Annual Reports in Organic Synthesis—1983

Annual Reports in Organic Synthesis

ANNUAL REPORTS IN ORGANIC SYNTHESIS—1970
John McMurry and R. Bryan Miller, Eds.

ANNUAL REPORTS IN ORGANIC SYNTHESIS—1971
John McMurry and R. Bryan Miller, Eds.

ANNUAL REPORTS IN ORGANIC SYNTHESIS—1972
John McMurry and R. Bryan Miller, Eds.

ANNUAL REPORTS IN ORGANIC SYNTHESIS—1973
R. Bryan Miller and Louis S. Hegedus, Eds.
John McMurry, Series Editor

ANNUAL REPORTS IN ORGANIC SYNTHESIS—1974
Louis S. Hegedus and Stephen R. Wilson, Eds.
R. Bryan Miller, Series Editor

ANNUAL REPORTS IN ORGANIC SYNTHESIS—1975
R. Bryan Miller and L. G. Wade, Jr., Eds.

ANNUAL REPORTS IN ORGANIC SYNTHESIS—1976
R. Bryan Miller and L. G. Wade, Jr., Eds.

ANNUAL REPORTS IN ORGANIC SYNTHESIS—1977
R. Bryan Miller and L. G. Wade, Jr., Eds.

ANNUAL REPORTS IN ORGANIC SYNTHESIS—1978
L. G. Wade, Jr., and Martin J. O'Donnell, Eds.

ANNUAL REPORTS IN ORGANIC SYNTHESIS—1979
L. G. Wade, Jr., and Martin J. O'Donnell, Eds.

ANNUAL REPORTS IN ORGANIC SYNTHESIS—1980
L. G. Wade, Jr., and Martin J. O'Donnell, Eds.

ANNUAL REPORTS IN ORGANIC SYNTHESIS—1981
L. G. Wade, Jr., and Martin J. O'Donnell, Eds.

ANNUAL REPORTS IN ORGANIC SYNTHESIS—1982
L. G. Wade, Jr., and Martin J. O'Donnell, Eds.

ANNUAL REPORTS IN ORGANIC SYNTHESIS—1983
Martin J. O'Donnell and Louis Weiss, Eds.

Annual Reports in Organic Synthesis—1983

edited by
Martin J. O'Donnell
Department of Chemistry, Indiana University–Purdue University at Indianapolis
Indianapolis, Indiana

Louis Weiss
Research and Development
Hoffmann–La Roche Inc.
Nutley, New Jersey

ACADEMIC PRESS   1984

(Harcourt Brace Jovanovich, Publishers)

ORLANDO   SAN DIEGO   NEW YORK   LONDON
TORONTO   MONTREAL   SYDNEY   TOKYO

Academic Press Rapid Manuscript Reproduction

Copyright © 1984, by Academic Press, Inc.
ALL RIGHTS RESERVED.
NO PART OF THIS PUBLICATION MAY BE REPRODUCED OR
TRANSMITTED IN ANY FORM OR BY ANY MEANS, ELECTRONIC
OR MECHANICAL, INCLUDING PHOTOCOPY, RECORDING, OR ANY
INFORMATION STORAGE AND RETRIEVAL SYSTEM, WITHOUT
PERMISSION IN WRITING FROM THE PUBLISHER.

ACADEMIC PRESS, INC.
Orlando, Florida 32887

United Kingdom Edition published by
ACADEMIC PRESS, INC. (LONDON) LTD.
24/28 Oval Road, London NW1 7DX

Library of Congress Catalog Card Number: 17-167779
ISBN 0-12-040814-7

PRINTED IN THE UNITED STATES OF AMERICA

84 85 86 87   9 8 7 6 5 4 3 2 1

CONTENTS

PREFACE.. ix
JOURNALS ABSTRACTED xi
GLOSSARY OF ABBREVIATIONS xiii

I. **CARBON–CARBON BOND FORMING REACTIONS**........... 1
    Carbon–Carbon Single Bonds (*see also:* I.E., I.F, I.G).......... 1
    1. Alkylation of Aldehydes, Ketones, and Their Derivatives 1
    2. Alkylations of Nitriles, Acids, and Acid Derivatives 7
    3. Alkylation of β-Dicarbonyl and β-Cyanocarbonyl Systems
       and Other Active Methylene Compounds.................. 11
    4. Alkylation of N-, S-, and Se-Stabilized Carbanions 18
    5. Alkylation of Organometallic Reagents (*see also:* I.F, I.G).. 23
    6. Other Alkylation Procedures and Reviews................. 33
    7. Nucleophilic Addition to Electron-Deficient Carbon 36
       a. 1,2-Additions 36
          (1) Aldol-Type Condensations...................... 36
             (a) Intermolecular 36
             (b) Intramolecular 51
          (2) Addition of N-, S-, or Se-Stabilized Carbanions 53
          (3) Grignard-Type Additions....................... 58
       b. Conjugate Additions................................. 70
          (1) Enolate-Type Carbanions....................... 70
          (2) Organometallic Reagents....................... 77
          (3) Other Conjugate Additions 82
    8. Other Carbon–Carbon Single Bond Forming Reactions 84
   B. Carbon–Carbon Double Bonds (*see also:* I.E.1, III.G,
    VI.A.16) 107
    1. Wittig-Type Olefination Reactions 107
    2. Eliminations 116
       a. Alcohols and Derivatives 116
       b. Halides 118
       c. Other Eliminations 119
    3. Other Carbon–Carbon Double Bond Forming Reactions ... 121
    4. Allene Forming Reactions.............................. 134
   C. Carbon–Carbon Triple Bonds (*see also:* VI.A.16)............ 137

v

D. Cyclopropanations ... 141
   1. Carbene or Carbenoic Additions to a Multiple Bond
      (*see also:* VI.A.7) ... 141
   2. Other Cyclopropanations ... 144
E. Thermal Reactions ... 153
   1. Cycloadditions ... 153
   2. Other Thermal Reactions ... 179
F. Aromatic Substitutions Forming a New Carbon–Carbon
   Bond ... 187
   1. Friedel–Crafts-Type Reactions ... 187
   2. Coupling Reactions ... 197
   3. Other Aromatic Substitutions ... 201
G. Synthesis via Organometallics ... 219
   1. Organoboranes ... 219
   2. Carbonylation Reactions ... 221
   3. Other Synthesis *via* Organometallics ... 225
   4. Reviews ... 227

II. OXIDATIONS ... 230
A. C–O Oxidations ... 230
   1. Alcohol → Ketone, Aldehyde ... 230
   2. Alcohol, Aldehyde → Acid, Acid Derivative ... 234
B. C–H Oxidations ... 236
   1. C–H → C–O ... 236
   2. C–H → C–Hal ... 241
   3. Other C–H Oxidations ... 247
C. C–N Oxidations ... 251
D. Amine Oxidations ... 255
E. Sulfur Oxidations ... 257
F. Oxidative Additions to C–C Multiple Bonds ... 260
   1. Epoxidations ... 260
   2. Hydroxylation ... 263
   3. Other ... 264
G. Phenol → Quinone Oxidation ... (1982, 265)
H. Oxidative Cleavages ... 269
I. Photosensitized Oxygenations ... 272
J. Dehydrogenation ... 273
K. Other Oxidations and Reviews ... 274

III. REDUCTIONS ... 276
A. C = O Reductions (*see also:* III.F.1) ... 276
B. C–N Multiple Bond Reductions ... 282
   1. Nitrile Reductions ... (1982, 279)
   2. Imine Reductions ... 282

CONTENTS

- C. Reduction of Sulfur Compounds 284
- D. N–O Reductions 284
- E. C–C Multiple Bond Reductions........................... 286
  1. C = C Reductions..................................... 286
  2. C ≡ C Reductions..................................... 289
  3. Reduction of Aromatic Rings 290
- F. Hydrogenolysis of Hetero Bonds 291
  1. C–O → C–H 291
  2. C–Hal → C–H 297
  3. C–S → C–H 299
  4. C–N → C–H 302
- G. Reductive Eliminations..................................... 304
- H. Reductive Cleavages.. 306
- I. Hydroboration (reduction only) 309
- J. Other Reductions and Reviews............................. 309

IV. SYNTHESIS OF HETEROCYCLES........................... 312
- A. Aziridines 312
- B. Furans, etc.. 313
- C. Indoles... 318
- D. Lactams.. 320
- E. Lactones... 330
- F. Pyridines, Quinolines, etc................................. 340
- G. Pyrroles, etc... 343
- H. Other Heterocycles with One Heteroatom
  (see also: II.F.1, VI.A.9).................................. 346
- I. Heterocycles with Two or More Heteroatoms................ 350
  1. Heterocycles with 2 Ns 350
     a. 5-Membered 350
     b. 6-Membered 352
     c. Other 353
  2. Heterocycles with 1 N and 1 O........................ 354
  3. Heterocycles with 1 N and 1 S........................ 357
  4. Heterocycles with 1 S and 1 O................ (1982, 342)
  5. Heterocycles with 3 Ns 361
  6. Other Heterocycles................................... 362
- J. General Reviews... 364

V. PROTECTING GROUPS 366
- A. Hydroxyl (see also: VI.A.10, VI.A.11) 366
- B. Amine (see also: VI.A.4)................................... 372
- C. Sulfhydryl (see also: VI.A.19)............................. 379
- D. Carboxyl (see also: VI.A.4, VI.A.10) 380
- E. Ketone, Aldehyde (see also: VI.A.18) 382

F.	Phosphate	386
G.	Pi Bond	387
H.	Miscellaneous Protecting Groups	388

VI. USEFUL SYNTHETIC PREPARATIONS ... 390

A. Functional Group Preparations ... 390
  1. Acids, Acid Halides, Anhydrides (*see also:* II.A.2) ... 390
  2. Alcohols, Phenols (*see also:* II.B.1, III.A., III.F.1) ... 392
  3. Alkyl, Aryl Halides (*see also:* II.B.2) ... 397
  4. Amides (*see also:* IV.D, VI.A.17) ... 400
  5. Amines (*see also:* III.B.2, III.D) ... 407
  6. Amino Acids and Derivatives (*see also:* III.E.1, VI.A.4, VI.A.10) ... 412
  7. Carbenes (*see also:* I.D) ... (1982, 389)
  8. Enamines ... 415
  9. Epoxides (*see also:* II.F.1, IV.H) ... 415
  10. Esters (*see also:* IV.E, V.D) ... 416
  11. Ethers (*see also:* V.A) ... 422
  12. Ketones and Aldehydes (*see also:* I.A.2, II.A.1, III.F.1) ... 423
  13. Nitriles ... 426
  14. Nitro ... 429
  15. Nucleotides, etc. (*see also:* IV.I.1a, b, V.F) ... 430
  16. Olefins, Acetylenes (*see also:* I.B, I.C, II.J, III.G) ... 431
  17. Peptides (*see also:* V.B, V.C, V.D, VI.A.4) ... 433
  18. Vinyl Halides, Vinyl Ethers, Vinyl Esters ... 434
  19. Sulfur Compounds (*see also:* II.E, III.C) ... 436

B. Ring Enlargement and Contraction ... 445
  1. Enlargement ... 445
  2. Contraction ... 446

C. Multistep Transformations ... (1982, 449)
  1. Masked Carbonyl Systems ... (1982, 450)
  2. Other Multistep Transformations ... (1982, 451)

VII. MISCELLANEOUS REVIEWS ... 447

AUTHOR INDEX ... 459

PREFACE

One of the most difficult problems facing chemists today is that of "keeping up with the literature." For several reasons, the problem is particularly severe for the synthetic organic chemist. Bits of information of potential use are scattered throughout common chemistry journals and can be found in any paper, not just those dealing strictly with synthesis. Thus, synthetic chemists must read a large number of journals and must organize and index what they read to make the information available for future reference. All synthetic chemists do this; but the task is becoming more difficult each year as the flow of information increases.

The problem, however, is shared to some extent by all. Most organic chemists are at some time faced with the problem of synthesizing a desired material, and for many the problems are formidable. Nonspecialists faced with the synthetic problem are not likely to have kept pace with the developments in synthetic chemistry that may well solve their problems, and they will not have the necessary information in their files.

Thus, we felt that an organized annual review of synthetically useful information would prove beneficial to nearly all organic chemists, both specialist and nonspecialist in synthesis. It should help relieve some of the information-storage burden of the specialist and should enable the nonspecialist who is seeking help with a specific problem to become rapidly aware of recent synthetic advances. Ideally also, it should appear as promptly as possible after the close of the abstracting period. This year we have placed particular emphasis on keeping the abstracts as concise as possible, while indicating the generality of the reactions involved. We have tried to combine similar publications into inclusive abstracts, particularly in Chapter I. This practice has allowed us to include a larger number of references without a substantial increase in the book's length.

In producing *Annual Reports in Organic Synthesis—1983*, we have abstracted 47 primary chemistry journals, selecting useful synthetic advances. We have tried to present the information in an organized manner, emphasizing rapid visual retrieval. Only the common journals received by our libraries have been abstracted. Any journal received after March 1, 1984 will be covered in the next volume. We have also exercised selectivity in choosing which papers to abstract. Our general guidelines have been to include all reactions and methods that are new, synthetically useful, and reasonably general. Each entry is comprised primarily of structures, accompanied by very few comments. The purpose of this emphasis is to aid the reader in scanning the book. The mind is capable of absorbing a whole picture in an instant, but is considerably slowed by having to read sentences. If the pictures presented catch the reader's interest, he or she should then seek details from the original paper.

For the eighth year we have included a principal author index to aid the user. No subject index is included because to do so would greatly increase both the cost of the book and the lead time for publication. Instead, we have chosen to use an extensive table of contents. Chapters I–III are organized by reaction type and constitute a major part of the book. The organization of these sections is self-explanatory; thus, there should be no difficulty in locating a new method of oxidation or a new cyclo-propanation procedure. Chapter IV deals with methods of synthesizing heterocyclic systems and Chapter V covers the use of new protecting groups. Chapter VI is divided into three main parts and covers those synthetically useful transformations that do not fit easily into the first three chapters. The first part deals only with functional group synthesis; the second covers ring expansion and contraction; and the third involves useful multistep sequences, the individual steps of which may be well known. Future volumes of this series will maintain the present table of contents as much as possible. If no entry is found for a particular section, the last volume in which one appears will be cited in the table of contents.

Any undertaking of this type involves a series of compromises. We have chosen to emphasize reasonable cost, rapid publication, and rapid visual retrieval of information at the admitted expense of detail and beauty.

The arduous task of drawing the multitude of structures appearing in this review was carried out by Ms. Katy Krupa and Ms. Carol P. Bertram. We thank them very much for their efforts. We also thank William Bennett, Tonette Tucker, and Carol P. Bertram for aid in proofreading the manuscript.

MARTIN J. O'DONNELL
LOUIS WEISS

JOURNALS ABSTRACTED

Accounts of Chemical Research
Acta Chemica Scandinavica
Aldrichimica Acta
Angewandte Chemie International Edition in English
Australian Journal of Chemistry
Bulletin of the Chemical Society of Japan
Bulletin de Sociétés Chimiques Belges
Bulletin de la Société Chimique de France
Canadian Journal of Chemistry
Chemical Communications
Chemical and Pharmaceutical Bulletin
Chemical Reviews
Chemical Society Reviews
Chemische Berichte
Chemistry and Industry
Chemistry Letters
Collection of Czechoslovakian Chemical Communications
Comptes Rendus Hebdomadaires de Seances de l'Academie des Sciences (C)
Gazzetta Chimica Italiana
Helvetica Chimica Acta
Indian Journal of Chemistry
Journal of the American Chemical Society
Journal of Chemical Research
Journal of the Chemical Society (Perkin I)
Journal of the Chemical Society (Perkin II)
Journal of General Chemistry (USSR)
Journal of Heterocyclic Chemistry
Journal of Medicinal Chemistry
Journal of Organic Chemistry
Journal of Organic Chemistry (USSR)
Journal of Organometallic Chemistry
Journal für Praktische Chemie
Liebig's Annalen der Chemie
Monatschefte für Chemie
Nouveau Journal de Chimie
Organic Preparations and Procedures International
Organic Syntheses
Organometallics
Pure and Applied Chemistry
Recueil des Travaux Chimiques des Pays-bas
Russian Chemical Reviews
Synthesis
Synthetic Communications
Tetrahedron
Tetrahedron Letters
Topics in Current Chemistry
Zeitschrift für Chemie

GLOSSARY OF ABBREVIATIONS

Ac	acetyl
AIBN	azobisisobutyronitrile
Ar	aryl
9-BBN	9-borabicyclo[3.3.1]nonane
BOC (t-Boc)	t-butyloxycarbonyl
Bu	butyl
Bz	benzyl
Cbz	benzyloxycarbonyl
COD	1,5-cyclooctadiene
Cp	cyclopentadienyl
CSA	camphorsulfonic acid
DABCO	1,4-diazabicyclo[2.2.2]octane
DBN	1,5-diazabicyclo[4.3.0]non-3-ene
DBU	1,5-diazabicyclo[5.4.0]undec-5-ene
DCC	dicyclohexylcarbodiimide
DDQ	2,3-dichloro-5,6-dicyanobenzoquinone
de	diasteriomeric excess
DEAD	diethyl azodicarboxylate
DIBAH (DIBAL)	diisobutylaluminum hydride
DMAD	dimethyl acetylenedicarboxylate
DMAP	4-N,N-dimethylaminopyridine
DME	1,2-dimethoxyethane
DMF	dimethylformamide
DMSO	dimethyl sulfoxide
E+	general electrophile
ee	enantiomeric excess
Et	ethyl
Fp	η^5-C_5H_5Fe(CO)$_2$
Hex	hexyl
HMPA, HMPT	hexamethyl phosphoramide (hexamethylphosphoric triamide)
hν	irradiation with light
KAPA	potassium 3-aminopropylamide
L	triphenylphosphine ligand
LAH	lithium aluminum hydride
LDA	lithium diisopropylamide
LICA	lithium isopropylcyclohexylamide
LTA	lead tetraacetate
MCPBA	meta-chloroperbenzoic acid
Me	methyl
MEM	β-methoxyethoxymethyl
MOM	methoxymethyl
Ms	methanesulfonyl
MSA	methanesulfonic acid
MTM	methylthiomethyl
NBS	N-bromosuccinimide
NCS	N-chlorosuccinimide
NIS	N-iodosuccinimide
Ni(R)	Raney Nickel
[O]	general oxidation
Ⓟ	polymeric backbone
PCC	pyridinium chlorochromate
PDC	pyridinium dichromate
Ph	phenyl
(Phen)	1,10-phenanthroline
Phth	phthaloyl
PPA	polyphosphoric acid
PPE	polyphosphate ester
Pr	propyl
Py, pyr	pyridine
PTC	phase-transfer catalysis
Q+	quaternary ammonium
RT	room temperature
TBDMS	t-butyldimethylsilyl
TCNQ	7,7,8,8-tetracyanoquinodimethane
Tf	trifluoromethane sulfonate
TFA	trifluoroacetic acid
TFAA	trifluoroacetic anhydride
THF	tetrahydrofuran
THP	tetrahydropyranyl
TMEDA	tetramethylethylenediamine
TMP	2,2,6,6-tetramethylpiperidine
TMS	trimethylsilyl
Tol	tolyl
Tr	trityl
Ts, Tos	p-toluenesulfonyl
TSA	toluenesulfonic acid
Z	benzyloxycarbonyl; also used for electron-withdrawing groups such as -CN, -COOR, etc.
Δ	heat
φ	phenyl
18-C-6	18-crown-6

I
CARBON–CARBON BOND FORMING REACTIONS

I.A. Carbon-Carbon Single Bonds

   (see also: I.E, I.F, I.G).

I.A.1. Alkylations of Aldehydes, Ketones and Their Derivatives

I.A.1-1 G. S. Bates and S. Ramaswamy, Can. J. Chem., 61, 2000, 2466 (1983); S. M. Makin et al., J. Org. Chem. (USSR), 19, 1044 (1983).

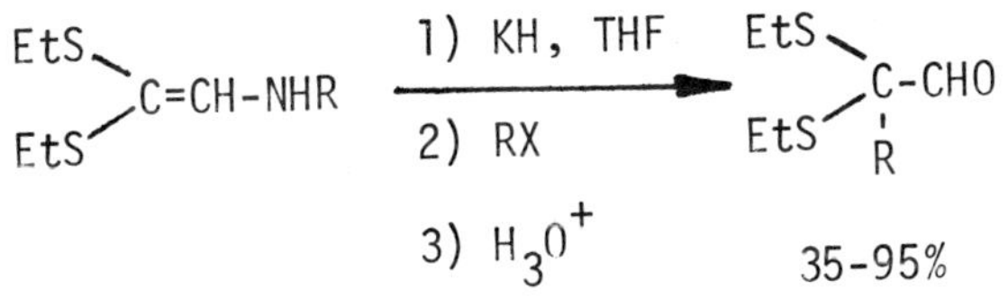

$$\text{EtS}_2\text{C=CH-NHR} \xrightarrow[\substack{\text{2) RX} \\ \text{3) H}_3\text{O}^+}]{\text{1) KH, THF}} \text{EtS}_2\text{C-CHO} \atop \text{R}$$

35-95%

I.A.1-2 E. I. Negishi and S. Chatterjee, Tetrahedron Lett., 24, 1341 (1983); M. A. Krafft and R. A. Holton, ibid, 24, 1345 (1983); J. d'Angelo and G. Revial, ibid, 24, 2103 (1983); E. I. Negishi et al., J. Org. Chem., 48, 2427 (1983).

[Reaction scheme: 2-methylcyclohexanone → 2,2-dimethylcyclohexanone using 1) KH, 2) Et₃B, 3) MeI]

79%

(90% 2,2-)

Highly regiosel. generation of "thermodynamic" enolates.

I.A.1-3 L. A. Paquette et al., J. Am. Chem. Soc., 105, 6975, 7352, 7358 (1983); J. Org. Chem., 48, 3282 (1983); A. G. Schultz and J. P. Dittami, ibid, 48, 2318 (1983); L. M. Jackman and B. C. Lange, ibid, 48, 4789 (1983); D. Gravel, R. Deziel and L. Bordeleau, Tetrahedron Lett., 24, 699 (1983); E. V. Vasil'eva, E. M. Auvinen and I. A. Favorskaya, J. Org. Chem. (USSR), 19, 1266 (1983).

[Reaction scheme with bicyclic ketone + =CHSnBu, treated with 1) LDA, THF and CH₂=C(SiMe₃)CH₂I; 2) 25% KOH, O(CH₂CH₂OH)₂, Δ → product with Me₃Si-allyl group, 45%]

I.A.1-4 J. A. M. van den Goorbergh and A. van der Gen, Rec. Trav. Chim., 102, 393 (1983); P. M. Booth, C. M. J. Fox and S. V. Ley, Tetrahedron Lett., 24, 5143 (1983); G. B. Trimitsis et al., J. Org. Chem., 48, 2957 (1983).

[Reaction scheme: methyl 5-methyl-3-oxo-hex-4-enoate + 1) 2 LDA, 2) E⁺ → α-substituted product, 60-94%]

E⁺ = RX, RCHO, R₂CO.

I.A.1-5 L. S. Liebeskind and M. E. Welker, Organometallics, 2, 194 (1983).

$$\text{Cp}\diagdown\underset{\underset{\text{Co}}{L}}{\text{Fe}}-\overset{\overset{\text{O}}{\|}}{\text{C}}-\text{CH}_3 \xrightarrow[\substack{\text{2) PhCH}_2\text{Br} \\ \text{3) NBS, CH}_2\text{Cl}_2 \\ \text{EtOH}}]{\text{1) LDA, THF}} \text{Et}\overset{\overset{\text{O}}{\|}}{\text{C}}\text{CH}_2-\text{CH}_2\text{Ph}$$

81%

I.A.1-6 E. I. Negishi and R. A. John, J. Org. Chem., 48, 4098 (1983).

(M=BFt$_3$K or ZnCl)
(Ph$_3$P)$_4$Pd

Also, study of countercation effects in Pd-catalyzed alkylation.

I.A.1-7 W. Kantlehner, W. M. Mergen and E. Haug, Liebigs Ann. Chem., 290 (1983).

$$\text{EtOC(NMe}_2)_3 \xrightarrow[\text{PhH, }\Delta]{\text{R}-\overset{\overset{\text{O}}{\|}}{\text{C}}-\text{CH}_3} \text{R}-\overset{\overset{\text{O}}{\|}}{\text{C}}-\text{CH}=\text{C}\diagup^{\text{NMe}_2}_{\diagdown\text{NMe}_2}$$

66-84%

I.A.1-8 J. K. Whitesell and M. A. Whitesell, Synthesis, 517 (1983).

Review: "Alkylation of Ketones and Aldehydes via their Nitrogen Derivatives."

I.A.1-9 K. Saigo et al., Tetrahedron Lett., 24, 511 (1983); D. Enders and U. Baus, Liebigs Ann. Chem., 1439 (1983).

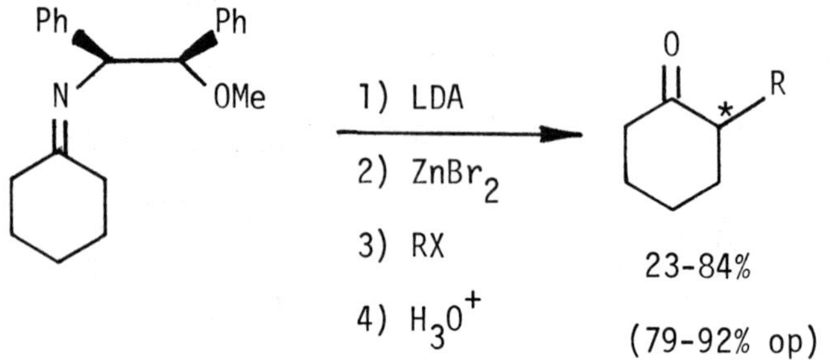

cyclohexanone imine of Ph-CH(OMe)-CH(Ph)-NH₂

1) LDA
2) ZnBr$_2$
3) RX
4) H$_3$O$^+$

→ 2-R-cyclohexanone (*)

23-84%
(79-92% op)

I.A.1-10 M. Newcomb, D. E. Bergbreiter, A. I. Meyers et al., Tetrahedron Lett., 24, 3559 (1983); J. Am. Chem. Soc., 105, 4396 (1983); J. F. Honek, M. L. Mancini and B. Belleau, Tetrahedron Lett., 24, 257 (1983).

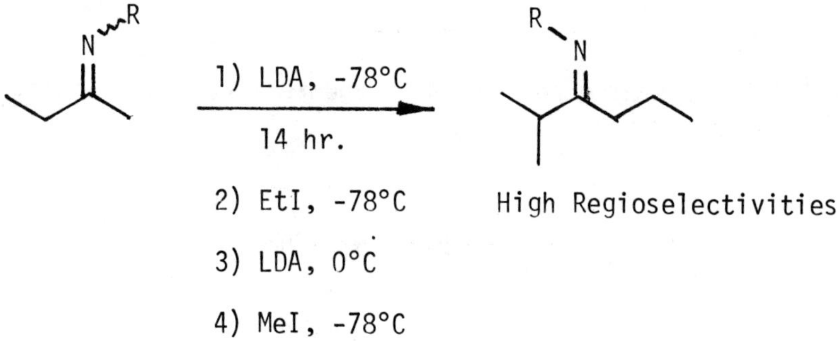

1) LDA, -78°C, 14 hr.
2) EtI, -78°C
3) LDA, 0°C
4) MeI, -78°C

High Regioselectivities

I.A.1-11 T. C. Flood et al., J. Organometal. Chem., 244, 61 (1983).

[Reaction scheme: CpFe(CO)(o-BP)-CH$_2$Cl + cyclohexenyl pyrrolidine enamine, 1) then 2) H$_2$O, gives CpFe(CO)(o-BP)-CH$_2$-(2-oxocyclohexyl)* in 68% (60% ee)]

Chiral Electrophile

o-BP = tri(o-biphenyl)phosphite

I.A.1-12 R. Noyori, I. Nishida and J. Sakata, J. Am. Chem. Soc., 105, 1598 (1983).

[Reaction: Ph-C(OSiMe$_3$)=CHMe, 1) (Et$_2$N)$_3$S$^+$ Me$_3$SiF$_2^-$, 2) MeI, gives PhC(O)CH(Me)Me]

Also, erythro selective aldol.

I.A.1-13 T. V. Lee and J. O. Okonkwo, Tetrahedron Lett., 24, 323 (1983); I. Fleming and J. Iqbal, ibid, 24, 327, 2913 (1983); D. J. Ager, ibid, 24, 419 (1983); M. A. Ibragimov and W. A. Smit, ibid, 24, 961 (1983); L. Duhamel, J. Chauvin and C. Goument, ibid, 24, 2095 (1983).

[Reaction: 1-(trimethylsilyloxy)cyclohexene + PhS-CHCl-CO$_2$Me, ZnBr$_2$, gives 2-(1-(phenylthio)-1-(methoxycarbonyl)methyl)cyclohexanone, 75%]

Products transformed to γ-ketoesters or unsaturated γ-ketoesters.

I.A.1-14 M. T. Reetz and M. Sauerwald, Tetrahedron Lett., 24, 2837 (1983); J. Tsuji, I. Minami and I. Shimizu, ibid, 24, 4713 (1983); Chem. Lett., 1325 (1983); R. R. Schmidt and M. Hoffmann, Angew Chem.,Int. Ed. Engl., 22, 406 (1983).

[Ar-CH(OAc)-CH$_3$]·Cr(CO)$_3$ + CH$_2$=CH-OSiMe$_3$ →(ZnCl$_2$, CH$_2$Cl$_2$, RT)→ [Ar-CH(CH$_3$)-CH$_2$-C(O)CH$_3$]·Cr(CO)$_3$ 84%

I.A.1-15 H. Sakurai, K. Sasaki and A. Hosomi, Bull. Chem. Soc. Jpn., 56, 3195 (1983); I. Kuwajima et al., Chem. Commun., 796 (1983); D. Kellner and H. J. E. Loewenthal, Tetrahedron Lett., 24, 3397 (1983); M. T. Reetz and H. Muller-Starke, Liebigs Ann. Chem., 1726 (1983).

cyclopentenyl-OSiMe$_3$ + PhCHC(OMe)$_2$ →(cat. Me$_3$SiI, CH$_2$Cl$_2$, -78°C)→ 2-[CH(Ph)(OMe)]cyclopentanone 87%

(99% Erythro)

I.A.1-16 M. T. Reetz and H. Heimbach, Helv. Chim. Acta, 66, 3702 (1983); M. T. Reetz, R. E. Schmidt et al., ibid, 116, 3708 (1983).

1,2-bis(OSiMe$_3$)cyclohexene →(1) ^{t}BuCl, ZnCl$_2$, RT; 2) Hydrol.)→ 2-tBu-2-hydroxycyclohexanone 66%

Also, t-butylation of carboxylic acid and carboxylic ester derivatives.

I.A.2. Alkylations of Nitriles, Acids and Acid Derivatives

I.A.2-1   K. Sukata et al., Bull. Chim. Soc. Jpn., 56, 3306 (1983); Y. Kimura, P. Kirszensztejn and S. L. Regen, J. Org. Chem., 48, 385 (1983); T. Balakrishnau and W. T. Ford, ibid, 48, 1029 (1983); S. Takano et al., Chem. Commun., 760 (1983); G. A. Garcia, H. Munoz and J. Tamariz, Synth. Commun., 13, 569 (1983).

$$PhCH_2CN \xrightarrow[NaOH/Al_2O_3]{RX, PhH \atop KOH/Al_2O_3 \; or} PhCHCN\atopR$$

48-94% (GC)

(0-6% Dialkyl)

I.A.2-2   J. D. Albright, Tetrahedron, 39, 3207 (1983).

Review: "Reactions of Acyl Anion Equivalents Derived from Cyanohydrins, Protected Cyanohydrins and α-Dialkylaminonitriles."

I.A.2-3   T. Takahashi, J. Tsuji, et al., Tetrahedron Lett., 24, 2005, 3485, 3489, 4695 (1983).

1) NaN(SiMe$_3$)$_2$, THF, 50°C
2) Acid

63%

I.A.2-4 E. Hebert, N. Maigrot and Z. Welvart, Tetrahedron Lett., 24, 4683 (1983); K. Takahashi et al., J. Org. Chem., 48, 1909 (1983).

Et$_2$N-CH-CN $\xrightarrow{\text{1) } \diagup\!\!\diagdown\!\!\diagup\!\!\diagdown\text{I} \quad \text{2) Hydrol.}}$ CH$_2$=CH-CH$_2$-CH$_2$-CH$_2$-CHO

70%

Also, alkylations of O-protected cyanohydrins.

I.A.2-5 I. H. Sanchez et al., Synth. Commun., 13, 35, 43 (1983).

ArCH$_2$CN $\xrightarrow[\text{KOH}]{\text{1) PhSSPh, Br}\diagup\!\!\diagdown\!\!\diagup}$ Ar-CH(CHO)-CH$_2$-CH=CH$_2$

2) Ni, Me$_2$C=O

3) DIBAL, 5°C

55%

I.A.2-6 M. F. Semmelhack and J. W. Herndon, Organometallics, 2, 363 (1983); J. Am. Chem. Soc., 105, 2497 (1983); A.J. Pearson and I. C. Richards, Tetrahedron Lett., 24, 2465 (1983); A. J. Pearson, T. R. Perrior and D. A. Griffin, J. Chem. Soc., Perkin I, 625 (1983).

[cyclohexadiene-Fe(CO)$_3$] $\xrightarrow[\text{THF, HMPA}]{\text{1) LiC(Me)}_2\text{CO}_2\text{Li}}$ [cyclohexadiene-C(Me)$_2$CO$_2$H]

2) CF$_3$CO$_2$H

3) Workup

82%

Various anions studied.

I.A.2-7 J. S. Bajwa and M. J. Miller, *J. Org. Chem.*, **48**, 1114 (1983).

$$\begin{array}{c} CO_2H \\ HO-\!\!\!\!\!-\!\!\!\!\!-H \\ CH_2 \\ CO_2H \end{array} \xrightarrow[\text{via chiral lactone}]{\text{Several Steps}} \begin{array}{c} CO_2H \\ H-\!\!\!\!\!-\!\!\!\!\!-H \\ Me-\!\!\!\!\!-\!\!\!\!\!-H \\ CO_2H \end{array}$$

Substituted and functionalized optically active succinic acid fragments from malic acid.

I.A.2-8 H. Ahlbrecht and H. Simon, *Synthesis*, 58, 61 (1983); P. A. Aristoff and C. L. Nelson, *Org. Prep. Proced. Int.*, **15**, 149 (1983).

$$\underset{H}{\overset{Me}{\diagdown}}C=C\underset{CO_2Me}{\overset{N(Me)-Ph}{\diagup}} \xrightarrow[\text{2) } E^+]{\text{1) LDA, THF, } -78°C} \underset{H}{\overset{E-CH_2}{\diagdown}}C=C\underset{CO_2Me}{\overset{N(Me)-Ph}{\diagup}} \quad 42-85\%$$

E^+ = RX or $R^1R^2C=O$

I.A.2-9 M. Gaudemar, *Tetrahedron Lett.*, **24**, 2749 (1983); M. Visnick, L. Strekowski and M. A. Battiste, *Synthesis*, 284 (1983); F. Orsini and F. Pelizzoni, *Synth. Commun.*, **13**, 523 (1983).

$$BrZn\!\!-\!\!CH_2\!\!-\!\!CO_2Et \xrightarrow[\text{5\% Cu(acac)}_2]{CH_2=CH\!-\!CH_2\!-\!Br} CH_2=CH\!-\!CH_2\!-\!CH_2\!-\!CO_2Et \quad 82\%$$

I.A.2-10 M. N. Danchenko and Y. G. Gololobov, J. Org. Chem. (USSR), 19, 632 (1983); T. Fujisawa, T. Itoh and T. Sato, Chem. Lett., 1901 (1983).

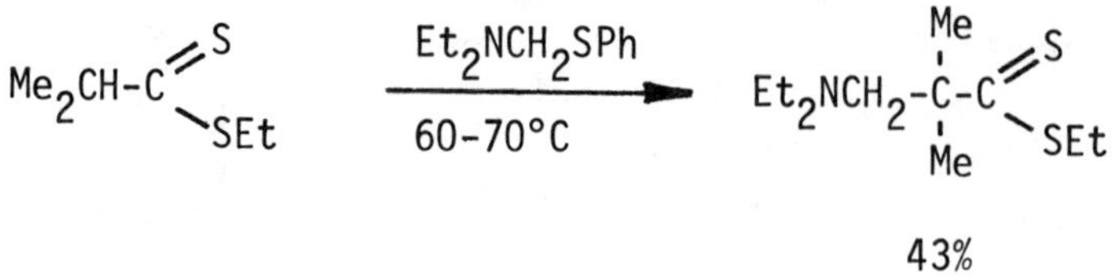

$Me_2CH-C(=S)SEt$ →[Et_2NCH_2SPh][60-70°C] $Et_2NCH_2-C(Me)(Me)-C(=S)SEt$

43%

I.A.2-11 G. Helmchen et al., Tetrahedron Lett., 24, 3213, 1235 (1983); G. J. McGarvey et al., ibid, 24, 2733 (1983).

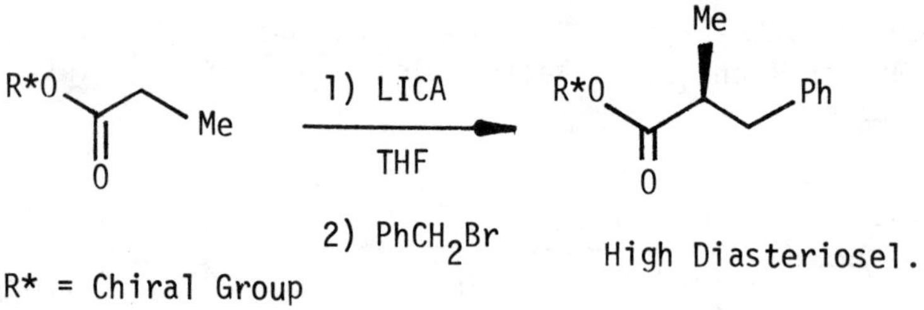

R*O-C(=O)-CH_2Me →[1) LICA, THF][2) PhCH_2Br] R*O-C(=O)-CH(Me)-CH_2Ph

R* = Chiral Group

High Diasteriosel.

I.A.2-12 D. Seebach et al., Liebigs Ann. Chem., 1930 (1983); J. Am. Chem. Soc., 105, 5390 (1983); T. R. Hoye and M. J. Kurth, Tetrahedron Lett., 24, 4769 (1983).

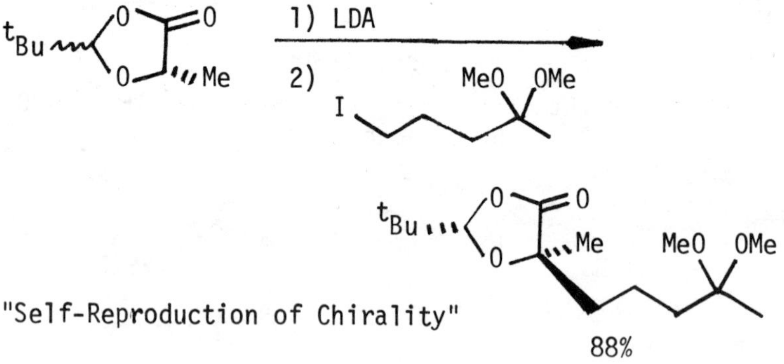

[tBu-dioxolanone with Me] →[1) LDA][2) I-CH_2CH_2CH_2-C(OMe)_2Me]

[product dioxolanone with tBu, Me, and -(CH_2)_3-C(OMe)_2Me chain]

"Self-Reproduction of Chirality"

88%

I.A.2-13 K. Ikeda, K. Achiwa and M. Sekiya, Tetrahedron Lett., 24, 913 (1983); S. K. Patel and I. Paterson, ibid, 24, 1315 (1983); S. Tanimoto et al., Synthesis, 787 (1983); Bull. Chem. Soc. Jpn., 56, 645 (1983).

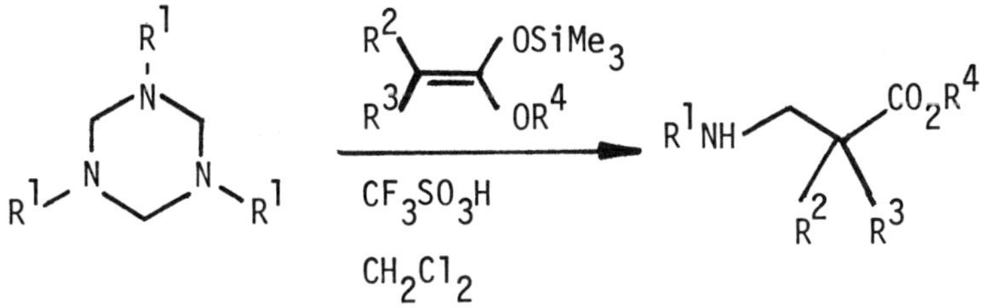

I.A.2-14 Z. Yoshida et al., J. Org. Chem., 48, 3631 (1983).

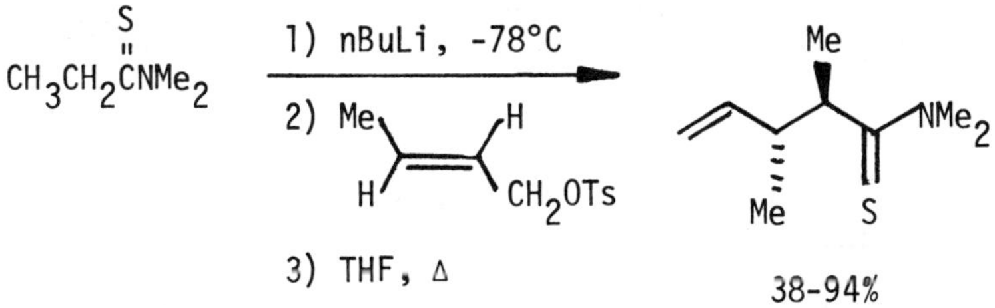

Trans gives erythro, cis gives threo.

I.A.3. Alkylation of β-Dicarbonyl, β-Cyanocarbonyl Systems and Other Active Methylene Compounds

I.A.3-1 M. E. Garst and B. J. McBride, J. Org. Chem., 48, 1362 (1983).

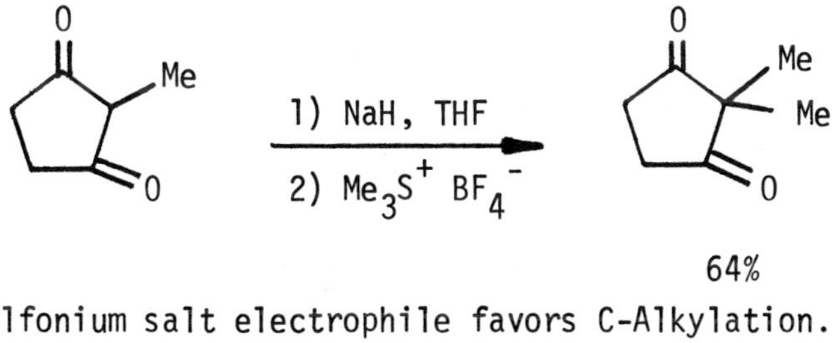

Sulfonium salt electrophile favors C-Alkylation.

I.A.3-2 D. Henderson, K. A. Richardson and R. J. K. Taylor, Synthesis, 996 (1983); K. K. Pivnitskii, et al., J. Org. Chem. (USSR), 19, 259, 261 (1983).

cyclohexanone-2-CO$_2^t$Bu
1) NaH, DMF, PhH
2) RBr, NaI
3) CF$_3$CO$_2$H
4) Δ

→ 2-R-cyclohexanone   80-91%

I.A.3-3 S. D. Barker and R. K. Norris, Aust. J. Chem., 36, 527 (1983); S. E. Drewes et al., J. Chem. Soc., Perkin I, 2293 (1983); K. S. Rumyantseva et al., J. Org. Chem.(USSR), 19, 1205 (1983).

Ar(H)C=CH–CH^{t}Bu(Cl) $\xrightarrow[hv]{Et\bar{C}(CN)_2\ Na^+}$ Ar(CH)–C(CN)$_2$Et ... CH=CH^{t}Bu

98%

Allylic Halides Reactions with Various Nucleophiles

I.A.3-4 A. J. Pearson, Pure Appl. Chem., 55, 1767 (1983).

Review: "Natural Products Synthesis using Organoiron Complexes."

I.A.3-5 D. Astruc, Tetrahedron, 39, 4027 (1983).

Review: "Organo-iron Complexes of Aromatic Compounds. Applications in Synthesis."

I.A.3-6 T. C. T. Chang and M. Rosenblum, Tetrahedron Lett., 24, 695 (1983).

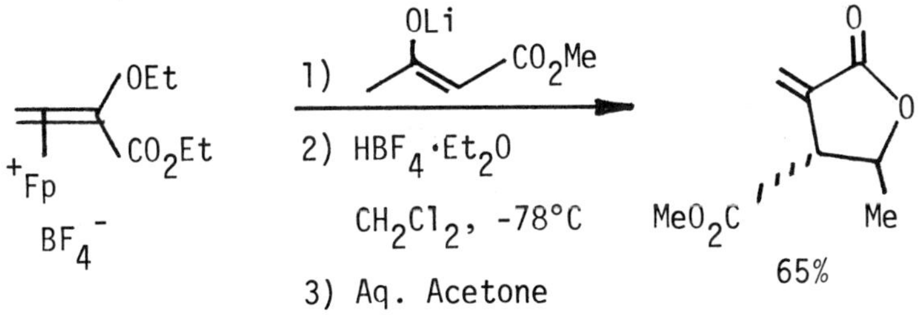

I.A.3-7 E. Ghera and Y. Ben-David, Tetrahedron Lett., 24, 3533 (1983); S. Niwas and A. P. Bhaduri, Synthesis, 110 (1983); T. L. Ho, Synth. Commun., 13, 341, (1983); N. S. Zefirov et al., J. Org. Chem. (USSR), 19, 474 (1983).

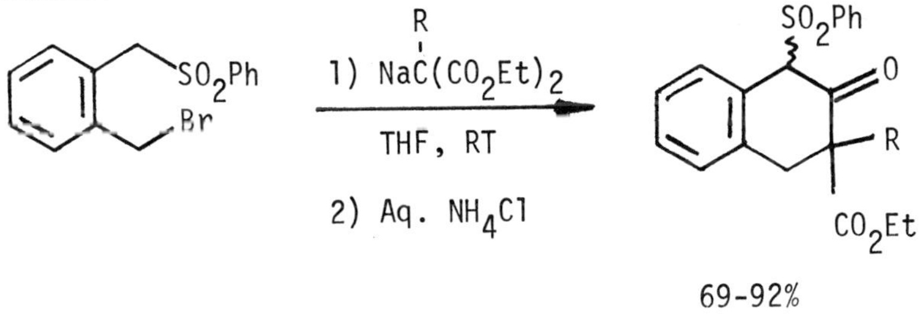

I.A.3-8 J. C. Blazejewski, R. Dorme and C. Wakselman, Chem. Commun., 1050 (1983).

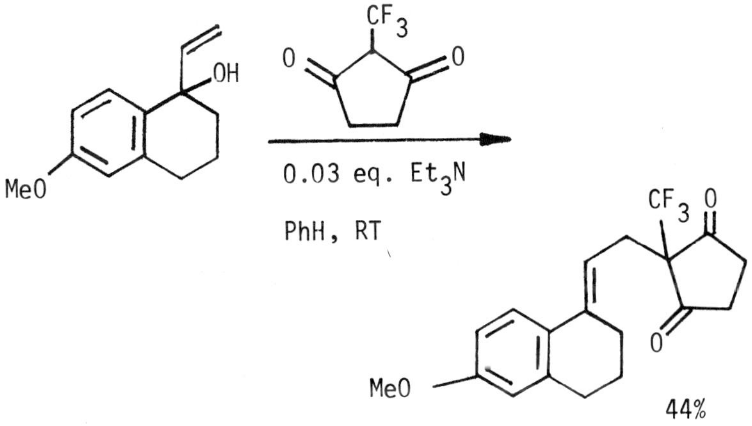

I.A.3-9 B. M. Trost, J. Cossy and J. Burks, J. Am. Chem. Soc., 105, 1052 (1983); H. Stamm and J. Budny, J. Chem. Res. (S), 54 (1983).

[cyclopropane with two SO_2Ph groups] $\xrightarrow[\text{2) MeI}]{\text{1) KCH(CO}_2\text{Et})_2, \text{ DMF, 80°C}}$ EtO_2C–CH(–CO_2Et)–CH_2CH_2–C(Me)(SO_2Ph)(SO_2Ph) 64%

I.A.3-10 Y. Nagao, S. Yamada and E. Fujita, Tetrahedron Lett., 24, 2287, 2291 (1983); B. M. Trost and A. C. Lavoie, J. Am. Chem. Soc., 105, 5075 (1983).

[4-(^{t}BuO$_2$C)-2-methyloxazole] $\xrightarrow[\text{hv}]{\text{1) NBS, CCl}_4}$

2) Nu$^-$

(KCN, NaCH(CO$_2$Et)$_2$, etc.)

[4-(^{t}BuO$_2$C)-2-(CH$_2$-Nu)oxazole]

2-Benzenesulfonylmethyl starting reagent deprotonated and reacts with electrophiles.

I.A.3-11 T. Mukaiyama, T. Oriyama and M. Murakami, Chem. Lett., 985 (1983).

Me_3Si–C(CN)(H)–CO_2Me $\xrightarrow[\text{2) RCH}_2\text{X}]{\text{1) LDA}}$ $\underset{\text{3) Ac}_2\text{O}}{}$ R–CH=C(OAc)–CO_2Me

45-74%

I.A.3-12 S. Yongai and T. Miwa, Chem. Commun., 68 (1983).

$$\text{AcO} \cdots \overset{\text{O}}{\underset{\text{AcO}}{\bigcirc}} \xrightarrow[\text{BF}_3 \cdot \text{Et}_2\text{O}]{\text{CH}_3\overset{\text{O}}{\text{C}}\text{CH}_2\overset{\text{O}}{\text{C}}\text{CH}_3} \text{AcO} \cdots \overset{\text{O}}{\bigcirc}\text{—CH}(\overset{\text{O}}{\text{C}}\text{CH}_3)_2$$

72-81%

(α : β = 5 : 1)

I.A.3-13 E. Keinan and Z. Roth, J. Org. Chem., 48, 1769 (1983); H.A. Dieck et al., ibid, 48, 807 (1983); J. E. Nystrom and J. E. Backvall, ibid, 48, 3947 (1983); D. E. Bergbreiter and B. Chen, Chem. Commun., 1238 (1983); L. S. Hegedus et al., Organometallics, 2, 1658 (1983); B. M. Trost and C. R. Self, J. Am. Chem. Soc., 105, 5942 (1983); T. Cuvigny and M. Julia, J. Organometal. Chem., 250, C21 (1983).

$$\text{Ph} \diagup\hspace{-0.5em}= \hspace{-0.5em}\diagdown \underset{\text{CN}}{\overset{\text{OAc}}{\diagdown}} \xrightarrow[(\text{Ph}_3\text{P})_4\text{Pd}]{\text{Nu}^-} \quad \text{Regiochemical Classification of Nucleophiles}$$

Nu⁻ = Various Nucleophiles Ranging from Malomates to Grignards.

I.A.3-14 B. M. Trost and M. Lautens, J. Am. Chem. Soc., 105, 3343 (1983); Organometallics, 2, 1687 (1983); T. Tatsumi et al., J. Organometal. Chem., 252, 105 (1983).

$$\underset{\text{PhCH}_3}{\overset{\text{OAc}}{\bigcirc}\hspace{-0.3em}\diagdown} \xrightarrow[\text{NaH, 5\% Mo(CO)}_6]{\text{Me}\diagup\overset{\text{CO}_2\text{Me}}{\diagdown}\text{CO}_2\text{Me}}$$

$$\bigcirc\hspace{-0.3em}=\hspace{-0.3em}\diagdown\hspace{-0.3em}\diagup\hspace{-0.3em}\diagdown\text{C(CH}_3\text{)}_2\text{(CO}_2\text{Me)}_2$$

73%

I.A.3-15 B. M. Trost and M. H. Hung, J. Am. Chem. Soc., 105, 7757 (1983).

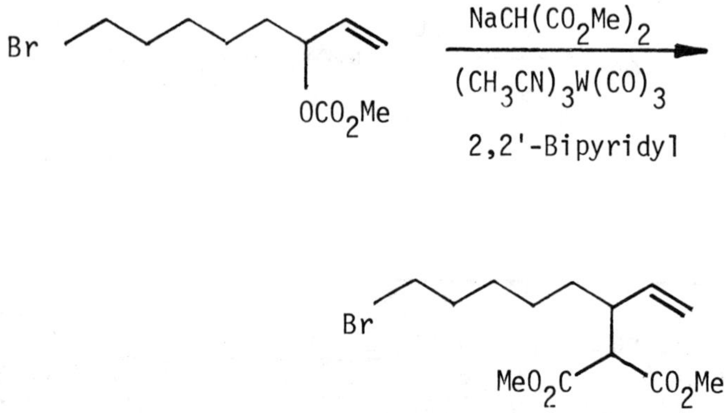

$$\text{Br}\diagdown\diagdown\diagdown\diagup\text{CH(OCO}_2\text{Me)CH=CH}_2 \xrightarrow[\substack{(CH_3CN)_3W(CO)_3 \\ 2,2'\text{-Bipyridyl}}]{\text{NaCH(CO}_2\text{Me})_2} \text{Br}\diagdown\diagdown\diagdown\diagup\text{CH(CH(CO}_2\text{Me})_2)\text{CH=CH}_2$$

57%

I.A.3-16 U. Schollkopf, Pure Appl. Chem., 55, 1799 (1983).

Review: "Asymmetric Syntheses of Amino Acids via Metallated Bis-Lactim Ethers of 2,5-Diketopiperazines."

I.A.3-17 U. Schollkopf et al., Synthesis, 37, 673, 675 (1983); D. H. Rich et al., Tetrahedron Lett., 24, 5305, 5309 (1983).

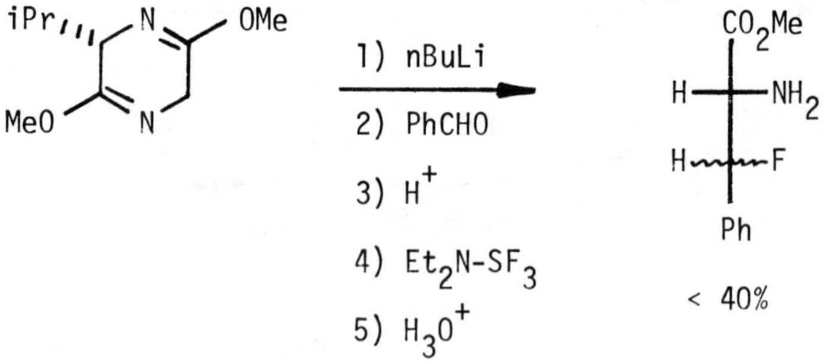

Reagents:
1) nBuLi
2) PhCHO
3) H^+
4) $Et_2N\text{-}SF_3$
5) H_3O^+

< 40%

I.A.3-18 D. Seebach et al., Tetrahedron Lett., 24, 3311, 3315 (1983); I. Hoppe, U. Schollkopf and R. Tolle, Synthesis, 789 (1983); M.Kolb and J. Barth, Liebigs Ann. Chem., 1668 (1983).

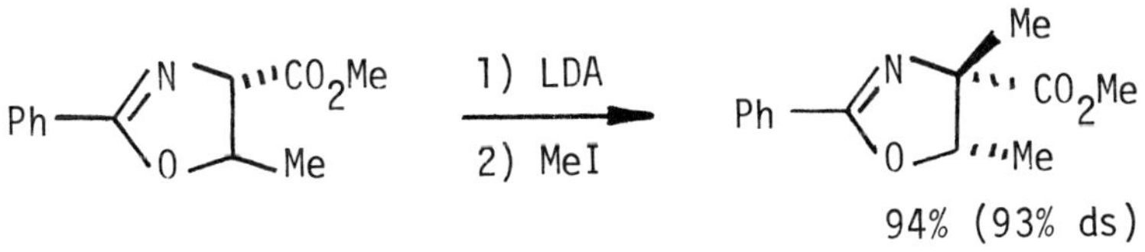

94% (93% ds)

Starting material from L-threonine.

I.A.3-19 T. Takahashi, H. Kataoka and J. Tsuji, J. Am. Chem. Soc., 105, 147 (1983); B. M. Trost and S. J. Brickner, ibid, 105, 568 (1983); J. Wicha and M. M. Kabat, Chem. Commun., 985 (1983).

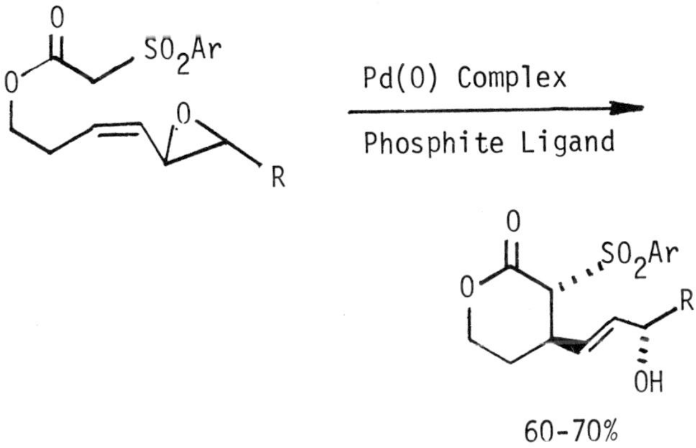

60-70%

I.A.3-20 J. S. Yadav et al., Synth. Commun., 13, 331, 379, 1067 (1983); H. Sasaki and T. Kitagawa, Chem. Pharm. Bull., 31, 2868 (1983).

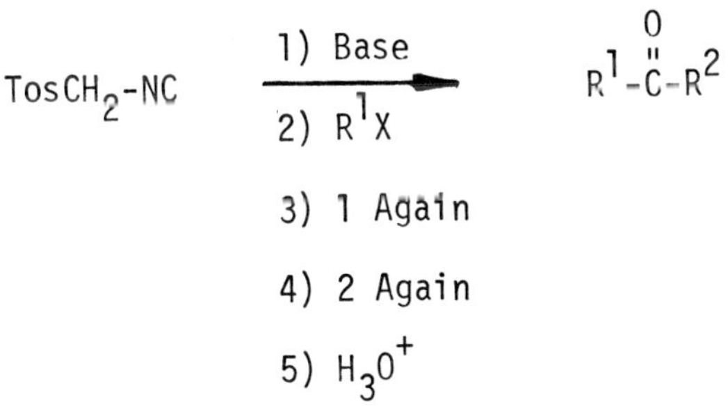

$TosCH_2-NC$

1) Base
2) R^1X
3) 1 Again
4) 2 Again
5) H_3O^+

$R^1-\overset{O}{\underset{\|}{C}}-R^2$

I.A.3-21 R. S. Torr and S. Warren, *J. Chem. Soc. Perkin I*, 1173 (1983).

$$\underset{\underset{O=C-R^2}{|}}{Ph_2\overset{O}{\overset{\|}{P}}-CH-R^1} \xrightarrow[2)\ R^3X]{1)\ Base} \underset{\underset{O=C-R^2}{|}}{Ph_2\overset{O}{\overset{\|}{P}}-\overset{R^3}{\underset{|}{C}}-R^1}$$

25-80%

Also, Michael additions.

I.A.4. Alkylation of N-, S- and Se-Stabilized Carbanions

I.A.4-1 A. E. Feiring, *J. Org. Chem.*, <u>48</u>, 347 (1983); S. I. Al-Khalil and W. R. Bowman, *Tetrahedron Lett.*, <u>24</u>, 2517 (1983); R. Beugelmans, A. Lechevallier and H. Rousseau, *ibid*, <u>24</u>, 1787 (1983).

$$CH_3-\underset{\underset{}{CH_3}}{\overset{}{CH}}-NO_2 \xrightarrow[\underset{2)\ R_fI,\ hv}{\phi H\ (-H_2O)}]{1)\ nBu_4N^+\ OH^-} CH_3-\underset{\underset{R_f}{|}}{\overset{\overset{CH_3}{|}}{C}}-NO_2$$

51-79%

R_f = Perfluoroalkyl.

I.A.4-2 H. Kurosawa et al., *J. Organometal. Chem.*, <u>250</u>, 83 (1983); A. R. Katritzky et al., *J. Am. Chem. Soc.*, <u>105</u>, 90 (1983).

$$Me_2\overset{-}{C}NO_2 \xrightarrow[MeOH\ or\ DMSO]{RTl(OAc)_2} R-\underset{\underset{Me}{|}}{\overset{\overset{Me}{|}}{C}}-NO_2$$

68-92%

I.A.4-3 A. Krowczynski and L. Kozerski, Synthesis, 489 (1983).

$$R^1CH_2NO_2 \xrightarrow[\Delta]{\substack{1) R^3R^4NH, HC(OEt)_3 \\ 2) R^2NH_2 \text{ (Excess)}}} \underset{O_2N}{\overset{R^1}{>}}=\underset{NHR^2}{\overset{H}{<}}$$

46-59%

I.A.4-4 J. E. Saavedra, J. Org. Chem., 48, 2388 (1983).

$$\underset{\underset{NO}{|}}{\overset{OMe}{\underset{N}{\diagdown}}}\hspace{-2pt}\diagup \xrightarrow{\substack{1) \text{ LDA} \\ 2) E^+ \\ 3) \text{ Deprotection}}} E-CH_2-NH_2$$

37-71%

Starting material from $CH_3NH_2 + CH_3CHO + MeOH + HONO$.

I.A.4-5 S. Warren et al., Tetrahedron Lett., 24, 2113, 3391 (1983).

$$\underset{R^2}{\overset{PhS}{\diagdown}}\hspace{-2pt}C=C\hspace{-2pt}\underset{R^1}{\overset{CO_2R^3}{\diagup}} \xrightarrow{\substack{1) KO^tBu, THF \\ -78°C \text{ to } 0°C. \\ 2) R^4X}} \underset{R^2}{\overset{PhS}{\diagdown}}\hspace{-2pt}\underset{R^1}{\overset{R^4 \, CO_2R^3}{\diagup}}$$

73-98%

I.A.4-6 T. Hayashi, I. Hori and T. Oishi, J. Am. Chem. Soc., 105, 2909 (1983).

$$\xrightarrow{\substack{1) \text{ LDA} \\ 2) R^1X \\ 3) \Delta \\ 2 \times [3,3] \\ 4) \text{ LDA} \\ 5) R^2X}}$$

79-97%

I.A.4-7 D. J. Ager, J. Chem. Soc., Perkin I, 1131 (1983);
T. Takeda et al., Bull. Chem. Soc. Jpn., 56, 967 (1983); Chem. Lett., 1285 (1983); F. Ogura, T. Otsubo and N. Ohira, Synthesis, 1006 (1983).

$$PhSCH_2SiMe_3 \xrightarrow[\text{Hexane}]{\text{1) nBuLi, TMEDA}} \xrightarrow{\text{2) RX}} PhS-\underset{R}{CH}-SiMe_3$$

0-99%

Full paper. Products converted to RCHO by oxidation, thermal rearrangement and hydrolysis.

I.A.4-8 V. Reutrakul and V. Rukachaisirikul, Tetrahedron Lett., 24, 725 (1983).

$$PhS-\underset{\underset{Li}{|}}{\overset{\overset{O}{\|}\ F}{C}}-H \xrightarrow[\substack{\text{THF} \\ -78 \text{ to } -20°C}]{\text{1) RCH}_2X} \xrightarrow{\text{2) }\Delta} RCH=CHF$$

49-79%

Reaction with aldehydes gives fluoromethyl ketones.

I.A.4-9 D. Scholz, Liebigs Ann. Chem., 98 (1983).

$$R^1-\overset{O}{\underset{\|}{C}}-\underset{R^2}{CH}-SO_2CH_2R^3 \xrightarrow[\text{2) }R^4X]{\text{1) NaH, BuLi}} R^1-\overset{O}{\underset{\|}{C}}-\underset{R^2}{CH}-CO_2CH\underset{R^4}{\overset{R^3}{<}}$$

15-94%

Bromination then Ramberg-Backlund of products gives olefins.

I.A.4-10 H. Kotake et al., Bull. Chem. Soc. Jpn., 56, 2539 (1983); G. Ourisson et al., Bull. Soc. Chim. Fr. II, 189 (1983); K. Tanaka et al., Chem. Lett., 633 (1983); V. Reutrakul and K. Herunsalee, Tetrahedron Lett., 24, 527 (1983).

$$RCH_2SO_2Ar \xrightarrow[\begin{array}{c}1)\ nBuLi \\ 2)\ R^2X \\ 3)\ nBuLi \\ 4)\ (MeS)_2 \\ 5)\ CuCl_2,\ SiO_2\end{array}]{} R^1-\overset{O}{\underset{\|}{C}}-R^2$$

I.A.4-11 H. Takayama et al., Chem. Lett., 1003 (1983); K. Sato et al., ibid, 725 (1983); A. Jonczyk and T. Radwan-Pytlewski, ibid, 1557 (1983); J. Org. Chem., 48, 910 (1983); M. Ochiai, T. Ukita and E. Fujita, Chem. Pharm. Bull., 31, 1641 (1983).

Reagents: 1) $(Me_3Si)_2NLi$ 2) R^1I 3) $(Me_3Si)_2NLi$ 4) R^2I

32-39%

Thermolytic desulfonylation gives E,Z-dienes.

I.A.4-12 B. M. Trost and M. Shimizu, J. Am. Chem. Soc., 105, 6757 (1983); B. M. Trost and R. W. Warner, ibid, 105, 5940 (1983); B. M. Trost and R. Remuson, Tetrahedron Lett., 24, 1129 (1983); F. Benedetti and C. J. M. Stirling, Chem. Commun., 1374 (1983).

Reagent: $CH_2(SO_2Ph)_2$, NaH, DMF

69-96%

Dibromide Prepared from Bis(silane).

I.A.4-13 K. Ogura et al., Chem. Lett., 767 (1983); H. Ikehira, S. Tanimoto et al., J. Org. Chem., 48, 1120 (1983).

$$CH_2 \begin{array}{c} SMe \\ SO_2Ar \end{array} \xrightarrow[\begin{array}{c} PhCH_3, RX \\ 2)\ H_2O_2,\ AcOH \\ 3)\ H^+,\ MeOH \end{array}]{1)\ 50\%\ Aq.\ NaOH} R-CO_2Me$$

Also, reactions with ArCHO.

I.A.4-14 J. V. Comasseto, J. Organometallic Chem., 253, 131 (1983).

Review: "Vinylic Selenides."

I.A.4-15 S. Warren et al., J. Chem. Soc., Perkin I, 2879, 2893 (1983); G. M. Blackburn and M. J. Parratt, Chem. Commun., 886 (1983).

$$Ph_2\overset{O}{\overset{\|}{P}}CH_2R^1 \xrightarrow[\begin{array}{c} 1)\ BuLi \\ 2)\ CuI \\ 3)\ Cl\diagdown\!\!\overset{Cl}{\diagup}\!\!= \\ 4)\ H_2SO_4,\ H_2O \end{array}]{} Ph_2\overset{O}{\overset{\|}{P}}\diagdown\!\!\underset{R^1}{\diagup}\!\!\diagdown\!\!=O \quad 52\%$$

Nine routes to β-(Diphenylphosphinoyl) Ketones.

I.A.5. Alkylations of Organometallic Reagents.

(see also: I.F, I.G).

I.A.5-1 G. Courtois and P. Miginiac, Bull. Soc. Chim. Fr. II, 148 (1983); L. Poncini, Bull. Soc. Chim. Belg., 92, 215 (1983).

$$nBuO-CH_2-NEt_2 \xrightarrow[\text{THF, 0°C}]{Cl(CH_2)_4MgBr} Cl(CH_2)_4-CH_2NEt_2$$
$$53-90\%$$

Also, reaction with immonium salts.

I.A.5-2 K. Tamao, N. Ishida and M. Kumada, J. Org. Chem., 48, 2120 (1983); R. A. Volkmann, J. T. Davis and C. N. Meltz, ibid, 48, 1767 (1983).

$$(iPrO)_2MeSiCH_2MgCl \xrightarrow[\substack{\text{Cat., THF} \\ \text{2) 30\% H}_2O_2}]{\text{1) RX}} HO-CH_2-R$$
$$47-87\%$$

RX = Alkyl, vinyl and aryl halides.

Cat. = CuI, $NiCl_2$ (dppp) or $PdCl_2$ (dppf).

I.A.5-3 M. Julia et al., Tetrahedron, 39, 3283, 3289 (1983).

<chemical reaction: allylic sulfone with OH group + nHexMgBr, 1% Cu(acac)$_2$, THF, 20°C → allylic alcohol with nHex substituent, 90%>

I.A.5-4 M. F. Semmelhack and R. Tamura, *J. Am. Chem. Soc.*, **105**, 4099 (1983); J. E. Baldwin et al., *Chem. Commun.*, 1040 (1983); K. J. H. Kruithof, R. F. Schmitz and G. W. Klumpp, *Tetrahedron*, **39**, 3073 (1983); J. Klaveness and K. Undheim, *Acta Chem. Scand. B*, **37**, 258 (1983).

$$(CO)_4Fe=C\begin{matrix}OLi\\Ph\end{matrix} \quad \xrightarrow[\text{2) FeCl}_3]{\text{1) EtI, THF, HMPA}} \quad Et-\underset{\underset{O}{\|}}{C}-Ph \quad 53\%$$

Control of Site of Alkylation Based on Hard-Soft Acid-Base Theory.

I.A.5-5 V. Calo, L. Lopez and W. F. Carlucci, *J. Chem. Soc., Perkin I*, 2953 (1983); S. Suzuki et al., *Tetrahedron Lett.*, **24**, 5103 (1983); *Synthesis*, 804 (1983); S. Araki and Y. Butsugan, *Bull. Chem. Soc. Jpn.*, **56**, 1446 (1983); H. J. Liu and L. K. Ho, *Can. J. Chem.*, **61**, 632 (1983); G. Consiglio, F. Morandini and O. Piccolo, *Chem. Commun.*, 112 (1983).

benzimidazole-2-S-geranyl $\xrightarrow[\text{THF, -30°C}]{\text{4 nBuMgBr, 1 CuBr}}$ nBu-geranyl 98% (GC)

I.A.5-6 P. Mangeney, A. Alexakis and J. F. Normant, *Tetrahedron Lett.*, **24**, 373 (1983); M. Huche et al., *ibid*, **24**, 585 (1983); F. J. Sardina, A. Mourino and L. Castedo, *ibid*, **24**, 4477 (1983); N. R. Schmuff and B. M. Trost, *J. Org. Chem.*, **48**, 1404 (1983).

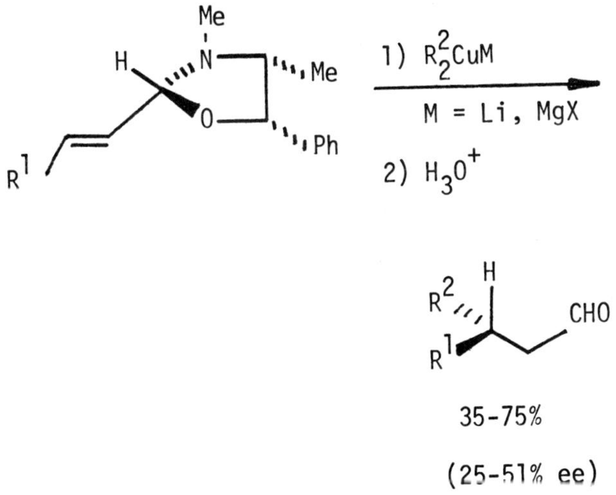

CARBON–CARBON BOND FORMING REACTIONS

1) R_2^2CuM $\quad M = Li, MgX$

2) H_3O^+

35-75%

(25-51% ee)

I.A.5-7 H. L. Goering et al., *J. Org. Chem.*, **48**, 715, 721, 1531, 3986 (1983).

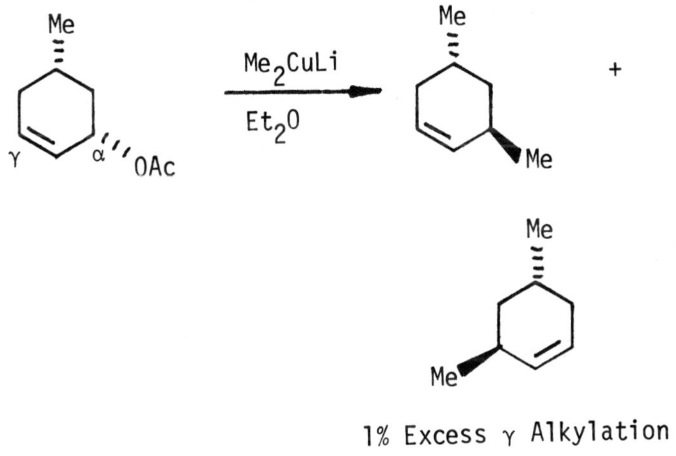

Me_2CuLi / Et_2O

1% Excess γ Alkylation

I.A.5-8 T. Fujisawa et al., *Tetrahedron Lett.*, **24**, 5745 (1983).

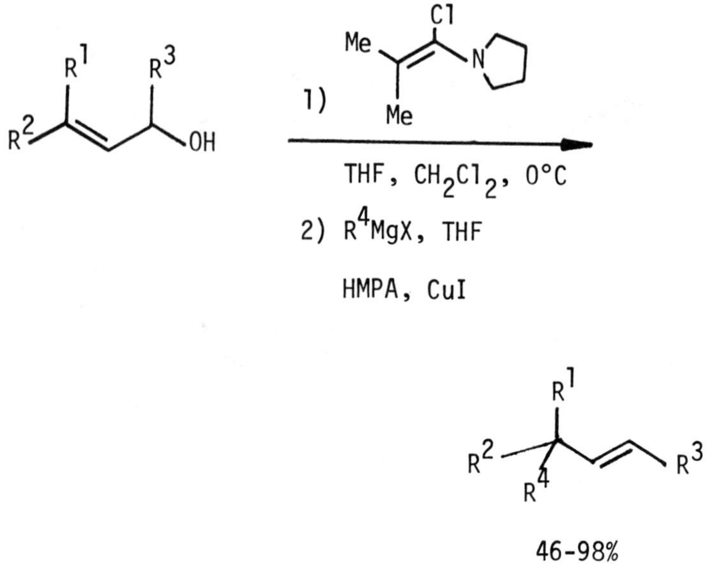

$R^2\text{-}CR^1\text{=}CH\text{-}CH(OH)R^3$

1) Me-C(Me)=C(Cl)-N(pyrrolidine), THF, CH_2Cl_2, 0°C
2) R^4MgX, THF, HMPA, CuI

→ $R^2R^1R^4C\text{-}CH\text{=}CH\text{-}R^3$

46-98%

I.A.5-9   J. Klein, Tetrahedron, 39, 2733 (1983).

Review: "Directive Effects in Allylic and Benzylic Polymetalations: the Question of U-stabilization, Y-Aromaticity and Cross-Conjugation."

I.A.5-10   P. Beak, J. E. Hunter and Y. M. Jun, J. Am. Chem. Soc., 105, 6350 (1983).

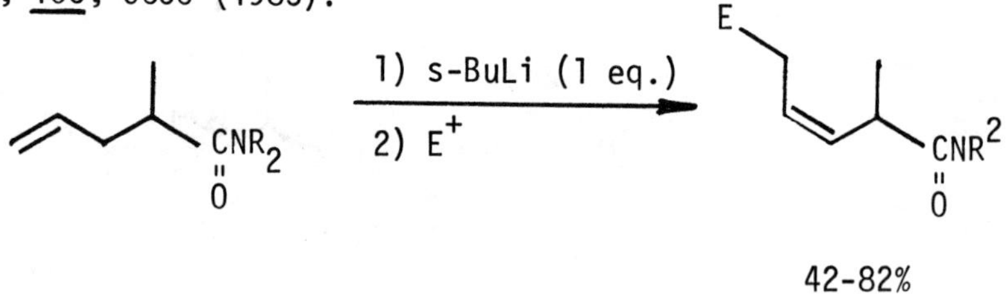

$CH_2\text{=}CH\text{-}CH_2\text{-}CH(Me)\text{-}C(=O)NR_2$

1) s-BuLi (1 eq.)
2) E^+

→ $E\text{-}CH_2\text{-}CH\text{=}CH\text{-}CH(Me)\text{-}C(=O)NR_2$ (Z-alkene)

42-82%

E^+ = MeI, $Me_2C\text{=}O$, Me_3SiCl.

I.A.5-11 A. Nakamura et al., Organometallics, 2, 21 (1983).

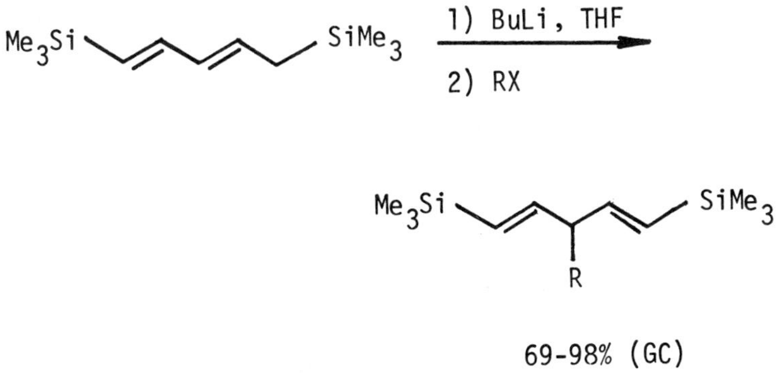

Me₃Si–CH=CH–CH=CH–CH₂–SiMe₃ → 1) BuLi, THF 2) RX

Me₃Si–CH=CH–CH(R)–CH=CH–SiMe₃

69-98% (GC)

I.A.5-12 R. M. Carlson and L. L. White, Synth. Commun., 13, 237 (1983); A. A. Kandil and K. N. Slessor, Can. J. Chem., 61, 1166 (1983); H. L. Holland and Jahangir, Tetrahedron Lett., 24, 1577 (1983); T. L. Ho, Synth. Commun., 13, 769 (1983).

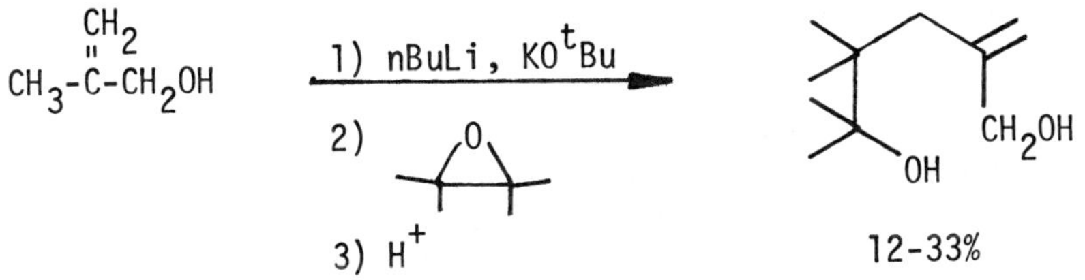

CH₃-C(=CH₂)-CH₂OH → 1) nBuLi, KOᵗBu 2) epoxide 3) H⁺ → product, CH₂OH, OH

12-33%

I.A.5-13 W. R. Roush, M. A. Adam and S. M. Peseckis, Tetrahedron Lett., 24, 1377 (1983); R. S. Matthews et al., J. Org. Chem., 48, 409 (1983).

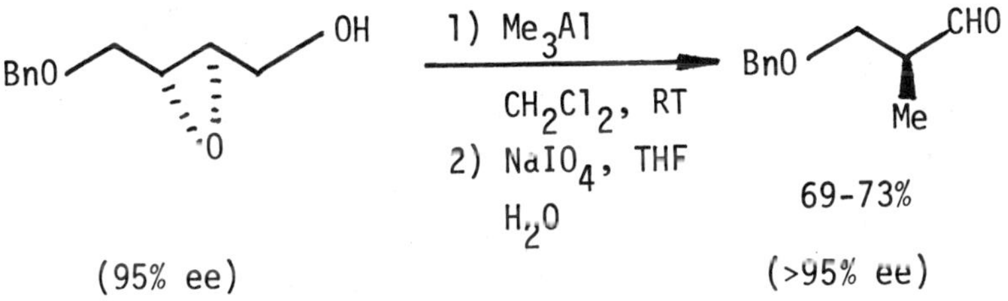

BnO–CH₂–CH(O)–CH₂–CH₂–OH → 1) Me₃Al, CH₂Cl₂, RT 2) NaIO₄, THF, H₂O → BnO–CH₂–CH(Me)–CHO

(95% ee)   69-73%
                                  (>95% ee)

I.A.5-14 B. H. Lipshutz et al., J. Org. Chem., 48, 546, 3306, 3334 (1983).

"Chemistry of Higher Order Mixed Cuprates. Choice of the Cu(I) Salt for the Formation of R_2CuLi."

I.A.5-15 M. E. Jung and P. K. Lewis, Synth. Commun., 13, 213 (1983); P. Viallefont et al., Tetrahedron Lett., 24, 3717, 3721 (1983).

$$R^1CHO \xrightarrow[\text{Et}_2O]{\begin{array}{l}1) \text{ Me}_3\text{SiI}\\ 2) \text{ R}^2\text{CuLi}\\ 3) \text{ H}_2\text{O}\end{array}} R^1-\underset{R^2}{C}H-OH$$

48-77%

I.A.5-16 L. Hamon and J. Levisalles, J. Organometal. Chem., 251, 133; 253, 259 (1983); A. E. Greene et al., J. Org. Chem., 48, 4763 (1983).

$$R^1\underset{X}{C}H\overset{O}{\underset{\|}{C}}R^2 \xrightarrow[\text{Et}_2O, \text{ DMF}]{\text{MeCuCNLi}} R^1\underset{Me}{C}H\overset{O}{\underset{\|}{C}}R^2$$

Comparison with Me_2CuLi.

I.A.5-17 A. J. Pearson, S. L. Kole and B. Chen, J. Am. Chem. Soc., 105, 4483 (1983).

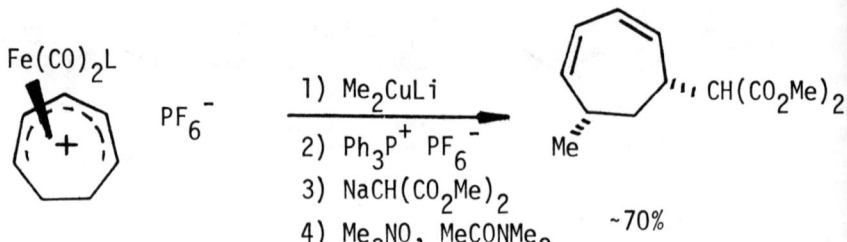

~70%

I.A.5-18 K. B. Sharpless et al., Pure Appl. Chem., 55, 589 (1983).

Review: "Stereo and Regioselective Openings of Chiral 2,3-Epoxy Alcohols, Versatile Routes to Optically Pure Natural Products and Drugs, Unusual Kinetic Resolutions."

I.A.5-19 G. Teutsch and G. Costerousse, J. Chem. Res. (S), 294 (1983); J. A. Marshall et al., J. Am. Chem. Soc., 105, 3360, 6515 (1983).

$\xrightarrow{\text{R}_2\text{CuMgBr} \cdot \text{Me}_2\text{S}}{\text{THF}}$

52-97%

I.A.5-20 T. Fujisawa et al., Bull. Chem. Soc. Jpn., 56, 345 (1983); T. Kato, M. Sato et al., Chem. Pharm. Bull., 31, 1108, 4346 (1983).

$\text{BrMg(CH}_2)_8\text{MgBr}$

1) 2 [β-propiolactone], THF
   Li_2CuCl_4
2) H_3O^+
3) EtOH, $-\text{H}_2\text{O}$

$\longrightarrow \text{EtO}_2\text{C-(CH}_2)_{12}\text{-CO}_2\text{Et}$

60-68%

I.A.5-21 C. Mioskowski, S. Manna and J. R. Falck, Tetrahedron Lett., 24, 5521 (1983); P. A. Wender and G. B. Dreyer, ibid, 24, 4543 (1983); R. F. Newton et al., J. Chem. Soc., Perkin I, 683 (1983).

cyclopropyl methyl ketone + R$_2$CuCNLi$_2$·BF$_3$, Et$_2$O, Hexane, −78°C → R–CH$_2$CH$_2$C(O)CH$_3$ 0-78%

I.A.5-22 M. Koreeda and S. G. Mislankar, J. Am. Chem. Soc., 105, 7203 (1983); J. Barluenga, J. Florez and M. Yus, J. Chem. Soc., Perkin I, 3019 (1983); J. G. Smith, E. Oliver and T. J. Boettger, Organometallics, 2, 1577 (1983).

3-(OiBu)-cyclopent-2-enone
1) 2.5 LDA, THF, −78°C
2) ICH$_2$CMe$_2$CHO
3) MeOCH$_2$Cl
→ bicyclic product 65%

I.A.5-23 A. I. Meyers and L. M. Fuentes, J. Am. Chem. Soc., 105, 117 (1983); G. W. Klumpp et al., Rec. Trav. Chim., 102, 542 (1983).

1,2,3,4-tetrahydroisoquinoline N-CH=N-R*
1) LDA, −78°C
2) RX, −78°C or −100°C
3) NH$_2$NH$_2$, HOAc
→ 1-R-1,2,3,4-tetrahydroisoquinoline 65-97% (10→99% ee)

I.A.5-24 R. H. Mitchell, T. W. Dingle and R. V. Williams, J. Org. Chem., 48, 903 (1983); D. Wilhelm, T. Clark and P. v. R. Schleyer, Chem. Commun., 211 (1983); D. K. Ellison and R. T. Iwamoto, Tetrahedron Lett., 24, 31 (1983); T. A. Engler and H. Schechter, ibid, 24, 4645 (1983); M. Takagi, M. Nojima and S. Kusabayashi, J. Am. Chem. Soc., 105, 4676 (1983).

Br_2 or Me_3SiCl gave ring substitution (C-1).

I.A.5-25 M. Braun and E. Ringer, Tetrahedron Lett., 24, 1233 (1983); R. D. Wood and B. Ganem, ibid, 24, 4391 (1983); T. A. Oster and T. M. Harris, J. Org. Chem., 48, 4307 (1983).

E^+ = RX or PhCHO

I.A.5-26 H. Shechter et al., J. Am. Chem. Soc., 105, 6096, 6104 (1983); A. R. Katritsky et al., Tetrahedron, 39, 4133 (1983).

18-66%

E^+ = MeI, CH_3COCl, CO_2, Me_3SiCl, $\underset{CH_2-CH_2}{\overset{O}{\triangle}}$.

I.A.5-27 S. G. Davies et al., Chem. Commun., 1316 (1983).

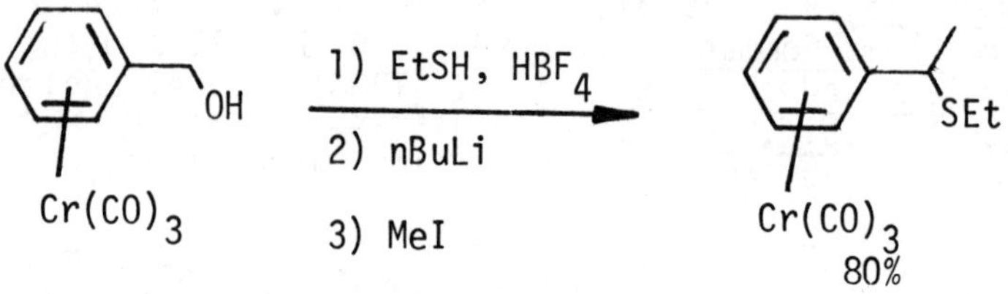

PhCH$_2$OH·Cr(CO)$_3$ → 1) EtSH, HBF$_4$ 2) nBuLi 3) MeI → PhCH(SEt)CH$_3$·Cr(CO)$_3$ 80%

Wittig rearrangement suppressed.

I.A.5-28 M. Bourhis, J. J. Bosc and R. Golse, J. Organometal. Chem., 256, 193 (1983).

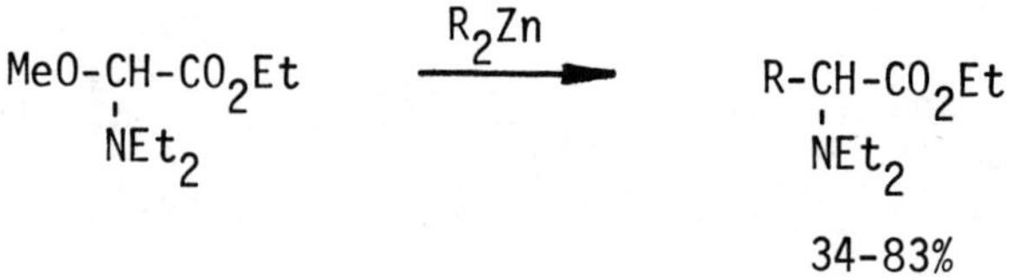

MeO-CH(NEt$_2$)-CO$_2$Et $\xrightarrow{R_2Zn}$ R-CH(NEt$_2$)-CO$_2$Et

34-83%

I.A.5-29 J. P. Godschalx and J. K. Stille, Tetrahedron Lett., 24, 1905 (1983); E. Kleinan and M. Peretz, J. Org. Chem., 48, 5302 (1983).

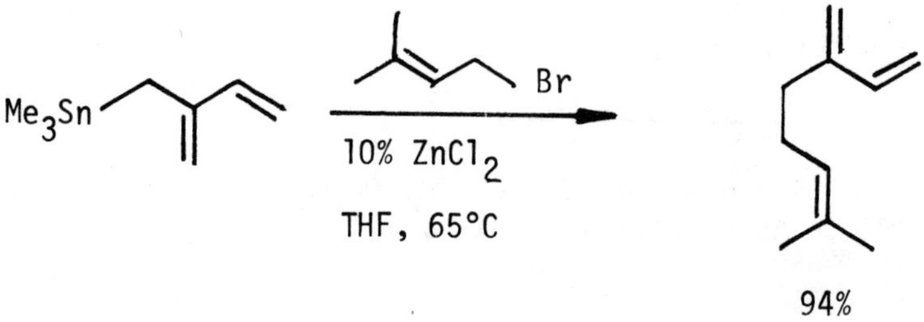

Me$_3$Sn-CH$_2$-C(=CH$_2$)-CH=CH$_2$ + (CH$_3$)$_2$C=CH-CH$_2$-Br $\xrightarrow[\text{THF, 65°C}]{10\% \text{ ZnCl}_2}$ product

94%

CARBON-CARBON BOND FORMING REACTIONS

I.A.6. Other Alkylation Procedures and Reviews

I.A.6-1 K. Akiba et al., <u>Tetrahedron Lett.</u>, <u>24</u>, 1711, 1715 (1983); R. P. Alexander and I. Patterson, <u>ibid</u>, <u>24</u>, 5911 (1983).

$$\text{PhS-CH(Cl)-(CH}_2)_n\text{-C(O)-R}^1 \quad \xrightarrow[\text{2) } R^2R^3C=CHCH_2SiMe_3]{\text{1) Lewis Acid}}$$

(n = 0-2)

$$\text{PhS-CH[CH(R}^3)\text{C(R}^2)=CH_2]\text{-(CH}_2)_n\text{-C(O)-R}^1$$

20-91%

I.A.6-2 A. Hosomi, Y. Sakata and H. Sakurai, <u>Chem. Lett.</u>, 405, 409 (1983); K. Itoh et al., <u>J. Org. Chem.</u>, <u>48</u>, 1557 (1983); A. P. Kozikowski et al., <u>Tetrahedron Lett.</u>, <u>24</u>, 1563 (1983).

$$\text{Me}_3\text{SiCH}_2\text{CH=CH}_2 \quad \xrightarrow[\text{Me}_3\text{SiI or Me}_3\text{SiOTf}]{\text{PhCH}_2\text{OCH}_2\text{Cl}} \quad \text{CH}_2\text{=CHCH}_2\text{CH}_2\text{OCH}_2\text{Ph}$$

78%

Also, reactions of TMS enol ethers.

I.A.6-3 P. A. Bartlett, W. S. Johnson and J. D. Elliott, J. Am. Chem. Soc., 105, 2088 (1983); S. J. Hathaway and L. A. Paquette, J. Org. Chem., 48, 3351 (1983); T. Hayashi et al., Tetrahedron Lett., 24, 5661 (1983).

$$\text{dioxane} \xrightarrow[\text{TiCl}_4, \text{CH}_2\text{Cl}_2]{1) \diagdown \!\!\!\text{SiMe}_3} \text{RC(OH)(H)CH}_2\text{CH=CH}_2$$

1) ⟶ SiMe$_3$ / TiCl$_4$, CH$_2$Cl$_2$
2) [O]
3) Na, Et$_2$O

59-94%
(>65% ee)

I.A.6-4 A. G. Schultz and J. P. Dittami, Tetrahedron Lett., 24, 1369 (1983); S. Bhattacharyya, B. Basu and D. Mukherjee, Tetrahedron, 39, 4221 (1983); K. Pramod, H. Ramanathan and G. S. R. Subba Rao, J. Chem. Soc., Perkin I, 7 (1983); D. F. Taber, B. P. Gunn and I. C. Chiu, Org. Syn., 61, 59 (1983).

(o-methoxy methyl benzoate)
1) Reductive Alkylation / RX
2) Bromination
3) Dehydrobromination

⟶ cyclohexadienone with R, CO$_2$Me

I.A.6-5 F. Bickelhaupt et al., Tetrahedron Lett., 24, 3935 (1983).

Cp$_2$Ti(CH$_2$)(Cl)(MgBr)
1) R^{1}CH=CR^{2}R^3
2) Br$_2$
⟶ R^{1}CH(Br)–C(R^2)(CH$_2$Br)–R^3

I.A.6-6   R. Baker, V. B. Rao and E. Erdik, J. Organometal. Chem., 243, 451 (1983);   P. Lennon and M. Rosenblum, J. Am. Chem. Soc., 105, 1233 (1983).

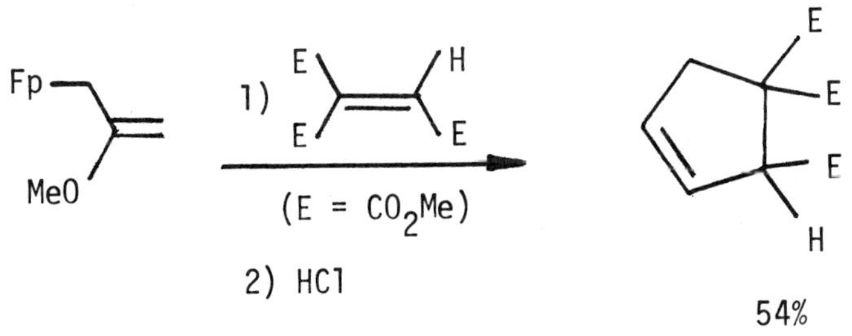

[Scheme: Fp-C(=CH2)(OMe) + 1) (E)(H)C=C(E)(E) where E = CO2Me; 2) HCl → cyclopentene bearing two quaternary centers with E,E and E,H substituents, 54%]

Fp = Cp(CO)FeL

I.A.6-7   B. DePoorter, J. Muzart and J. P. Pete, Organometallics, 2, 1494 (1983).

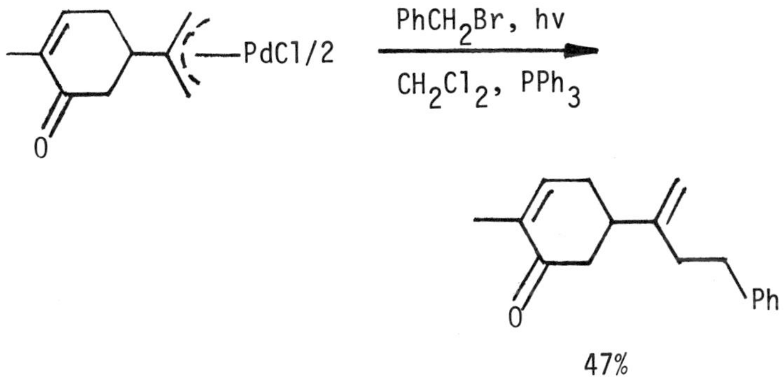

[Scheme: (methylcyclohexenone)-allyl–PdCl/2 complex + PhCH2Br, hv, CH2Cl2, PPh3 → methylcyclohexenone with CH2=C(CH2CH2Ph)– substituent, 47%]

I.A.6-8   G. G. Yakobson and N. E. Akhmetova, Synthesis, 169 (1983).

Review: "Alkali Metal Fluorides in Organic Synthesis."

I.A.6-9 H. W. Pinnick, Org. Prep. Proceed. Int., 15, 199 (1983).

Review: "Potassium Hydride in Organic Synthesis."

I.A.6-10 P. Brownbridge, Synthesis, 1, 85 (1983).

Review: "Silyl Enol Ethers in Synthesis - Parts I and II."

I.A.6-11 Tetrahedron 39, Issue 6 (1983).

Reviews: "Recent Developments in the Use of Silicon in Organic Synthesis." (Tetrahedron Report - 19 Papers).

I.A.6-12 I. Fleming and N. K. Terrett, Pure Appl. Chem., 55, 1707 (1983).

Review: "Some Uses of Silicon Compounds in Organic Synthesis."

I.A.7. Nucleophilic Addition to Electron Deficient Carbon

I.A.7.a.1a. Intermolecular Aldol-Type 1,2 Additions

I.A.7.a.1a-1 T. Mukaiyama, Pure Appl. Chem., 55, 1749 (1983).

Review: "Metal Enolates in Organic Synthesis."

I.A.7.a.1a-2 F. M. Menger, Tetrahedron, 39, 1013 (1983).

Review: "Directionality of Organic Reactions in Solution."

I.A.7.a.1a-3 T. Ando et al., Bull. Chem. Soc. Jpn., 56, 1885 (1983).

KF-Alumina - Solid Base for Elimination, Michael Addition, Aldol Condensation, and Darzens Condensation.

I.A.7.a.1a-4 S. H. Mashraqui and R. M. Kellogg, J. Am. Chem. Soc., 105, 7792 (1983); C. Chuit, R. J. P. Corriu and C. Reye, Synthesis, 294 (1983).

$$\text{Dihydropyridine} \xrightarrow[\substack{\text{Mg(ClO}_4)_2 \cdot 2\text{H}_2\text{O} \\ \text{CH}_3\text{CN, RT} \\ \text{2) H}_2\text{O}}]{\text{1) RCHO}} \text{Ph-CO-CH}_2\text{-CH(OH)-R} \quad 37\text{-}90\%$$

Aldol Under Unusually Mild Conditions.

I.A.7.a.1a-5 S. Thaisrivongs and D. Seebach, J. Am. Chem. Soc., 105, 7407 (1983); C. H. Wong and G. M. Whitesides, J. Org. Chem., 48, 3199 (1983).

$$R^1\text{-CH(OH)-*CH-CO}_2\text{Me} \xrightarrow{\substack{1) \alpha\text{-Alkylation} \\ (R_2X) \\ 2) \text{Wittig Olefination} \\ 3) \text{Benzyloxymercuration} \\ 4) \text{Demercuration}}}$$

New Strategy for Aldol-Type Products.

$$R^1 \overset{*}{-} \underset{R^2}{\overset{OR}{\overset{|}{C}}} \overset{*}{-} \overset{OH}{\overset{|}{C}} \overset{*}{-} CO_2Me$$

I.A.7.a.1a-6 P. Ballesteros, B. W. Roberts and J. Wong, J. Org. Chem., 48, 3603 (1983); N. Ono, A. Kaji et al., Tetrahedron Lett., 24, 3477 (1983); G. Schneider et al., Synthesis, 665 (1983); B. W. Disanayaka and A. C. Weedon, ibid, 952 (1983).

$$CH_2(CO_2{}^tBu)_2 \xrightarrow[\substack{KOAc, HOAc \\ Cu(OAc)_2 \cdot H_2O \\ 100°C}]{(CH_2O)_n} CH_2=C(CO_2{}^tBu)_2$$

53%

I.A.7.a.1a-7 G. A. Kraus and P. Gottschalk, J. Org. Chem., 48, 2111 (1983).

$$\underset{BrCH_2}{\overset{H}{>}}C=C\underset{H}{\overset{CH_2Br}{<}} \xrightarrow[\substack{\text{2) } Ph_3P, 0°C \\ \text{3) } Nu^-}]{\text{1) } O_3, CH_2Cl_2, -78°C} \underset{Nu}{\overset{OH}{\underset{|}{C}}}\!\!\!\diagup\!\!\!\diagdown Br$$

60-100%

$Nu^- = PhC(OLi)=CH_2$, PhLi, RLi, $CH_2=C(OLi)-O^tBu$, etc.

I.A.7.a.1a-8 R. Noyori et al., J. Org. Chem., 48, 932 (1983).

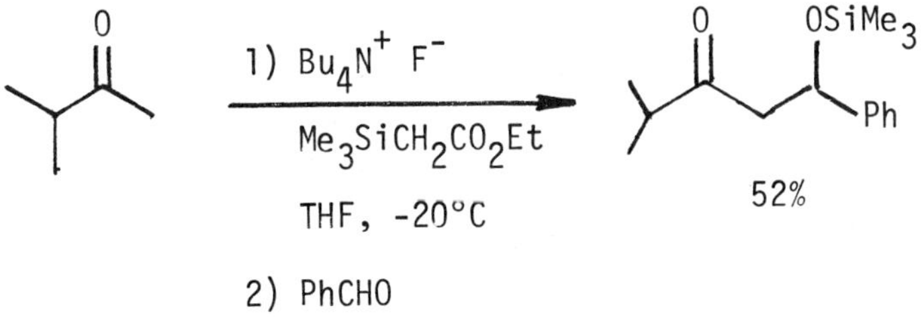

[Reaction: 3-methyl-2-butanone + 1) Bu$_4$N$^+$ F$^-$, Me$_3$SiCH$_2$CO$_2$Et, THF, -20°C; 2) PhCHO → product with OSiMe$_3$ and Ph, 52%]

Full paper.

I.A.7.a.1a-9 M. T. Reetz, B. Wenderoth and R. Peter, Chem. Commun., 406 (1983).

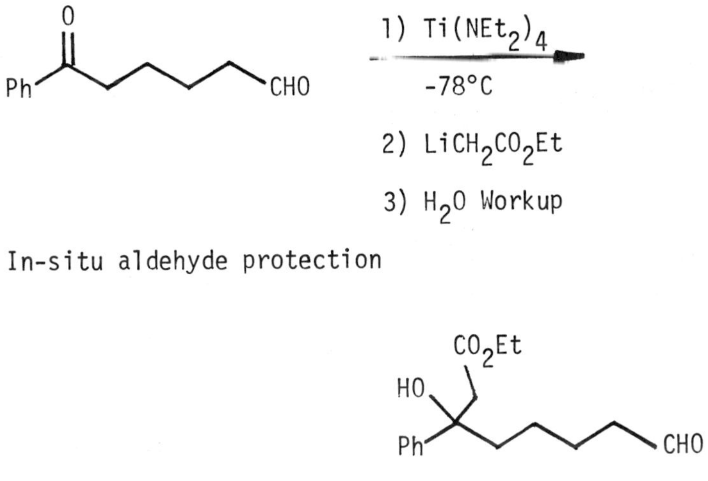

[Reaction: PhCO(CH$_2$)$_3$CHO + 1) Ti(NEt$_2$)$_4$, -78°C; 2) LiCH$_2$CO$_2$Et; 3) H$_2$O workup → HO, Ph, CO$_2$Et, CHO product]

In-situ aldehyde protection

I.A.7.a.1a-10 M. T. Reetz et al., Angew. Chem., Int. Ed. Engl., 22, 989 (1983); J. Am. Chem. Soc., 105, 4833 (1983).

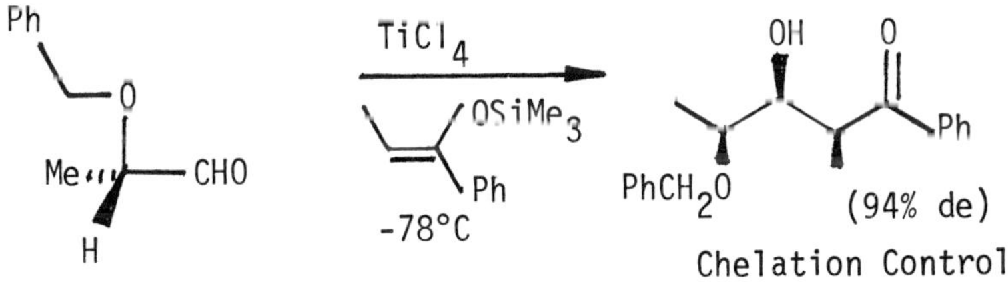

[Reaction: PhCH$_2$O-protected α-methyl aldehyde + TiCl$_4$, CH$_2$=C(OSiMe$_3$)Ph, -78°C → product with OH and C(=O)Ph, PhCH$_2$O (94% de)]

Chelation Control

I.A.7.a.1a-11 H. Urabe and I. Kuwajima, Tetrahedron Lett., 24, 5001 (1983); M. Yamana, T. Ishihara and T. Ando, ibid, 24, 507 (1983).

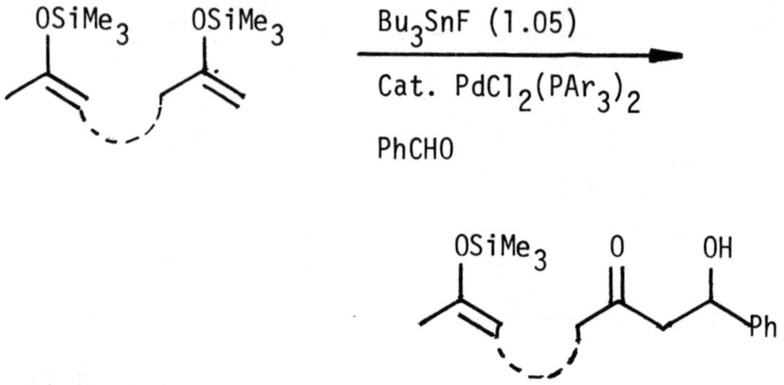

$$\text{(OSiMe}_3\text{)(OSiMe}_3\text{)-diene} \xrightarrow[\text{Cat. PdCl}_2(\text{PAr}_3)_2 \quad \text{PhCHO}]{\text{Bu}_3\text{SnF (1.05)}} \text{OSiMe}_3\text{-CH=C(Me)-CH}_2\text{-CO-CH}_2\text{-CH(OH)-Ph}$$

71%

I.A.7.a.1a-12 Y. Yamamoto, K. Maruyama and K. Matsumoto, J. Am. Chem. Soc., 105, 6963 (1983); I. Kuwajima et al., Tetrahedron Lett., 24, 3341, 3343, 3347 (1983).

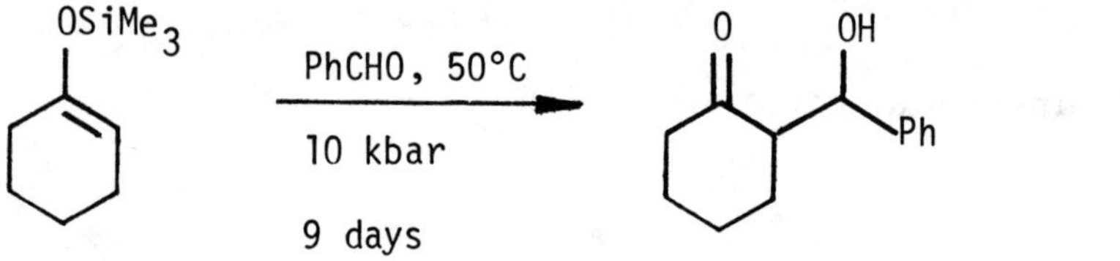

$$\text{1-OSiMe}_3\text{-cyclohexene} \xrightarrow[\text{10 kbar, 9 days}]{\text{PhCHO, 50°C}} \text{2-(CH(OH)Ph)-cyclohexanone}$$

90%

(Erythro:Threo = 75:25)

I.A.7.a.1a-13 K. Yamamoto and Y. Tomo, Chem. Lett., 531 (1983);
C. H. Heathcock and L. A. Flippin, J. Am. Chem. Soc., 105,
1667 (1983); C. H. Heathcock and B. L. Finkelstein, Chem.
Commun., 919 (1983); C. Goasdoue, N. Goasdoue and M. Gaudemar,
Tetrahedron Lett., 24, 4001 (1983); Y. E. Rhodes et al., Synth.
Commun., 13, 449 (1983).

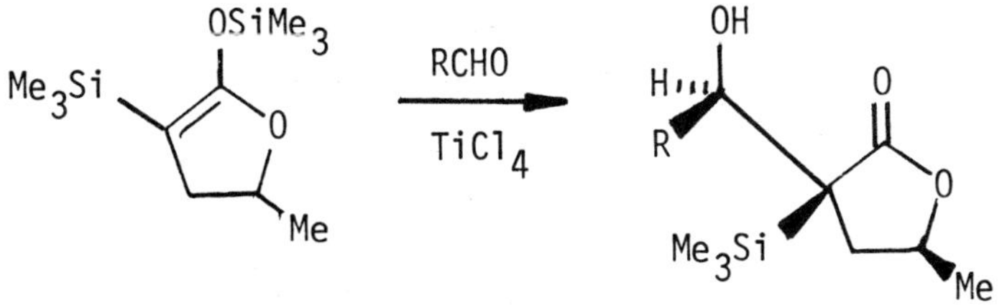

82-98%

(100% Threo)

I.A.7.a.1a-14 T. Mukaiyama et al., Chem. Lett., 297, 595,
1727, 1799, 1825 (1983); S. Kozima et al., ibid, 851 (1983).

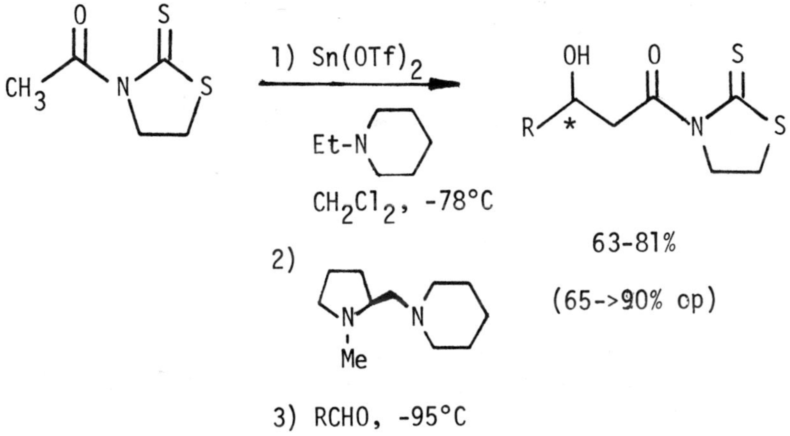

63-81%

(65->90% op)

I.A.7.a.1a-15 R. Baker and J. A. Devlin, Chem. Commun., 147 (1983); I. Patterson, S. K. Patel and J. R. Porter, Tetrahedron Lett., 24, 3395 (1983); M. Braun and R. Devant, Angew. Chem., Int. Ed. Engl., 22, 788 (1983).

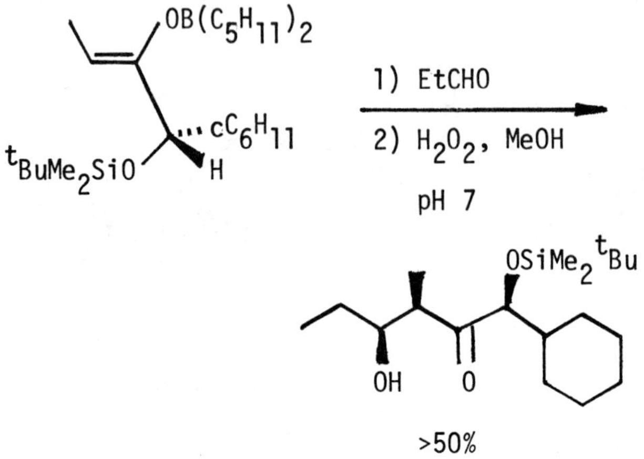

1) EtCHO
2) H_2O_2, MeOH
pH 7

>50%

I.A.7.a.1a-16 T. Imamoto, T. Kusumoto and M. Yokoyama, Tetrahedron Lett., 24, 5233 (1983).

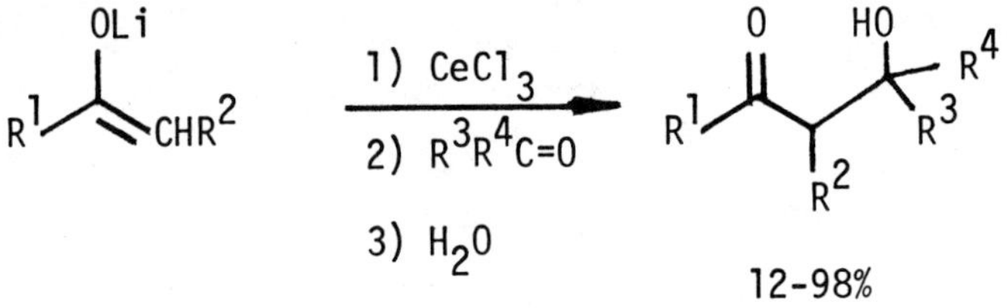

1) $CeCl_3$
2) $R^3R^4C=O$
3) H_2O

12-98%

I.A.7.a.1a-17 C. H. Heathcock and J. Lampe, J. Org. Chem., 48, 4330 (1983); F. Babudri, L. DiNunno and S. Florio, Tetrahedron Lett., 24, 3883 (1983); R. F. Newton et al., J. Chem. Soc., Perkin I, 1809 (1983).

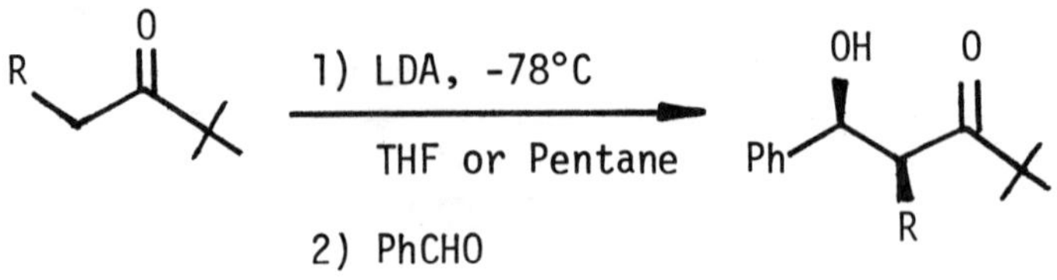

1) LDA, -78°C
   THF or Pentane
2) PhCHO

R = Me, Et, nPr, nBu.

I.A.7.a.1a-18 F. Germain, Y. Chapleur and B. Castro, Synthesis, 119 (1983).

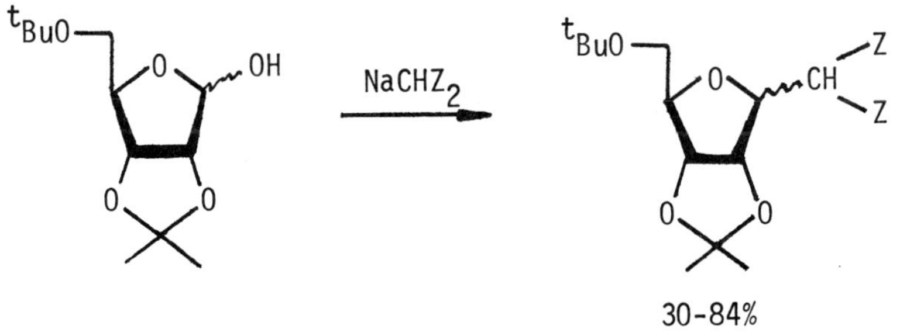

30-84%

I.A.7.a.1a-19 J. Baghdadchi and C. A. Panetta, J. Org. Chem., 48, 3852 (1983); R. J. Crawford, J. Org. Chem., 48, 1366 (1983); J. Guyot and A. Kergomard, Tetrahedron, 39, 1161, 1167 (1983); C. Degrand, P. L. Compagnon and F. Gasquez, Chem. Commun., 383 (1983); O. Attanasi, P. Filippone and A. Mei, Synth. Commun., 13, 1203 (1983); H. Yamanaka et al., Heterocycles, 20, 1541 (1983).

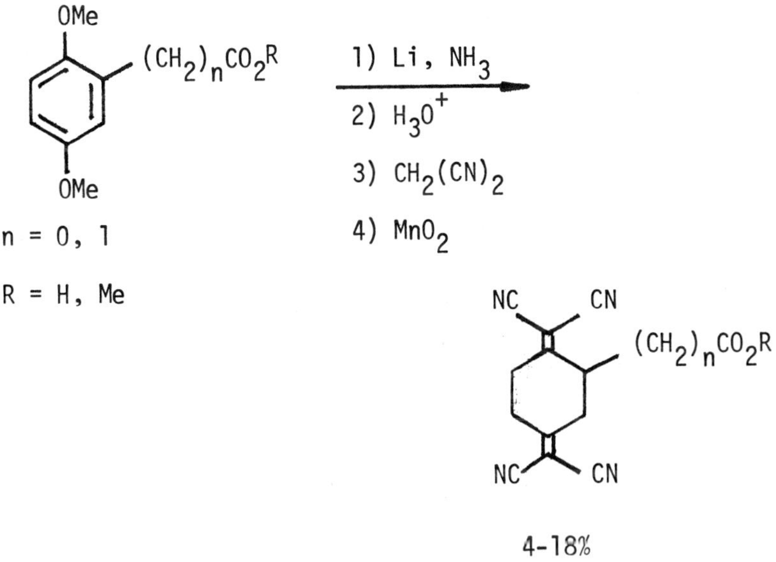

n = 0, 1

R = H, Me

1) Li, NH_3
2) H_3O^+
3) $CH_2(CN)_2$
4) MnO_2

4-18%

I.A.7.a.1a-20   R. Tanikaga et al., Synthesis, 134 (1983).

$$\underset{R^2}{\overset{R^1}{\diagdown}}CH-CHO \xrightarrow[\text{piperidine-NH}]{MeO_2C-CH_2-\overset{O}{\underset{\|}{S}}-Ar} \underset{R^2}{\overset{R^1}{\diagdown}}\underset{OH}{\overset{}{C}}-CH=CH-CO_2Me$$

64-84%

Knoevenagel, double-bond shift, [2,3]-sigmatropic rearrangement, then hydrolysis.

I.A.7.a.1a-21   J. O. Karlsson and T. Frejd, J. Org. Chem., 48, 1921 (1983); K. Annen et al., Liebigs Ann. Chem., 712 (1983).

<chemical reaction: 3-methylcyclohexanone + POCl$_3$, DMF, Cl$_2$C=CHCl, 60°C, 3 hr, (52% yield) → 2-chloro-1-formyl-4-methylcyclohexene (90) + 2-chloro-1-formyl-5-methylcyclohexene (10)>

Regioselectivity of Vilsmeier-Haack reaction.

I.A.7.a.1a-22   A. A. Croteau and J. Termini, Tetrahedron Lett., 24, 2481 (1983); N. Matsumura, A. Kunugihara, S. Yoneda, ibid, 24, 3239 (1983); N. Y. Grigorieva, I. M. Avrutov and A. V. Semenovsky, ibid, 24, 5531 (1983).

$$Me_3Si\underset{R^1}{\diagdown}\overset{N-{}^tBu}{\underset{\|}{C}}-CH_3 \xrightarrow[\substack{2)\ R^2R^3C=O\\ -78°\ to\ 0°C\\ 3)\ Hydrolysis}]{1)\ LDA,\ THF,\ 0°C} \underset{R^1}{\overset{R^2}{\diagdown}}C=\underset{R^3}{\overset{}{C}}-CO-CH_3$$

40-86%

I.A.7.a.1a-23 U. Schollkopf et al., Liebigs Ann. Chem., 1133 (1983); D. M. Hrubowchak and F. X. Smith, Tetrahedron Lett., 24, 4951 (1983); D. Villemin, Chem. Ind. (London), 478 (1983).

$$\text{iPr}_{\text{,,}}\diagdown\underset{\text{MeO}}{\overset{\text{N}}{\diagup}}\diagdown\text{OMe} \quad \xrightarrow[\text{3) H}^+]{\begin{array}{l}\text{1) BuLi}\\ \text{2) R}^1\text{COR}^2\end{array}} \quad \text{iPr}_{\text{,,}}\diagdown\underset{\text{MeO}}{\overset{\text{N}}{\diagup}}\diagdown\overset{\text{OMe}}{\underset{R^1\;R^2}{\overset{\text{C-OH}}{\diagdown}}}$$

92-95%

(>85% de)

I.A.7.a.1a-24 Y. Sato and S. Takeuchi, Synthesis, 734 (1983); I. Paterson, Tetrahedron Lett., 24, 1311 (1983); G. L. Larson, F. Quiroz and J. Suarez, Synth. Commun., 13, 833 (1983).

$$\underset{\text{Me}_3\text{Si}}{\overset{\text{Me}_3\text{Si}}{\diagdown}}\text{C=C}\underset{\text{O}^t\text{Bu}}{\overset{\text{OLi}}{\diagup}} \quad \xrightarrow[\text{3) R}^2\text{CHO}]{\begin{array}{l}\text{1) R}^1\text{CHO}\\ \text{2) Q}^+\text{ F}^-\end{array}} \quad R^1\text{CH=C}\underset{\text{CH-R}^2}{\overset{\text{CO}_2{}^t\text{Bu}}{\diagup}} \;\; \underset{\text{OH}}{}$$

35-84%

I.A.7.a.1a-25 J. d'Angelo et al., Tetrahedron Lett., 24, 5869 (1983); A. Pelter et al., ibid, 24, 523 (1983); W. Ladner, Chem. Ber., 116, 3413 (1983); S. Matui, K. Tanaka and A. Kaji, Synthesis, 127 (1983).

$$\underset{\text{Me}}{\overset{\text{CH}_2\text{-CO}_2{}^t\text{Bu}}{\underset{\text{O}}{|}\;\;\text{Ph}}} \quad \xrightarrow[\text{3) H}_2,\text{ Pd(OH)}_2]{\begin{array}{l}\text{1) LDA}\\ \text{2) CH}_3\text{CHO (-120°C)}\end{array}} \quad \overset{\text{OH}}{\underset{\text{OH}}{\diagup\diagdown}}\text{CO}_2{}^t\text{Bu}$$

80-90%

(>90% Erythro)

I.A.7.a.1a-26   C. Scolastico et al., <u>Chem. Commun.</u>, 1112 (1983)

$Me_2N\diagdown\diagdown\diagdown CO_2{}^tBu$ $\xrightarrow{\begin{array}{l}1)\ LDA,\ THF\\ 2)\ \underset{R^1\diagup\diagdown CHO}{\overset{OR^2}{|}}\\ 3)\ MeI\\ 4)\ DBU\end{array}}$ $R^1\underset{\overline{\equiv}}{\overset{OR^2}{|}}\diagdown\underset{OH}{\overset{}{|}}\diagdown\overset{\diagup\!\!\!\diagdown}{C}CO_2{}^tBu$

55-65%

(~80% Indicated Diast.)

I.A.7.a.1a-27   B. Castro et al., <u>Bull. Soc. Chim. Fr. II</u>, 207, 230 (1983);   A. Takeda, S. Sato et al., <u>Tetrahedron Lett.</u>, <u>24</u>, 2393 (1983).

$Cl_3CCO_2Me \xrightarrow[\begin{array}{c}(Me_2N)_3P\\ MgCl_2\end{array}]{PhCHO} \underset{OH\ \ Cl}{\overset{Cl}{PhCH-C-CO_2Me}}$

64-80%

I.A.7.a.1a-28   Y. Tanura et al., <u>Chem. Pharm. Bull.</u>, <u>31</u>, 52 (1983);   A. Fischer and G. N. Henderson, <u>Tetrahedron Lett.</u>, <u>24</u>, 131 (1983);   C. Wakselman et al., <u>Synthesis</u>, 322 (1983).

[cyclohexenone with Ph C=O and Cl] $\xrightarrow[THF,\ -78°C]{LiCH_2CO_2Et}$ [product with EtO_2CCH_2, OH, Ph C=O, Cl]

50%

I.A.7.a.1a-29 S. M. Hannick and Y. Kishi, J. Org. Chem., 48, 3833 (1983); C. Trombini, A. Umani-Ronchi et al., ibid, 48, 4108 (1983).

$$R^1O_2C-CHR^2-Br \xrightarrow[\text{THF, }\Delta]{\begin{array}{c}1)\ R^3CN \\ \text{Active Zn} \\ 2)\ H_3O^+\end{array}} R^1O_2C-CHR^2-C(=O)-R^3$$

43-81%

I.A.7.a.1a-30 P. J. Cowan and M. W. Rathke, Synth. Commun., 13, 183 (1983); S. N. Kulkarni et al., Ind. J. Chem., 22B, 684 (1983); R. G. Bhandari and G. V. Bhide, ibid, 22B, 331 (1983); R. M. Weinstein, W. L. Wang and D. Seyferth, J. Org. Chem., 48, 3367 (1983).

$$CH_3CO_2SiMe_3 \xrightarrow[\begin{array}{c}2)\ RCOCl \\ 3)\ H_3O^+, \Delta\end{array}]{\begin{array}{c}1)\ LDA,\ THF \\ -78°C\end{array}} R-\overset{O}{\underset{\|}{C}}-CH_3$$

26-99% (GC)

I.A.7.a.1a-31 R. W. Saalfrank and W. Rost , Angew Chem.,Int. Ed. Engl., 22, 321 (1983).

$$(EtO)_2C=C=C(OEt)_2 \xrightarrow{Cl_2C=O} O=C=C(CO_2Et)_2$$

I.A.7.a.1a-32 R. M. DiPardo and M. G. Bock, Tetrahedron Lett., 24, 4805 (1983); T. A. Oster and T. M. Harris, ibid, 24, 1851 (1983).

propionyl-(4S)-isopropyl-oxazolidinone

1) LDA, THF
   $-78°C$
2) $(^tBoc\text{-}D\text{-}Ala)_2O$

^{t}BocNH—CH(CH$_3$)—C(O)—CH(CH$_3$)—C(O)—N(oxazolidinone with isopropyl)

55%

I.A.7.a.1a-33 W. Wierenga and H. I. Skulnick, Org. Syn., 61, 5 (1983); D. V. Rao and F. A. Stuber, Synthesis, 308 (1983).

$CH_2(CO_2Et)(CO_2H)$

1) 2 nBuLi, THF
2) $Me_2CHCOCl$

$Me_2CH\text{-}\underset{\underset{O}{\|}}{C}\text{-}CH_2\text{-}CO_2Et$

80%

I.A.7.a.1a-34 J. L. LaMattina and C. J. Mularski, Tetrahedron Lett., 24, 2059 (1983); G. Guanti et al., ibid, 24, 817 (1983); S. Ohta et al., Synthesis, 715 (1983).

$BrCH_2\underset{\underset{OEt}{|}}{\overset{\overset{OEt}{|}}{C}}\text{-}CO_2\text{-}C_6H_4\text{-}NO_2$

$\xrightarrow{LiCH_2CN}$

$BrCH_2\text{-}\underset{\underset{OEt}{|}}{\overset{\overset{OEt}{|}}{C}}\text{-}\underset{\underset{}{\|}}{\overset{\overset{O}{\|}}{C}}\text{-}CH_2CN$

92%

I.A.7.a.1a-35 L. N. Mander and S. P. Sethi, Tetrahedron Lett., 24, 5425 (1983); P. E. Eaton and P. G. Jobe, Synthesis, 796 (1983).

[Cyclopentanone with R substituent]
1) LDA, THF, -78°C
2) HMPA
3) MeO$_2$C-CN
4) H$_2$O
→ [Cyclopentanone with R and CO$_2$Me substituents]   65-96%

I.A.7.a.1a-36 K. Takahashi et al., J. Org. Chem., 48, 3566 (1983); K. Takahashi et al., Chem. Lett., 859 (1983); K. Takahashi et al., Synthesis, 1043 (1983); J. Simonet et al., Tetrahedron, 39, 1551 (1983).

Ph,Me-N-CH$_2$-CN
1) R^{1}R^{2}C=O
2) KH
3) H$_3$O$^+$, Δ
→ R^{1}R^{2}CH-CO$_2$H   24-82%

I.A.7.a.1a-37 K. Kobayashi and T. Hiyama, Tetrahedron Lett., 24, 3509 (1983); M. Larcheveque, P. Perriot and Y. Petit, Synthesis, 297 (1983).

Li-CR^1(CN) + R^{2}R^{3}C(OR^4)(CN)
THF, -78°C
→ R^{2}R^{3}C(OR^4)-C(=C(R^1)(CN))-NH$_2$ 51-99%

I.A.7.a.1a-38 J. P. Celerier, M. G. Michaud and G. Lhommet, Synthesis, 195 (1983); R. N. Renaud, D. Berube and C. J. Stephens, Can. J. Chem., 61, 1379 (1983); J. Liebscher, B. Neumann and H. Hartmann, J. Prakt. Chem., 325, 915 (1983).

$$\text{piperidinium chloride} \xrightarrow[\text{Et}_3\text{N, CHCl}_3]{\text{Meldrum's acid}} \text{product}$$

55-97%

I.A.7.a.1a-39 T. V. Khenkina and V. B. Mochalin, J. Org. Chem. (USSR), 19, 1109 (1983); T. Mukaiyama, Y. Goto and S. I. Shoda, Chem. Lett., 671 (1983).

$$\text{cyclohexanone} \xrightarrow[\substack{\text{BF}_3 \cdot \text{Et}_2\text{O, Et}_2\text{O} \\ 5^\circ\text{C}}]{\text{PhCH=NCO}_2\text{Et}} \text{product}$$

73%

I.A.7.a.1a-40 N. Matsumura, N. Asai and S. Yoneda, Chem. Commun., 1487 (1983); N. Matsumura, T. Ohba and S. Yoneda, Chem. Lett., 317 (1983).

$$R^1\overset{O}{\underset{}{C}}CH_2R^2 \xrightarrow[\substack{\text{DMF} \\ 2)\ \text{H}_2\text{O}^+}]{1)\ \text{BrMgO}_2\text{C-N}\underset{\text{S}}{\overset{}{=}}\text{C-N-CO}_2\text{MgBr}} R^1\overset{O}{\underset{}{C}}\underset{\underset{\text{CO}_2\text{H}}{|}}{C}HR^2$$

49-80%

I.A.7.a.1b. Intramolecular Aldol-Type 1,2-Additions

I.A.7.a.1b-1 D. Caine, E. Crews and J. M. Salvino, Tetrahedron Lett., 24, 2083 (1983); D. Bottger and P. Welzel, ibid, 24, 5201 (1983).

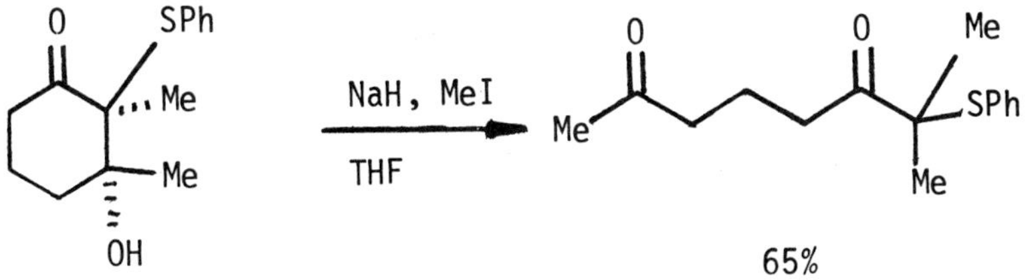

I.A.7.a.1b-2 D. P. Curran, Tetrahedron Lett., 24, 3443 (1983); A. Dormond, A. El Bouadili and C. Moise, ibid, 24, 3087 (1983); K. B. G. Torsell et al., Tetrahedron, 39, 2227, 2231, 2237, 2241, 2247 (1983); S. Innocenti et al., Synthesis, 124 (1983), G. Cainelli et al., J. Org. Chem., 48, 123 (1983); T. L. Ho, Synth. Commun., 13, 435 (1983); F. X. Kohl and P. Jutzi, J. Organometal. Chem., 243, 119 (1983).

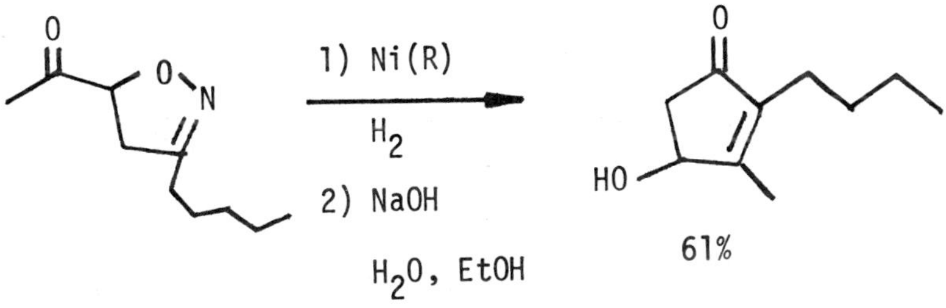

I.A.7.a.1b-3 R. J. Pariza, F. Kuo and P. L. Fuchs, Synthesis, 13, 243 (1983); P. R. R. Costa, C. C. Lopes and A. V. Pinto, Synth. Commun., 13, 691 (1983); S. Niwas, S. Kumar and A. P. Bhaduri, Ind. J. Chem., 22B, 524 (1983).

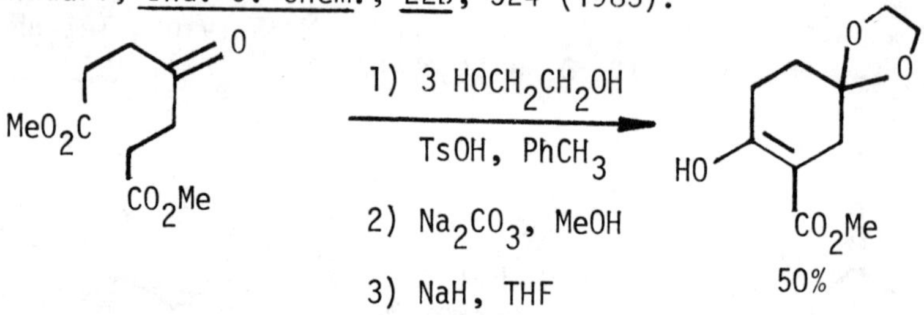

O
MeO_2C
CO_2Me
1) 3 $HOCH_2CH_2OH$
TsOH, $PhCH_3$
2) Na_2CO_3, MeOH
3) NaH, THF
O
O
HO
CO_2Me
50%

Second Route to product also.

I.A.7.a.1b-4 H. O. House et al., J. Org. Chem., 48, 5285 (1983).

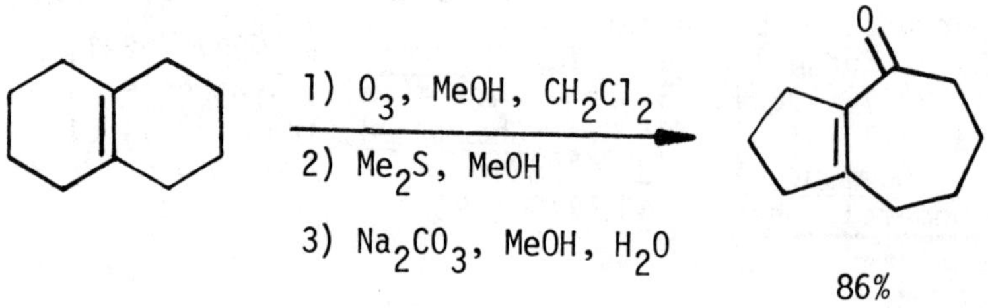

1) O_3, MeOH, CH_2Cl_2
2) Me_2S, MeOH
3) Na_2CO_3, MeOH, H_2O
O
86%

I.A.7.a.1b-5 J. Ficini, G. Stork et al., Tetrahedron Lett., 24, 907 (1983).

O
O
NEt_2
O
1) 1% Aq. NaOH
2) 5% Aq. HCO_2H
H
$CONEt_2$
O
36%

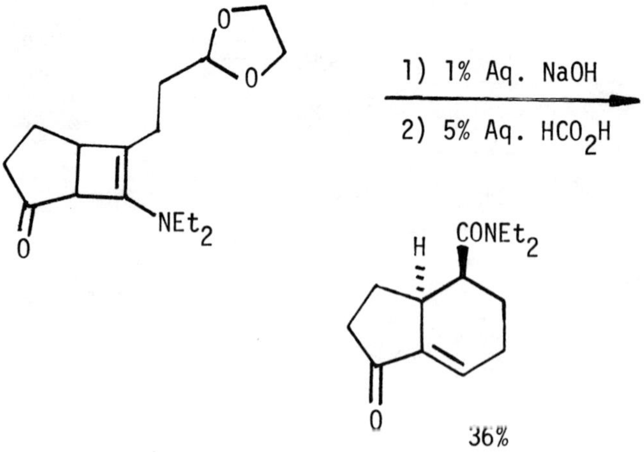

I.A.7.a.2. 1,2-Additions of N-, S- or Se-Stabilized Carbanions.

I.A.7.a.2-1 G. Rosini, R. Ballini and P. Sorrenti, <u>Synthesis</u>, 1014 (1983); D. H. R. Barton, W. B. Motherwell and S. Z. Zaid, Bull. Soc. Chim. Fr. II, 61 (1983);   T. Fujisawa, Y. Kurita and T. Sato, <u>Chem. Lett.</u>, 1537 (1983).

$$\underset{R^2}{\overset{R^1}{>}}CH-NO_2 \quad \xrightarrow[\text{Dry } Al_2O_3]{R_3CHO} \quad R^1-\underset{\underset{R^2}{|}}{\overset{\overset{NO_2}{|}}{C}}-\underset{OH}{\overset{|}{C}}H-R^3$$

69-86%

I.A.7.a.2-2 K. S. Petrakis, G. Batu and J. Fried, <u>Tetrahedron Lett.</u>, <u>24</u>, 3063 (1983); G. Rosini, R. Ballini and P. Sorrenti, <u>Tetrahedron</u>, <u>39</u>, 4127 (1983).

[Reaction scheme: PhCH=CH-C(O)-CH(OMe)(OH) + O₂N-CH₂CH₂-OSiMe₂ᵗBu, THF, H₂O, Me₃N, 0°C → PhCH=CH-C(O)-C(O)-CH₂CH₂-OSiMe₂ᵗBu]

90%

I.A.7.a.2-3 S. Warren et al., Tetrahedron Lett., 24, 3927, (1983); A. de Groot and B. J. M. Jansen, Synth. Commun., 13, 985 (1983).

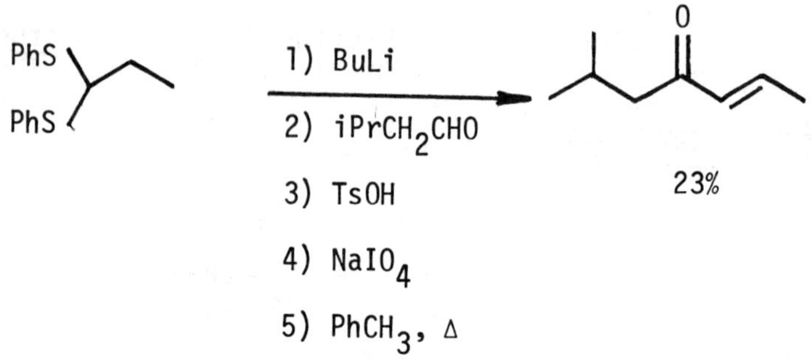

PhS−CH(CH₃)−SPh

1) BuLi
2) iPrCH$_2$CHO
3) TsOH
4) NaIO$_4$
5) PhCH$_3$, Δ

→ iPr-CH$_2$-C(=O)-CH=CH-CH$_3$ 23%

I.A.7.a.2-4 Y. Ohtsuka and T. Oishi, Chem. Pharm. Bull., 31, 443, 454 (1983).

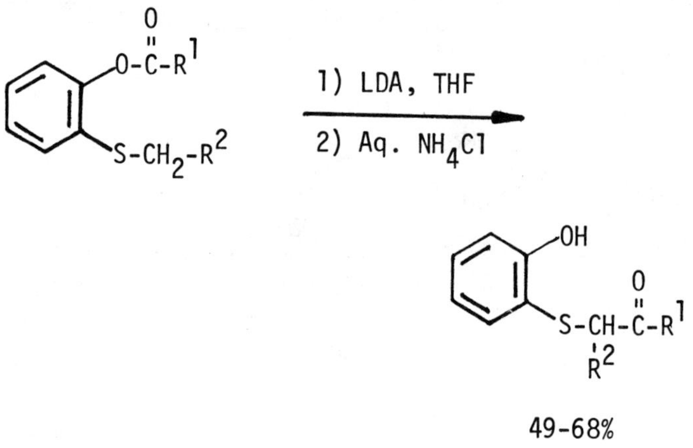

[o-(acyloxy)phenyl] S-CH$_2$-R^2, O-C(=O)-R^1

1) LDA, THF
2) Aq. NH$_4$Cl

→ o-hydroxyphenyl S-CH(R^2)-C(=O)-R^1

49-68%

I.A.7.a.2-5 K. Fuji, M. Ueda and E. Fujita, Chem. Commun., 49 (1983); J. Thiem et al., Liebigs Ann. Chem., 2185 (1983).

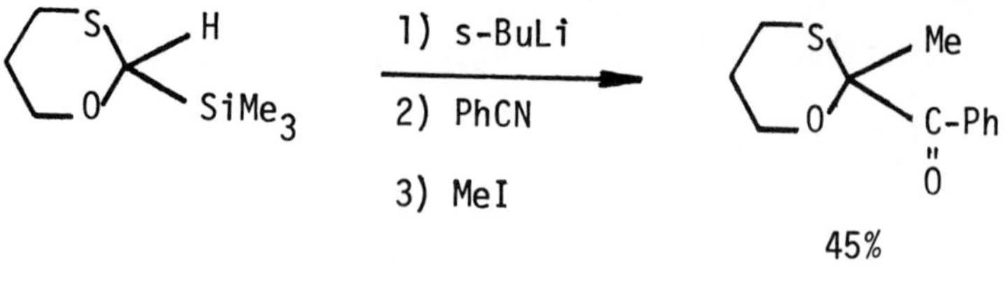

1,3-oxathiane-2-H, 2-SiMe$_3$

1) s-BuLi
2) PhCN
3) MeI

→ 1,3-oxathiane-2-Me, 2-C(=O)-Ph

45%

I.A.7.a.2-6 M. Rosenberger, N. Cohen et al., J. Am. Chem. Soc., 105, 3656, 3661 (1983); B. Zwanenburg et al., Synthesis, 628 (1983); M. H. H. Nkunya and B. Zwanenburg, Rec. Trav. Chim., 102, 461 (1983); F. Fischer and P. Palitzsch, J. Prakt. Chem., 325, 835 (1983).

$$R-C\equiv C-CH_2-CH=CH-CH_2-\overset{+}{S}(CH_2CH_2CH_2CH_2)\ Br^- \xrightarrow[\text{Aq. NaOH, } CH_2Cl_2]{OHC(CH_2)_3CO_2Me}$$

$$R-C\equiv C-CH=CH-\underset{O}{\overset{H}{C-C}}-(CH_2)_3CO_2Me$$

72%

I.A.7.a.2-7 C. N. Hsiao and H. Schechter, Tetrahedron Lett., 24, 2371 (1983); P. Kocienski and M. Todd, J. Chem. Soc., Perkin I, 1777, 1783 (1983).

$$Me_3SiCH_2CH_2\overset{O}{\underset{\|}{S}}Ph \xrightarrow[\text{2) } R^1R^2C=O]{\text{1) MeLi}} \underset{H}{\overset{Me_3Si}{C}}=\underset{R^1\ R^2}{\overset{H}{C-C-OH}}$$

81-89%

Also, reactions with epoxides.

I.A.7.a.2-8 R. Annunziata, M. Cinquini and A. Gilardi, Synthesis, 1016 (1983); C. Scolastico, M. Cinquini et al., Chem. Commun., 403 (1983); M. Cinquini, F. Montanari et al., ibid, 1138 (1983); G. Solladie, R. Zimmerman and R. Bartsch, Tetrahedron Lett., 24, 755 (1983).

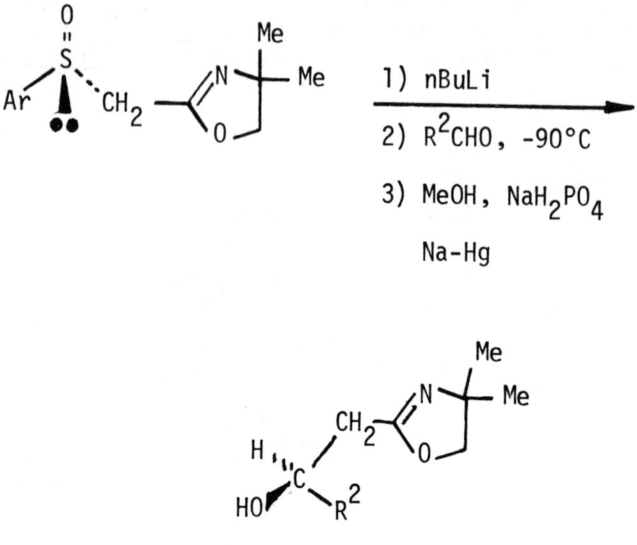

Ar—S(=O)—CH₂ + oxazoline(Me,Me)

1) nBuLi
2) R²CHO, −90°C
3) MeOH, NaH₂PO₄
   Na-Hg

→ HO-C(H)(R²)-CH₂-oxazoline

60-85%

(26-53% ee)

I.A.7.a.2-9 K. S. Kyler and D. S. Watt, *J. Am. Chem. Soc.*, **105**, 619 (1983); H. Uda et al., *Chem. Commun.*, 1065 (1983); D. D. Ridley and M. A. Smal, *Aust. J. Chem.*, **36**, 1049 (1983); M. Julia et al., *Tetrahedron Lett.*, **24**, 4311, 4315, 4319 (1983).

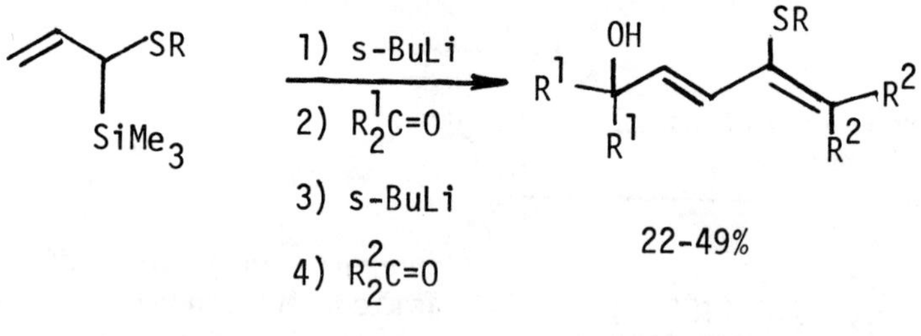

CH₂=CH-CH(SR)(SiMe₃)

1) s-BuLi
2) R¹₂C=O
3) s-BuLi
4) R²₂C=O

→ R¹₂C(OH)-CH=CH-C(SR)=CR²₂

22-49%

I.A.7.a.2-10 L. L. Vasil'eva, V. I. Mel'nikova and K. K. Pivnitskii, J. Org. Chem. (USSR), 19, 581 (1983); C. Fehr, Helv. Chim. Acta, 66, 2519 (1983).

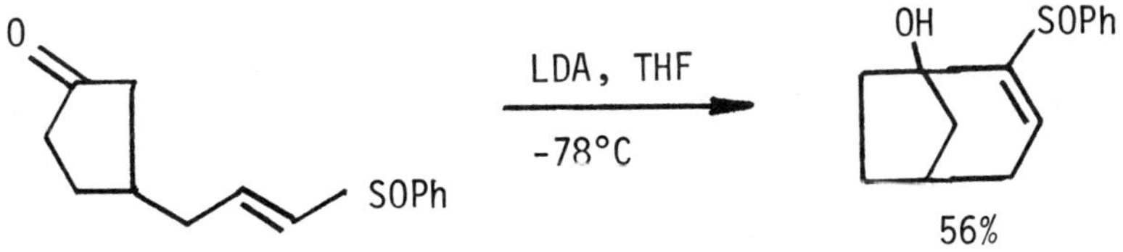

56%

I.A.7.a.2-11 A. F. Ermolov and A. F. Eleev, J. Org. Chem. (USSR), 19, 1200 (1983).

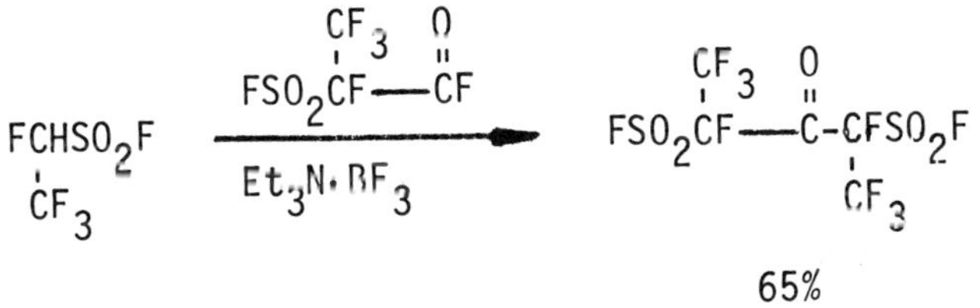

65%

I.A.7.a.2-12 Y. Yamamoto, Y. Saito and K. Maruyama, J. Org. Chem., 48, 5408 (1983); T. Sakakibara, M. d. Manandhar and Y. Ishido, Synthesis, 920 (1983).

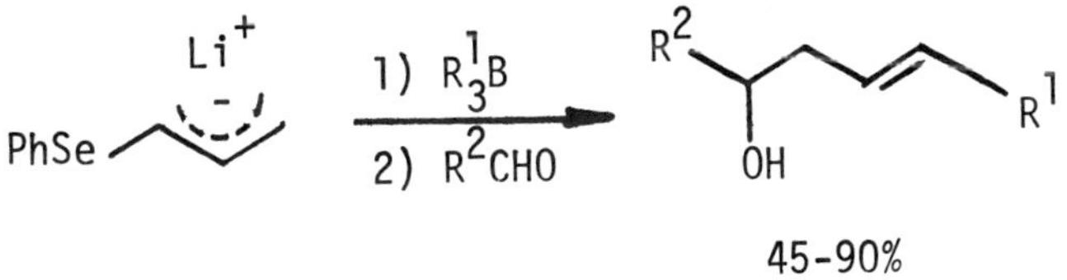

45-90%

I.A.7.a.3. 1,2-Additions of Grignard-type Carbanions

I.A.7.a.3-1 T. Shono, Y. Matsumura and T. Kanazawa, Tetrahedron Lett., 24, 4577 (1983); P. Coutrot, J. R. Dormoy and A. Moukimou, J. Organometal. Chem., 258, C25 (1983).

$$\underset{\underset{CO_2Me}{|}}{\overset{}{N}}-\overset{}{CH}-\overset{\overset{O}{\|}}{P}Ph_2 \quad \xrightarrow[\substack{2)\ PhCHO \\ 3)\ \Delta \\ 4)\ H_2,\ PtO_2 \\ 5)\ LAH}]{1)\ LDA} \quad \underset{\underset{|}{N}}{\overset{}{}} -CH(CH_3)-CH(OH)-Ph$$

25%

I.A.7.a.3-2 E. E. Aboujaoude, N. Collignon and P. Savignac, Synthesis, 634 (1983); K. C. Tang and J. K. Coward, J. Org. Chem., 48, 5001 (1983); H. J. Bestmann, K. Kumar and L. Kisielowski, Chem. Ber., 116, 2378 (1983); L. Capuano, T. Triesch and A. Willmes, ibid, 116, 3767 (1983).

$$(R^1O)_2\overset{\overset{O}{\|}}{P}CH\underset{R^3}{\overset{R^2}{<}} \quad \xrightarrow[\substack{2)\ Me_2NCHO \\ 3)\ H^+}]{1)\ nBuLi} \quad (R^1O)_2\overset{\overset{O}{\|}}{P}-\underset{R^3}{\overset{R^2}{C}}-CHO$$

51-88%

I.A.7.a.3-3 E. J. Corey and T. M. Eckrich, Tetrahedron Lett., 24, 3163, 3165 (1983).

$$^tBuOCH_3 \quad \xrightarrow[\substack{KO^tBu \\ 2)\ E^+}]{1)\ s\text{-}BuLi} \quad ^tBuOCH_2E$$

63-93%

$E^+ = CO_2$, RCHO, R^1R^2CO, RCOCl, α,β-enones (+CuBr).

I.A.7.a.3-4   D. Hoppe and A. Bronneke, Tetrahedron Lett., 24, 1687 (1983);   J. Armand et al., J. Org. Chem., 48, 2847 (1983); K. Smith et al., J. Chem. Res. (S), 30, 31 (1983).

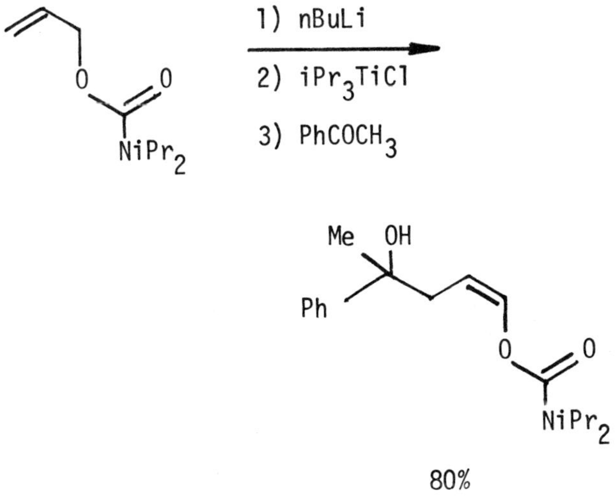

I.A.7.a.3-5   N. H. Wertiuk, Tetrahedron, 39, 205 (1983).

Review: "Homoenolate Anions and Homoenolate Anion Equivalents."

I.A.7.a.3-6   J. B. Stothers et al., Chem. Commun., 204 (1983).

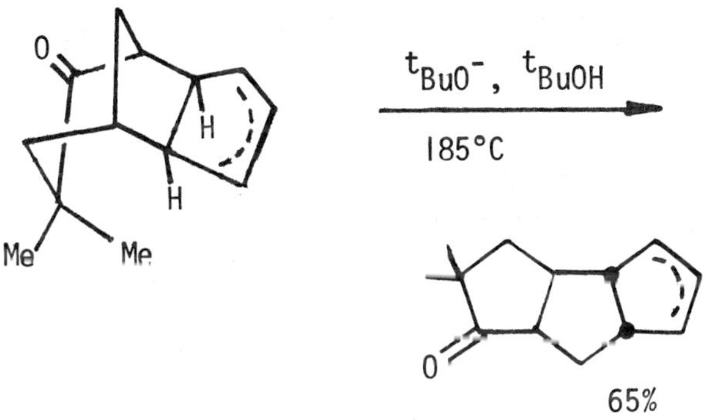

I.A.7.a.3-7   D. Seebach et al., Org. Syn., 61, 24, 42 (1983).

Ph-CHO $\xrightarrow[\text{-140°C}]{\text{nBuLi, Isopentane, DBB}}$ Ph-CH(OH)-nBu

80-90% (30% oy)

DBB = Chiral Covalent (Prepared on p. 24).

I.A.7.a.3-8   T. Hiyama, M. Obayashi and M. Sawahata, Tetrahedron Lett., 24, 4113 (1983);   D. Seyferth et al., J. Am. Chem. Soc., 105, 4634 (1983).

CF_3-C(R)=CH$_2$
1) Disilane, $Bu_4N^+ F^-$ or Me_3SiLi
2) KO^tBu or $(Me_2N)_3S^+ Me_3F_2Si^-$
   $R^1R^2C=O$

→ $R^1R^2C(OH)-CF_2-C(R)=CH_2$

74-85%

I.A.7.a.3-9   J. Sepulveda et al., Bull Soc. Chim. Fr. II, 233, 237, 240 (1983);   T. Nishio, T. Tokunaga and Y. Omote, Chem. Ind. (London), 243 (1983);   R. Perez-Ossorio et al., J. Chem. Soc., Perkin II, 1645 (1983);   I. G. Vasi and R. H. Acharya, Ind. J. Chem. 22B, 67 (1983);   J. W. Blunt et al., Aust. J. Chem., 36, 565 (1983).

$\xrightarrow[\text{-70°C}]{R^2MgBr}$

80-95%
(100% trans)

R^2 = CH_3CH_2-, $CH_2=CH-$, $HC\equiv C-$

I.A.7.a.3-10 M. Kawana, T. Koresawa and H. Kuzuhara, <u>Bull. Chem. Soc. Jpn.</u>, <u>56</u>, 1095 (1983); I. Hoppe and U. Schollkopf, <u>Liebigs Ann. Chem.</u>, 372 (1983); P. Canonne and M. Akssira, <u>Tetrahedron Lett.</u>, <u>24</u>, 5519 (1983); G. M. Strunz, P. Giguere and M. Ebacher, <u>Synth. Commun.</u>, <u>13</u>, 823 (1983).

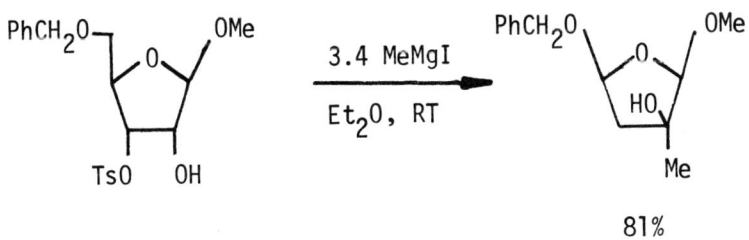

I.A.7.a.3-11 Y. Suzuki and H. Takahashi et al., <u>Chem. Pharm. Bull.</u>, <u>31</u>, 31, 2183, 2895 (1983); M. Asami and T. Mukaiyama, <u>Chem. Lett.</u>, 93 (1983); T. R. Kelly and P. N. Kaul, <u>J. Org. Chem.</u>, <u>48</u>, 2775 (1983); J. Leder, H. Fujioka and Y. Kishi, <u>Tetrahedron Lett.</u>, <u>24</u>, 1463 (1983).

ArCH=N—CHR(OH) → 1) PhCH₂MgCl (xs), THF, -5°C; 2) Aq. NH₄Cl → PhCH₂—CH(Ar)—NH—CH(R)—OH

44-85%

(40-100% de)

I.A.7.a.3-12 Y. Yamamoto and K. Maruyama, <u>J. Org. Chem.</u>, <u>48</u>, 1564 (1983); Y. Naruta and K. Maruyama, <u>Chem. Commun.</u>, 1264 (1983); J. M. Coxon et al., <u>Tetrahedron Lett.</u>, <u>24</u>, 1427 (1983); P. Wieland and J. Kalvoda, <u>ibid</u>, <u>24</u>, 5603 (1983); J. d'Angelo et al.,<u>Tetrahedron Lett.</u>, <u>24</u>, 895, 899 (1983).

RCHO + CH₂=CHCH₂CH=CH—MgCl → AlCl₃, -78°C → R—CH(OH)—CH₂—CH=CH—CH₃

Predominant Product is α-Adduct.

I.A.7.a.3-13 P. Albaugh-Robertson and J. A. Katzenellenbogen, J. Org. Chem., 48, 5288 (1983); T. Hayashi, M. Konishi and M. Kumada, ibid, 48, 281 (1983); P. E. Bauer, K. S. Kyler and D. S. Watt, ibid, 48, 34 (1983); A. Hosomi, Y. Araki and H. Sakurai, ibid, 48, 3122 (1983); T. Hayashi, M. Kumada et al., Tetrahedron Lett., 24, 2865 (1983); S. I. Koyooka and C. H. Heathcock, ibid, 24, 4765 (1983).

$$\text{Me-C(=CH}_2\text{)-CH(SiMe}_3\text{)-CO}_2\text{Et} \xrightarrow[\text{Lewis Acid}]{E^+} \text{E-CH}_2\text{-C(Me)=C(H)-CO}_2\text{Et}$$

22-86%

E^+ = RCHO, RCOCl, R_2C=O, Acetals, Ketals, Chloro Thioethers

---

I.A.7.a.3-14 Y. Yamamoto, et al., Chem. Commun., 489, 742, 774 (1983); J. Otera et al., Chem. Lett., 1529 (1983); T. Mukaiyama, T. Yamada and K. Suzuki, ibid, 5 (1983); W. A. Nugent and B. E. Smart, J. Organometal. Chem., 256, C9 (1983); J. Nokami et al., Organometallics, 2, 191 (1983); J. P. Quintard, B. Elissondo and M. Pereyre, J. Org. Chem., 48, 1559 (1983).

$$\text{CH}_3\text{-CH=CH-CH}_2\text{-SnBu}_3 \xrightarrow[\text{AlCl}_3, \text{ iPrOH}]{\text{RCHO}} \text{CH}_3\text{-CH=CH-CH}_2\text{-CH(OH)-R}$$

68-80%

(58-98% α Selective)

Use of $TiCl_4$, $SnCl_4$ or $BF_3 \cdot Et_2O$ leads to γ-adduct.

I.A.7.a.3-15 I. Matsuda, H. Okada and Y. Izumi, Bull. Chem. Soc. Jpn., 56, 528 (1983); Chem. Lett., 97 (1983); W. Ando and H. Tsumaki, Chem. Lett., 1409 (1983).

$$\text{(Me}_3\text{Si)(R}^3\text{)C=C=N-SiMe}_3 \xrightarrow[\text{BF}_3 \cdot \text{Et}_2\text{O}]{R^1R^2C=O} \text{R}^1\text{R}^2\text{C=C(SiMe}_3\text{)(CN)}$$

45-99%

I.A.7.a.3-16 K. Mikami, N. Kishi and T. Nakai, Tetrahedron Lett., 24, 795 (1983); S. E. Denmark and E. J. Weber, Helv. Chim. Acta, 66, 1655 (1983).

[Reaction: acid chloride with allylsilane]

1) 3 AlCl$_3$, CH$_2$Cl$_2$, 0°C
2) Aq. NaHCO$_3$

76%

Both α and β-cyclizations observed in other cases.

I.A.7.a.3-17 W. Amaratunga and J. M. J. Frechet, Tetrahedron Lett., 24, 1143 (1983).

[Piperidine-N-N(Me)-CHO hydrazone]

1) RMgX
2) H$_3$O$^+$

RCHO

80-92%

I.A.7.a.3-18 S. R. Ramadas et al., Synthesis, 605 (1983).

Review: "Methods of Synthesis of Dithiocarboxylic Acids and Esters."

I.A.7.a.3-19 T. P. Burns and R. D. Rieke, J. Org. Chem., 48, 4141 (1983); F. Bickelhaupt et al., Synthesis, 721 (1983).

PhO(CH$_2$)$_n$Br
(n = 3-6)

1) Mg*, THF, −78° to 20°C
2) CO$_2$

PhO(CH$_2$)$_n$CO$_2$H

71-89%

Highly reactive magnesium avoids cyclopropane formation.

I.A.7.a.3-20 G. M. Rubottom and C. W. Kim, J. Org. Chem., 48, 1550 (1983); C. G. Knudsen and H. Rapoport, ibid, 48, 2260 (1983); V. Fiandanese, G. Marchese and L. Ronzini, Tetrahedron Lett., 24, 3677 (1983); T. Fujisawa et al., Chem. Lett., 1267, 1791 (1983).

$$R-CO_2H \xrightarrow[\text{2) Me}_3\text{SiCl Quench}]{\text{1) MeLi, THF}} R-\overset{O}{\underset{}{C}}-CH_3$$

65-95%

Aq. NH_4Cl quench gives substantial carbinol.

I.A.7.a.3-21 N. K. Kochetkov et al., Tetrahedron Lett., 24, 4359 (1983); S. Kim and J. I. Lee, J. Org. Chem., 48, 2608 (1983).

$$R^1-\overset{O}{\underset{}{C}}-X \xrightarrow[\substack{\text{THF, Me}_2\text{S, HMPA} \\ -25°C}]{R^2Cu \cdot MgBr} R^1-\overset{O}{\underset{}{C}}-R^2$$

33-96%

(X = SeMe or Cl)

I.A.7.a.3-22 J. Barluenga et al., Synthesis, 378, 736 (1983); J. Org. Chem., 48, 609 (1983).

$$ClCH_2CH_2CH_2\overset{O}{\underset{}{C}}Cl \xrightarrow[\substack{\text{2) Li} \\ \text{3) E}^+}]{\text{1) 2 RMgBr}} E-CH_2CH_2CH_2-\underset{R}{\overset{OH}{\underset{|}{C}}}-R$$

51-91%

E^+ = RCHO, $R_2C=O$, RX, Me_3SiCl, Imines, H_2O, D_2O, O_2.

I.A.7.a.3-23 J. K. Stille et al., J. Am. Chem. Soc., 105, 669, 6129 (1983); J. Org. Chem., 48, 4634 (1983); A. Gambaro, V. Peruzzo and D. Marton, J. Organometal. Chem., 258, 291 (1983).

$$R^1\overset{O}{\overset{\|}{C}}Cl \xrightarrow[\text{LnPd(0)}]{R^2_3SnR^3} R^1-\overset{O}{\overset{\|}{C}}-R^2$$

Full Paper Includes Mechanistic Studies.

I.A.7.a.3-24 Y. Akiyama, T. Kawasaki and M. Sakamoto, Chem. Lett., 1231 (1983); J. Leyendecker, U. Niewohner and W. Steglich, Tetrahedron Lett., 24, 2375 (1983); L. R. Krepski, S. M. Heilmann and J. K. Rasmussen, ibid, 24, 4075 (1983).

$$RMgBr \xrightarrow[\substack{\text{2) NCCO}_2\text{Et, ZnCl}_2 \\ \text{3) H}_3\text{O}^+}]{\text{1) CdBr}_2} R\overset{O}{\overset{\|}{C}}CO_2Et \quad 24\text{-}66\%$$

I.A.7.a.3-25 A. C. Regan and J. Staunton, Chem. Commun., 764 (1983); H. Hamana and T. Sugasawa, Chem. Lett., 333 (1983); H. B. Kagan et al., J. Organometal. Chem., 250, 227 (1983); B. Bennetau and J. Dunogues, Tetrahedron Lett., 24, 4217 (1983).

41% (53% ee)

I.A.7.a.3-26 T. Shono et al., Chem. Lett., 1311 (1983); K. Krohn et al., Liebigs Ann. Chem., 1818, 2151 (1983); M. Saljoughian et al., Monat. Chem., 114, 813 (1983).

$$\text{pyrrolidinone} \xrightarrow[\substack{2) \text{ RCHO} + \text{CHCl}_3 \\ \text{DMF, } -78°C}]{\substack{1) +e^-, \text{ DMF} \\ \text{QOTs}}} R\underset{\text{OH}}{\overset{}{\text{CH}}}\text{CCl}_3 \quad 49\text{-}80\%$$

Electrogenerated base.

I.A.7.a.3-27 B. Weidmann and D. Seebach. Angew. Chem., Int. Ed. Engl., 22, 31 (1983).

Review: "Organometallic Compounds of Titanium and Zirconium as Selective Nucleophilic Reagents in Organic Synthesis."

I.A.7.a.3-28 D. Seebach et al., Pure Appl. Chem., 55, 1807 (1983).

Review: "Some Recent Advances in the Use of Titanium Reagents for Organic Synthesis."

I.A.7.a.3-29 J. R. Stille and R. H. Grubbs, J. Amer. Chem. Soc., 105, 1664 (1983); T. S. Chou and S. B. Huang, Tetrahedron Lett., 24, 2169 (1983); J. J. Eisch and A. Piotrowski, ibid, 24, 2043 (1983); F. Sato et al., Chem. Commun., 921 (1983).

$$R-\overset{O}{\underset{}{C}}-Cl \xrightarrow[\substack{2) \text{ PhCHO}}]{\substack{1) \text{ Cp}_2\text{Ti}=\text{CH}_2}} R\overset{O}{\overset{}{C}}CH_2\overset{OH}{\overset{}{CH}}Ph$$

53-69%

I.A.7.a.3-30 A. Clerici and O. Porta, J. Org. Chem., 48, 1690 (1983); M. T. Reetz and J. Westermann, ibid, 48, 254 (1983); M. Schiess and D. Seebach, Helv. Chim. Acta, 66, 1618 (1983).

$$R^1-\overset{O}{\underset{}{C}}-Z \xrightarrow[R^2CH=CH-CHO]{2\ TiCl_3/H_3O^+} R^1-\overset{Z}{\underset{OH}{C}}-\underset{OH}{CH}-CH=CHR^2$$

THF   27-78%

Z = -CN, -CO$_2$H, -CO$_2$Me, 2 and 4-Pyridyl.

I.A.7.a.3-31 A. Nakamura et al., Chem. Lett., 219 (1983); G. Erker et al., Angew. Chem., Int. Ed. Engl., 22, 494, 777, (1983).

$$Cp_2ZrCl_2 \xrightarrow[\substack{2)\ RCHO \\ 3)\ H^+}]{1)\ CH_3CH=CHCH_2MgCl}$$

Threo selective

I.A.7.a.3-32 B. B. Snider et al., J. Org. Chem., 48, 2789 (1983); J. Am. Chem. Soc., 105, 2364 (1983).

$$\underset{Me}{\diagup}\!\!\!\diagdown^{OEt} \xrightarrow[Me_3Al]{CH_2O} \underset{Me}{\overset{Me}{\diagup}}\!\!\!\overset{OEt}{\diagdown}_{CH_2OH}$$

61%

I.A.7.a.3-33 T. Hiyama, M. Sawahata and M. Obayashi, Chem. Lett., 1237 (1983); W. C. Still and D. Mobilio, J. Org. Chem., 48, 4785 (1983); B. Cazes, C. Verniere and J. Gore, Synth. Commun., 13, 73 (1983).

$$\diagup\!\!\!\diagdown\!\!\text{Br} \xrightarrow[R^1R^2C=O]{\text{Mn Powder}} \begin{array}{c} R^1 \\ R^2\!\!-\!\!\text{C}\!\!-\!\!\text{CH}_2\text{CH=CH}_2 \\ \text{OH} \end{array}$$

58-97%

I.A.7.a.3-34 P. G. M. Wutz, P. A. Thompson and G. R. Callen, J. Org. Chem., 48, 5398 (1983).

50-97%

I.A.7.a.3-35 W. R. Roush, et al., Tetrahedron Lett., 24, 2227, 2231 (1983); R. W. Hoffmann and B. Landmann, ibid, 24, 3209 (1983); R. W. Hoffmann, G. Eichler and A. Endesfelder, Liebigs Ann. Chem., 2000 (1983); H. C. Brown and P. K. Jadhav, J. Am. Chem. Soc., 105, 2092 (1983); Y. Yamamoto, T. Komatsu and K. Maruyama, Chem. Commun., 191 (1983); P. G. M. Wuts and S. S. Bigelow, J. Org. Chem., 48, 3489 (1983).

70%

(>95% Diästeriösel.)

I.A.7.a.3-36 C. Fuganti, P. Grasselli and G. Pedrocchi-Fantoni, J. Org. Chem., 48, 909 (1983); T. Shono et al., ibid, 48, 1621 (1983); N. Oguni et al., Chem. Lett., 841 (1983); G. Rousseau and J. Drouin, Tetrahedron, 39, 2307 (1983).

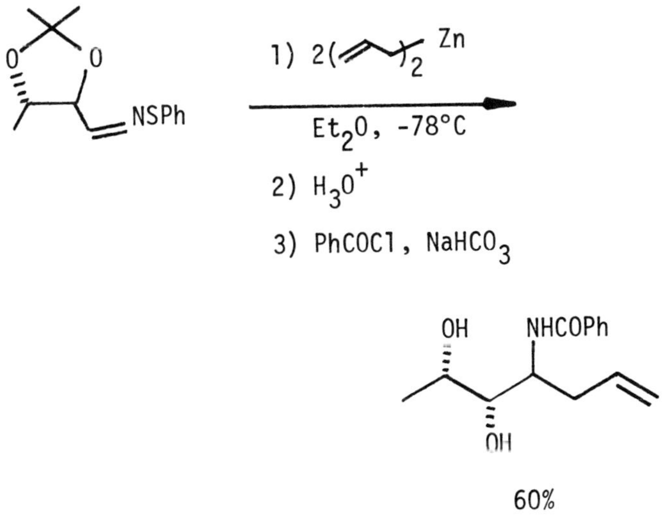

$$\text{(acetonide-CH=NSPh)} \xrightarrow[\substack{\text{Et}_2\text{O, } -78°\text{C} \\ 2) \text{ H}_3\text{O}^+ \\ 3) \text{ PhCOCl, NaHCO}_3}]{1) \ 2(\text{CH}_2\text{=CHCH}_2)_2\text{Zn}} \text{product}$$

60%

I.A.7.a.3-37 E. I. Negishi et al., Tetrahedron Lett., 24, 5181 (1983); S. I. Inaba and R. D. Rieke, ibid, 24, 2451 (1983).

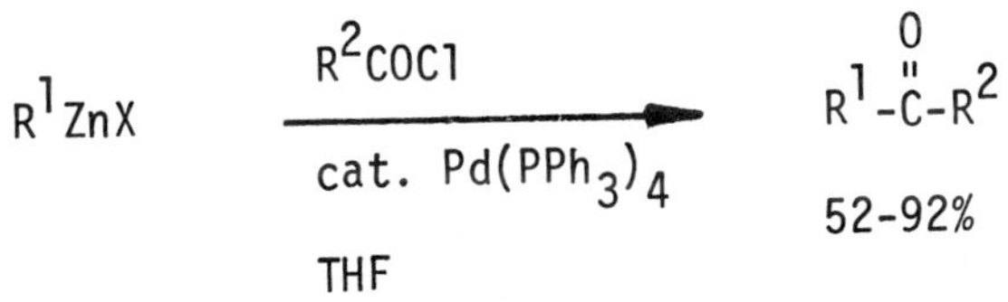

$$R^1\text{ZnX} \xrightarrow[\substack{\text{cat. Pd(PPh}_3)_4 \\ \text{THF}}]{R^2\text{COCl}} R^1\text{-}\underset{\underset{\text{O}}{\|}}{\text{C}}\text{-}R^2$$

52-92%

R^1 = Alkyl, alkenyl, alkynyl, aryl.

I.A.7.b.1. Conjugate Additions of Enolate-type Carbanions

I.A.7.b.1-1 W. J. Thompson and C. A. Buhr, J. Org. Chem., 48, 2769 (1983); N. Ono, H. Miyake and A. Kaji, Chem. Commun., 875 (1983).

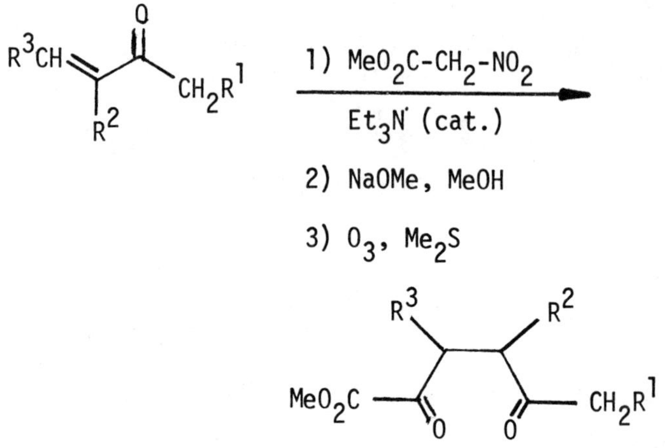

$R^3CH=C(R^2)-C(O)-CH_2R^1$

1) $MeO_2C-CH_2-NO_2$, Et_3N (cat.)
2) NaOMe, MeOH
3) O_3, Me_2S

→ $MeO_2C-CH(R^3)-C(O)-CH(R^2)-C(O)-CH_2R^1$

73-85%

I.A.7.b.1-2 H. Ahlbrecht and H. M. Kompter, Synthesis, 645 (1983); I. C. Ivanov, P. B. Sulay and D. K. Dantchev, Liebigs Ann. Chem., 753 (1983); I. Crossland and S. I. Hommeltoft, Acta Chem. Scand. B, 37, 21 (1983).

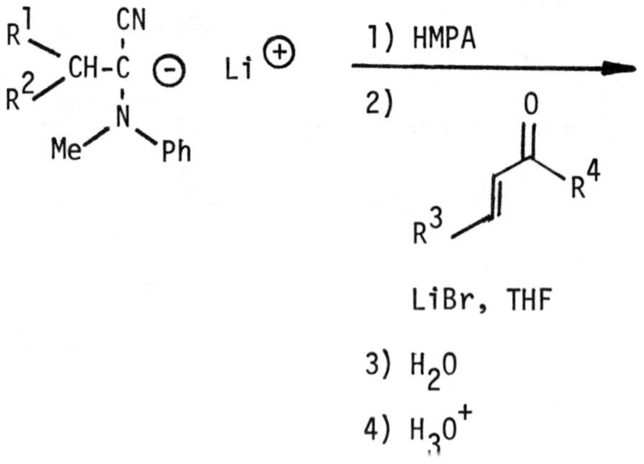

$R^1R^2CH-C(CN)(N(Me)Ph)^- Li^+$

1) HMPA
2) $R^3CH=CH-C(O)R^4$, LiBr, THF
3) H_2O
4) H_3O^+

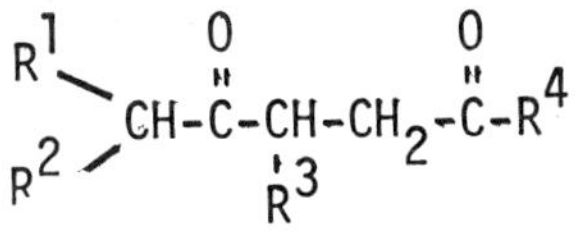

$$R^1_{R^2}CH-\overset{O}{\underset{}{C}}-\overset{}{\underset{R^3}{CH}}-CH_2-\overset{O}{\underset{}{C}}-R^4$$

42-87%

I.A.7.b.1-3 J. Mulzer et al., Chem. Commun., 869 (1983).

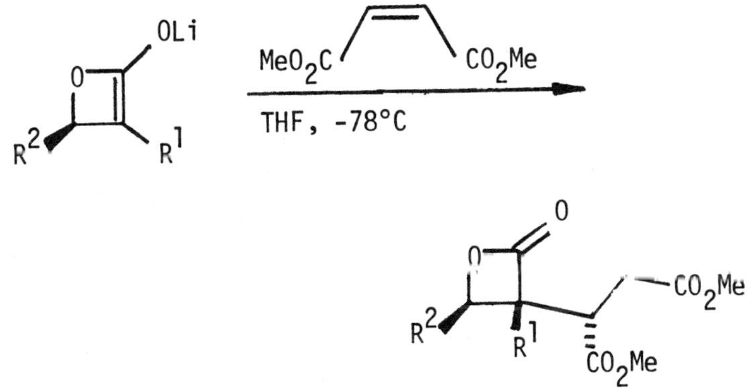

45-73%

(82-100% indicated diasteriomer)

I.A.7.b.1-4 W. Oppolzer, M. Guo and K. Baettig, Helv. Chim. Acta, 66, 2140 (1983); J. C. Barriere et al., ibid, 66, 1392 (1983); F. E. Ziegler and J. J. Mencel, Tetrahedron Lett., 24, 1859 (1983); S. H. Bertz, L. W. Jelinski and G. Dabbagh, Chem. Commun., 388 (1983).

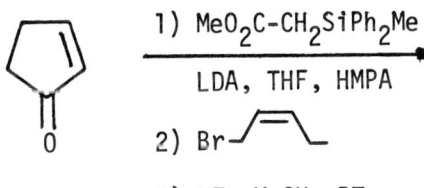

1) $MeO_2C-CH_2SiPh_2Me$
   LDA, THF, HMPA

2) Br⟋=⟍

3) KF, MeOH, RT

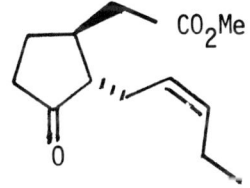

50%

I.A.7.b.1-5 R. J. P. Corriu et al., Tetrahedron, 39, 117 (1983); Z. I. Dyusenova et al., J. Org. Chem. (USSR), 19, 1112 (1983); G. Arsenault, A. D. Broadbent and P. Hutten-Czapski, Chem. Commun., 437 (1983).

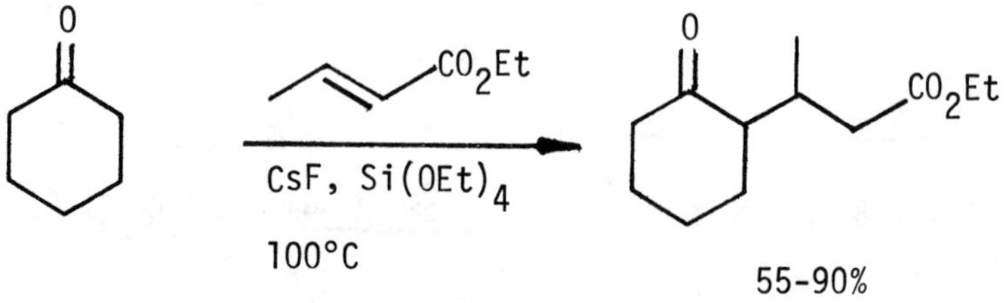

cyclohexanone + CH$_3$CH=CHCO$_2$Et $\xrightarrow{\text{CsF, Si(OEt)}_4, 100°C}$ 2-(3-ethoxycarbonyl-1-methylpropyl)cyclohexanone 55-90%

I.A.7.b.1-6 S. J. Blarer and D. Seebach, Chem. Ber., 116, 2250, 3086 (1983); F. Benedetti, S. Fabrissin and A. Risaliti, Tetrahedron, 39, 3887 (1983); V. V. Kane and M. Jones, Jr., Org. Syn., 61, 129 (1983).

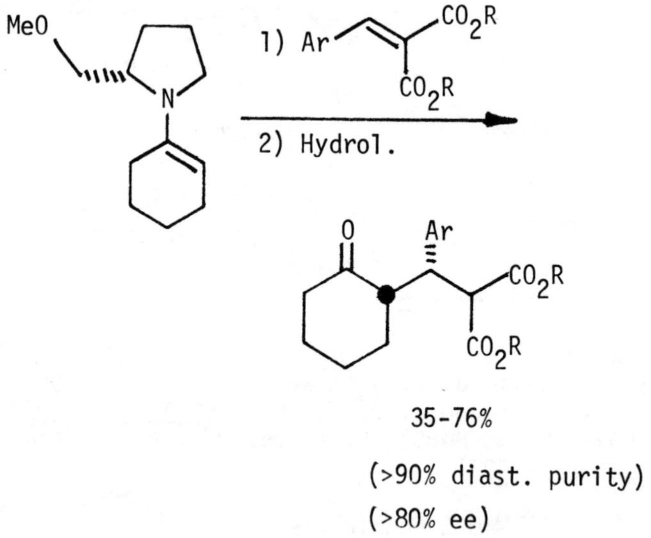

2-methoxymethyl-1-(1-cyclohexenyl)pyrrolidine $\xrightarrow{\text{1) ArCH=C(CO}_2\text{R)}_2 \text{ 2) Hydrol.}}$ 2-substituted cyclohexanone 35-76%

(>90% diast. purity)
(>80% ee)

I.A.7.b.1-7 W. G. Dauben, C. H. Heathcock et al., Tetrahedron Lett., 24, 4943, 3841 (1983); A. Koskinen and M. Lounasmaa, ibid, 24, 1951 (1983).

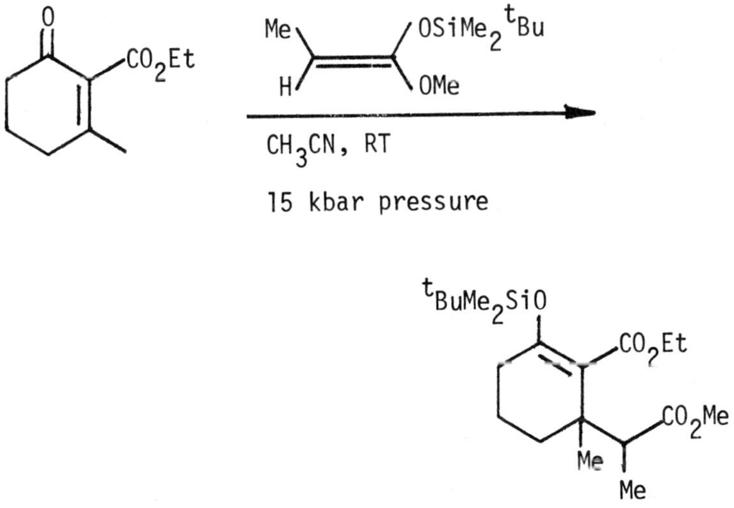

[Reaction: 2-methyl-6-carbethoxy-cyclohex-2-enone + Me(H)C=C(OMe)(OSiMe$_2^t$Bu), CH$_3$CN, RT, 15 kbar pressure → cyclohexene product with ^{t}BuMe$_2$SiO, CO$_2$Et, and CH(Me)CO$_2$Me substituents, 81%]

I.A.7.b.1-8 E. Hatzigrigoriou and L. Wartski, Synth. Commun., 13, 319 (1983); Bull. Soc. Chim. Fr. II, 313 (1983); J. Seyden-Penne et al., Tetrahedron, 39, 3415 (1983); Synthesis, 494 (1983).

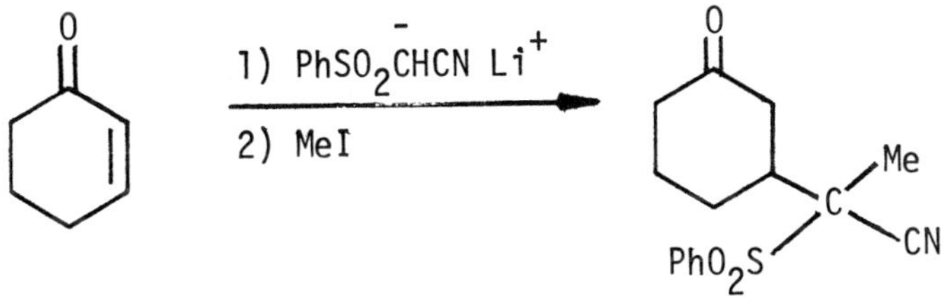

[Reaction: cyclohex-2-enone, 1) PhSO$_2$CHCN$^-$ Li$^+$, 2) MeI → 3-substituted cyclohexanone bearing -C(Me)(CN)(SO$_2$Ph) group, 80%]

I.A.7.b.1-9 F. E. Ziegler and K. J. Hwang, J. Org. Chem., 48, 3349 (1983); R. J. Pariza and P. L. Fuchs, ibid, 48, 2306 (1983); W. G. Dauben and R. A. Bunce, ibid, 48, 4642 (1983); E. Y. Chen, Synth. Commun., 13, 927 (1983); A. Jellal and M. Santelli, Tetrahedron Lett., 24, 2847 (1983).

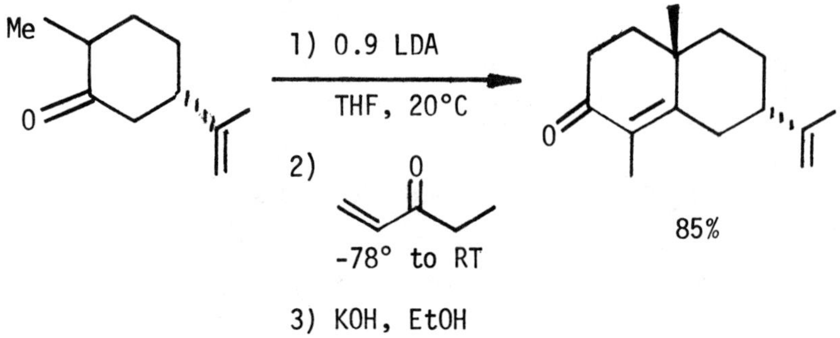

Aprotic Robinson Annelation.

I.A.7.b.1-10 S. Danishefsky et al., J. Org. Chem., 48, 3615 (1983); L. Lombardo and L. N. Mander, ibid, 48, 2298 (1983); P. G. Baraldi et al., Tetrahedron Lett., 24, 5669 (1983); K. Maruyama, H. Uno and Y. Naruta, Chem. Lett., 1767 (1983).

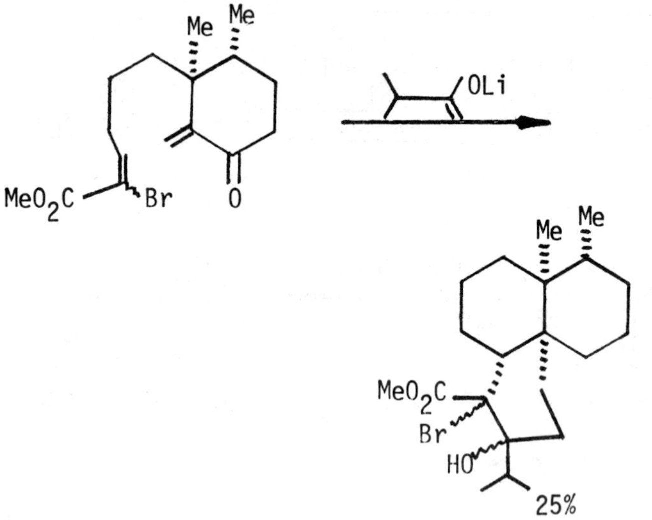

Multiple annulation via sequential Michael and Aldol.

I.A.7.b.1-11 H. Takei et al., Tetrahedron Lett., 24, 5127 (1983); M. Horton and G. Pattenden, ibid, 24, 2125 (1983); P. A. Grieco et al., ibid, 24, 3807 (1983).

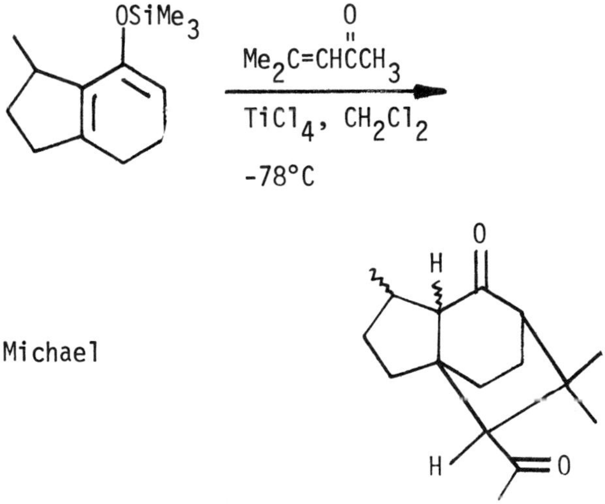
OSiMe3
Me2C=CHCCH3
O
TiCl4, CH2Cl2
-78°C
O
H
H
O

Double Michael

64%

I.A.7.b.1-12 G. Stork, J. D. Winkler and N. A. Saccomano, Tetrahedron Lett., 24, 465 (1983); W. A. Nugent and F. W. Hobbs, Jr., J. Org. Chem., 48, 5364 (1983); K. Schank and W. Lorig, Liebigs Ann. Chem., 112 (1983).

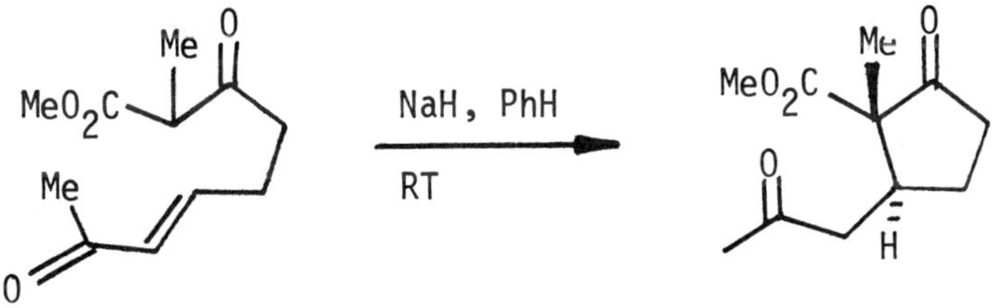
Me
O
MeO2C
Me
O
NaH, PhH
RT
Me
O
MeO2C
O
H

90%

I.A.7.b.1-13 P. R. Hamann and P. L. Fuchs, J. Org. Chem., 48, 914 (1983); A. T. Hewson and D. T. MacPherson, Tetrahedron Lett., 24, 5807 (1983); D. N. Jones et al., ibid, 24, 405 (1983); D. J. Ager, ibid, 24, 95 (1983); A. G. Cameron and A. T. Hewson, J. Chem. Soc., Perkin I, 2979 (1983); Y. K. Rao and M. Nagarajan, Ind. J. Chem., 22B, 519 (1983).

$$\text{OSiMe}_2{}^t\text{Bu} \quad \xrightarrow[\text{2) MeI}]{\text{1) Nu}^- \text{K}^+ \ \ \text{THF}} \quad \text{OSiMe}_2{}^t\text{Bu}, \text{Nu}, \text{SO}_2{}^t\text{Bu}, \text{Me}$$

51-98%

NuH = CH_3CN, CH_3CONMe_2, CH_3CO_2Et, CH_3COPh, $CH_3CO_2{}^tBu$.

I.A.7.b.1-14 H. Yoshida et al., Bull. Chem. Soc. Jpn., 56, 3015 (1983); Chem. Lett., 155 (1983).

1) MeBr
2) [acetylacetone]
3) Et_3N, PhH

73%

I.A.7.b.1-15 P. Hodge, E. Khoshdel and J. Waterhouse, J. Chem. Soc., Perkin I, 2205 (1983); D. Enders and K. Papadopoulos, Tetrahedron Lett., 24, 4967 (1983).

$$\xrightarrow[\text{PhCH}_3, \text{ Base}]{\text{CH}_2=\text{CHCCH}_3,\ \text{(P)}^*-\overset{+}{\text{NR}}_3\ X^-}$$

—CO_2Me

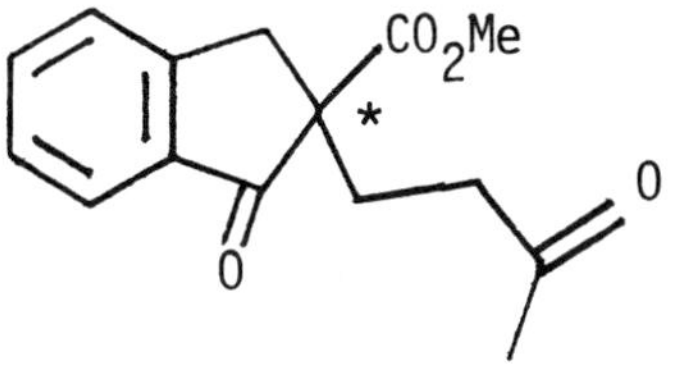

61-100% (≤27% ee)

*Ammonium salt prepared from cinchona alkaloids.

I.A.7.b.2. Conjugate Additions of Organometallic Reagents

I.A.7.b.2-1 H. G. Richey, Jr. et al., J. Org. Chem., 48, 3821, 3822 (1983); T. Mukaiyama and T. Ohsumi, Chem. Lett., 875 (1983).

$$PhCH=CHCH_2NMe_2 \xrightarrow[\text{2) MeOH}]{\text{1) RLi, Hexane} \atop (R = nBu \text{ or } {}^tBu)} PhCH_2\underset{R}{CH}CH_2NMe_2$$

52-56%

Also, Fragmentation of Metalated 1° Amines.

I.A.7.b.2-2 B. H. Lipshutz, Tetrahedron Lett., 24, 127 (1983).

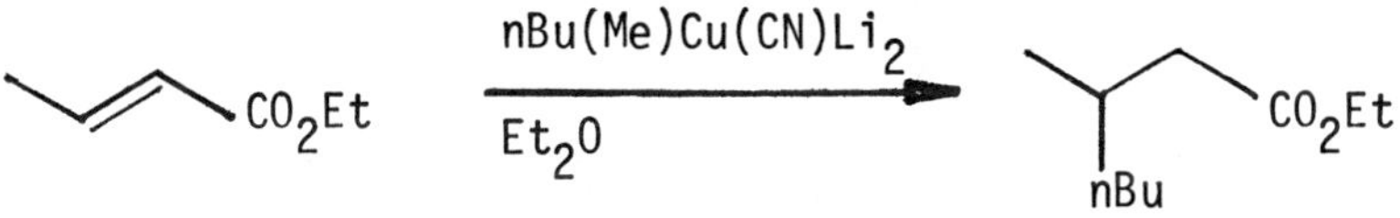

86%

I.A.7.b.2-3 M. Mitani, I. Kato and K. Koyama, J. Amer. Chem. Soc., 105, 6719 (1983).

cyclohexyl-Br + CH$_2$=CHCN $\xrightarrow[h\nu]{\text{CuCl, Bu}_3\text{P}, \text{ THF, RT}}$ cyclohexyl-CH$_2$CH(CN)(Br)

35-73%

I.A.7.b.2-4 F. Barbot, A. Kadib-Elban and P. Miginiac, Tetrahedron Lett., 24, 5089 (1983); F. Barbot, A. Kadib-Elban and P. Miginiac, J. Organometal. Chem. 255, 1 (1983); G. Hallnemo and C. Ullenius, Tetrahedron, 39, 1621 (1983); C. Kashima, T. Tajima and Y. Omote, Heterocycles, 20, 1811 (1983).

CH$_3$-CH=CH-CH=CH-C(O)-NEt$_2$ $\xrightarrow[20°C]{\text{RMgBr}}$ CH$_3$-CH=CH-CH(R)-CH$_2$-C(O)-NEt$_2$

35-80%

R$_2$CuLi reagents give 1,6-addition.

I.A.7.b.2-5 T. E. Goodwin et al., J. Org. Chem., 48, 376 (1983); R. K. Dieter and J. W. Dieter, Chem. Commun., 1378 (1983).

^{t}BuMe$_2$SiO-CH$_2$-(2-MeO-pyranone) $\xrightarrow[\text{THF, -78°C}]{\text{R(PhS)CuLi}}$

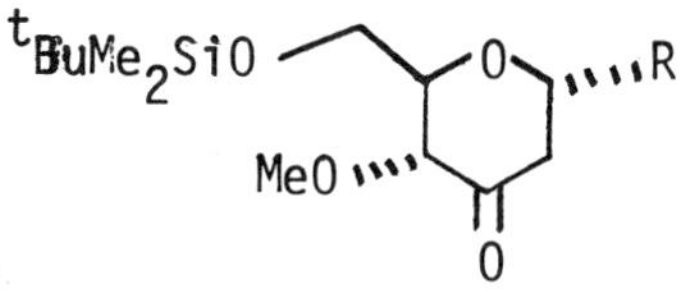

$^tBuMe_2SiO$
O
R
MeO
O

65%

I.A.7.b.2-6 W. Oppolzer et al., Tetrahedron Lett., 24, 4971, 4975 (1983); K. Soai, A. Ookawa and Y. Nohara, Synth. Commun., 13, 27 (1983); J. Berlan et al., J. Organometal. Chem., 256, 181 (1983); L. Jalander et al., Acta Chem. Scand. B, 37, 15, 173 (1983).

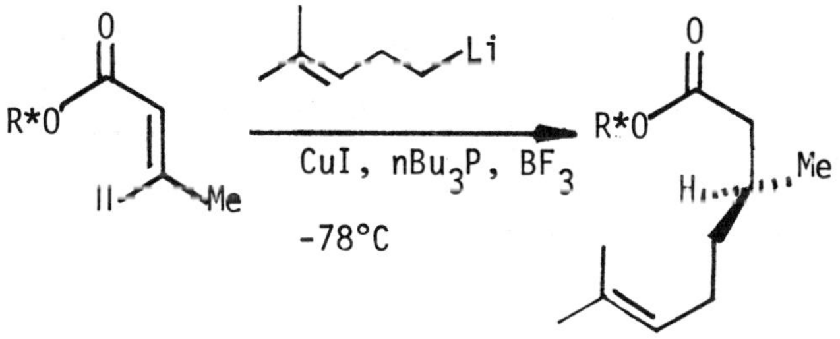

O
R*O
H
Me
Li
CuI, nBu$_3$P, BF$_3$
-78°C
O
R*O
H
Me

81% (98.5% de)

R* contains chiral neopentylether group derived from camphor.

I.A.7.b.2-7 F. Leyendecker et al., Tetrahedron Lett., 24, 3513, 3517, (1983).

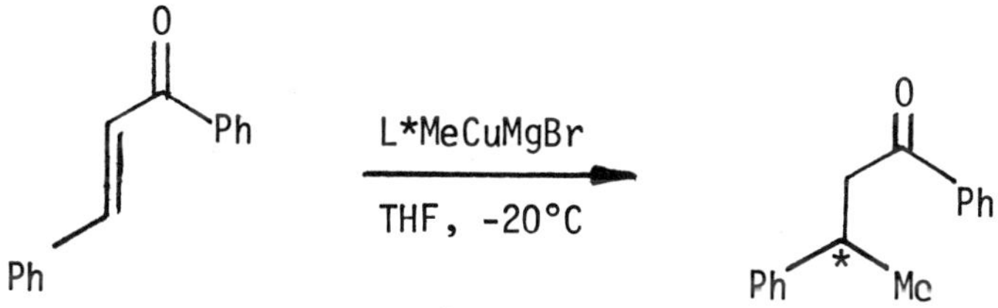

O
Ph
Ph
L*MeCuMgBr
THF, -20°C
O
Ph
Ph
*
Mc

Study of Ligand Effect

I.A.7.b.2-8 Z. Florjanczyk and U. Iwaniak, J. Organometal. Chem., 252, 275 (1983).

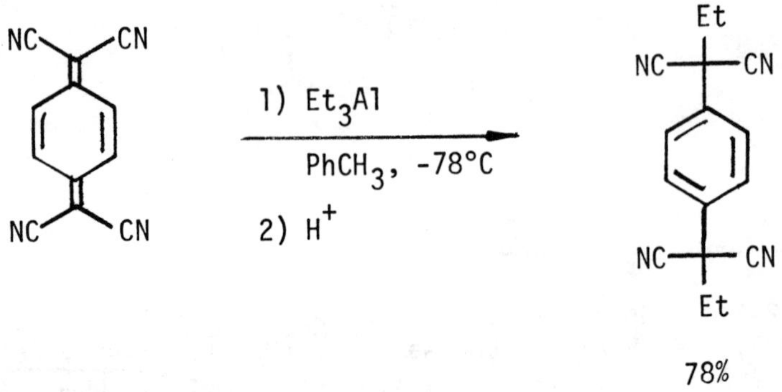

1) Et_3Al / $PhCH_3$, -78°C
2) H^+

78%

I.A.7.b.2-9 W. T. Monte, M. M. Baizer and R. D. Little, J. Org. Chem., 48, 803 (1983); J. H. Clark et al., Chem. Lett., 1145 (1983); A. T. Hewson and D. T. MacPherson, Tetrahedron Lett., 24, 647 (1983); S. Tomoda, Y. Nomura et al., ibid, 24, 2795 (1983); J. H. Clark, D. G. Cork and H. W. Gibbs, J. Chem. Soc., Perkin I, 2253 (1983).

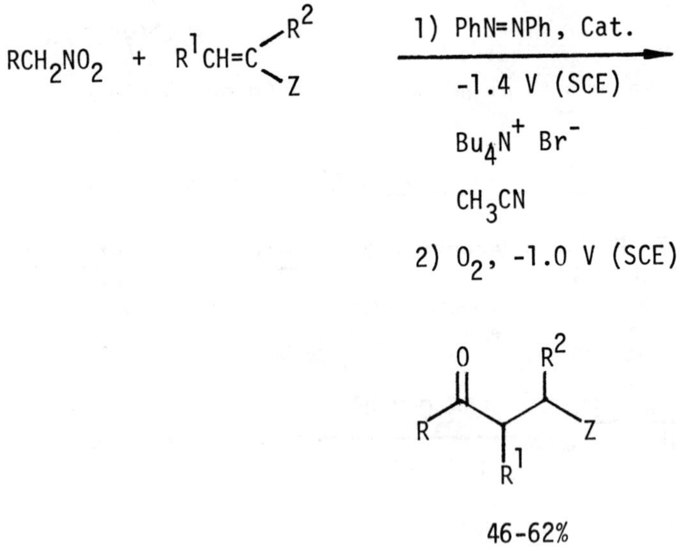

RCH_2NO_2 + $R^1CH=C\begin{smallmatrix}R^2\\Z\end{smallmatrix}$

1) PhN=NPh, Cat.
   -1.4 V (SCE)
   Bu_4N^+ Br^-
   CH_3CN
2) O_2, -1.0 V (SCE)

46-62%

I.A.7.b.2-10 J. Gonzalez, F. Sanchez and T. Torres, Synthesis, 911 (1983); A. Pelter, R. S. Ward et al., J. Chem. Soc., Perkin I, 643 (1983); W. Dumont, J. Lucchetti and A. Krief, Chem. Commun., 66 (1983); L. L. Vasil'eva et al., J. Org. Chem. (USSR), 19, 835 (1983).

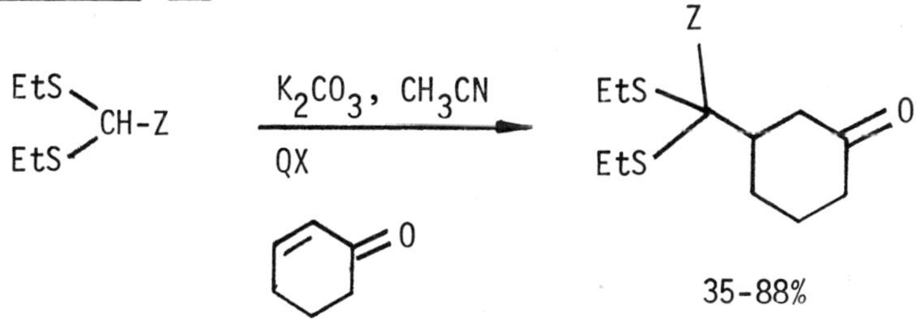

$$\text{(EtS)}_2\text{CH-Z} + \text{cyclohexenone} \xrightarrow[\text{QX}]{K_2CO_3,\ CH_3CN} \text{product}$$

35-88%

I.A.7.b.2-11 D. S. Watt et al., J. Org. Chem., 48, 383 (1983); R. K. Dieter and L. A. Silks, ibid, 48, 2786 (1983); E. Fujita et al., Chem. Pharm. Bull., 31, 3346 (1983); A. J. Bridges and J. W. Fischer, J. Chem. Soc., Perkin I, 2359 (1983).

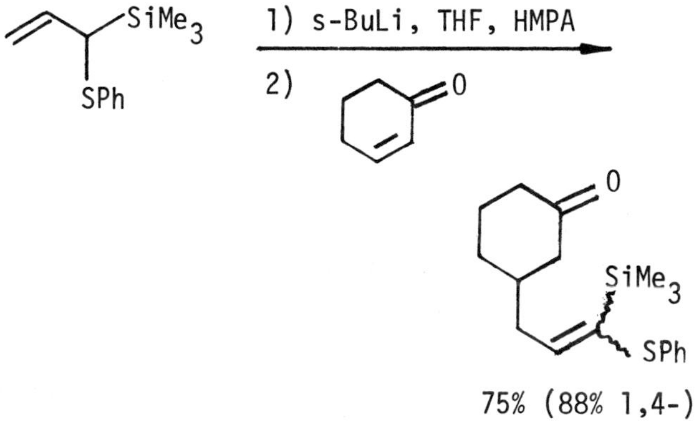

1) s-BuLi, THF, HMPA
2) cyclohexenone

75% (88% 1,4-)

Exclusive γ selectivity.

I.A.7.b.3. Other Conjugate Additions

I.A.7.b.3-1 K. Maruyama et al., <u>Chem. Lett.</u>, 1683, 1687 (1983); T. A. Blumenkopf and C. H. Heathcock, <u>J. Am. Chem. Soc.</u>, 105, 2354 (1983); G. Majetich et al., <u>Tetrahedron Lett.</u>, 24, 1909, 1913 (1983).

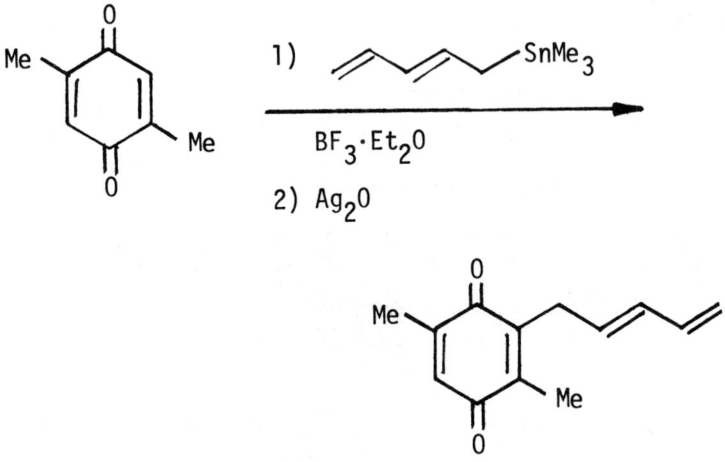

$$\text{Me-quinone} \xrightarrow[\text{2) Ag}_2\text{O}]{\text{1) } \diagup\!\!\diagdown\!\!\diagup\!\!\diagdown\text{SnMe}_3,\ BF_3\cdot Et_2O} \text{product}$$

86%

(no Diels-Alder)

I.A.7.b.3-2 A. P. Kozikowski and J. Scripko, <u>Tetrahedron Lett.</u> 24, 2051 (1983); R. Henning and H. Urbach, <u>ibid</u>, 24, 5343 (1983).

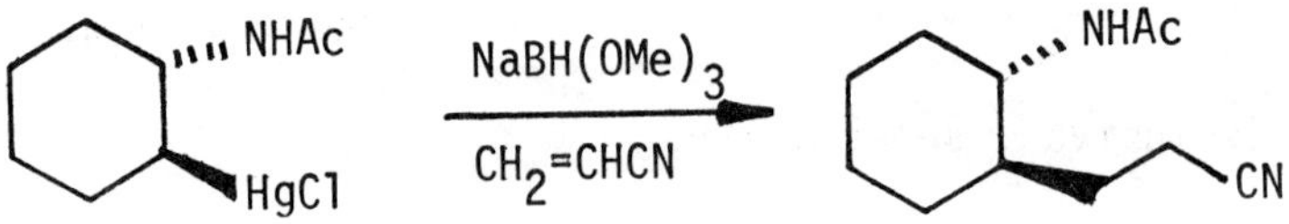

$$\text{cyclohexane-NHAc, HgCl} \xrightarrow[\text{CH}_2=\text{CHCN}]{\text{NaBH(OMe)}_3} \text{cyclohexane-NHAc, CH}_2\text{CH}_2\text{CN}$$

78%

(60:40 - trans, cis)

I.A.7.b.3-3 T. Minami et al., Tetrahedron Lett., 24, 767 (1983); N. Minowa et al., ibid, 24, 2391 (1983).

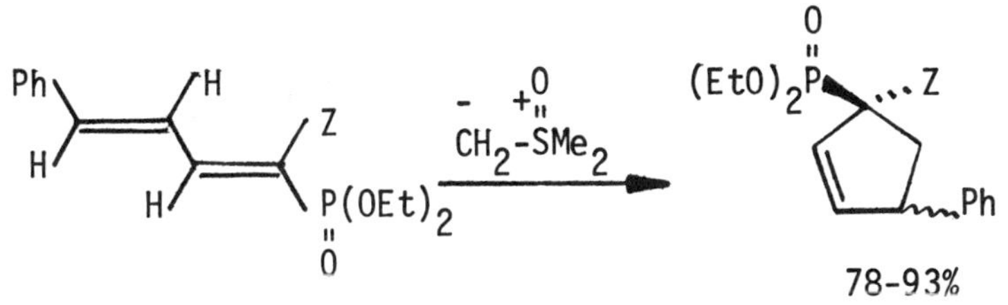

[Reaction: Ph-CH=CH-CH=C(Z)(P(OEt)₂=O) + CH₂⁻–S⁺Me₂ → cyclopentene with (EtO)₂P(=O), Z, and Ph substituents]

78-93%

I.A.7.b.3-4 M. E. Garst et al., J. Org. Chem., 48, 8, 16 (1983).

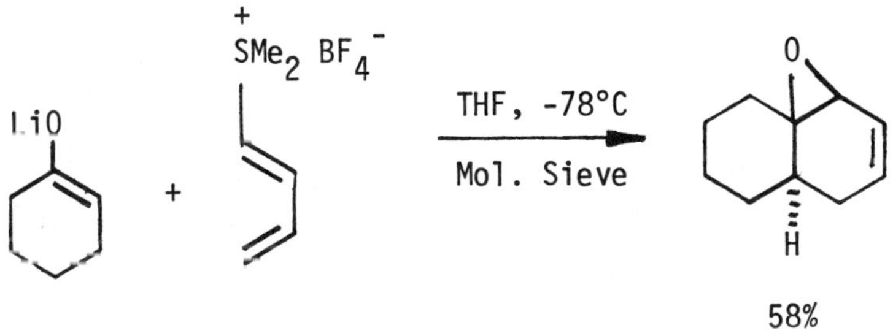

[Reaction: LiO-cyclohexenyl + SMe₂⁺BF₄⁻ allyl + butadiene, THF, -78°C, Mol. Sieve → bicyclic decalone]

58%

I.A.7.b.3-5 S. Bozzini et al., Tetrahedron, 39, 3409 (1983); V. Dryanska and C. Ivanov, Synthesis, 143 (1983).

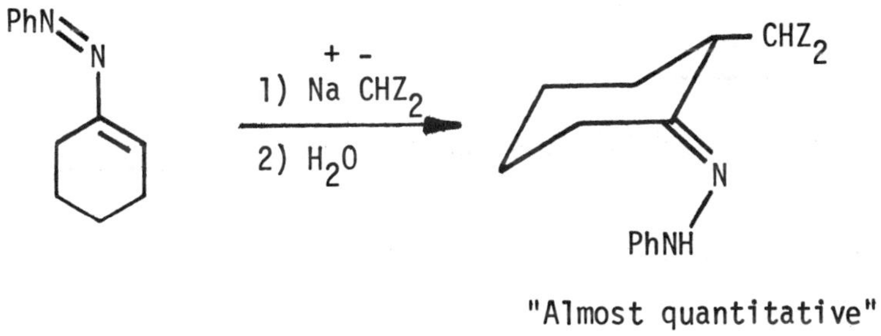

[Reaction: PhN=N-cyclohexenyl + 1) Na⁺ CHZ₂⁻ 2) H₂O → cyclohexane with =N-PhNH and CHZ₂ substituents]

"Almost quantitative"

I.A.8. Other Carbon-Carbon Single Bond Forming Reactions

I.A.8-1   I. Cutting and P. J. Parson, Chem. Commun. , 1435 (1983); F. Sato, Y. Tanaka and H. Kanbara, ibid, 1024 (1983); R. F. Newton et al., ibid, 932 (1983).

$$R^2\text{-CH=C}(R^1)\text{-CH}(SiMe_3)\text{-CH}_2\text{-epoxide} \xrightarrow{SnCl_4} R^2\text{-CH=C}(R^1)\text{-CH}_2\text{-C}(=CH_2)\text{-CH}_2\text{-OH}$$

26-45%

I.A.8-2   C. Giordano et al., J. Org. Chem., **48**, 4658 (1983); I. K. Moiseev and R. I. Doroshenko, J. Org. Chem. (USSR), **19**, 999 (1983);   E. Suarez et al., J. Chem. Soc., Perkin I, 2757 (1983);   S. Durani and R. S. Kapil, ibid, 211 (1983).

$$Ar\text{-C}(OMe)(OMe)\text{-CH}(R^2)(X) \xrightarrow[\substack{PhCH_3, \Delta \\ 2)\ OH^- \\ 3)\ H^+}]{1)\ Cat.\ ZnBr_2} ArCH(R^2)CO_2H$$

25-98%

I.A.8-3   C. Santelli-Rouvier and M. Santelli, Synthesis, 429 (1983).

Review: "The Nazarov Cyclization."

I.A.8-4 E. E. van Tamelen, J. R. Hwu and T. M. Leiden, Chem. Commun., 62 (1983); G. Stork and K. S. Atwal, Tetrahedron Lett., 24, 3819 (1983); S. H. Liu, ibid, 24, 439 (1983); H. Mayr, H. Klein and E. Sippel, Chem. Ber., 116, 3624 (1983).

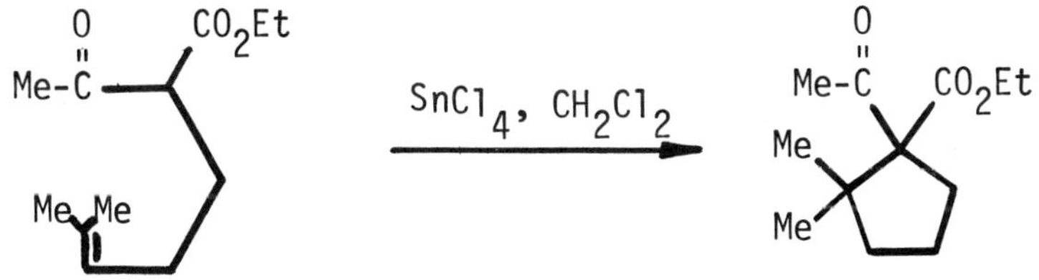

51%

I.A.8-5 J. K. Sutherland et al., J. Chem. Soc., Perkin I, 747, 751, 755, 759 (1983); S. Dev et al., Ind. J. Chem., 22B, 189, 193, 200, 206, 212 (1983); R. J. Armstrong and L. Weiler, Can. J. Chem., 61, 214 (1983).

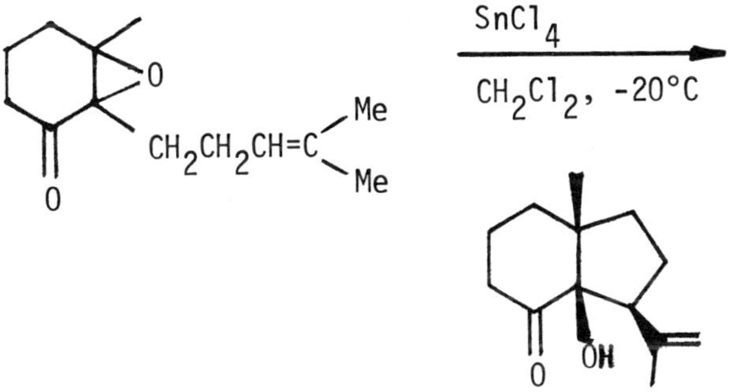

94%

I.A.8-6 M. A. Boaventura, J. Drouin and J. M. Conia, Synthesis, 801 (1983); M. L. Roumestant, B. Cavallin and M. Bertrand, Bull. Soc. Chim. Fr. II, 309 (1983).

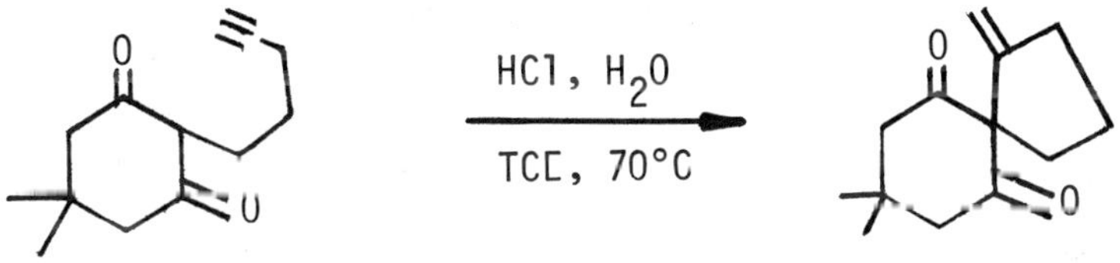

88%

I.A.8-7 R. J. Armstrong and L. Weiler, Can. J. Chem., 61, 2530 (1983); H. Itokawa, H. Nakanishi and S. Mihashi, Chem. Pharm. Bull., 31, 1991 (1983).

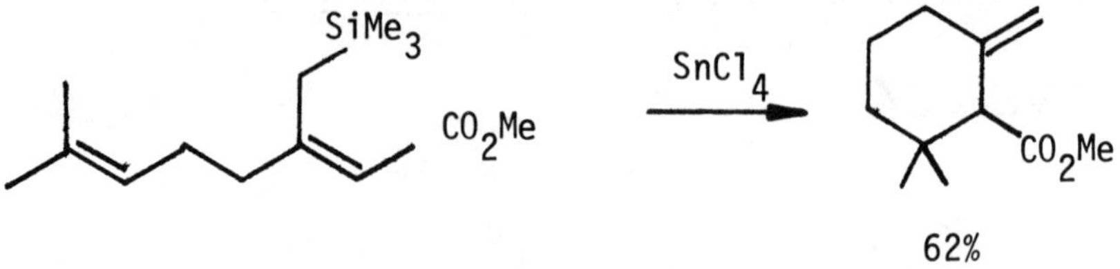
SiMe3
CO2Me
SnCl4
CO2Me
62%

I.A.8-8 G. S. Cockerill and P. Kocienski, Chem. Commun., 705 (1983).

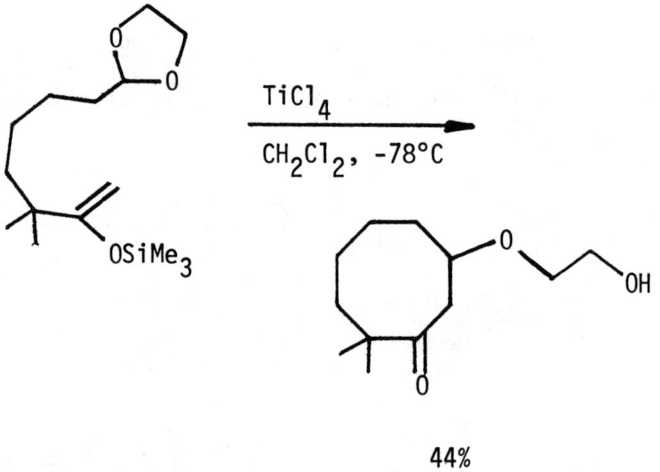
O
O
OSiMe3
TiCl4
CH2Cl2, -78°C
O
OH
O
44%

I.A.8-9 J. Nishimura et al., J. Am. Chem. Soc., 105, 4758 (1983).

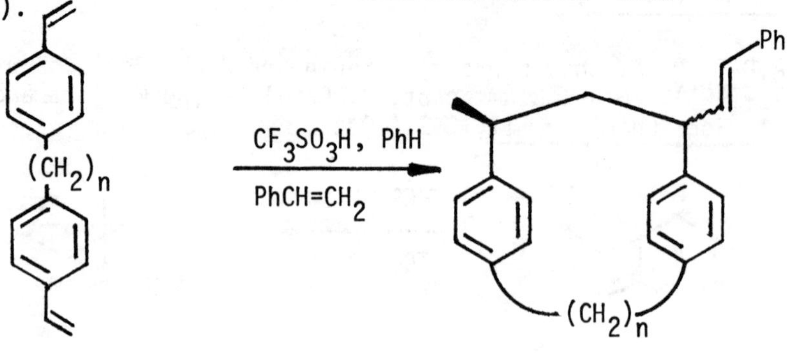
(CH2)n
CF3SO3H, PhH
PhCH=CH2
Ph
(CH2)n

I.A.8-10 H. Shirahama et al., <u>Tetrahedron Lett.</u>, 24, 2869 (1983); A. G. Gonzalez et al., <u>ibid</u>, 24, 969 (1983); T. Ohtsuka, H. Shirahama and T. Matsumoto, <u>ibid</u>, 24, 3851 (1983).

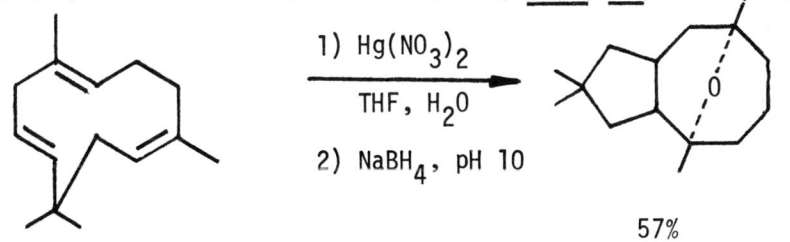

57%

I.A.8-11 W. Kirmse and J. Streu, <u>Synthesis</u>, 994 (1983); K. Takeda, Y. Shimono and E. Yoshii, <u>J. Am. Chem. Soc.</u>, 105, 563 (1983).

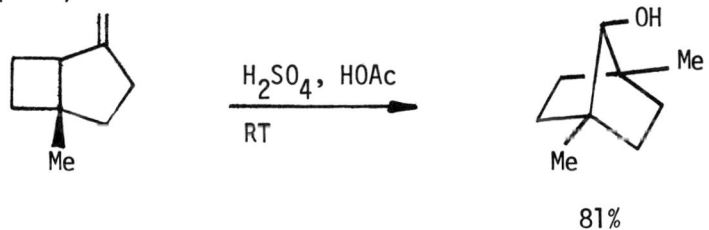

81%

I.A.8-12 J. Hoflack and P. J. De Clercq, <u>Bull. Soc. Chim. Belg.</u>, 92, 407 (1983); R. D. Gupta, B. C. Ranu and U. R. Ghatak, <u>Ind. J. Chem.</u>, 22B, 619 (1983); M. Ishitsuka, T. Kusumi and H. Kakisawa, <u>Chem. Lett.</u>, 999 (1983); R. M. Bettolo, A. Lupi et al., <u>Helv. Chim. Acta</u>, 66, 760, 1922 (1983).

Via an Electronically Disfavored Course.   50%

I.A.8-13 M. Nishizawa et al., Tetrahedron Lett., 24, 2581 (1983); P. Gosselin and F. Rouessac, ibid, 24, 5515 (1983); M. Nishizawa, H. Takenaka and Y. Hayashi, Chem. Lett., 1459 (1983); P. F. Vlad, N. D. Ungur and M. N. Koltsa, Tetrahedron, 39, 3947 (1983).

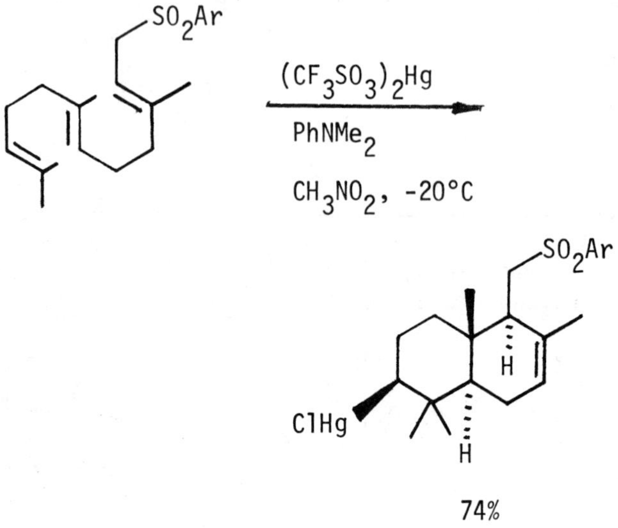

74%

I.A.8-14 W. S. Johnson, Y. Q. Chen and M. S. Kellogg, J. Am. Chem. Soc., 105, 6653 (1983); E. E. van Tamelen and J. R. Hwu, ibid, 105, 2490 (1983); G. Mikhail and M. Demuth, Helv. Chim. Acta, 66, 2362 (1983); P. M. Bishop, J. R. Pearson and J. K. Sutherland, Chem. Commun., 123 (1983).

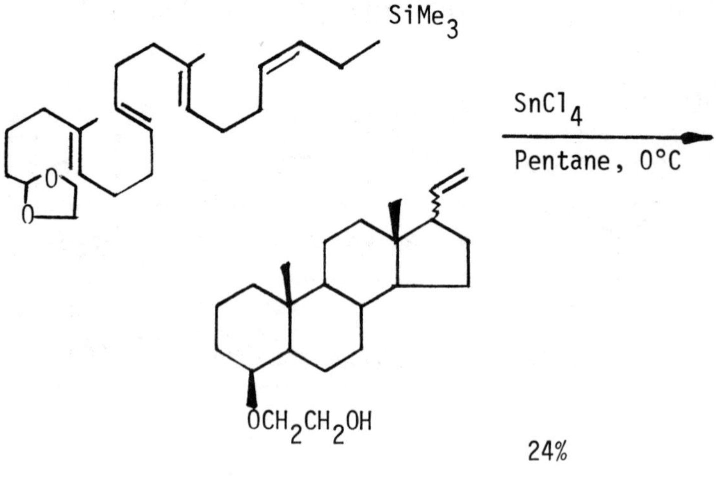

24%

I.A.8-15 J. V. Cooney, J. Het. Chem., 20, 823 (1983).

Review: "Reissert Compounds and Their Open-Chain Analogs in Organic Synthesis."

I.A.8-16 J. R. Gibson and W. Reusch, Tetrahedron, 39, 55 (1983); C. Agami and M. Fadlallah, ibid, 39, 777 (1983); S. A. Jacobs and R. G. Harvey, J. Org. Chem., 48, 5134 (1983).

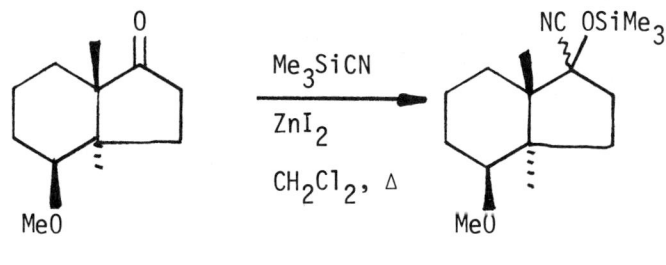

93%

I.A.8-17 J. D. Elliott, V. M. F. Choi and W. S. Johnson, J. Org. Chem., 48, 2294 (1983); G. A. Olah et al., Synthesis, 498 (1983); T. Harayama et al., Tetrahedron Lett., 24, 5241 (1983).

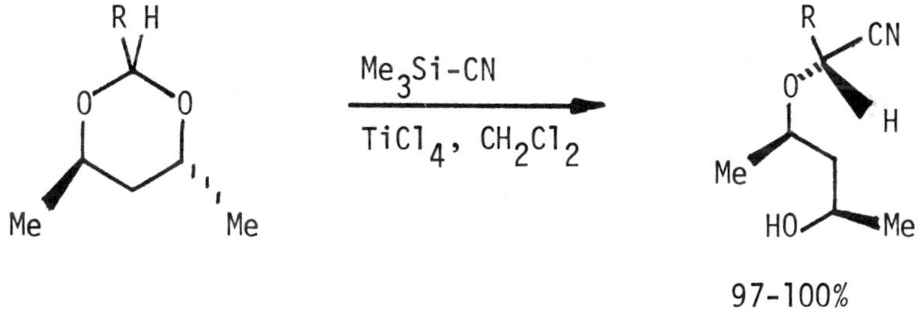

97-100%

Products converted to cyanohydrins, β-amino alcohols and α-hydroxy acids with op > 90%.

I.A.8-18 G. A. Olah, M. Arvanaghi and G. K. S. Prakash, Synthesis, 636 (1983); T. Ando et al., ibid, 637 (1983); M. Sakamoto et al., Chem. Pharm. Bull., 31, 2623 (1983).

$$\text{Ar-C(=O)-Cl} \xrightarrow[\text{SnCl}_4,\ \text{CH}_2\text{Cl}_2]{\text{Me}_3\text{SiCN}} \text{Ar-C(=O)-CN}$$

46-94%

I.A.8-19 H. Bohme, G. Braun and A. Ingendoh, Liebigs Ann. Chem., 717 (1983); M. Taddei et al., Tetrahedron Lett., 24, 2311 (1983).

$$\text{Ph}_2\text{C=N-CH}_2\text{-NMe}_2 \xrightarrow[\substack{2)\ \text{NaCN, H}_2\text{O} \\ \Delta}]{1)\ \text{MeI}} \text{Ph}_2\text{C=N-CH}_2\text{-CN}$$

38%

I.A.8-20 F. Duboudin et al., Tetrahedron Lett., 24, 4335 (1983); V. S. Russkikh and G. G. Abashev, J. Org. Chem., (USSR), 19, 742 (1983).

$$\text{MeSiCl}_3 \xrightarrow[\substack{\text{Pyr, CH}_3\text{CN} \\ 2)\ R^1R^2\text{C=O} \\ 3)\ \text{LAH}}]{1)\ \text{KCN, NaI}} R^2\text{-}\underset{\text{OH}}{\overset{R^1}{\text{C}}}\text{-CH}_2\text{NH}_2$$

64-75%

I.A.8-21 R. Chenevert, R. Plante and N. Voyer, Synth. Commun., 13, 403 (1983); W. J. Greenlee and D. G. Hangauer, Tetrahedron Lett., 24, 4559 (1983).

$$R^1\text{-C(=O)-}R^2 \xrightarrow[\substack{\text{KCN} \\ \text{CH}_2\text{Cl}_2,\ \text{18-C-6}}]{\text{PhCOCl}} R^1\text{-}\underset{\text{CN}}{\overset{\text{OC(=O)Ph}}{\text{C}}}\text{-}R^2$$

30-94%

I.A.8-22 W. A. Davis and M. P. Cava, J. Org. Chem., 48, 2774 (1983).

$$Ar^1CH_2CN \xrightarrow[\substack{2\ Ar^2CH_2SCN \\ (Ar^2 = 2\text{-}ClC_6H_4\text{-}) \\ 2)\ H_3O^+}]{1)\ 2\ R_2NLi} Ar^1CH(CN)_2 \quad 56\text{-}100\%$$

I.A.8-23 T. Funabiki, Y. Sato and S. Yoshida, Bull. Chem. Soc. Jpn., 56, 2863 (1983).

$$HC\equiv C(CH_2)_nC\equiv CH \xrightarrow[\substack{KCN,\ NaBH_4\ or \\ Zn \\ \text{Ethylene Glycol or } H_2O}]{K_2Ni(CN)_4} CH_3\underset{CN}{CH}(CH_2)_n\underset{CN}{CH}CH_3 \quad 28\text{-}85\%$$

I.A.8-24 T. L. Ho and S. H. Liu, Synth. Commun., 13, 1125 (1983); K. Karimian, F. Mohanazadeh and S. Rezai, J. Het. Chem., 20, 1119 (1983); H. Stetter and H. T. Leinen, Chem. Ber., 116, 254 (1983).

<chemical reaction: CH2=CH-C(=O)-CH2-OAc + 1) CH3CHO, (P)-thiazolium salt → CH3-C(=O)-CH2-CH2-C(=O)-CH2-OAc with additional ketone; 65%>

I.A.8-25 B. Giese, Angew Chem., Int. Ed. Engl., 22, 753 (1983).

Review: "Formation of CC Bonds by Addition of Free Radicals to Alkenes."

I.A.8-26 B. Giese et al., Chem. Ber., 116, 1240, 1264 (1983);
Tetrahedron Lett., 24, 3221 (1983); Synthesis, 733 (1983);
Angew Chem. Int. Ed. Engl., 22, 622 (1983).

$$R^1-\underset{\underset{R^2}{\|}}{C}-R^2 \xrightarrow[\text{2) } \diagup\!\!\!\diagdown CN, NaBH_4]{\text{1) HgO, Hg(OAc)}_2} R^1-\underset{\underset{R^2}{|}}{\overset{\overset{OAc}{|}}{C}}-CH_2CH_2CN$$

25-77%

I.A.8-27 A. L. J. Beckwith, D. M. O'Shea and D. H. Roberts,
Chem. Commun., 1445 (1983); D. L. J. Clive and P. L. Beaulieu,
ibid, 307 (1983); N. N. Marinovic and H. Ramanathan, Tetrahedron Lett., 24, 1871 (1983); M. Bertrand, P. Teisseire and
G. Pelerin, Nouv. J. Chem., 7, 61 (1983).

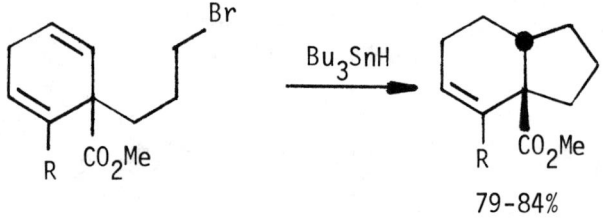

79-84%

I.A.8-28 S. K. Chung and L. B. Dunn, Jr., J. Org. Chem., 48,
1125 (1983).

$$(CH_2)_n\underset{CO_2R}{\overset{CO_2R}{\diagup}} \xrightarrow[\substack{\text{THF} \\ -78°C \to RT}]{\text{2.2 LDA, 2 CuBr}_2} (CH_2)_{n-2}\underset{CO_2R}{\overset{CO_2R}{\diagup}}$$

88-99%

I.A.8-29 A. Citterio, F. Ferrario and S. De Bernardinis, J.
Chem. Res. (S), 310 (1983); J. J. Villenave et al., Synth.
Commun., 13, 361 (1983); I. Tabbaa, M. Cazaux and R. Lalande,
Bull. Soc. Chim. Belg., 92, 1011 (1983).

$$R^1CH_2\underset{\underset{O}{\|}}{C}CH_2R^1 \xrightarrow[\text{H}_2\text{O}]{\underset{\text{AgNO}_3,\ \text{Na}_2\text{S}_2\text{O}_8}{R^2CH=CH_2}} R^1CH_2\underset{\underset{O}{\|}}{C}\underset{R^1}{\overset{|}{C}}H-CH_2CH_2R^2$$

5-68%

I.A.8-30 H. G. Viehe et al., J. Organometal. Chem., **250**, 197 (1983); T. Fujita et al., Chem. Ind. (London), 167 (1983).

$$\overset{CO_2Me}{\diagup\!\!\!\!=} \xrightarrow[\substack{\text{1) PhSeBr} \\ \text{2) Et}_3\text{N} \\ \text{3) AIBN}}]{}$$

$$\underset{\text{Me}}{\overset{\text{Me}}{\underset{|}{\text{R-C-CH}_2}}}\diagdown\underset{\text{MeO}_2\text{C}}{=}\diagup\overset{CO_2Me}{\diagdown}\underset{\text{CH}_2\text{-C-R}}{\diagup}\overset{\text{Me}}{\underset{\text{Me}}{|}}$$

43%

I.A.8-31 J. D. Wuest et al., J. Org. Chem., **48**, 3810 (1983).

$$\text{Ph}_3\text{Sn}\diagdown\!\!\diagup\!\!\diagdown\text{SnPh}_3 \xrightarrow[\text{AIBN}]{\text{CCl}_4,\ \Delta} \text{Cl}_3\text{C}\diagdown\!\!\diagup\!\!\diagdown\text{CCl}_3$$

65%

I.A.8-32 S. Wolff, W. C. Agosta, et al., *J. Am. Chem. Soc.*, **105**, 1292, 1299 (1983); *J. Org. Chem.*, **48**, 1718 (1983).

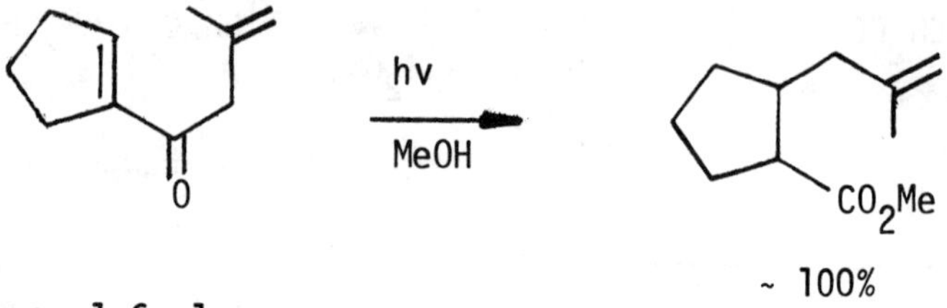

1,5 vs. 1,6 closures.

I.A.8-33 M. Pasternak and A. Morduchowitz, *Tetrahedron Lett.*, **24**, 3439, 4275 (1983); K. Tsujimoto, Y. Inaba and M. Ohashi, *Chem. Lett.*, 1113 (1983).

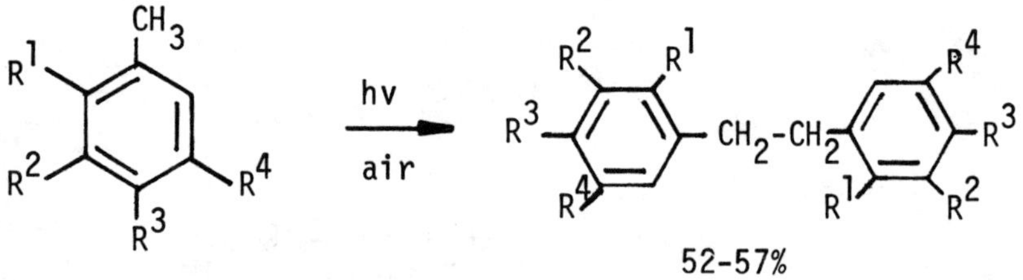

I.A.8-34 P. G. Gassman and J. L. Smith, *J. Org. Chem.*, **48**, 4438 (1983).

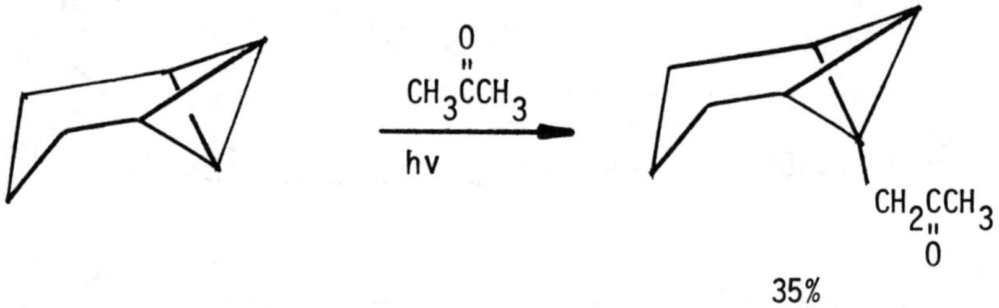

I.A.8-35 M. A. Meador and P. J. Wagner, *J. Am. Chem. Soc.*, **105**, 4484 (1983); T. H. Kim, Y. Hayase and S. Isoe, *Chem. Lett.*, 1421 (1983); H. Seto et al., *ibid*, 989 (1983); N. K. Hamer, *J. Chem. Soc., Perkin I*, 61 (1983); P. A. Wender and J. J. Howbert, *Tetrahedron Lett.*, **24**, 5325 (1983).

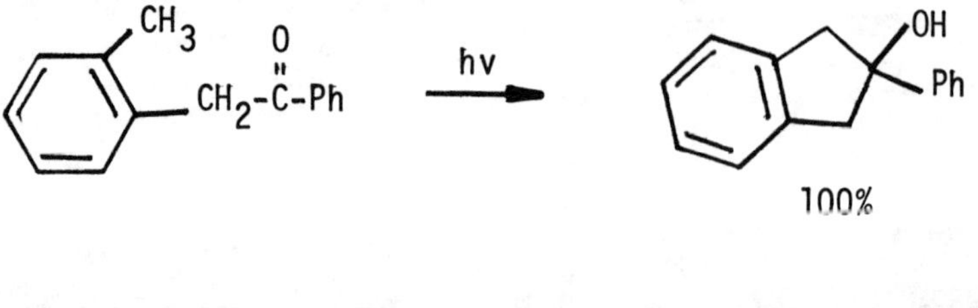

I.A.8-36 S. L. Schreiber, A. H. Hoveyda and H. J. Wu, J. Am. Chem. Soc., 105, 660 (1983); J. P. Morizur and J. Tortajada, Bull. Soc. Chim. Fr. II, 175 (1983).

Me-furan
1) RCHO, hv
2) 0.01 N HCl
THF
→ product 88-92%

I.A.8-37 J. J. Bonet et al., Chem. Commun., 718 (1983); P. K. Chowdhury, R. P. Sharma and J. N. Baruah, Tetrahedron Lett., 24, 5429 (1983).

hv, H_2SO_4 → 53%

I.A.8-38 J. S. Swenton, Acct. Chem. Res., 16, 74 (1983).

Review: "Quinone Bis- and Monoketals via Electrochemical Oxidation. Versatile Intermediates for Organic Synthesis."

I.A.8-39 R. Scheffold and R. Orlinski, J. Am. Chem. Soc., 105, 7200 (1983); S. Satoh, H. Suginome and M. Tokuda, Bull. Chem. Soc. Jpn., 56, 1791 (1983).

$(C_7H_{15}\overset{O}{\overset{\|}{C}})_2O$ $\xrightarrow[\text{DMF, Cat. Vitamin B}_{12}]{CH_2=CHCHO, \text{ e}, hv}$ $C_7H_{15}\overset{O}{\overset{\|}{C}}-CH_2CH_2CHO$ 30-80%

I.A.8-40 N. Muller, J. Org. Chem., 48, 1370 (1983); T. Davies R. N. Haszeldine and A. E. Tipping, J. Chem. Soc., Perkin I, 1353 (1983).

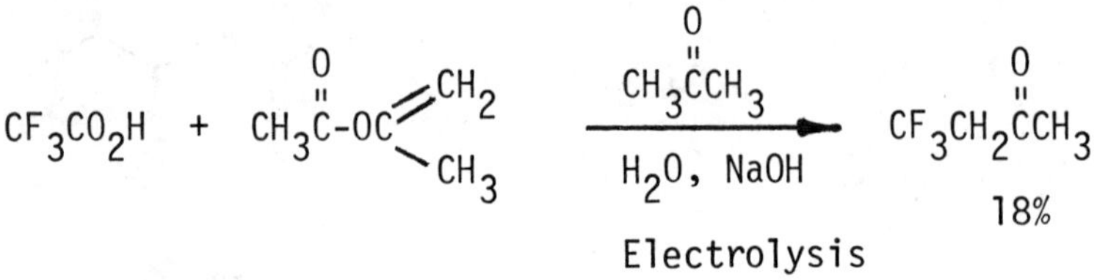

CF3CO2H
CH3C-OC
CH2
CH3
CH3CCH3
H2O, NaOH
Electrolysis
CF3CH2CCH3
18%

I.A.8-41 S. Torii, T. Inokuchi and R. Oi, J. Org. Chem., 48, 1944 (1983); D. Lelandais et al., Bull. Soc. Chim. Belg., 92, 305 (1983).

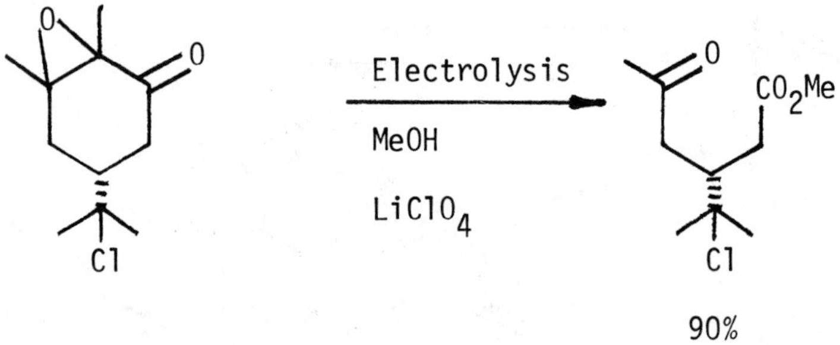

Electrolysis
MeOH
LiClO4
CO2Me
Cl
Cl
90%

I.A.8-42 T. H. Black, Aldrichimica Acta, 16, 3 (1983).

Review: "The Preparation and Reactions of Diazomethane."

I.A.8-43 E. C. Taylor and H. M. L. Davies, Tetrahedron Lett., 24, 5453 (1983).

OEt
N2
Ph
H
1) Rh(OAc)3
PhF
2) H2O, Δ
CO2Et
Ph
67%

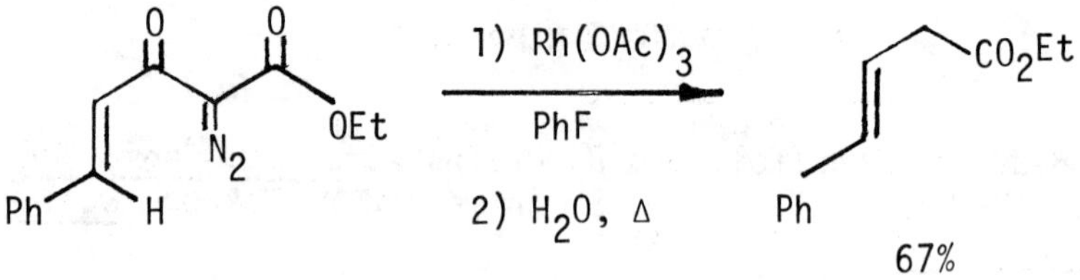

I.A.8-44 M. P. Doyle, M. L. Trudell and J. W. Terpstra, J. Org. Chem., 48, 5146 (1983); J. P. Anselme et al., ibid, 48, 4407 (1983); T. Aoyama, T. Shioiri et al., Chem. Pharm. Bull., 31, 2957 (1983); K. Nagao, M. Chiba and S. W. Kim, Synthesis, 197 (1983).

Ph–CH=CH–CH(OMe)$_2$ $\xrightarrow[\text{CH}_2\text{Cl}_2,\ 0°\text{C}]{\text{N}_2\text{CHCO}_2\text{Et} \atop \text{BF}_3\cdot\text{Et}_2\text{O}}$ Ph–CH=CH–C(CO$_2$Et)H–CH(OMe)$_2$

70%

I.A.8-45 D. F. Taber and K. Raman, J. Am. Chem. Soc., 105, 5935 (1983); T. Hudlicky et al., J. Org. Chem., 48, 4453 (1983); S. V. Govidan, T. Hudlicky and F. J. Koszyk, ibid, 48, 3581 (1983); T. Hudlicky et al., ibid, 48, 3422 (1983); J. C. Gilbert, D. H. Giamalva and U. Weerasooriya, ibid, 48, 5251 (1983).

[α-diazo-β-ketoester with pentyl chain] $\xrightarrow[\text{CH}_2\text{Cl}_2,\ \text{RT}]{\text{Rh}_2\text{OAc}_4}$ [cyclopentanone with CO$_2$R* and pentyl substituents]

R* – Camphor-Derived Residue. 60%

(74% de)

I.A.8-46 D. Mukherjee, et al., Tetrahedron Lett., 24, 3921, 5919 (1983).

[tricyclic aromatic substrate with CH$_2$COCHN$_2$ side chain, MeO and OMe substituents] $\xrightarrow[\text{CH}_2\text{Cl}_2,\ -25°\text{C}]{\text{CF}_3\text{CO}_2\text{H}}$

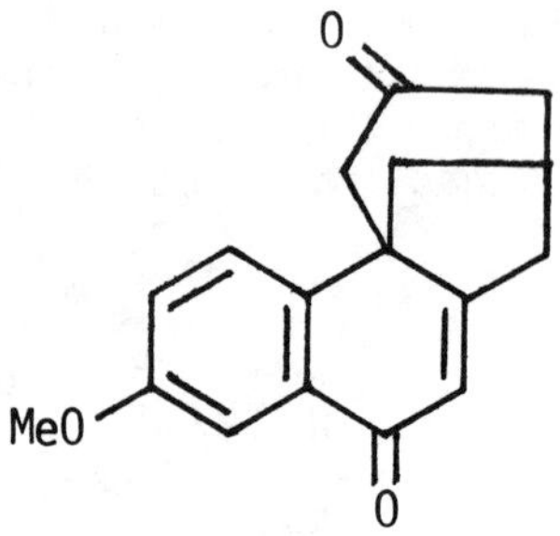

50%

I.A.8-47 R. D. Little et al., J. Am. Chem. Soc., 105, 928, 6976 (1983); J. Org. Chem., 48, 3139, 4487 (1983).

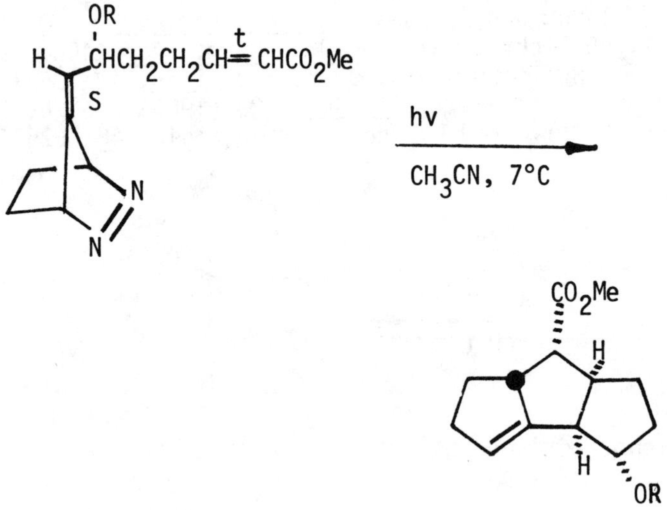

$$\text{structure with OR, CHCH}_2\text{CH}_2\text{CH} \overset{t}{=} \text{CHCO}_2\text{Me}$$

$$\xrightarrow{h\nu}{CH_3CN, 7°C}$$

≥ 85%

(93% de)

I.A.8-48 T. Harada, E. Akiba and A. Oku, J. Am. Chem. Soc., 105, 2771 (1983).

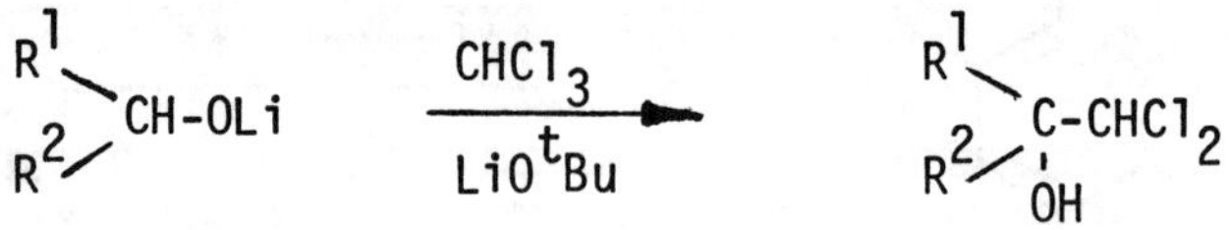

$$R^1R^2\text{CH-OLi} \xrightarrow[\text{LiO}^t\text{Bu}]{\text{CHCl}_3} R^1R^2\text{C(OH)-CHCl}_2$$

THF, Hexane   32-91%

I.A.8-49 J. Barluenga et al., J. Org. Chem., 48, 3116 (1983);
A. Sekiguchi, T. Sato and W. Ando, Chem. Lett., 1083 (1983).

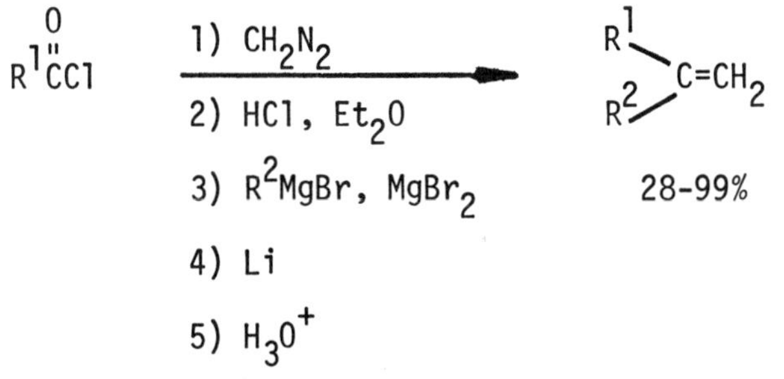

$R^1\overset{O}{\underset{\|}{C}}Cl$
1) CH_2N_2
2) HCl, Et_2O
3) R^2MgBr, $MgBr_2$
4) Li
5) H_3O^+

→ $R^1{\atop R^2}\!\!\!>\!\!C=CH_2$

28-99%

I.A.8-50 J. E. McMurry, Acct. Chem. Res., 16, 405 (1983).

Review: "Titanium-Induced Dicarbonyl-Coupling Reactions."

I.A.8-51 J. E. McMurry and D. D. Miller, J. Am. Chem. Soc., 105, 1660 (1983); Tetrahedron Lett., 24, 1885 (1983); P. C. Anderson, D. L. J. Clive and C. F. Evans, ibid, 24, 1373 (1983); Z. I. Yoshida et al., ibid, 24, 3469, 3473 (1983).

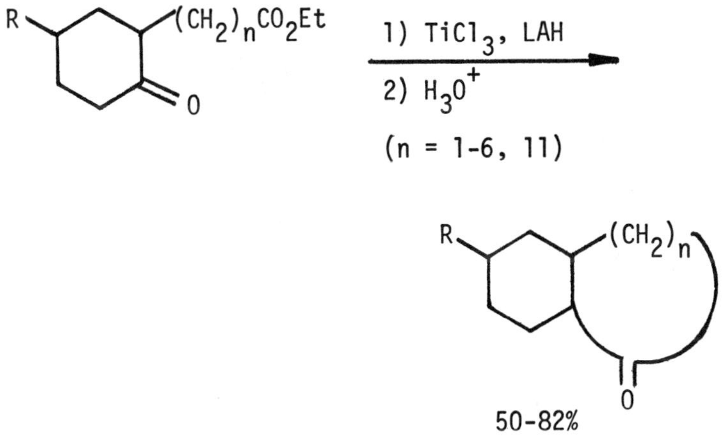

[cyclohexanone with R substituent and $(CH_2)_n CO_2Et$ side chain]

1) $TiCl_3$, LAH
2) H_3O^+

(n = 1-6, 11)

[bicyclic product with R substituent and $(CH_2)_n$ bridge containing C=O]

50-82%

I.A.8-52 H. Rupp, W. Schwarz and H. Musso, Chem. Ber., 116, 2554 (1983); J. L. Namy, J. Souppe and H. B. Kagan, Tetrahedron Lett., 24, 765 (1983).

$$O=\square-CO_2Et \xrightarrow{\text{Low Valent Ti Reagents}} EtO_2C-\square(OH)-\square(OH)-CO_2Et$$

13%

I.A.8-53 K. Hirai and I. Ojima, Tetrahedron Lett., 24, 785 (1983).

2 $R^1O\text{-}C(OSiMe_3)\text{=}C(R^2)\text{-}C(R^3)\text{=}CH_2$ $\xrightarrow[CH_2Cl_2, 0°C]{TiCl_4}$ Study of regiochemistry of Homo-coupling.

I.A.8-54 S. Murai et al., J. Am. Chem. Soc., 105, 7192 (1983).

[Me$_3$SiO-cyclopropanated cyclohexane] $\xrightarrow[Cu(BF_4)_2]{AgBF_4 \text{ or}}$ [bis(2-oxocyclohexyl)ethane]

42-87%

I.A.8-55 C. Ruchardt et al., Chem. Ber., 116, 3235 (1983).

$$Ar\text{-}\underset{R^2}{\overset{R^1}{C}}\text{-}X \xrightarrow[(X = OH, Cl, Br, H)]{\text{Redn.}} Ar\text{-}\underset{R^2}{\overset{R^1}{C}}\text{-}\underset{R^2}{\overset{R^1}{C}}\text{-}Ar$$

I.A.8-56 K. Itoh et al., Tetrahedron Lett., 24, 4021 (1983);
T. T. Takahashi, K. Nomura and J. Y. Satoh, Chem. Commun.,
1441 (1983); K. Ando, K. Tajima and K. Takase, J. Org. Chem.,
48, 1210 (1983); M. G. Pettett and A. B. Holmes, J. Chem. Soc.,
Perkin I, 1243 (1983).

$$\text{AcO-N=cyclohexylidene-CH}_2\text{SiMe}_3 \xrightarrow{\text{cat. TMSOTf.}} \text{N≡C-CH}_2\text{CH}_2\text{CH}_2\text{CH=CH}_2$$

89%

I.A.8-57 R. L. Snowden, Helv. Chim. Acta, 66, 1031 (1983);
W. Oppolzer, F. Zutterman and K. Battig, Helv. Chim. Acta, 66,
522 (1983).

$$\text{norbornenyl-R, OH} \xrightarrow[\text{2) Aq. NH}_4\text{Cl, 0°C}]{\text{1) KH, HMPA, 30°C}} \text{cyclopentenyl-CH}_2\text{C(O)R}$$

8-85%

I.A.8-58 J. Tsuji, et al., Tetrahedron Lett., 24, 1793, 1797,
3865 (1983); J. E. Backvall, R. E. Nordberg and J. Vagberg,
ibid, 24, 411 (1983); J. H. Babler and K. P. Spina, ibid, 24,
3835 (1983); M. Chandler, P. J. Parsons and E. Mincione, ibid,
24, 5781 (1983); E. C. Taylor and J. S. Skotnicki, Synth.
Commun., 13, 1137 (1983).

$$\text{cyclohexenyl-OCO}_2\text{CH}_2\text{CH=CH}_2 \xrightarrow[\text{Ph}_3\text{P, DME}]{\text{Pd(dba)}_3 \cdot \text{CHCl}_3} \text{2-allylcyclohexanone}$$

91%

I.A.8-59 J. S. Tou and A. A. Schleppnik, J. Org. Chem., 48, 753 (1983).

$$\underset{Z}{\overset{CO_2Me}{R-C-CH_2-CO_2Me}} \xrightarrow[\text{(Z = CO_2Me or CN)}]{KH, \text{ Glyme}} \underset{Z}{R-CH-CH}\begin{matrix}CO_2Me\\CO_2Me\end{matrix}$$

80-88%

I.A.8-60 T. Shioiri et al., Chem. Pharm. Bull., 31, 2564, 3139 (1983).

$$CH_3CH_2\overset{O}{\overset{\|}{C}}\text{-}\langle\text{Ar}\rangle\text{-}CH_2CHMe_2 \xrightarrow[\substack{\text{2) (PhO)}_2PON_3\\\text{THF}\\\text{3) KOH}}]{\text{1) pyrrolidine}} HO_2C\text{-}\underset{CH_3}{CH}\text{-}\langle\text{Ar}\rangle\text{-}CH_2CHMe_2$$

79%

I.A.8-61 S. Sebti and A. Foucaud, Synthesis, 546 (1983).

$$\underset{Cl}{\overset{O}{R^1\overset{\|}{C}CHR^2}} \xrightarrow[R^4NC]{R^3CO_2H} \underset{\underset{R^2}{CH\text{-}Cl}}{\overset{O}{R^3\overset{\|}{C}O}\text{-}\overset{R^1}{\underset{|}{C}}\text{---}\overset{O}{\overset{\|}{C}}\text{-}NH\text{-}R^4}$$

53-98%

Product + CsF yields 3-acyloxy-2-azetidinones.

I.A.8-62 H. Yamamoto and K. Maruoka, Pure Appl. Chem., 55, 1853 (1983).

Review: "Selective Reactions using Organoaluminum Reagents."

I.A.8-63 H. Yamamoto et al., J. Am. Chem. Soc., 105, 672, 2831, 6154, 6312 (1983).

$$R^1\underset{\underset{OSO_2R}{N}}{\diagup\!\!\!\diagdown}R^2 \xrightarrow{R_2AlX} R^1\underset{\underset{X}{N=}}{\diagup\!\!\!\diagdown}R^2$$

46-97%

Organoaluminum-promoted Beckmann Rearrangement to Form Imino Thioethers and Selenoethers, Imino Nitriles, α-Alkylated and α,α-Dialkylated Amines.

I.A.8-64 H. G. Richey, Jr., L. M. Moses and J. J. Hangeland, Organometallics, 2, 1545 (1983).

1) ROH, CH_2Cl_2
   Ti(acac)$_2Cl_2$
2) R_2AlCl

95% (GC)

I.A.8-65 A. Solladie-Cavallo and J. L. Haesslein, Helv. Chim. Acta, 66, 1760 (1983); A. S. Kende and P. J. Sanfilippo, Synth. Commun., 13, 715 (1983).

$$\diagup\!\!\!\diagdown R \xrightarrow[Pd(II)]{^-Nu^*} CH_3\underset{*}{\diagup\!\!\!\diagdown}\overset{Nu}{R}$$

$$^-\text{Nu}^* = \text{Ar}\overset{\overset{O}{\|}}{S}\text{-CH}_2\text{Li}, \text{LiCH}_2\text{-}\underset{N}{\overset{O}{\diagup}}\hspace{-0.1cm}\underset{\text{'''}CH_2OMe}{\diagdown}^{Ph}$$

Palladium-promoted alkylation of alkenes.

I.A.8-66 H. S. Mosher and J. D. Morrison, Science, 221, 1013 (1983).

Review: "Current Status of Asymmetric Synthesis."

I.A.8-67 J. Lieto et al., CHEMTECH, 46 (1983).

Review: "Polymeric Supports for Catalysts."

I.A.8-68 P. Cocagne, J. Elguero and R. Gallo, Heterocycles, 20, 1379 (1983).

Review: "The Present Use and the Possibities of Phase-Transfer Catalysis in Drug Synthesis."

I.A.8-69 G. M. Whitesides and C. H. Wong, Aldrichimica Acta, 16, 27 (1983).

Review: "Enzymes as Catalysts in Organic Synthesis."

I.A.8-70 V. Jager et al., Bull. Soc. Chim. Belg., 92, 1039 (1983).

Review: "Isoxazolines-Key Intermediates for Synthesis of Some Naturally Occuring Amino Compounds."

I.A.8-71 P. J. Stang and M. R. White, Aldrichimica Acta, 16, 15 (1983).

Review: "Triflic Acid and Its Derivatives."

I.A.8-72 E. F. V. Scriven, Chem. Soc. Rev., 12, 129 (1983).

Review: "4-Dialkylaminopyridines: Super Acylation and Alkylation Catalysts."

I.A.8-73 A. J. L. Cooper, J. Z. Ginos and A. Meister, Chem. Rev., 83, 321 (1983).

Review: "Synthesis and Properties of α-Keto Acids."

I.A.8-74 N. De Kimpe and N. Schamp, Org. Prep. Proced. Int., 15, 71 (1983).

Review: "Reactivity of β-Haloenamines."

I.A.8-75 K. N. Houk, Pure Appl. Chem., 55, 277 (1983).

Review: "Theoretical Studies of the Stereoselectivities of Organic Reactions."

I.A.8-76 U. Schollkopf, Top. Curr. Chem., 109, 65 (1983).

Review: "Enantioselective Synthesis of Nonproteinogenic Amino Acids."

I.A.8-77 L. Rossa and F. Vogtle, Top. Curr. Chem., 113, 1 (1983).

Review: "Synthesis of Medio- and Macrocyclic Compounds by High Dilution Principle Techniques."

I.A.8-78 B. M. Trost, Science, 219, 245 (1983).

Review: "Selectivity: A Key to Synthetic Efficiency."

I.A.8-79 L. N. Mander, Acct. Chem. Res., 16, 48 (1983).

Review: "New Strategies for the Construction of Highly Functionalized Organic Molecules: Applications to the C_{19} Gibberellin Synthesis."

I.A.8-80 J. Redpath and F. J. Zeelen, Chem. Soc. Rev., 12, 75 (1983).

Review: "Stereoselective Synthesis of Steroid Side-Chains."

I.A.8-81 G. Quinkert and H. Stark, Angew. Chem.,Int. Ed. Engl., 22, 637 (1983).

Review: "Stereoselective Synthesis of Enantiomerically Pure Natural Products - Estrone as Example."

I.A.8-82 D. Seebach and A. Hidber, Chimia, 37, 449 (1983).

Review: "Synthesis at Temperatures Below -80°C."

I.A.8-83 V. Gold, Pure Appl. Chem., 55, 1281 (1983).

Review: "Glossary of Terms Used in Physical Organic Chemistry."

I.B. Carbon-Carbon Double Bonds

(See also: I.E.1, III.G, VI.A.16).

I.B.1. Wittig-Type Olefination Reactions

I.B.1-1 H. J. Bestmann and O. Vostrowsky, Top. Curr. Chem., 109, 85 (1983).

Review: "Selected Topics of the Wittig Reaction in the Synthesis of Natural Products."

I.B.1-2 H. Pommer and P. C. Thieme, Top. Curr. Chem., 109, 165 (1983).

Review: "Industrial Applications of the Wittig Reaction."

I.B.1-3 M. Schlosser et al., Chimia, 37, 10 (1983).

Review: "SCOOPY and Oxirane Reactions: α-Lithio-Ylids vs. Conventional Ylids."

I.B.1-4 H. Schmidbaur, Angew Chem., Int. Ed. Engl., 22, 907 (1983).

Review: "Phosphorous Ylides in the Coordination Sphere of Transition Metals: An Inventory."

I.B.1-5 R. H. Grubbs et al., Pure Appl. Chem., 55, 1733 (1983).

Review: "Cp_2TiCl_2 Complexes in Synthetic Applications."

I.B.1-6 T. Kawashima, T. Ishii and N. Inamoto, Tetrahedron Lett., 24, 739 (1983).

$$Me_3SiCH-\underset{R^1}{\overset{O}{\overset{\|}{P}}}(OMe)_2 \xrightarrow[\text{2) } R^2R^3C=O]{\text{1) } F^-} \underset{R^3}{\overset{R^2}{>}}C=CHR^1$$

20-85%

I.B.1-7 E. Angeletti, P. Tundo and P. Venturello, Chem. Commun., 269 (1983).

$$Ph_3\overset{+}{P}CH_2R^1 \ X^- \xrightarrow[\substack{K_2CO_3 \\ \text{Carbowax 600 (liquid)} \\ 150-170°C}]{R^2CHO \text{ (gas)}} R^1CH=CHR^2 \text{ (gas)}$$

(solid)   8-86%

Gas-Liquid PTC Wittig.

I.B.1-8 Y. LeBigot, M. Delmas and A. Gaset, Tetrahedron Lett., 24, 193 (1983); A. S. Sarma and A. K. Gayen, ibid, 24, 3385 (1983); A. Krief et al., ibid, 24, 3413 (1983); A. P. Kozikowski and A. K. Ghosh, ibid, 24, 2623 (1983); F. Ricciardi and M. M. Joullie, Org. Prep. Proced. Int., 15, 17, 296 (1983); E. Schaumann and S. Fittkau, Synthesis, 449 (1983).

salicylaldehyde + $Ph_3\overset{+}{P}-CH_2R \ Br^-$ $\xrightarrow[\text{MeOH, RT}]{K_2CO_3}$ 2-hydroxystyryl product (CH=CHR)

74-88%

I.B.1-9 D. W. Knight and B. Ojhara, J. Chem. Soc., Perkin I, 955 (1983); H. A. Tolstikov, A. M. Moisecnkov et al., Org. Prep. Proced. Int., 15, 283 (1983); R. Zamboni and J. Rokach, Tetrahedron Lett., 24, 999 (1983); K. Takabe and J. D. White, ibid, 24, 3709 (1983); H. T. Toh and W. H. Okamura, J. Org. Chem., 48, 1414 (1983).

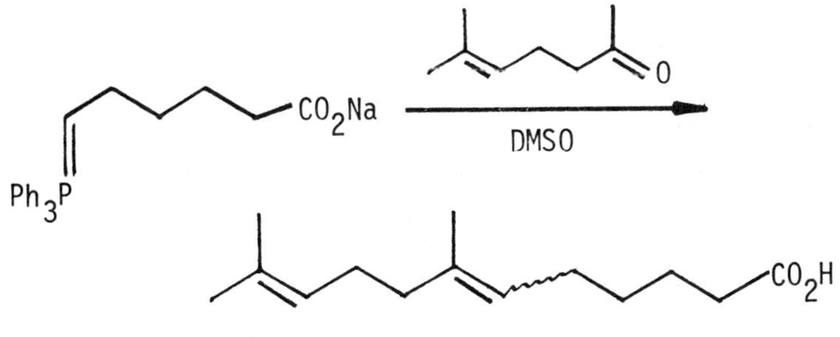

I.B.1-10 S. Warren et al., Tetrahedron Lett., 24, 111, 295, 2603, 3931, 5293 (1983).

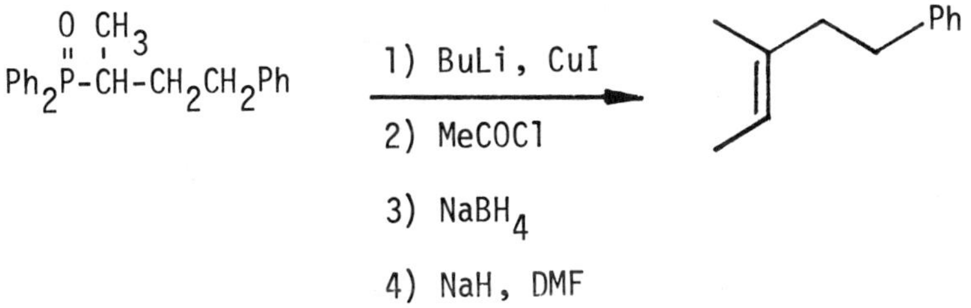

Erythio and threo alcohols containing phosphine oxide can be separated chromatographically.

I.B.1-11 H. Yamamoto et al., Tetrahedron Lett., 24, 4029 (1983); B. E. Maryanoff, A. B. Reitz and B. A. Duhl-Emswiler, ibid, 24, 2477 (1983); W. E. McEwen and J. V. Cooray, J. Org. Chem., 48, 983 (1983).

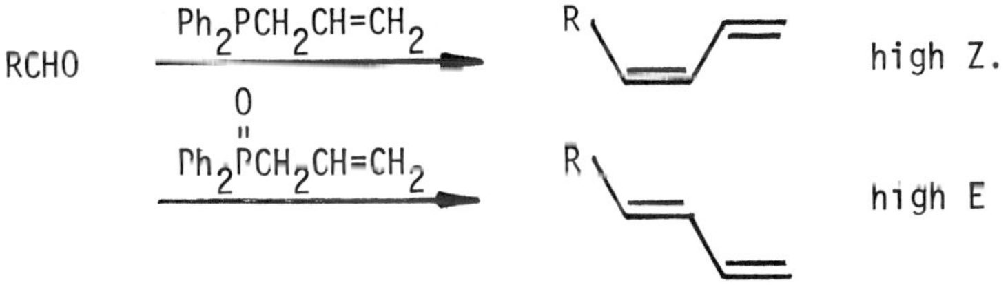

I.B.1-12 F. Buzzetti, N. Barbugian and C. A. Gandolfi, <u>Tetrahedron Lett.</u>, <u>24</u>, 2505 (1983).

$$Ph_3P=CH-\overset{O}{\underset{\|}{C}}-CH\overset{X}{\underset{R}{\diagdown}}$$

Prep and reactions of chiral Wittig reagents.

X = Me(S), Me(R), F(S).

I.B.1-13 A. J. H. Labuschagne and D. F. Schneider, <u>Tetrahedron Lett.</u>, <u>24</u>, 743 (1983); A. B. Reitz and B. E. Maryanoff, <u>Synth. Commun.</u>, <u>13</u>, 845 (1983); D. Beeman, C. Wenger and H. D. Perlmutter, <u>ibid</u>, 853 (1983).

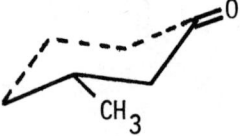

I.B.1-14 M. Duraisamy and H. M. Walborsky, <u>J. Am. Chem. Soc.</u>, <u>105</u>, 3252 (1983); D. F. Murray, <u>J. Org. Chem.</u>, <u>48</u>, 4860 (1983).

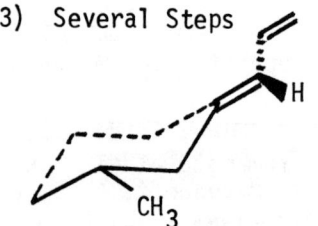

1) $(EtO)_2\overset{O}{\underset{\|}{P}}CH_2CO_2Et$ / Base

2) Saponification

3) Several Steps

Chiral Dienes of Known Configuration

I.B.1-15 Y. Okamoto et al., Synthesis, 916, 917 (1983); A. Minsky and M. Rabinovitz, ibid, 497 (1983); Y. Le Bigot, M. Delmas and A. Gaset, Synth. Commun., 13, 177 (1983).

$$(EtO)_2\overset{O}{\overset{\|}{P}}CH_2CH\overset{SiMe_3}{\underset{CN}{\diagdown}} \xrightarrow[\substack{2) R^1R^2C=O \\ 3) LDA \\ 4) R^1R^2C=O}]{1) LDA} \underset{R^2}{\overset{R^1}{\diagdown}}C=CH-\underset{CN}{\overset{}{\underset{|}{C}}}=C\underset{R^2}{\overset{R^1}{\diagup}}$$

34-89%

I.B.1-16 W. T. Ford et al., J. Org. Chem., 48, 326, 3164 (1983).

$$\underset{Me}{\overset{+}{\textcircled{P}}-PPh_2} \ I^-$$

Polymer Bound Wittig Reagents.

I.B.1-17 A. Osuka, H. Suzuki et al., Tetrahedron Lett., 24, 2599, 5109 (1983); J. B. Ousset, C. Mioskowski and G. Solladie, ibid, 24, 4419 (1983); Synth. Commun., 13, 1193 (1983); F. W. Hartner, Jr., J. Schwartz and S. M. Clift, J. Am. Chem. Soc., 105, 640 (1983).

$$R_2Te\overset{+}{-}\overset{-}{CHCO_2Et} \xrightarrow{R^1R^2C=O} \underset{R^2}{\overset{R^1}{\diagdown}}C=CH-CO_2Et$$

52-90%

I.B.1-18 R. J. Linderman and A. I. Meyers, Tetrahedron Lett., 24, 3043 (1983); T. N. De Castro Dantas, J. P. Laval and A. Lattes, Tetrahedron, 39, 3337 (1983).

$$R^1NHCH_2CH_2\overset{+}{-}PPh_3 \ Br^- \xrightarrow[\substack{2) R^2CHO \\ 3) Workup}]{1) 2\ nBuLi} R^1NHCH_2CH=CHR^2$$

40-90%

I.B.1-19 H. J. Bestmann and P. Ermann, Chem. Ber., 116, 3264 (1983).

$$RCH=PPh_3 \xrightarrow[\text{2) Hydrolysis}]{\text{1) OHC-CH(OEt)}_2} RCH=CHCHO$$

48-80%

I.B.1-20 R. C. Ronald and C. J. Wheeler, J. Org. Chem., 48, 138 (1983); H. J. Bestmann and K. Kumar, Chem. Ber., 2708 (1983); Y. Ueno, L. D. S. Yadav and M. Okawara, Chem. Lett., 831 (1983).

$$\phi_3P=\text{CH-CO-CH}_2\text{CH}_2\text{-CO}_2\text{Me} \xrightarrow[\text{DMF}]{\text{RCHO}} R\text{-CH=CH-CO-CH}_2\text{CH}_2\text{-CO}_2\text{Me}$$

70-93%

I.B.1-21 W. C. Still and C. Gennari, Tetrahedron Lett., 24, 4405 (1983); S. Tsuboi, H. Fukumoto and A. Takeda, Chem. Lett., 1219 (1983); J. Nokami et al., ibid, 1249, 1251 (1983); J. Villieras and M. Rambaud, Synthesis, 300 (1983); J. Villieras and M. Rambaud, Compt. Rend. (II), 296, 1175 (1983); N. Ono, A. Kaji et al., J. Org. Chem., 48, 3678 (1983).

$$R\text{-CHO} \xrightarrow[\text{THF}]{\underset{(Me_3Si)_2NK,\ 18\text{-}C\text{-}6}{(CF_3CH_2O)_2\overset{O}{\overset{\|}{P}}CHMeCO_2Me}} R\overset{Me}{\underset{CO_2Me}{\diagup\!\!\!\diagdown}}$$

79->95%

(Z:E=>30:1)

I.B.1-22 K. Schonauer and E. Zbiral, Tetrahedron Lett., 24, 573 (1983); Y. Vo-Quang et al., Chem. Commun., 1505 (1983); J. C. Gilbert and U. Weerasooriya, J. Org. Chem., 48, 448 (1983); T. A. M. van Schaik and A. van der Gen, Rec. Trav. Chim., 102, 465 (1983).

PhCHO → 1) Ph$_3$P=CHOCH$_2$CH$_2$SiMe$_3$ 2) 5% HF, CH$_3$CN → PhCH$_2$CHO

I.B.1-23 T. A. M. van Schaik, A. V. Henzen and A. van der Gen, Tetrahedron Lett., 24, 1303 (1983).

$$Ph_2\overset{O}{\overset{\|}{P}}-CH(OR^3)_2 \xrightarrow[\text{Et}_2\text{O, }-100°C]{\text{1) LDA, THF}} \begin{array}{c} R^1 \\ R^2 \end{array}\!\!\!\!>\!\!=\!\!<\!\!\!\!\begin{array}{c} OR^3 \\ OR^3 \end{array}$$

2) R^{1}R^{2}C=O   45-85%

3) H$_2$O

4) KO^{t}Bu

I.B.1-24 J. B. P. A. Wijnberg, J. Vader and A. de Groot, J. Org. Chem., 48, 4380 (1983); S. V. Ley and N. S. Simpkins, Chem. Commun., 1281 (1983).

[Decalone structure with R group and exocyclic methylene] → 1) Li-C(OMe)(SiMe$_3$)(SPh), THF, -80°C   2) HgCl$_2$, H$^+$, MeOH → [Decalin product with MeO$_2$C and R groups], 82%

I.B.1-25 E. Ohler, E. Zbiral and M. El-Badawi, Tetrahedron Lett., 24, 5599 (1983).

[Imidazo-pyrimidinone with OHC and Me substituents] → 2 NaCH(PO(OiPr)$_2$)$_2$

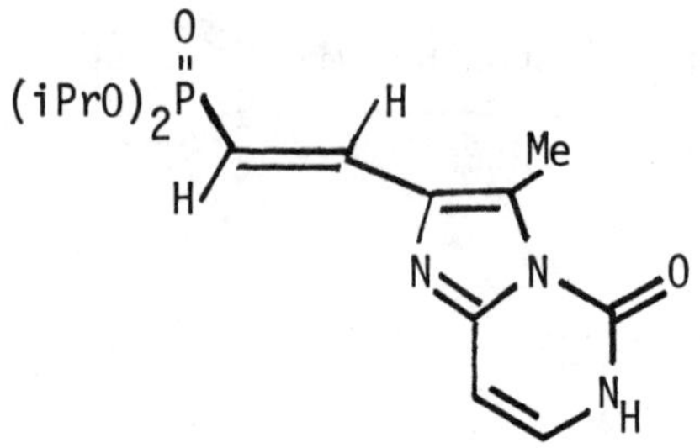

57%

I.B.1-26 R. Kober and W. Steglich, <u>Liebigs Ann. Chem.</u>, 599 (1983).

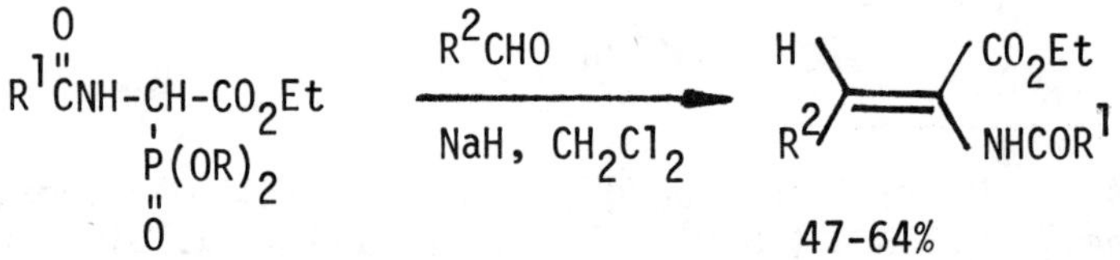

$$R^1\overset{O}{\underset{}{C}}NH-CH(P(OR)_2(=O))-CO_2Et \xrightarrow[\text{NaH, CH}_2\text{Cl}_2]{R^2CHO} \underset{R^2}{\overset{H}{>}}=\underset{NHCOR^1}{\overset{CO_2Et}{<}}$$

47-64%

I.B.1-27 A. Hosomi, M. Inaba and H. Sakurai, <u>Tetrahedron Lett</u>. $\underline{24}$, 4727 (1983); F. Camps et al., <u>ibid</u>, $\underline{24}$, 3387 (1983); M. C. Clingerman and J. A. Secrist, III, <u>J. Org. Chem.</u>, $\underline{48}$, 3141 (1983); M. D. Crenshaw and H. Zimmer, <u>ibid</u>, $\underline{48}$, 2782 (1983).

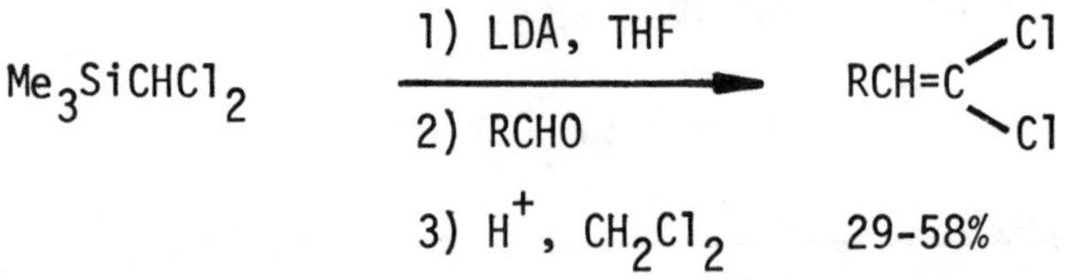

$$Me_3SiCHCl_2 \xrightarrow[\text{2) RCHO}]{\text{1) LDA, THF}} RCH=C(Cl)(Cl)$$

3) H^+, CH_2Cl_2 29-58%

I.B.1-28 D. J.Burton and D. G. Cox, J. Am. Chem. Soc., 105, 650 (1983); G. A. Wheaton and D. J. Burton, J. Org. Chem., 48, 917 (1983); T. Ishihara, T. Maekawa and T. Ando, Tetrahedron Lett., 24, 4229, 5657 (1983); D. Cantacuzene et al., Synthesis, 1010 (1983).

$$CFCl_3 \xrightarrow[\substack{2) R_FCF \\ \parallel \\ O \\ 3) NaOH, H_2O}]{1) 3 Bu_3P} \underset{14-50\%}{\overset{H\diagup\diagdown F}{F\diagup\diagdown R_F}}$$

I.B.1-29 J. R. Rocca et al., Tetrahedron Lett., 24, 1893 (1983); P. Babin and J. Dunogues, ibid, 24, 3071 (1983); P. A. Aristoff, Synth. Commun., 13, 145 (1983).

61%

I.B.1-30 R. J. Pariza and P. L. Fuchs, J. Org. Chem., 48, 2304 (1983); K. M. Pietrusiewicz, J. Monkiewicz and R. Bodalski, ibid, 48, 788 (1983); G. A. Flynn, ibid, 48, 4125 (1983); G. W. J. Fleet and T. K. M. Shing, Chem. Commun., 849 (1983).

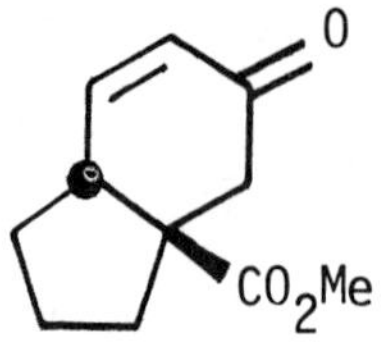
O
CO_2Me

64%

I.B.1-31 J. Auge and S. David, Tetrahedron Lett., 24, 4009 (1983); M. Moreno-Manas and A. Trius, Bull. Chem. Soc. Jpn., 56, 2154 (1983).

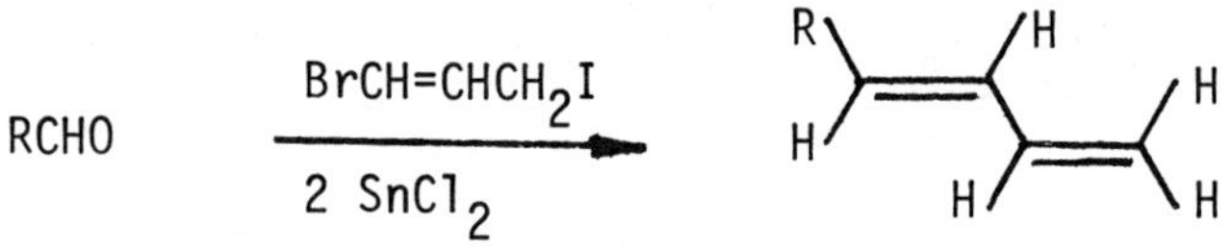
RCHO
$BrCH=CHCH_2I$
2 $SnCl_2$
R
H
H
H
H
H

54%

I.B.2.a. Eliminations of Alcohols and Derivatives to Form Double Bonds

I.B.2.a-1 L. Somekh and A. Shanzer, J. Org. Chem., 48, 907 (1983).

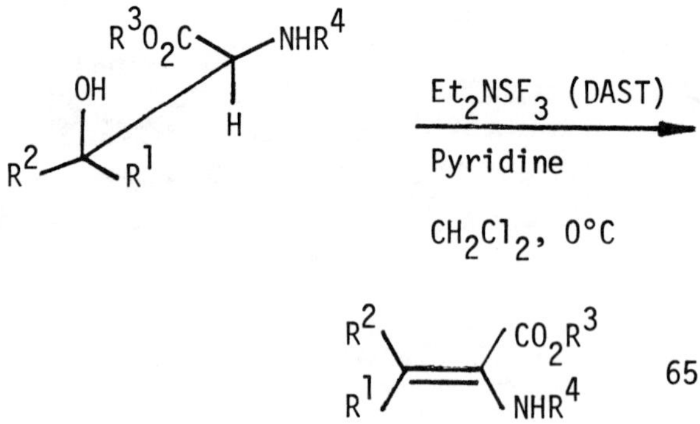
R^3O_2C
NHR^4
OH
H
R^2
R^1
Et_2NSF_3 (DAST)
Pyridine
CH_2Cl_2, 0°C
R^2
R^1
CO_2R^3
NHR^4

65-90%

I.B.2.a-2 T. Kolasa, Synthesis, 539 (1983).

$$\underset{\underset{OH}{|}}{\overset{\overset{CH_2R^1}{|}}{Ac-N-CH-CO_2R^2}} \xrightarrow[\substack{Et_3N \\ EtOAc \text{ or } CHCl_3}]{TosCl} \underset{82-96\%}{\overset{\overset{CHR^1}{\|}}{Ac-NH-C-CO_2R^2}}$$

I.B.2.a-3 B. M. Trost, M. Lautens and B. Peterson, Tetrahedron Lett., 24, 4525 (1983).

$$RCH_2\underset{\underset{}{}}{\overset{\overset{OH}{|}}{CH}}-CH=CH-CO_2Me \xrightarrow[\substack{2)\ Mo(CO)_6 \\ BSA, PhCH_3}]{1)\ Ac_2O,\ DMAP,\ Pyr.} R\frown\frown\frown CO_2Me$$

56%

BSA = O,N-bis(trimethylsilyl)acetamide

I.B.2.a-4 T. Kawashima, T. Ishii and N. Inamoto, Chem. Lett., 1375 (1983); J. L. Belletire et al., Synth. Commun., 13, 87, 589 (1983).

$$(MeO)_2\overset{\overset{O}{\|}}{P}-CH_2-\underset{\underset{OH}{|}}{\overset{\overset{R^1}{|}}{C}}-R^2 \xrightarrow{CsF,\ H_2O} CH_2=C\overset{R^1}{\underset{R^2}{\diagdown}}$$

46-85%

I.B.2.b. Eliminations of Halides to Form Double Bonds

I.B.2.b-1 S. Uemura and S. I. Fukuzawa, J. Am. Chem. Soc., 105, 2748 (1983); Y. Kimura and S. L. Regen, J. Org. Chem., 48, 195 (1983); D. R. Anton and R. H. Crabtree, Tetrahedron Lett., 24, 2449 (1983); G. G. Melikyan, K. A. Atanesyan and S. O. Badanyan, J. Org. Chem. (USSR), 19, 398 (1983).

$$RCH_2CH_2Br \xrightarrow[\begin{array}{c}\text{2) } Br_2, CCl_4 \\ \text{3) Aq. NaOH} \\ \text{4) Distill}\end{array}]{\text{1) } (PhTe)_2, NaBH_4} R-CH=CH_2$$
~50%

I.B.2.b-2 A. P. Croft and R. A. Bartsch, J. Org. Chem., 48, 876 (1983); Tetrahedron Lett., 24, 2737 (1983).

Dehydrochlorination favored (54-65%) over dehydrobromination with complex base (Syn. Elim.).

I.A.2.b-3 Z. Goren and I. Willner, J. Am. Chem. Soc., 105, 7764 (1983); V. Janout and P. Cefelin, Tetrahedron Lett., 24, 3913 (1983); J. Nakayama, H. Machida and M. Hoshino, ibid, 24, 3001 (1983); Y. Izawa, M. Takeuchi and H. Tomioka, Chem. Lett., 1297 (1983); F. Nome, M. C. Rezende and N. S. de Souza, J. Org. Chem., 48, 5357 (1983).

Debromination in 2-Phase System.

I.B.2.b-4 J. L. Belletire and D. R. Walley, <u>Tetrahedron Lett.</u>, <u>24</u>, 1475 (1983); J. Mulzer and O. Lammer, <u>Angew. Chem., Int. Ed. Engl.</u>, <u>22</u>, 628 (1983).

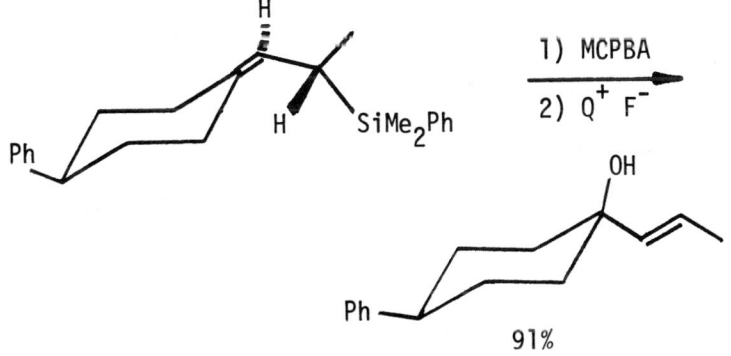

I.B.2.c. Other Eliminations to Form Double Bonds

I.B.2.c-1 I. Fleming and N. K. Terrett, <u>Tetrahedron Lett.</u>, <u>24</u>, 4151, 4153 (1983); T. Hayashi et al., <u>ibid</u>, <u>24</u>, 2665 (1983); K. Yamamoto and Y. Tomo, <u>ibid</u>, <u>24</u>, 1997 (1983).

I.B.2.c-2 E. Block et al., <u>J. Am. Chem. Soc.</u>, <u>105</u>, 6164, 6165 (1983); M. Julia et al., <u>Tetrahedron Lett.</u>, <u>24</u>, 1783 (1983); J. R. Schauder, J. N. Denis and A. Krief, <u>ibid</u>, <u>24</u>, 1657 (1983).

I.B.2.c-3 A. Hosomi, M. Inaba and H. Sakurai, Chem. Lett., 1763 (1983); M. Julia et al., Tetrahedron Lett., 24, 4331 (1983).

$$Me_3MCH_2CH_2\overset{\underset{|}{SOPh}}{C}HCH_2R \xrightarrow[PhCH_3]{Et_3N} Me_3MCH_2CH_2CH=CHR$$

44-84%

I.B.2.c-4 R. Block, D. Hassan and X. Mandard, Tetrahedron Lett., 24, 4691 (1983); H. Takayama et al., J. Org. Chem., 48, 3483 (1983); G. Schaden, ibid, 48, 5385 (1983).

[Structure showing bicyclic sulfone with OH, R^1, R^2, R^3, SO_2 groups] $\xrightarrow{650°C}$ $R^3\text{-CH=CH-CH=CH-}C(OH)(R^1)(R^2)$

73-78%

Starting material from anion of sulfone + $R^1R^2C=O$.

I.B.2.c-5 H. Takeshita and H. Mametsuka, Chem. Commun., 483 (1983).

[Tropone ether with $(CH_2)_n$ bridge] $\xrightarrow[180°C]{Me_2SO\ (d_6)}$ [cyclopentene with $(CH_2)_n$]

70-95%

(n = 1-4, 8)

Elimination by [s 8π + a 2σ + s 2σ].

I.B.3. Other Carbon-Carbon Double Bond Forming Reactions

I.B.3-1 R. M. Adlington and A. G. M. Barrett, Acct. Chem. Res., 16, 55 (1983).

Review: "Recent Applications of the Shapiro Reaction."

I.B.3-2 T. Baba, K. Avasthi and A. Suzuki, Bull. Chem. Soc. Jpn., 56, 1571 (1983); P. A. Butikofer and C. H. Eugster, Helv. Chim. Acta, 66, 1148 (1983); G. F. Cooper, A. R. Van Horn and D. Wren, Synth. Commun., 13, 1213 (1983); A. R. Chamberlin, E. L. Liotta and F. T. Bond, Org. Syn., 61, 141 (1983).

1) 2 BuLi, TMEDA
2) R_3B
3) I_2

67-98%

I.B.3-3 R. R. Schmidt, Bull. Soc. Chim. Belg., 92, 825 (1983).

Review: "Functionally Substituted Vinyl Carbanions Versatile Intermediates in Heterocyclic Synthesis."

I.B.3-4 C. B. B. Ekogha, O. Ruel and S. A. Julia, Tetrahedron Lett., 24, 4825, 4829 (1983); R. C. F. Jones and G. E. Peterson, ibid, 24, 4751, 4755, 4757 (1983); L. Duhamel, J. M. Poirier and N. Tedga, J. Chem. Res. (S), 222, 331 (1983); U. Melamed and B. A. Feit, J. Org. Chem., 48, 1928 (1983); R. R. Schmidt, J. Kast and H. Speer, Synthesis, 725 (1983).

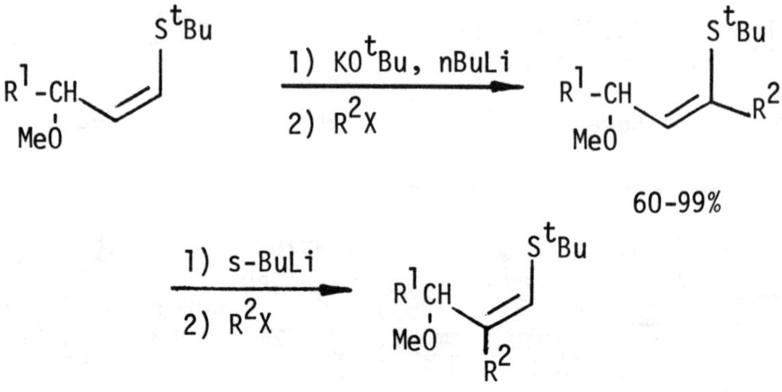

R¹-CH(MeO)-CH=CH-S^{t}Bu

1) KO^{t}Bu, nBuLi
2) R²X

→ R¹-CH(MeO)-CH=C(R²)-S^{t}Bu

60-99%

1) s-BuLi
2) R²X

→ R¹CH(MeO)-C(R²)=CH-S^{t}Bu

68-99%

I.B.3-5 R. B. Miller et al., J. Org. Chem., 48, 4113 (1983); Tetrahedron Lett., 24, 2055 (1983); Synth. Commun., 13, 969 (1983); W. Danikiewicz, T. Jaworski and S. Kwiatkowski, ibid, 13, 255 (1983); R. H. Smithers, J. Org. Chem., 48, 2095 (1983); A. B. Smith, III et al., Org. Syn., 61, 65 (1983).

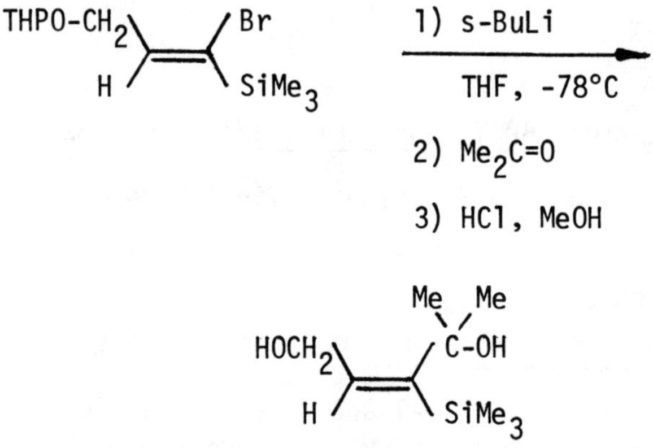

THPO-CH₂\\C(Br)=C(SiMe₃)/H

1) s-BuLi, THF, -78°C
2) Me₂C=O
3) HCl, MeOH

→ HOCH₂\\C=C(SiMe₃)(C(Me)(Me)-OH)/H

53%

I.B.3-6 Y. Sato and K. Hitomi, Chem. Commun., 170 (1983); T. Hirao et al., Bull. Chem. Soc. Jpn., 56, 1569 (1983).

$R^1CH=C(CN)SiMe_3$ → 1) Bu_4NF, THF; 2) $R^2R^3C=O$ → $R^1CH=C(CN)-C(R^2)(R^3)OH$

22-81%

R^1CHO plus $(Me_3Si)_2C=C=N-SiMe_3$, $BF_3 \cdot Et_2O$ gives starting material.

I.B.3-7 M. A. Tius and A. H. Fauq, J. Org. Chem., 48, 4131 (1983); S. Nunomoto, Y. Kawakami and Y. Yamashita, ibid, 48, 1912 (1983); T. Fujisawa et al., Chem. Lett., 1391 (1983).

$PhCH_2O$-epoxide-OH + isopropenyl-MgBr, THF, Et_2O, -25°C → $PhCH_2O$-CH$_2$-CH(OH)-CH(-C(=CH$_2$)CH$_3$)-CH$_2$OH

87%

I.B.3-8 F. K. Sheffy and J. K. Stille, J. Am. Chem. Soc., 105, 7173 (1983); M. Ochiai, T. Ukita and E. Fujita, Tetrahedron Lett., 24, 4025 (1983).

$CH_2=CH-CH_2-SnBu_3$ + $Br-CH_2-C(OMe)=CH-CO_2Me$, Cat. $Pd(dba)_2$, Cat. Ph_3P, THF, 50°C →

[structure: OMe-substituted diene ester] 75-90%

I.B.3-9 S. Martin, R. Sauvetre and J. F. Normant, Tetrahedron Lett., 24, 5615 (1983); A. S. Kende and P. Fludzinski, J. Org. Chem., 48, 1384 (1983); R. D. Chambers et al., J. Chem. Soc., Perkin I, 1235, 1239 (1983); M. M. Kremlev et al., J. Org. Chem. (USSR), 19, 814 (1983).

$CF_2=CFCl$

1) BuLi
2) Me_3SiCl
3) nBuLi
4) ^tBuCl, $AlCl_3$

→ nBu, F / F, ^tBu alkene   64%

I.B.3-10 H. M. R. Hoffmann and J. Rabe, Angew. Chem., Int. Ed. Engl., 22, 795, 796 (1983); K. Takai, T. Hiyama et al., Tetrahedron Lett., 24, 5281 (1983).

$CH_2=CHCO_2R^1$ —RCHO, cat. DABCO, RT→ $CH_2=C(CO_2R^1)CH(OH)R$ 33-95%

I.B.3-11 P. Vermeer et al., Rec. Trav. Chim., 102, 378 (1983); A. Kucerovy, K. Neuenschwander and S. M. Weinreb, Synth. Commun., 13, 875 (1983); A. J. Bridges and R. D. Thomas, Chem. Commun., 485 (1983); E. Giraudi and P. Teisseire, Tetrahedron Lett., 24, 489 (1983).

$CH_2=C=C(OMe)C(OAc)R^1R^1$ —R^2ZnCl, cat. $Pd(PPh_3)_4$, THF→ $H_2C=C(R^2)-C(OMe)=CR^1_2$ 50-95%

I.B.3-12 E. Piers and V. Karunaratne, J. Org. Chem., 48, 1774 (1983); M. P. Cooke, Jr., ibid, 48, 744 (1983); P. L. Fuchs et al., ibid, 48, 2167 (1983); S. Kurozumi et al., Tetrahedron Lett., 24, 4103 (1983).

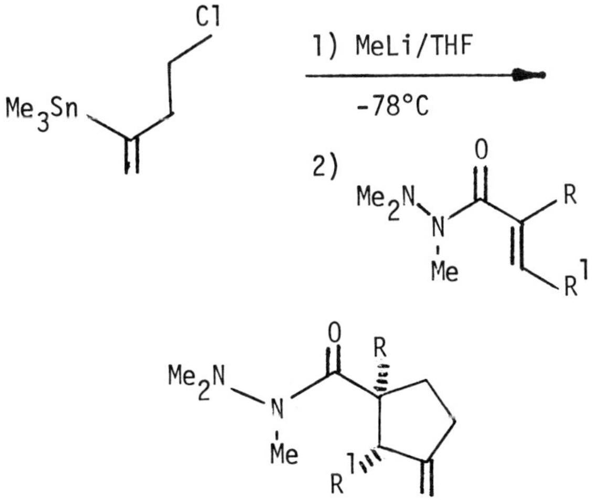

60-62%

Also, reaction with aldehydes and ketones.

I.B.3-13 K. J. Shea and P. Q. Pham, Tetrahedron Lett., 24, 1003 (1983); E. Piers et al., Chem. Commun., 935 (1983); Can. J. Chem., 61, 1226, 1239 (1983).

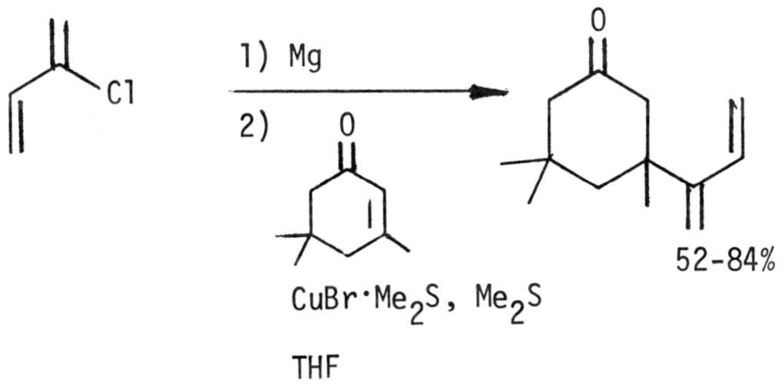

52-84%

I.B.3-14 W. G. Peet and W. Tam, Chem. Commun., 853 (1983); A. T. Hudson and M. J. Pether, J. Chem. Soc., Perkin I, 35 (1983); M. F. Aldersley, F. M. Dean and R. Nayyir-Mazhir, J. Chem. Soc., Perkin I, 1753 (1983); H. Brockmann and H. Laatsch, Liebigs Ann. Chem., 433, 1020 (1983); S. A. Russkikh, L. S. Klimenko and E. P. Fokin, J. Org. Chem. (USSR), 19, 144 (1983).

$$\text{2,3-dichloro-1,4-naphthoquinone} \xrightarrow{(nC_{12}H_{25})_3Al,\ ZnCl_2,\ THF,\ 25°C} \text{2-dodecyl-3-chloro-1,4-naphthoquinone}$$

82%

I.B.3-15 E. I. Negishi and J. A. Miller, J. Am. Chem. Soc., 6761 (1983); M. F. Semmelhack and R. Tamura, ibid, 105, 6750 (1983); J. W. Labadie and J. K. Stille, Tetrahedron Lett., 24, 4283 (1983); H. Alper, B. Despeyroux and J. B. Woell, ibid, 24, 5691 (1983).

$$\xrightarrow[\substack{1\ Pd(PPh_3)_4 \\ Et_3N,\ THF \\ 60°C}]{CO}$$

54%

I.B.3-16 E. J. Corey and S. G. Pyne, Tetrahedron Lett., 24, 2821 (1983); M. E. Jung and G. L. Hatfield, ibid, 24, 3175 (1983); G. Pattenden and G. M. Robertson, ibid, 24, 4617 (1983); G. Stork et al., J. Am. Chem. Soc., 105, 3720, 3741 (1983).

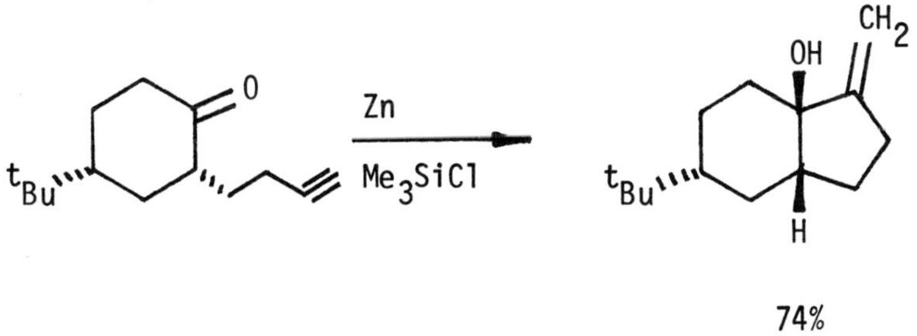

74%

I.B.3-17 R. F. Heck et al., J. Org. Chem., 48, 2792, 3894 (1983).

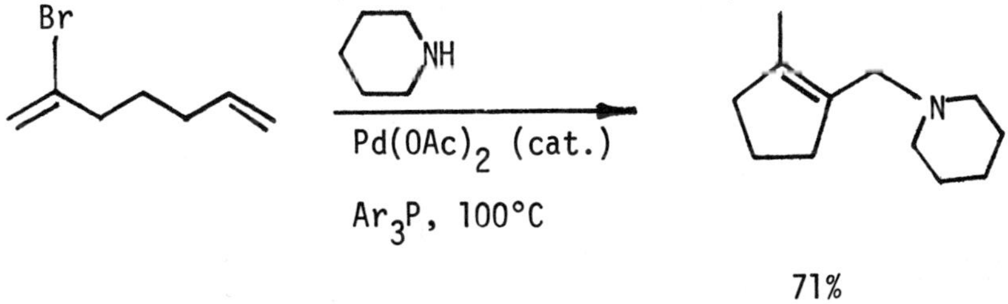

71%

I.B.3-18 E. Nakamura, K. Fukuzaki and I. Kuwajima, Chem. Commun., 499 (1983); K. Oshima et al., Tetrahedron Lett., 24, 2877 (1983); F. Henin and J. P. Pete, ibid, 24, 4687 (1983); T. K. Jones and S. E. Denmark, Helv. Chim. Acta, 66, 2377, 2397 (1983).

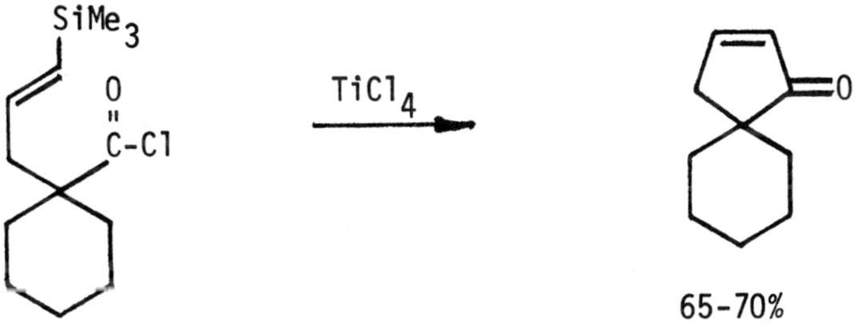

65-70%

I.B.3-19 R. A. Russell, R. N. Warrener et al., Chem. Commun., 994 (1983); T. Shono et al., J. Org. Chem., 48, 2503 (1983); H. Mayr and W. Striepe, ibid, 48, 1159 (1983); A. V. Shastin and E. S. Balenkova, J. Org. Chem. (USSR), 19, 579 (1983); S. O. Badanyan et al., ibid, 19, 1032 (1983); V. K. Singh and S. Dev, Ind. J. Chem., 22B, 319 (1983).

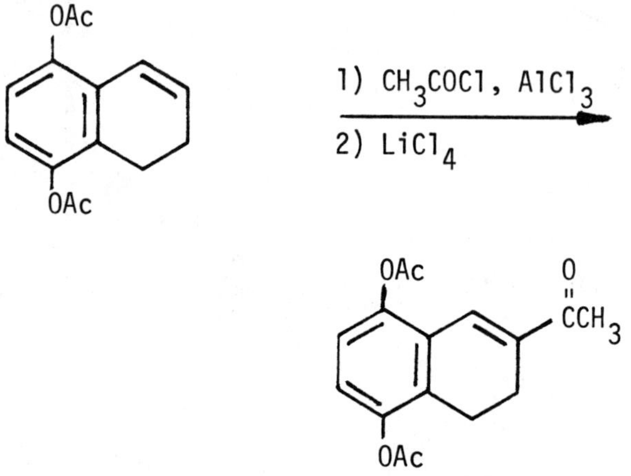
OAc
OAc
1) CH3COCl, AlCl3
2) LiCl4
OAc
O
CCH3
OAc

90%

I.B.3-20 P. Grenouillet, D. Neibecker and I. Tkatchenko, Chem. Commun., 542 (1983); R. F. Heck et al., J. Org. Chem., 48, 948 (1983); A. Sen and T. W. Lai, Organometallics, 2, 1059 (1983).

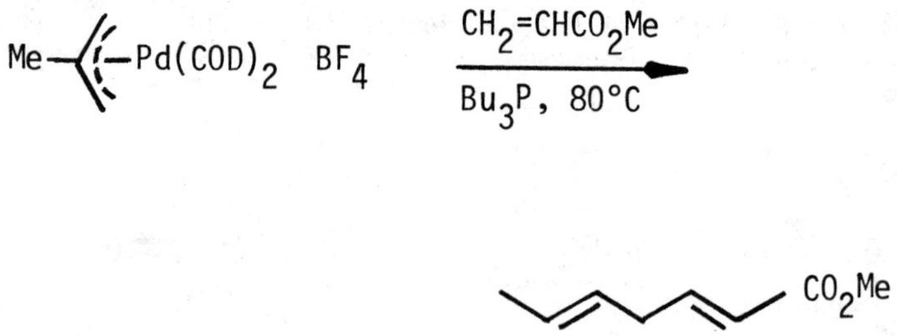
Me
Pd(COD)2 BF4
CH2=CHCO2Me
Bu3P, 80°C
CO2Me

83%

I.B.3-21 H. Kamogawa et al., Bull. Chem. Soc. Jpn., 56, 762
(1983); V. N. Gogte, A. A. Natu and V. S. Pandit, Tetrahedron
Lett., 24, 4131 (1983); T. Loerzer, R. Gerke and W. Luttke,
ibid, 24, 5861 (1983).

$$\text{(P)}-\text{C}_6\text{H}_4-\text{SO}_2\text{NHNH}_2 \quad \xrightarrow[\text{EtOH, }\Delta]{1)\ R^1R^2C=O} \quad R^1R^2C=CR^1R^2$$

2) NaOCH$_2$CH$_2$OH

HOCH$_2$CH$_2$OH, Δ

I.B.3-22 V. Galamb and H. Alper, Tetrahedron Lett., 24, 2965
(1983).

$$\underset{H}{\overset{Ar}{\diagdown}}C=C\underset{CO_2H}{\overset{H}{\diagup}} \quad \xrightarrow[\substack{2)\ \text{NaHCO}_3,\ \text{Me}_2\text{CO} \\ 3)\ \text{CO, Pd(PPh}_3)_4}]{1)\ \text{Br}_2,\ \text{CCl}_4} \quad \underset{H}{\overset{Ar}{\diagdown}}C=C\underset{H}{\overset{CO_2H}{\diagup}}$$

I.B.3-23 T. Funabiki et al., Bull. Chem. Soc. Jpn., 56, 649
(1983); N. De Kimpe et al., Chem. Ber., 116, 3846 (1983);
W. R. Jackson and C. G. Lovel, Aust. J. Chem., 36, 1975 (1983);
M. Prochazka and M. Siroky, Coll. Czeck. Chem. Commun., 48,
1765 (1983).

$$(CH_2)_n\text{-ring}=CHX \quad \xrightarrow[\substack{\text{KOH, H}_2\text{O} \\ (n = 3\text{-}6)}]{\text{KCN, CoCl}_2\ (\text{cat.})} \quad (CH_2)_n\text{-ring}=CHCN$$

68-90%

I.B.3-24 T. Takeda et al., Chem. Lett., 549 (1983); D. J. Hart and Y. M. Tsai, Tetrahedron Lett., 24, 4387 (1983).

$$\underset{\underset{Me_3Si}{PhS}}{\square}\text{-R, OH} \xrightarrow[\text{2) }\Delta,\text{ THF}]{\text{1) }H_2O_2,\text{ AcOH}} PhS\overset{O}{\diagup\!\!\!\diagdown}R$$

58-68%

I.B.3-25 K. B. Becker, Synthesis, 341 (1983).

Review: "Synthesis of Stilbenes."

I.B.3-26 M. Kumada et al., J. Org. Chem., 48, 2195 (1983); Bull. Chem. Soc. Jpn., 56, 363 (1983); C. Sahlberg, A. Quader and A. Claesson, Tetrahedron Lett., 24, 5137 (1983).

$$\underset{Me}{RCHMgCl} \xrightarrow[ML^*]{CH_2=CHBr} \underset{Me}{R\overset{*}{C}HCH=CH_2}$$

45->95% (GC)

6-83% ee

ML* = Chiral (β-Aminoalkyl) phosphines.

I.B.3-27 J. A. Soderquist and W. W. H. Leong, Tetrahedron Lett., 24, 2361 (1983); D. Caine and V. C. Ukachukwu, ibid, 24, 3959 (1983); M. Kosugi, I. Hagiwara and T. Migita, Chem. Lett., 839 (1983).

$$\underset{}{RCCl} + \underset{SnMe_3}{\overset{OMe}{=}} \xrightarrow{Pd(II)} \underset{R}{\overset{O}{\diagup\!\!\!\diagdown}}\overset{OMe}{=}$$

44-86%

I.B.3-28 N. Jabri, A. Alexakis and J. F. Normant, Bull. Soc. Chim. Fr. II, 321, 332 (1983); E. Negishi and F. T. Luo, J. Org. Chem., 48, 1560 (1983); K. Takagi and N. Hayama, Chem. Lett., 637 (1983).

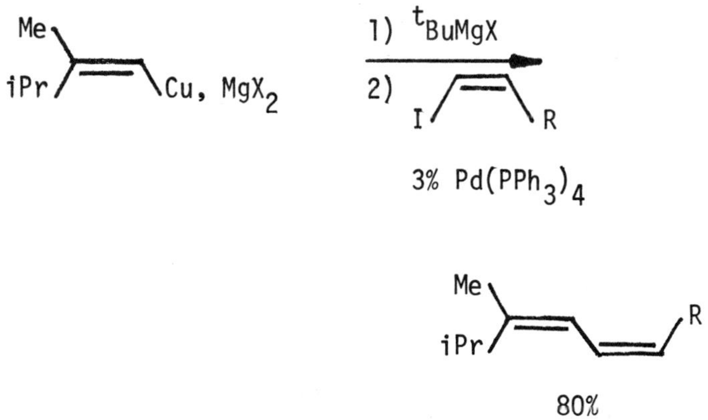

Me\\iPr/C=C\\Cu, MgX$_2$

1) ^{t}BuMgX
2) I/C=C\\R

3% Pd(PPh$_3$)$_4$

Me\\iPr/C=C–C=C/R

80%

I.B.3-29 G. Cassani, P. Massardo and P. Piccardi, Tetrahedron Lett., 24, 2513 (1983); N. Miyaura, H. Suginome and A. Suzuki, Tetrahedron, 39, 3271 (1983); M. Hoshi, Y. Masuda and A. Arase, Bull. Chem. Soc. Jpn., 56, 2855 (1983).

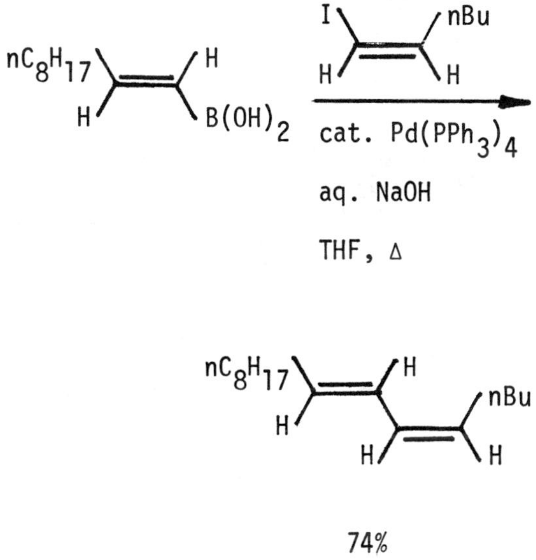

nC$_8$H$_{17}$\\H/C=C\\H/B(OH)$_2$ +   I\\H/C=C/nBu\\H

cat. Pd(PPh$_3$)$_4$
aq. NaOH
THF, Δ

nC$_8$H$_{17}$\\H/C=C/H\\H/C=C/nBu\\H

74%

I.B.3-30 H. Schubert and M. Regitz, *Angew. Chem., Int. Ed. Engl.*, **22**, 553 (1983).

[Triazole-N-CH=C(CN)₂ with 4-Me]

1) HC≡C-OMe
2) NaOEt, EtOH

[EtO,EtO-C=CH-CH=C(CN)₂ diene]

34%

Push-Pull Dienes

I.B.3-31 J. F. Normant et al., *Pure Appl. Chem.*, **55**, 1759 (1983).

Review: "Carbocupration of Acetylenic Acetals and Ketals: Synthesis of α,β-Ethylenic Acetals and of Dienals and Dienones."

I.B.3-32 M. Bourgain-Commercon, J. P. Foulon and J. F. Normant, *Tetrahedron Lett.*, **24**, 5077 (1983); N. Jabri, A. Alexakis and J. F. Normant, *ibid*, **24**, 5081 (1983); J. P. Marino and R. J. Linderman, *J. Org. Chem.*, **48**, 4621 (1983); G. A. Zheldubovskaya et al., *J. Org. Chem. (USSR)*, **19**, 1005 (1983).

$R^1C\equiv CH$

1) R^2Cu, MgX_2
2) R^3COX, THF
   cat. $(Ph_3P)_4Pd$

→ R^2, R^1 C=C, C(=O)R^3

58-84%

I.B.3-33 K. Oshima et al., J. Am. Chem. Soc., 105, 4491 (1983); F. Sato et al., Tetrahedron Lett., 24, 1041 (1983).

$$R-C\equiv C-H \xrightarrow[\text{2) }E^+]{\text{1) PhMe}_2\text{SiMgMe} \atop \text{PtCl}_2(\text{Bu}_3\text{P})_2 \text{ (cat.)}} \underset{\text{64-90\%}}{\overset{R \quad\quad H}{\underset{E \quad\quad SiMe_2Ph}{C=C}}}$$

E^+ = MeI, BuCHO, Me_3SiCl, I_2.

I.B.3-34 G. A. Molander, J. Org. Chem., 48, 5409 (1983); E. I. Negishi et al., Organometallics, 2, 563 (1983).

$$R^1C\equiv CSiMe_3 \xrightarrow[\text{2) H}^+]{\text{1) CH}_2=CHCH_2ZnBr}$$

$$\underset{\text{52-88\% (GC)}}{\overset{R^1 \quad\quad SiMe_3}{\underset{CH_2=CHCH_2 \quad H}{C=C}}}$$

I.B.3-35 E. I. Negishi et al., J. Am. Chem. Soc., 105, 6344 (1983); T. J. Zitzelberger, M. D. Schiavelli and D. W. Thompson, J. Org. Chem., 48, 4781 (1983).

Br-CH$_2$CH$_2$-C≡C-SiMe$_3$ $\xrightarrow[\text{(CH}_2\text{Cl)}_2\text{, 25°C}]{\text{2 Me}_3\text{Al} \atop \text{Cl}_2\text{ZrCp}_2}$ [cyclobutene with SiMe$_3$ and Me substituents]   92%

I.B.3-36 P. A. Wender, D. A. Holt and S. M. Sieburth, <u>J. Am. Chem. Soc.</u>, <u>105</u>, 3348 (1983); N. V. Komarov, O. I. Yurchenko and T. N. Dybova, <u>J. Org. Chem. (USSR)</u>, <u>19</u>, 401 (1983).

$$\text{2-chlorocyclohexanone} \xrightarrow[\substack{\text{2) LAH, NaOMe} \\ \text{THF, RT}}]{\substack{\text{1) R-}\equiv\text{-Li} \\ \text{THF, -78°C}}} \text{trans-2-(alkenyl)cyclohexanol}$$

66-86%

I.B.4. Allene Forming Reactions

I.B.4-1 W. Smadja, <u>Chem. Rev.</u>, <u>83</u>, 263 (1983).

Review: "Electrophilic Addition to Allenic Derivatives: Chemo-, Regio- and Stereochemistry and Mechanisms."

I.B.4-2 W. H. Okamura, <u>Acct. Chem. Res.</u>, <u>16</u>, 81 (1983).

Review: "Pericyclic Reactions of Vinylallenes: From Calciferols to Retinoids and Drimanes."

I.B.4-3 J. W. Patterson, Jr. et al., <u>J. Org. Chem.</u>, <u>48</u>, 2572 (1983).

$$Ph_3P=\!\!\!\diagdown\!\!\!\diagup\!\!=\!\!\bullet\!\!=\!\!\diagdown\!\!\!\diagup\!\!\!\diagdown\text{OSiMe}_2{}^t\text{Bu}$$

Wittig Rx

I.B.4-4 P. Vermeer et al., J. Org. Chem., 48, 1103 (1983).

$$HC\equiv C-C(Ph)(H)(OAc) \xrightarrow[\text{Et}_2O]{\text{PhZnCl, THF}} \underset{\text{cat. Pd(PPh}_3)_4}{} Ph(H)C=C=C(Ph)(H)$$

80%

82:18 = Anti:Syn Substitution

I.B.4-5 T. Hayashi, Y. Okamoto and M. Kumada, Tetrahedron Lett., 24, 807 (1983); T. J. Barton and G. P. Hussmann, J. Am. Chem. Soc., 105, 6316 (1983).

$$Ph-C\equiv C-C(SiMe_3)(Ph)(H) \xrightarrow[\text{CH}_2\text{Cl}_2, 0°C]{^t\text{BuCl, TiCl}_4} {}^tBu(Ph)C=C=C(Ph)(H)$$

23%

I.B.4-6 A. Claesson, A. Quader and C. Sahlberg, Tetrahedron Lett., 24, 1297 (1983).

[CH$_2$=CH-C(OP(O)(OEt)$_2$)=CH$_2$] $\xrightarrow[\text{10\% CuI}]{5 \text{ RMgX}}$ R-CH$_2$-CH=C=CH$_2$

23-72%

(6-12% Diene)

I.B.4-7 K. K. Wang, S. S. Nikam and C. D. Ho, J. Org. Chem., 48, 5376 (1983); C. Huynh and G. Linstrumelle, Chem. Commun., 1133 (1983).

$$Me_3SiC\equiv CCH_2R^1 \xrightarrow{\begin{array}{l}1)\ ^t\text{BuLi} \\ 2)\ \text{MeOB}\bigcirc \\ 3)\ 4/3\ \text{BF}_3\cdot\text{Et}_2\text{O} \\ 4)\ R^2R^3\text{CO} \\ 5)\ \text{NaOH, H}_2\text{O}_2\end{array}}$$

$$\underset{\underset{R^3}{R^2}}{\overset{Me_3Si}{\diagdown}}C=C=C\underset{H}{\overset{R^1}{\diagup}}$$
$$OH$$

71-95%

I.B.4-8 D. Hoppe and C. Riemenschneider, <u>Angew. Chem., Int. Ed. Engl.</u>, <u>22</u>, 54 (1983); A. C. Oehlschlager and E. Czyzewska, <u>Tetrahedron Lett.</u>, <u>24</u>, 5587 (1983); I. Cutting and P. J. Parsons, <u>ibid, 24</u>, 4463 (1983); <u>Chem. Commun.</u>, 1209 (1983).

$$\underset{\underset{Li}{|}\underset{O}{\|}}{Me-C\equiv C-CH-O-CNiPr_2} \xrightarrow[\text{2) } R^1R^2C=O]{\text{1) Ti(OiPr)}_4, -78°C}$$

$$\underset{\underset{R^2}{|}}{\overset{Me}{\underset{R^1-C-OH}{\diagdown}}}C=C=C\underset{\underset{O}{\|}}{\overset{H}{\diagup}}OCNiPr_2$$

59-89%

(de > 90%)

I.B.4-9 S. L. Buchwald and R. H. Grubbs, <u>J. Am. Chem. Soc.</u>, <u>105</u>, 5490 (1983).

$$\underset{R^1R^2}{Cp_2Ti\diagup\!\!\diagdown} \xrightarrow[\text{PhH, RT}]{R^3R^4C=O} \underset{R^1R^2}{\overset{R^3R^4}{C=C=C}}$$

53-80%

I.B.4-10 T. Jeffery-Luong and G. Linstrumelle, Synthesis, 32 (1983); T. Harada, Y. Nozaki and A. Oku, Tetrahedron Lett., 24, 5665 (1983).

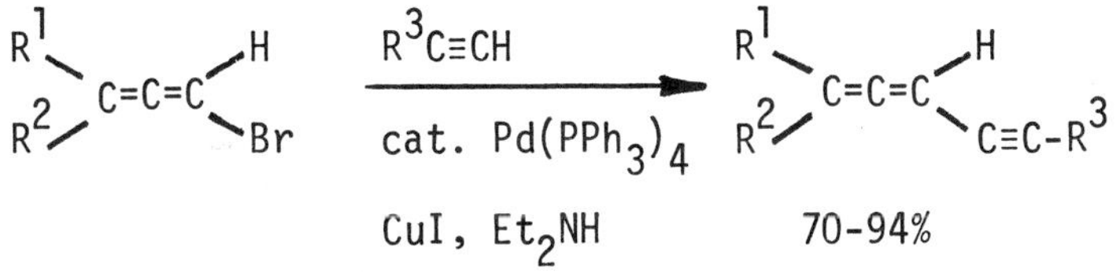

$$\underset{R^2}{\overset{R^1}{>}}C=C=C\underset{Br}{\overset{H}{<}} \quad \xrightarrow[\text{CuI, Et}_2\text{NH}]{R^3C\equiv CH \atop \text{cat. Pd(PPh}_3)_4} \quad \underset{R^2}{\overset{R^1}{>}}C=C=C\underset{C\equiv C-R^3}{\overset{H}{<}} \quad 70\text{-}94\%$$

I.C. Carbon-Carbon Triple Bonds

(See also: VI.A.16).

I.C-1 W. S. Johnson, R. Elliott and J. D. Elliott, J. Am. Chem. Soc., 105, 2904 (1983); A. Jellal, J. P. Zahra and M. Santelli, Tetrahedron Lett., 24, 1395 (1983); Y. Kawanami, T. Katsuki and M. Yamaguchi, ibid, 24, 5131 (1983).

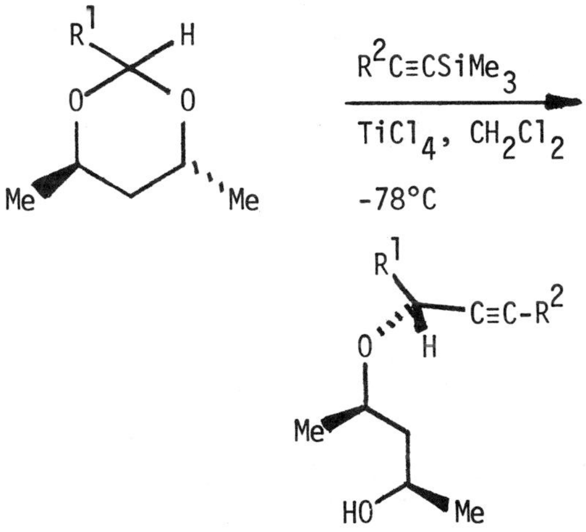

$$\xrightarrow[-78°C]{R^2C\equiv CSiMe_3 \atop TiCl_4,\ CH_2Cl_2}$$

81-98%

(72-93% de)

I.C-2   E. J. Corey et al., <u>Tetrahedron Lett.</u>, <u>24</u>, 4913 (1983); M. Yamaguchi and I. Hirao, <u>ibid</u>, <u>24</u>, 391 (1983); M. Yamaguchi, Y. Nobayashi and I. Hirao, <u>ibid</u>, <u>24</u>, 5121 (1983).

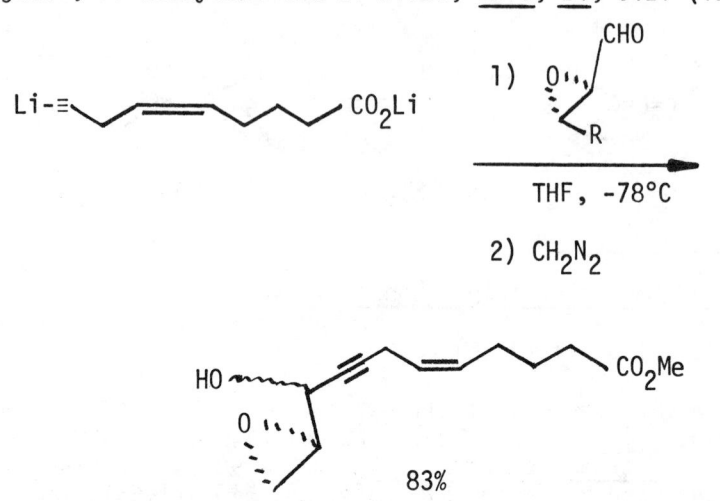

I.C-3   G. Courtois and P. Miginiac, <u>Bull. Soc. Chim. Fr. II</u>, 21 (1983).

BrMgOCH$_2$C≡CMgBr $\xrightarrow[\text{Et}_2\text{O, CH}_2\text{Cl}_2]{\text{1) nBuOCH}_2\text{NEt}_2}$ HOCH$_2$C≡C-CH$_2$NMe$_2$

2) Workup

75%

I.C-4   D. Villemin, <u>Chem. Commun.</u>, 1092 (1983);   G. Stork and J. M. Stryker, <u>Tetrahedron Lett.</u>, <u>24</u>, 4887 (1983);   J. M. Lancelin, P. H. A. Zollo and P. Sinay, <u>ibid</u>, <u>24</u>, 4833 (1983); G. Himbert and W. Schwickerath, <u>Liebigs Ann. Chem.</u>, 1185 (1983); F. Barbot and P. Miginiac, <u>Bull. Soc. Chim. Fr. II</u>, 41 (1983).

PhC≡CH $\xrightarrow[\text{"Dry Reaction"}]{\text{ArCHO, KF, Neutral Al}_2\text{O}_3}$ ArCH(OH)-C≡C-Ph

54%

I.C-5 M. Yamaguchi, T. Waseda and I. Hirao, Chem. Lett., 35 (1983); G. Giacomelli et al., J. Org. Chem., 48, 4887 (1983); H. Hoberg and H. J. Riegel, J. Organometal. Chem., 241, 245 (1983).

$$R^1-C\equiv C-H \xrightarrow[\text{3) }R^2CONMe_2]{\text{1) nBuLi} \atop \text{2) }BF_3\cdot Et_2O} R^1-C\equiv C-\overset{O}{\underset{\|}{C}}-R^2$$

77-97%

I.C-6 H. J. Bestmann, K. Kumar and W. Schaper, Angew. Chem., Int. Ed. Engl., 22, 167 (1983); S. Akiyama et al., Bull. Chem. Soc. Jpn., 56, 361 (1983); L. Horner and K. Dickerhof, Chem. Ber., 116, 1603, 1615 (1983).

$$2\ R^1CH=PPh_3 \xrightarrow[\text{3) 2\% Na/Hg}]{\text{1) }R^2COCl \atop \text{2) }(CF_3SO_2)_2O} R^1-C\equiv C-R^2$$

27-62%

THF

I.C-7 G. Hahn and G. Zweifel, Synthesis, 883 (1983); I. Kuwajima, S. Sugahara and J. Enda, Tetrahedron Lett., 24, 1061 (1983).

$$R^1CH=C=CH_2 \xrightarrow[\text{-90°C}]{\text{1) }^tBuLi,\ THF} \underset{E}{\overset{R^1}{>}}\!=\!\!\!=$$

2) iBu_3Al

3) E^+

58-97%

E^+ = RX, $R_2C=O$, CO_2.

I.C-8   D. G. Gillespie and B. J. Walker, J. Chem. Soc., Perkin I, 1689 (1983).

$$R^1C=CR^2 \atop \overset{|}{Br} \overset{|}{Br} \xrightarrow[(X = P \text{ or } As)]{2 \ Ph_2XLi} R^1-C\equiv C-R^2$$

37-80%

I.C-9   P. Sarti-Fantoni, J. Het. Chem., 20, 105 (1983); A. Svensson, J. O. Karlsson and A. Hallberg, ibid, 20, 729 (1983).

$$\underset{\text{Me—isoxazole—CH=CH-Ar}}{} \xrightarrow[\substack{1) \ Br_2 \\ 2) \ OH^- \\ 3) \ H^+}]{} HO_2C-C\equiv C-Ar$$

30-75%

I.C-10   A. Suzuki et al., Tetrahedron Lett., 24, 731, 735 (1983); R. Rossi and A. Carpita, Tetrahedron, 39, 287 (1983); S. R. Abrams, J. W. Quail and L. T. J. Delbaere, Can. J. Chem., 61, 2449 (1983).

$$R^1C\equiv CH \xrightarrow[\substack{1) \ X-B\triangleleft \\ X = Br \text{ or } I \\ 2) \ LiC\equiv CR^2 \\ 3) \ I_2}]{} \underset{X}{\overset{R^1}{}}C=C\underset{C\equiv C-R^2}{\overset{H}{}}$$

53-74%

I.C-11   J. A. Miller and G. Zweifel, Synthesis, 128 (1983); U. Stampfli, R. Galli and M. Neuenschwander, Helv. Chim. Acta, 66, 1631 (1983); V. Galamb, M. Gopal and H. Alper, Organometallics, 2, 801 (1983); S. Padmanabhan and K. M. Nicholas, Tetrahedron Lett., 24, 2239 (1983).

$$R-C\equiv CH \xrightarrow[\substack{1) \ nBuLi \\ 2) \ CuBr \quad 3) \ Me_3Si-C\equiv C-Br}]{} R-C\equiv C-C\equiv C-SiMe_3$$

70-93%

I.D. Cyclopropanations

I.D.1. Carbene or Carbenoic Additions to a Multiple Bond

(See also: VI.A.7.)

I.D.1-1 R. R. Kostikov, et al., J. Org. Chem. (USSR), 19, 215, 229, 1291 (1983); A. K. Khusid and N. Y. Sorokina, ibid, 19, 235 (1983); K. Kanematsu et al., Tetrahedron, 39, 1281 (1983); P. Weyerstahl et al., Chem. Ber., 116, 798, 808 (1983); T. L. Ho and S. H. Liu, Synth. Commun., 13, 685 (1983).

$$\text{Ph}\underset{R^1}{\overset{}{\rightthreetimes}}=\bullet=\underset{R^3}{\overset{R^2}{\leftthreetimes}} \quad \xrightarrow[\text{2) }\Delta]{\text{1) :CCl}_2} \quad \text{Ph-cyclopropane with =CCl}_2$$

30-60%

I.D.1-2 M. G. Banwell, Chem. Commun., 1453 (1983); W. Norden, V. Sander and P. Weyerstahl, Chem. Ber., 116, 3097 (1983); M. Ahmad et al., J. Chem. Res. (S), 301 (1983).

1) $CHCl_3$, QX
   50% NaOH

2) NaOH, MeOH

3) Pyridinium Chlorochromate

4) NaOMe, MeOH

75-100%

I.D.1-3 J. Hatem, J. P. Zahra and B. Waegell, Tetrahedron, 39, 2175 (1983); R. T. Taylor and L. A. Paquette, Org. Syn., 61, 39 (1983).

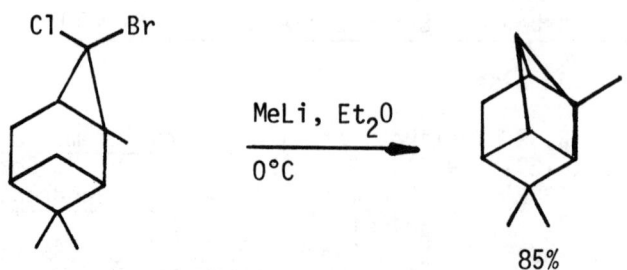

85%

I.D.1-4 M. W. Majchrzak, A. Kotelko and J. B. Lambert, Synthesis, 469 (1983); R. Carrie et al., Nouv. J. Chem., 7, 345 (1983); A. F. Noels et al., Tetrahedron, 39, 2169 (1983).

$$CH_2=C\begin{smallmatrix}R^1\\R^2\end{smallmatrix} \xrightarrow[40°C]{EtO_2C-CH=N_2 \atop Pd(OAc)_2, PhH}$$ [cyclopropane with R^3, EtO_2C, R^1, R^2]

30-85%

I.D.1-5 G. Rousseau and N. Slougui, Tetrahedron Lett., 24, 1251 (1983).

$$\begin{smallmatrix}R^1\\R^2\end{smallmatrix}\!\!\!\!\diagup\!\!\!\diagdown\begin{smallmatrix}OR^3\\OSiMe_3\end{smallmatrix} \xrightarrow[Et_2Zn, Et_2O]{CH_2I_2}$$ [cyclopropane with R^1, R^2, OR^3, $OSiMe_3$]

64-91%

I.D.1-6 C. H. Stammer et al., Tetrahedron Lett., 24, 3839
(1983); J. Org. Chem., 48, 4769 (1983); A. Oku, T. A. Yokoyama
and T. Harada, ibid, 48, 5333 (1983); C. D. Bedford and P. A.
S. Smith, ibid, 48, 4002 (1983).

$$CH_2 = C \begin{matrix} NHBoc \\ CO_2pNB \end{matrix} \quad \xrightarrow[\text{2) } PhCH_3,\ 90°C]{\text{1) } RCHN_2,\ Et_2O,\ -15°C}$$

cyclopropane with H, R on one carbon and NHBoc, CO_2pNB on adjacent carbon

46-90%

I.D.1-7 P. J. Stang, Pure Appl. Chem., 55, 369 (1983).

Review: "Properties and Reactions of Unsaturated Carbenes."

I.D.1-8 Y. Apeloig, P. J. Stang et al., J. Am. Chem. Soc.,
105, 4781 (1983); D. P. Fox, J. A. Bjork and P. J. Stang, J.
Org. Chem., 48, 3994 (1983).

$$\begin{matrix} ^tBu \\ Me \end{matrix} C=CHOTf \quad \xrightarrow[KO^tBu,\ -20°C]{Me_2C=CH_2} \quad$$

methylenecyclopropane product with tBu, Me on alkene carbon and gem-dimethyl on ring

47% (only E)

I.D.2. Other Cyclopropanations

I.D.2-1 J. Salaun, Chem. Rev., 83, 619 (1983).

Review: "Cyclopropanone Hemiacetals."

I.D.2-2 S. Murai, I. Ryu and N. Sonoda, J. Organometal. Chem., 250, 121 (1983).

Review: "Siloxycyclopropanes. Useful Synthetic Intermediates."

I.D.2-3 N. Ikekawa et al., Chem. Commun., 1180 (1983); N. De Kimpe et al., Tetrahedron Lett., 24, 2885 (1983); J. H. Babler and R. A. Haack, Synth. Commun., 13, 905 (1983).

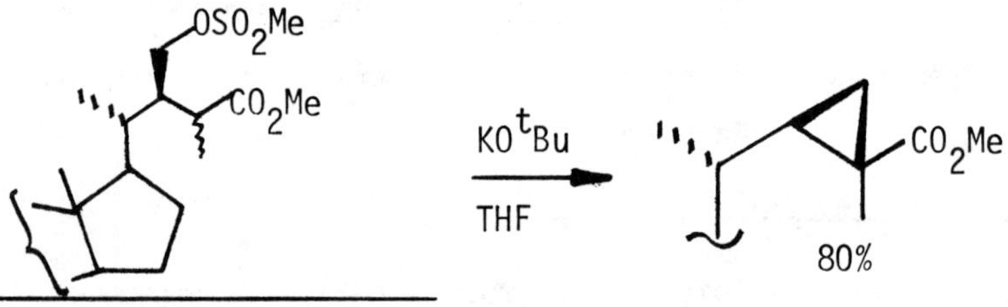

I.D.2-4 J. Barluenga, J. Florez and M. Yus, Synthesis, 647 (1983); P. Yates, P. H. Helferty and P. Mahler, Can. J. Chem., 61, 78, 936 (1983); C. Giomini, A. Inesi and E. Zeuli, J. Chem. Res. (S), 280 (1983).

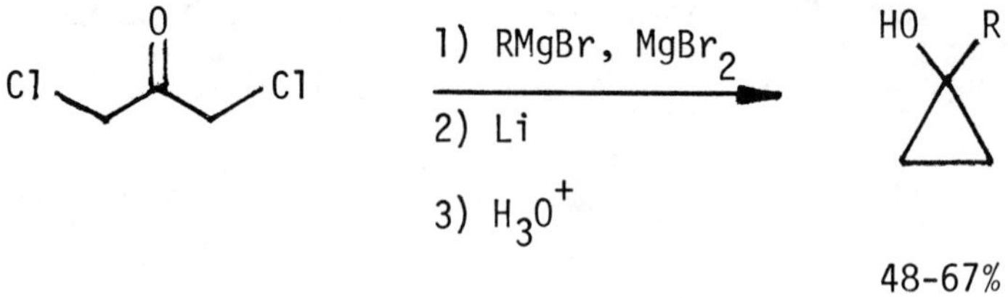

Product thermolysis yields ethyl ketones.

I.D.2-5 M. Apparu and M. Barrelle, Bull. Soc. Chim. Fr. II, 83 (1983); J. Ficini, S. Falou and J. d'Angelo, Tetrahedron Lett., 24, 375 (1983); N. De Kimpe et al., Bull. Soc. Chim. Belg., 92, 371 (1983); P. Prempree, S. Radviroongit and Y. Thebtaranonth, J. Org. Chem., 48, 3553 (1983).

$$\xrightarrow[\text{HMPA}]{\text{LiNEt}_2}$$

77%

I.D.2-6 R. Ando, T. Sugawara and I. Kuwajima, Chem. Commun., 1514 (1983); I. Kuwajima, R. Ando and T. Sugawara, Tetrahedron Lett., 24, 4429 (1983).

53-73%

I.D.2-7 T. Fujisawa et al., Chem. Lett., 1271, 1273 (1983).

1) 9-BBN
2) NaOMe

27-77%

I.D.2-8 W. T. Brady, S. J. Norton and J. Ko, Synthesis, 1002 (1983); P. Martin, Helv. Chim. Acta, 66, 1189 (1983); W. Huggenberg and M. Hesse, Helv. Chim. Acta, 66, 1519 (1983); Ch. R. Engel et al., J. Org. Chem., 48, 1954 (1983); A. Chatterjee et al., Tetrahedron, 39, 2965 (1983).

$$\xrightarrow{\text{KOH, H}_2\text{O}}$$

73%

I.D.2-9 I. Elphimoff-Felkin and P. Sarda, Tetrahedron Lett., 24, 4425 (1983); J. R. Christensen and W. Reusch, J. Org. Chem., 48, 3741 (1983); S. Stiver and P. Yates, Chem. Commun., 50 (1983); A. Fiecchi et al., Synthesis, 123 (1983).

$$\text{diene-OMe} \xrightarrow[\text{Et}_2\text{O}]{\text{Zn, HCl}} \text{bicyclic-OMe}$$

15-40%

I.D.2-10 I. Fleming and C. J. Urch, Tetrahedron Lett., 24, 4591 (1983).

1) Bu$_3$SnLi
2) MeI
3) 2 MeLi
4) BF$_3$·2AcOH

21%

I.D.2-11 D. J. Burton, B. E. Smart et al., J. Org. Chem., 48, 3616 (1983); F. S. Guziec, Jr. and F. A. Luzzio, ibid, 48, 2434 (1983); H. Abdallah, R. Gree and R. Carrie, Can. J. Chem., 61, 217 (1983); J. Mulzer and M. Kappert, Angew. Chem., Int. Ed. Engl., 22, 63 (1983); Y. Yamashita, Y. Miyauchi and M. Masumura, Chem. Lett., 489 (1983).

$$\xrightarrow[\text{Ph}_3\text{P, TME}]{\text{Ph}_3\overset{+}{\text{P}}\text{CF}_2\text{Br Br}^-} \quad \Delta$$

35%

I.D.2-12 T. Cohen and L. C. Yu, J. Am. Chem. Soc., 105, 2811 (1983); S. R. Landor, P. D. Landor and M. Kalli, J. Chem. Soc., Perkin I, 2921 (1983); J. C. Chalchat et al., Compt. Rend. (II), 296, 253 (1983).

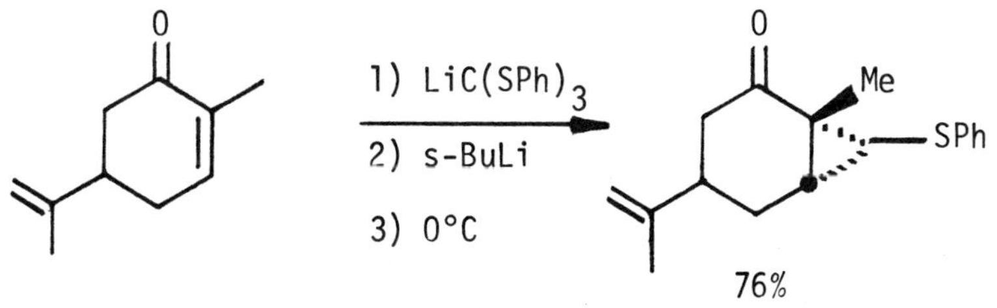

[Reaction scheme: 2-methyl-5-isopropenyl-cyclohex-2-enone + 1) LiC(SPh)$_3$ 2) s-BuLi 3) 0°C → product with Me and SPh groups, 76%]

I.D.2-13 M. Brookhart et al., J. Am. Chem. Soc., 105, 6721 (1983).

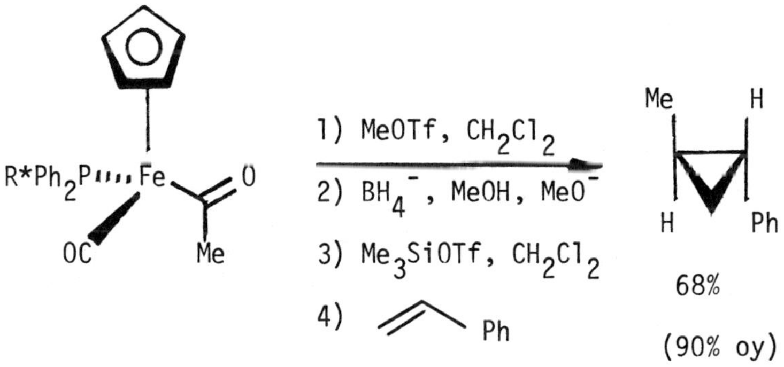

[Reaction scheme: Cp(R*Ph$_2$P)Fe(CO)(COMe) complex + 1) MeOTf, CH$_2$Cl$_2$ 2) BH$_4^-$, MeOH, MeO$^-$ 3) Me$_3$SiOTf, CH$_2$Cl$_2$ 4) CH$_2$=CHPh → cyclopropane with Me, H, H, Ph, 68% (90% oy)]

I.D.2-14 M. Reglier and S. A. Julia, Tetrahedron Lett., 24, 2387 (1983).

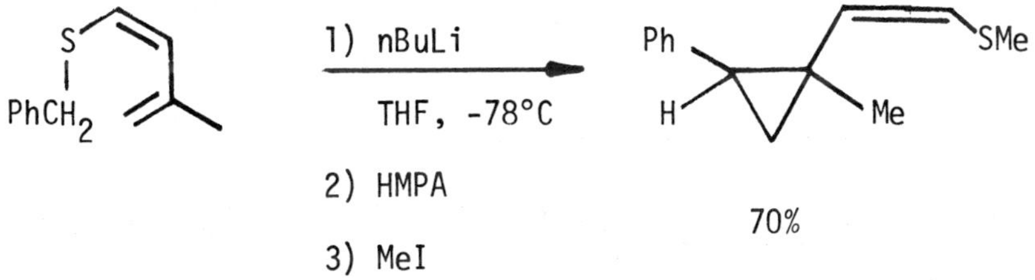

[Reaction scheme: PhCH$_2$-S-CH=CH-C(Me)=CH$_2$ + 1) nBuLi, THF, -78°C 2) HMPA 3) MeI → cyclopropane product with Ph, H, Me, and CH=CH-SMe, 70%]

I.D.2-15 M. Catellani and G. P. Chiusoli, <u>Tetrahedron Lett.</u>, <u>24</u>, 4493 (1983).

$R^1C\equiv CR^2$ / HCO_2NH_4 / Pd cat. → 29-42%

I.D.2-16 M. Mitani, Y. Yamamoto and K. Koyama, <u>Chem. Commun.</u>, 1446 (1983); A. Gilbert et al., <u>ibid</u>, 750 (1983); A. Padwa, G. R. Newkome et al., <u>J. Am. Chem. Soc.</u>, <u>105</u>, 137 (1983).

$CH_2=C\begin{smallmatrix}R^1\\R^2\end{smallmatrix}$ 1) CH_2Cl_2, hv / CuCl 2) $+e^-$ → 43-81%

I.D.2-17 T. Uyehara et al., <u>Tetrahedron Lett.</u>, <u>24</u>, 4445 (1983); T. Uyehara et al., <u>Chem. Commun.</u>, 17 (1983); A. Padwa, W. F. Rieker and R. J. Rosenthal, <u>J. Am. Chem. Soc.</u>, <u>105</u>, 4446 (1983); P. M. op den Brouw and W. H. Laarhoven, <u>J. Chem. Soc., Perkin II</u>, 1015 (1983).

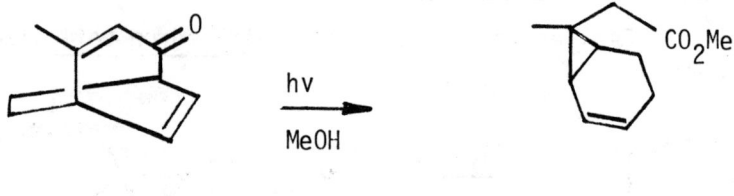

hv / MeOH → 80%

I.D.2-18 H. U. Reissig et al., <u>Chem. Ber.</u>, <u>116</u>, 3895 (1983); <u>Tetrahedron Lett.</u>, <u>24</u>, 715 (1983); E. G. E. Jahngen et al., <u>J. Org. Chem.</u>, <u>48</u>, 2472 (1983).

1) LDA, THF, -78°C 2) R^5X → 43-96%

I.D.2-19   D. J. Morgans, Jr. and G. B. Feigelson, <u>J. Am. Chem. Soc.</u>, <u>105</u>, 5477 (1983).

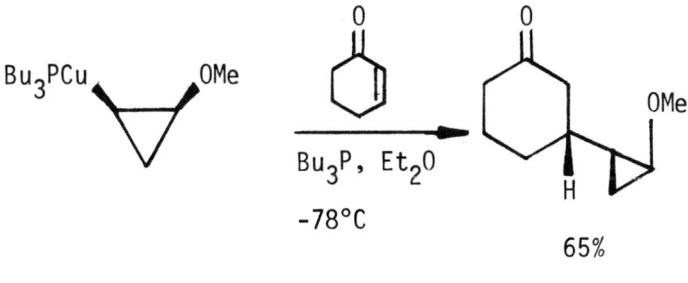

Various Organocopper Species.   (67% de)

I.D.2-20   B. M. Trost and P. L. Ornstein, <u>Tetrahedron Lett.</u>, <u>24</u>, 2833 (1983).

Use of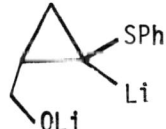

I.D.2-21   F. Scott, J. F. Normant et al., <u>Tetrahedron Lett.</u>, <u>24</u>, 5767 (1983); A. Jonczyk, M. Dabrowski and W. Wozniak, <u>ibid</u>, <u>24</u>, 1065 (1983); M. A. Fox, C. C. Chen and K. A. Campbell, J. Org. Chem., <u>48</u>, 321 (1983); J. P. Barnier, G. Rousseau and J. M. Conia, <u>Synthesis,</u> 915 (1983).

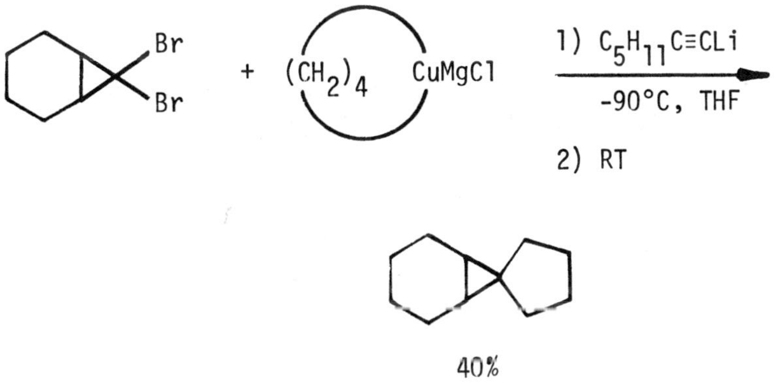

I.D.2-22 A. Lechevallier, F. Huet and J. M. Conia, Tetrahedron, 39, 3307, 3317, 3329 (1983); M. D. Taylor and A. B. Smith, III, Tetrahedron Lett., 24, 1867 (1983); T. Ibuka, E. Tabushi and M. Yasuda, Chem. Pharm. Bull., 31, 128 (1983).

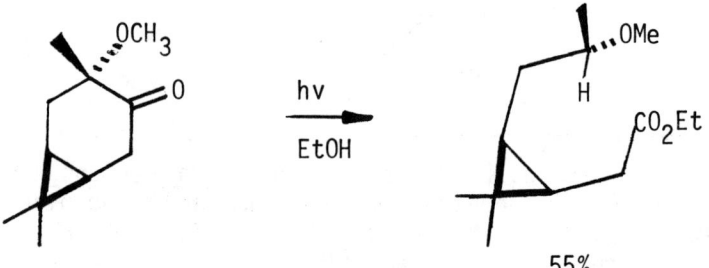

Preparation and reactions of α-cyclopropylidene ketones and aldehydes.

I.D.2-23 H. R. Sonawane, B. S. Nanjundiah and P. C. Purohit, Tetrahedron Lett., 24, 3917 (1983); H. R. Sonawane et al., ibid, 24, 3025 (1983).

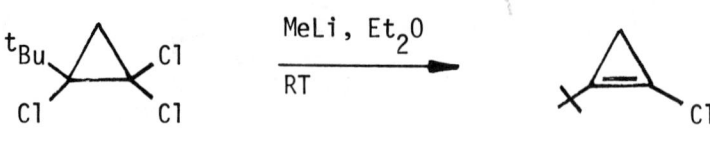

Epimer at C-2 undergoes completely different rx.

I.D.2-24 M. S. Baird and W. Nethercott, Tetrahedron Lett., 24, 605 (1983); A. Oku et al., J. Am. Chem. Soc., 105, 4400 (1983).

I.D.2-25 E. W. Thomas, Tetrahedron Lett., 24, 1467 (1983).

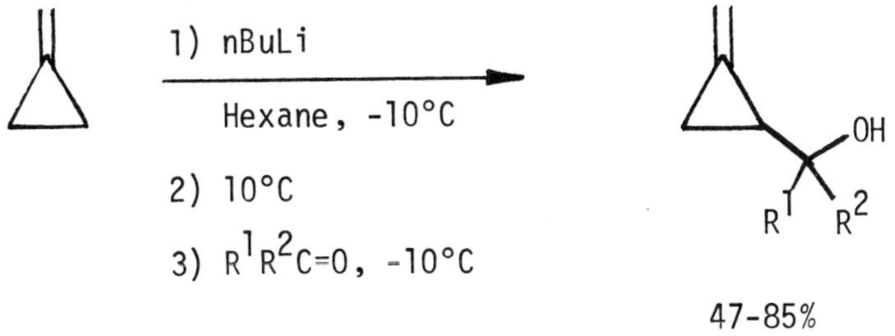
1) nBuLi
Hexane, -10°C
2) 10°C
3) R1R2C=O, -10°C
OH
R1
R2
47-85%

I.D.2-26 G. H. Kulkarni et al., Ind. J. Chem., 22B, 341, 355 (1983).

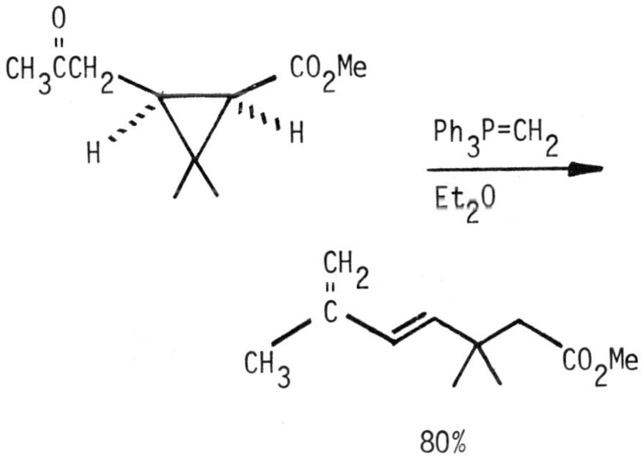
O
CH3CCH2
CO2Me
H
H
Ph3P=CH2
Et2O
CH2
C
CH3
CO2Me
80%

I.D.2-27 M. Nakayama et al., Chem. Lett., 147 (1983); J. M. Cook, S. H. Bertz et al., J. Org. Chem., 48, 139 (1983).

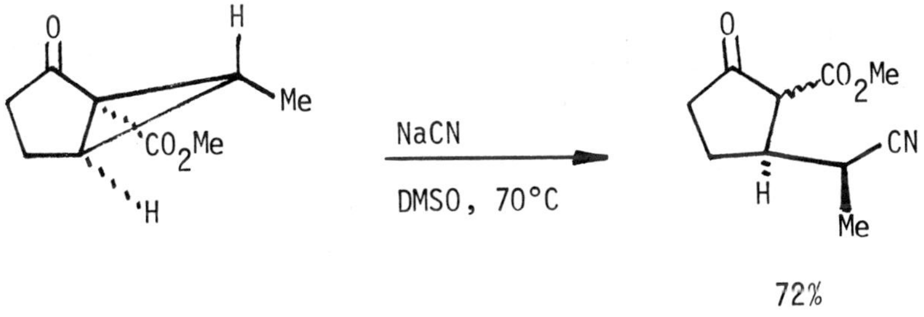
O
H
Me
CO2Me
H
NaCN
DMSO, 70°C
O
CO2Me
CN
H
Me
72%

I.D.2-28   E. Nakamura and I. Kuwajima, J. Am. Chem. Soc., 105, 651 (1983);   J. M. Conia and L. Blanco, Nouv. J. Chem., 7, 399 (1983);   R. A. Roberts, V. Schull and L. A. Paquette, J. Org. Chem., 48, 2076 (1983).

[cyclopropane with OSiMe$_3$ and OR substituents] 1) TiCl$_4$ / 2) PhCH$_2$CH$_2$CHO, CH$_2$Cl$_2$, 0°C →

Ph—CH$_2$CH$_2$—CH(OH)—CH$_2$CH$_2$—CO$_2$R

70-80%

I.D.2-29   E. W. Thomas, Tetrahedron Lett., 24, 2347 (1983); C. A. N. Catalan and C. Djerassi, ibid, 24, 3461 (1983);   H. Nemoto, T. Kametani et al., ibid, 24, 4257 (1983);   M. Ochiai, K. Sumi and E. Fujita, Chem. Pharm. Bull., 31, 3931 (1983); O. G. Kulinkovich et al., Synthesis, 383 (1983);   M. Demuth et al., Angew. Chem., Int. Ed. Engl., 22, 721 (1983).

[1-cyclopropyl-cyclopentanol] —HCO$_2$H→ [cyclopentylidene compound]—OCHO

61%

I.D.2-30   W. H. Tamblyn and R. E. Waltermire, Tetrahedron Lett., 24, 2803 (1983).

[cyclopropane with two CO$_2$Et groups] 1) Fe(CO)$_4^{-2}$, THF, RT / 2) CO / 3) HOAc → OHC—CH$_2$CH$_2$—CH(CO$_2$Et)$_2$

62%

I.E. Thermal Reactions

I.E.1. Cycloadditions

I.E.1-1 V. A. Mironov, A. D. Fedorovich and A. A. Akhrem, Russ. Chem. Rev., 52, 61 (1983).

Review: "Synthetic Methods for Cyclohexa-1,3-Dienes."

I.E.1-2 R. Gleiter and L. A. Paquette, Acct. Chem. Res., 16, 328 (1983).

Review: "σ/π Interaction as a Controlling Factor in the Stereoselectivity of Addition Reactions."

I.E.1-3 R. Gleiter and M. C. Bohm, Pure Appl. Chem., 55, 237 (1983).

Review: "Regio- and Stereoselectivity in Diels-Alder Reactions. Theoretical Considerations."

I.E.1-4 D. Ginsberg, Tetrahedron, 39, 2095 (1983).

Review: "The Role of Secondary Orbital Interactions in Control of Organic Reactions."

I.E.1-5 D. L. Boger, Tetrahedron, 39, 2869 (1983).

Review: "Diels-Alder Reactions of Azadienes."

I.E.1-6 N. L. Bauld et al., J. Am. Chem. Soc., 105, 2378 (1983).

Theory of Cation-Radical Pericyclic Reactions.

I.E.1-7 R. H. Schlessinger and J. A. Schultz, J. Org. Chem., 48, 407 (1983); J. A. Moore and E. M. Partain, III, ibid, 48, 1105 (1983); R. A. Pabon, D. J. Bellville and N. L. Bauld, J. Am. Chem. Soc., 105, 5158 (1983).

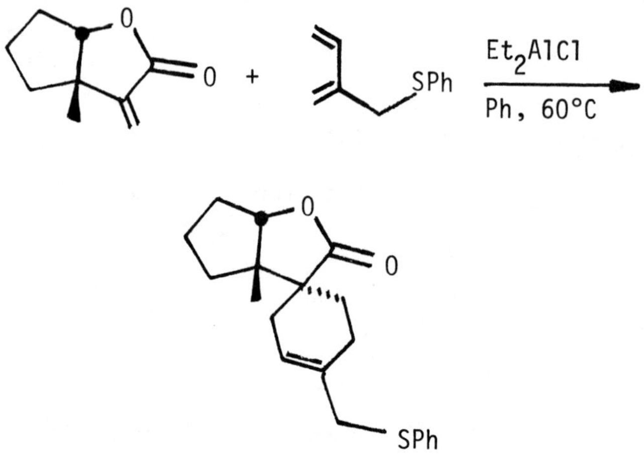

89%

I.E.1-8 K. Hayakawa, N. Hori and K. Kanematsu, Chem. Pharm. Bull., 31, 1809 (1983); J. H. Rigby and J. M. Sage, J. Org. Chem., 48, 3591 (1983).

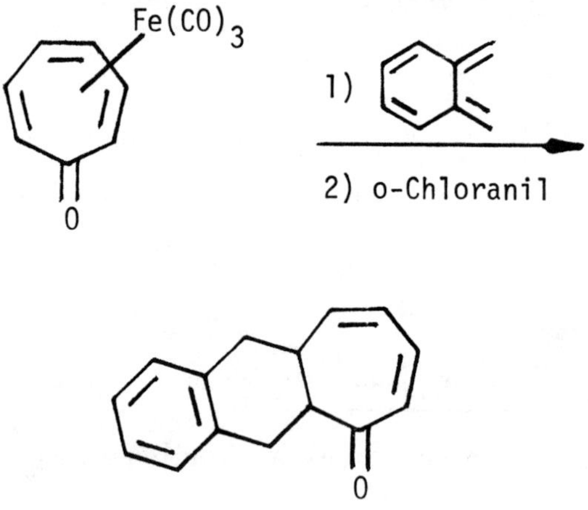

50%

[4+2] Cycloaddition Effected by Using $Fe(CO)_3$ as Masking Group.

I.E.1-9 W. D. Wulff and D. C. Yang, J. Am. Chem. Soc., 105, 6726 (1983).

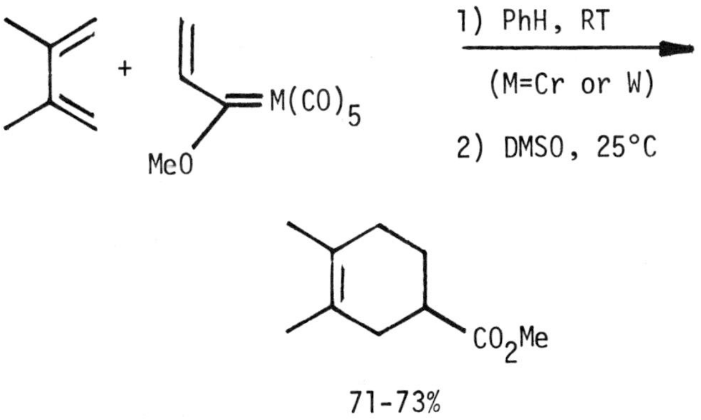
M(CO)5
MeO
1) PhH, RT
(M=Cr or W)
2) DMSO, 25°C
CO2Me
71-73%

I.E.1-10 P. DeShong and N. E. Lowmaster, Synth. Commun., 13, 537 (1983); Y. Okamoto, S. Giandinoto and M. C. Bochnik, J. Org. Chem., 48, 3830 (1983); J. Jurczak, C. H. Eugster et al., Helv. Chim. Acta, 66, 218, 222 (1983).

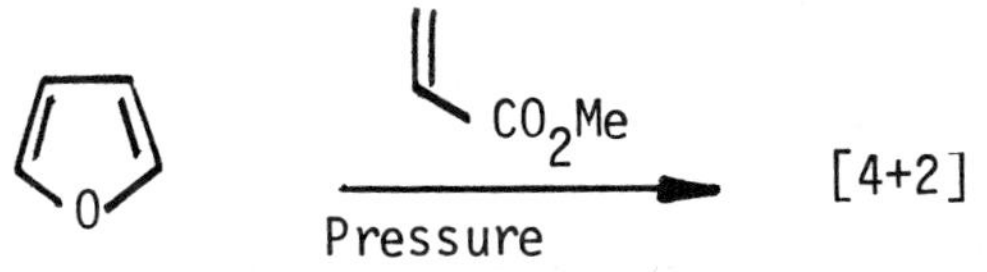
O
CO2Me
Pressure
[4+2]

Stereochem altered by varying temp.

I.E.1-11 J. F. W. Keana and D. D. Ward, Synth. Commun., 13, 729 (1983).

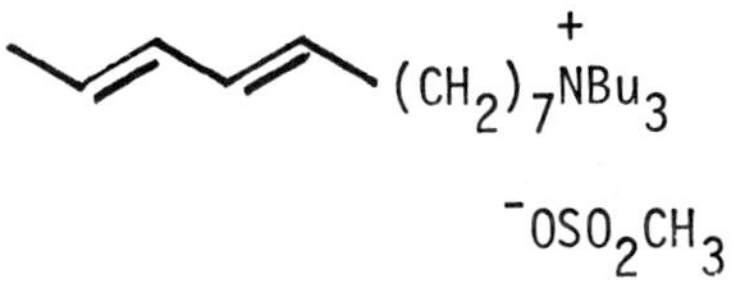
+
(CH2)7NBu3
−OSO2CH3

Phase-Transfer Catalyst which is removed by Diels-Alder rx.

I.E.1-12 L. A. Paquette and G. D. Crouse, J. Org. Chem., 48, 141 (1983).

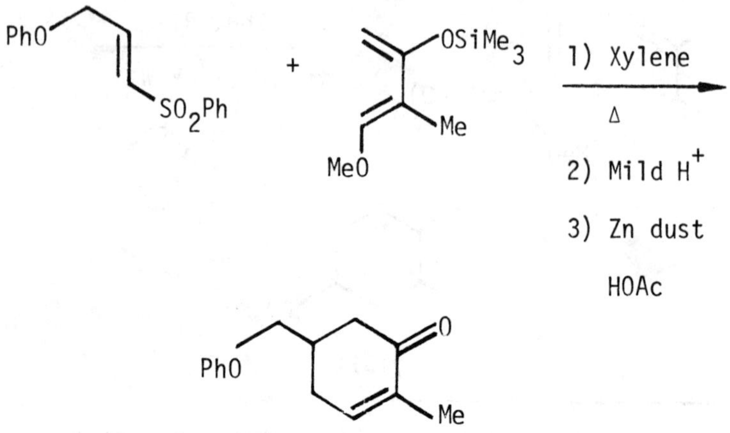

[Reaction scheme: PhO-CH=CH-SO₂Ph + diene with OSiMe₃, OMe, Me substituents; 1) Xylene, Δ; 2) Mild H⁺; 3) Zn dust, HOAc → cyclohexenone product with PhOCH₂ and Me substituents, 53%]

Seleno sulfonation and oxidation of terminal alkene gives starting dienophile.

I.E.1-13 P. A. Grieco et al., J. Org. Chem., 48, 3137 (1983); Tetrahedron Lett., 24, 1897 (1983); J. Am. Chem. Soc., 105, 1403 (1983).

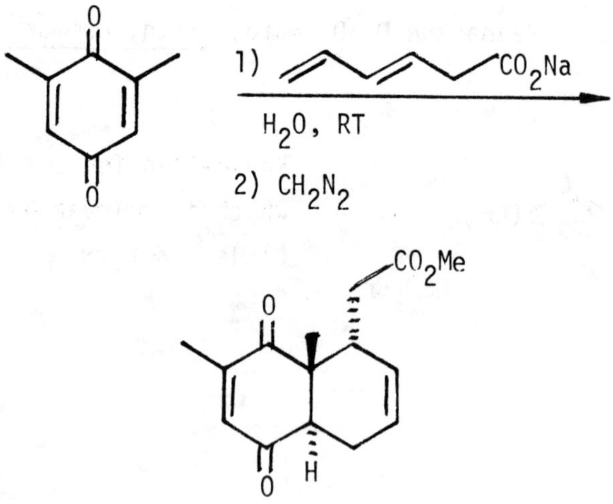

[Reaction scheme: 2,5-dimethylbenzoquinone + sodium hexadienoate; 1) H_2O, RT; 2) CH_2N_2 → bicyclic diketone with CO_2Me side chain, 77%]

I.E.1-14 F. Pelizzoni et al., J. Org. Chem., 48, 2866 (1983);
J. P. Gesson, J. C. Jacquesy and B. Renoux, Tetrahedron Lett.,
24, 2757, 2761 (1983); R. C. Gupta, P. A. Harland and R. J.
Stoodley, Chem. Commun., 754 (1983).

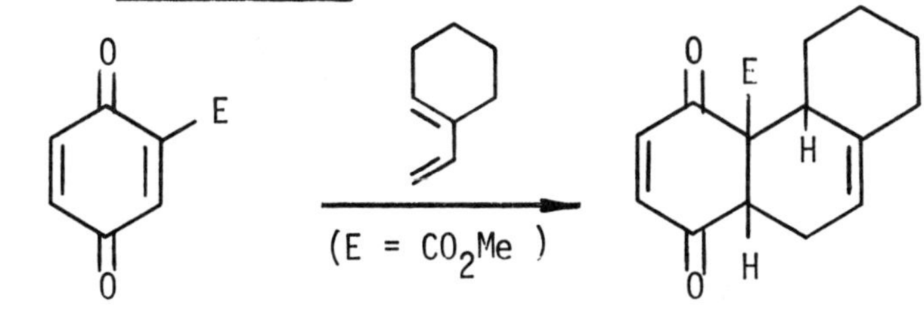

(E = CO_2Me)

92%

I.E.1-15 G. A. Kraus and S. Yue, Chem. Commun., 1198 (1983);
B. M. Trost et al., J. Org. Chem., 48, 3252 (1983); L.
Birkofer and V. Foremny, Z. Chem., 23, 250 (1983).

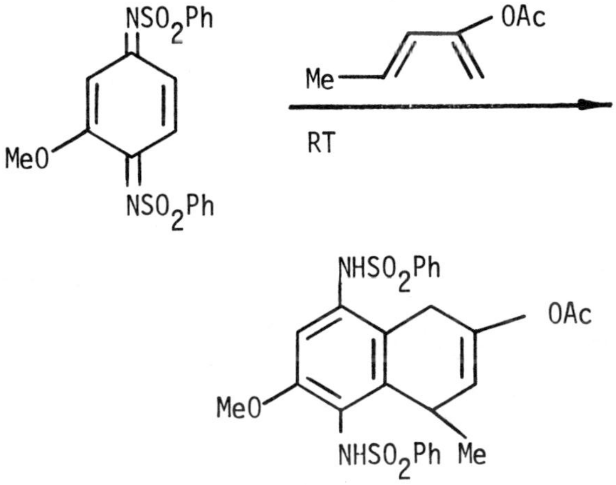

75%

I.E.1-16 P. Beak and C. W. Chen, Tetrahedron Lett., 24, 2945
(1983); C. Tamm et al., Helv. Chim. Acta, 66, 1796 (1983);
D. D. Weller and E. P. Stirchak, J. Org. Chem., 48, 4873 (1983).

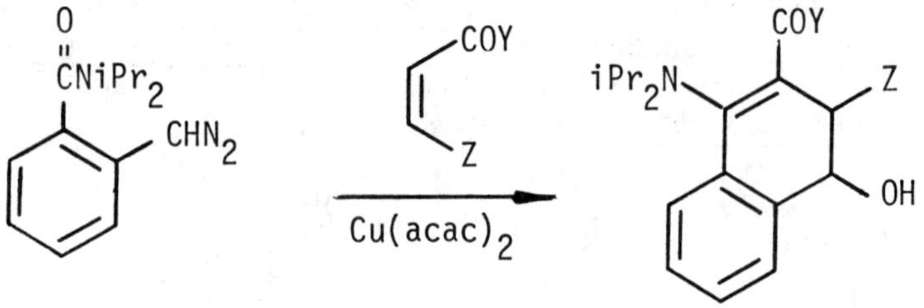

I.E.1-17 T. Minami et al., J. Org. Chem., 48, 2569 (1983); R. W. Franck and T. V. John, ibid, 48, 3269 (1983); T. Watabe, Y. Takahashi and M. Oda, Tetrahedron Lett., 24, 5623 (1983).

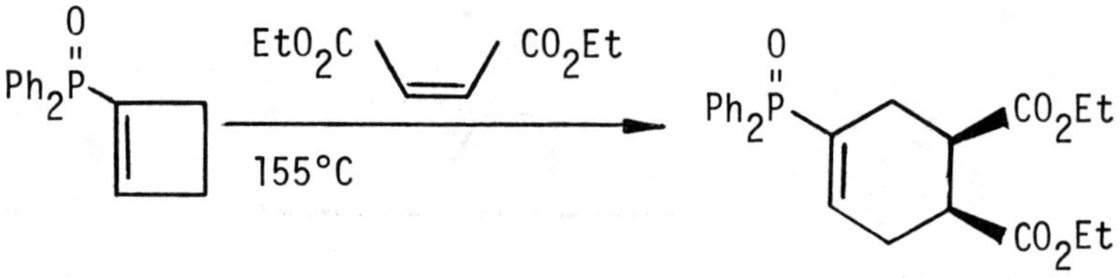

I.E.1-18 G. N. Fickes and R. S. Glass, Synth. Commun., 13, 721 (1983); N. S. Zefirov, M. A. Kirpichenok and T. G. Shestakova, J. Org. Chem. (USSR), 19, 221, 795 (1983); R. K. Geivandov, ibid, 19, 777 (1983); A. Oku et al., J. Org. Chem., 48, 4374 (1983).

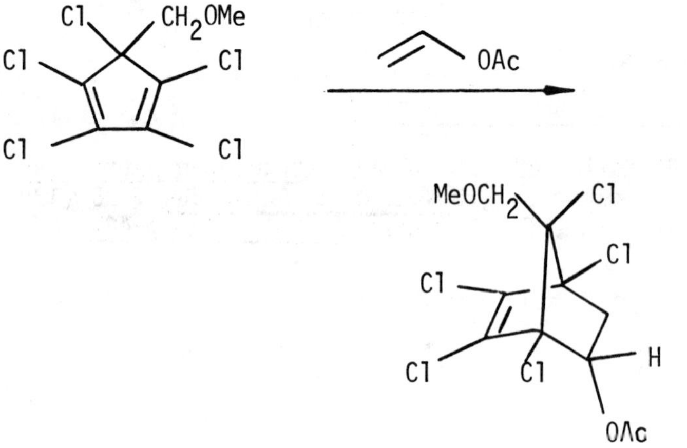

I.E.1-19 M. Iguchi, S. Yamamura et al., Chem. Pharm. Bull, 31, 2820, 2834, 2845, 2853 (1983); R. K. Razdan et al., J. Org. Chem., 48, 4137 (1983); S. Raucher and R. F. Lawrence, Tetrahedron Lett., 24, 2927 (1983).

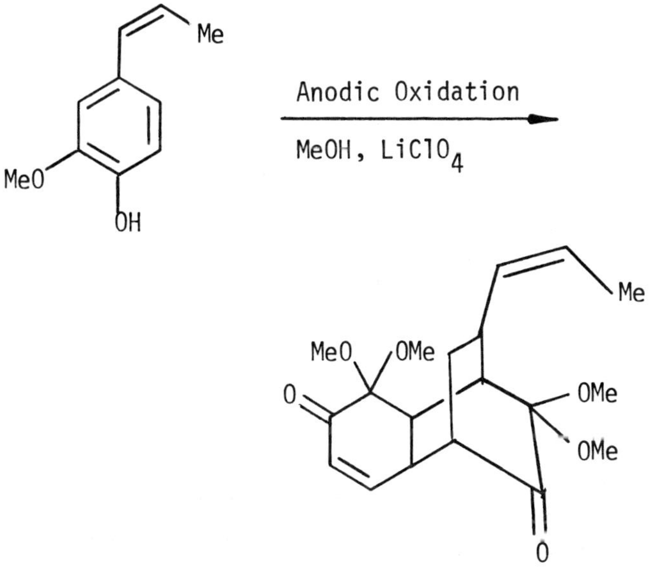

$$\text{Ar-CH=CH-Me} \xrightarrow[\text{MeOH, LiClO}_4]{\text{Anodic Oxidation}} \text{product}$$

14%

I.E.1-20 T. Tuschka, K. Naito and B. Rickborn, J. Org. Chem., 48, 70 (1983); J. Orban and J. V. Turner, Tetrahedron Lett., 24, 2697 (1983).

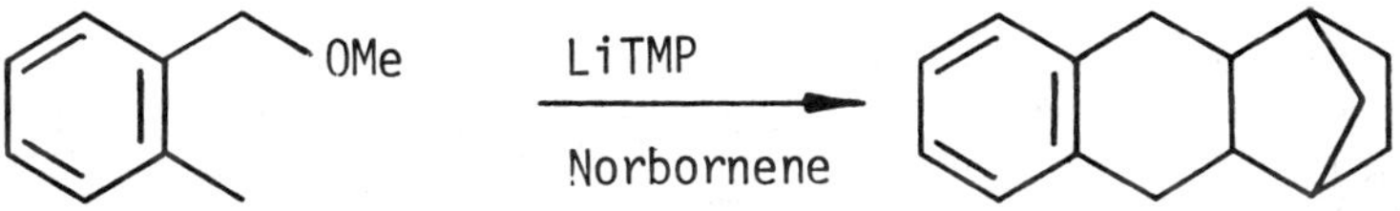

$$\text{o-MeC}_6\text{H}_4\text{CH}_2\text{OMe} \xrightarrow[\text{Norbornene}]{\text{LiTMP}} \text{product}$$

70%

I.E.1-21 K. G. Das et al., Synth. Commun., 13, 787 (1983); B. A. Keay and R. Rodrigo, Can. J. Chem., 61, 637 (1983).

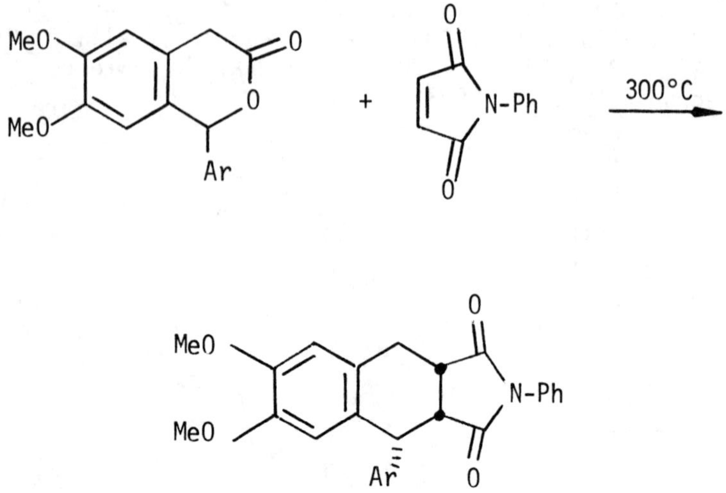

I.E.1-22 E. Bauml and H. Mayr, J. Org. Chem., 48, 2600 (1983); R. O. Angus, Jr. and R. P. Johnson, ibid, 48, 273 (1983); M. E. Wright, Organometallics, 2, 558 (1983); H. tom Dieck and R. Diercks, Angew. Chem., Int. Ed. Engl., 22, 778 (1983).

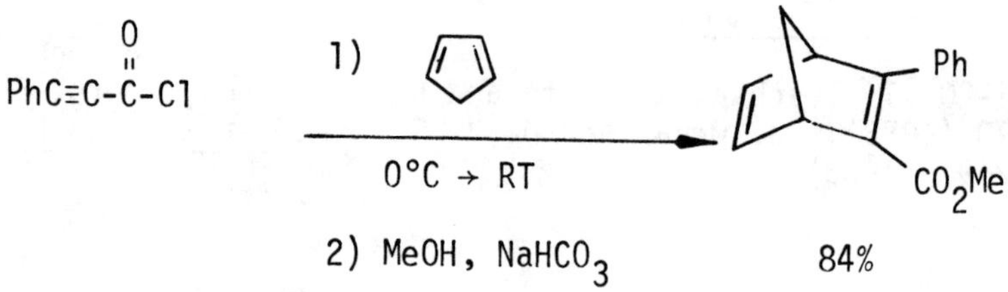

I.E.1-23 J. H. Markgraf et al., Tetrahedron Lett., 24, 241 (1983); S. V. Ley et al., J. Chem. Soc., Perkin I, 1579 (1983); M. Jalali-Naini, D. Guillerm and J. Y. Lallemand, Tetrahedron, 39, 749 (1983); E. Vedejs, W. H. Miller and J. R. Pribish, J. Org. Chem., 48, 3611 (1983).

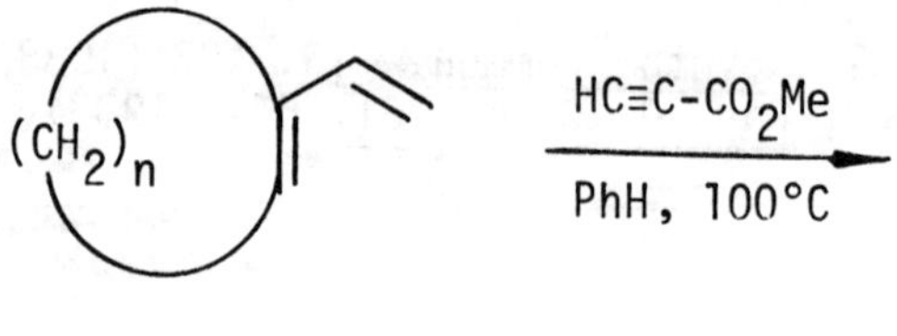

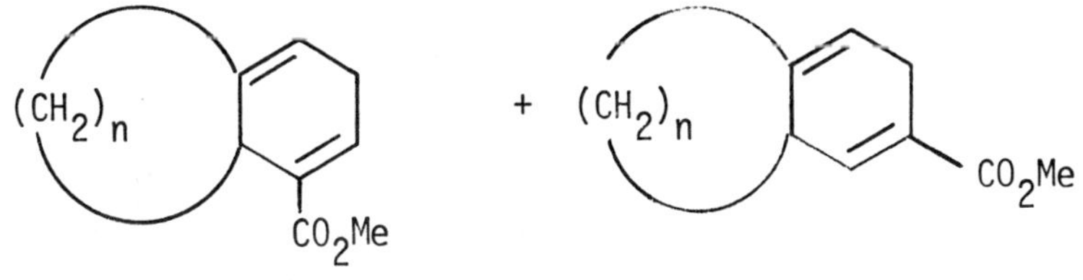

I.E.1-24 W. E. Noland et al., J. Org. Chem., 48, 2488 (1983); A. V. R. Rao et al., ibid, 48, 1552 (1983); F. Zutterman and A. Krief, ibid, 48, 1135 (1983).

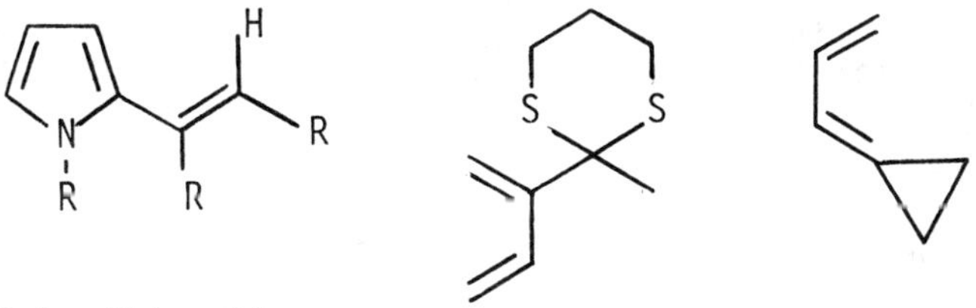

Diels-Alder Dienes.

I.E.1-25 S. M. Makin et al., J. Org. Chem. (USSR), 19, 640 (1983); J. Otera et al., Tetrahedron Lett., 24, 1171 (1983).

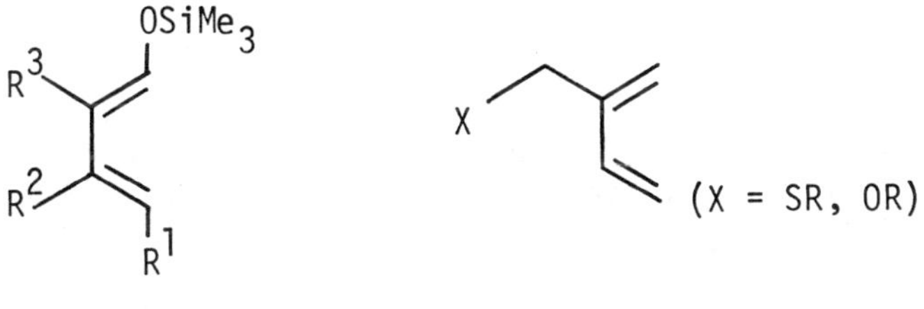

(X = SR, OR)

Diels-Alder Dienes.

I.E.1-26 S. Danishefsky, T. Kitahara and P. F. Schuda, Org. Syn., 61, 147 (1983); H. Hiranuma and S. I. Miller, J. Org. Chem., 48, 3096 (1983); O. Tsuge, E. Wada and S. Kanemasa, Chem. Lett., 1525 (1983).

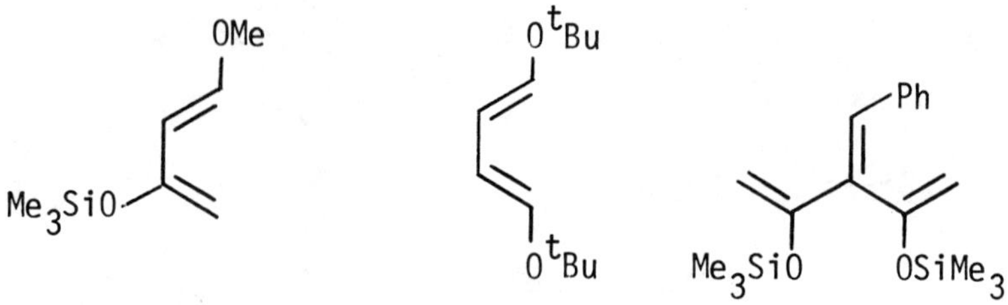

Diels-Alder Dienes.

I.E.1-27 L. E. Overman et al., <u>J. Am. Chem. Soc.</u>, <u>105</u>, 6335 (1983); A. J. Bridges and J. W. Fischer, <u>Tetrahedron Lett.</u>, <u>24</u>, 445, 447 (1983); P. V. Alston et al., <u>J. Org. Chem.</u>, <u>48</u>, 5051 (1983).

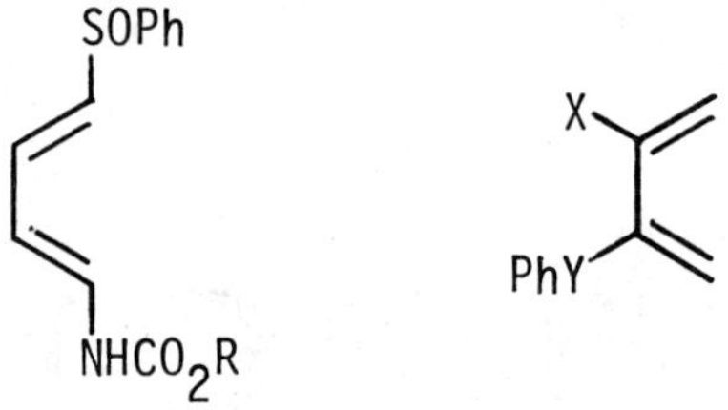

X = Cl or Br

Y = S or Se

Diels-Alder Dienes.

I.E.1-28 H. Schubert, I. Bast and M. Regitz, <u>Synthesis</u>, 661 (1983); V. Kvita, H. Sauter and G. Rihs, <u>Helv. Chim. Acta</u>, <u>66</u>, 2769 (1983).

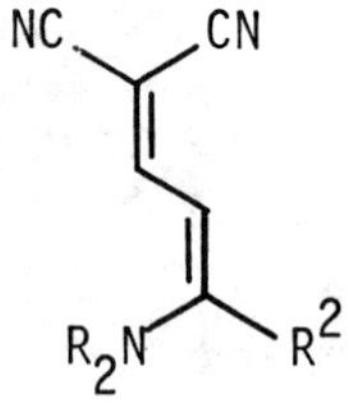

Push-Pull Diene. 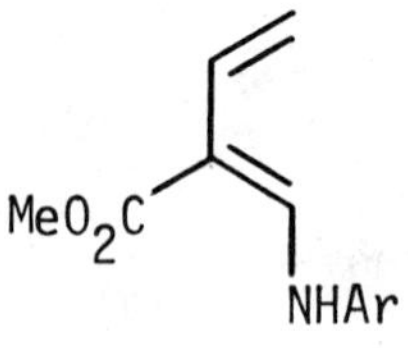   Diels-Alder Diene.

I.E.1-29 L. A. Paquette, R. Gleiter et al., J. Am. Chem. Soc., 105, 3126, 3136, 3148, 3642, 7364 (1983); J. Org. Chem., 48, 1250, 1257, 1262 (1983); D. Kaufmann and A. de Meijere, Chem. Ber., 116, 1897 (1983).

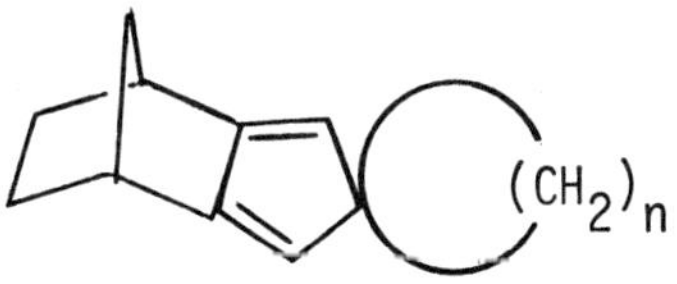

Diels-Alder Dienes.

I.E.1-30 P. Vogel et al., Tetrahedron Lett., 24, 1497, 3603 (1983); Helv. Chim. Acta, 66, 19, 1134, 1279, 2182 (1983); F. Lanzendorfer and M. Christl, Angew. Chem., Int. Ed. Engl., 22, 871 (1983); H. Ahlbrecht, M. Dietz and W. Raab, Synthesis, 231 (1983).

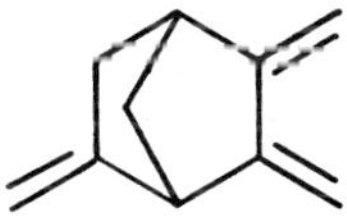

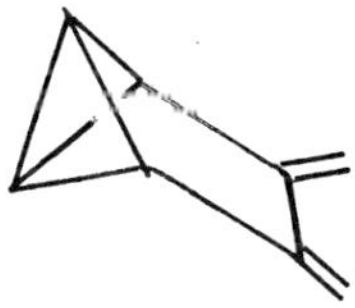

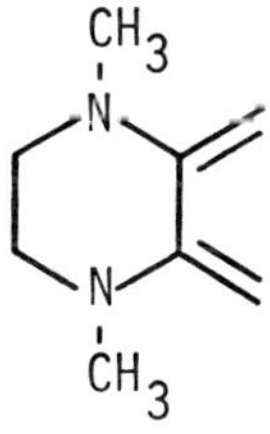

Diels-Alder Dienes.

I.E.1-31 S. Masamune et al., J. Org. Chem., 48, 1137, 4441 (1983); W. Oppolzer et al., Tetrahedron Lett., 24, 4665 (1983); Helv. Chim. Acta, 66, 2358 (1983); C. Maignan and R. A. Raphael, Tetrahedron, 39, 3245 (1983); D. Horton et al., Chem. Commun., 1164 (1983).

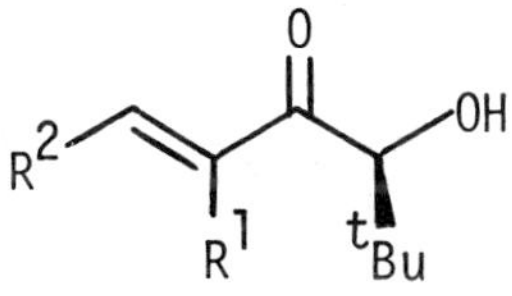

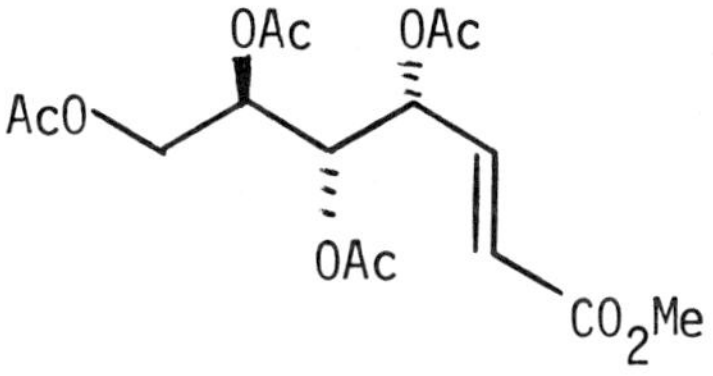

Various Chiral Diels-Alder Dienophiles.

I.E.1-32 B. Gaede and T. M. Balthazor, J. Org. Chem., 48, 276 (1983); A. Padwa and J. G. MacDonald, ibid, 48, 3189 (1983).

X = H, CO$_2$H, CO$_2$Et, CH$_2$OH   Y = Br, SAr, COCH$_3$, CO$_2$Me, CO$_2$H, SiMe$_3$

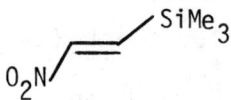

Diels-Alder Dienophiles.

I.E.1-33 O. De Lucchi and G. Modena, Tetrahedron Lett., 24, 1653 (1983); L. A. Paquette et al., J. Org. Chem., 48, 4976, 4986 (1983).

PhSO$_2$\\=/SO$_2$Ph   /=\\SO$_2$Ph   Me$_3$Si\\=/SO$_2$Ph

Maleic Anhydride
  Alternative

Diels-Alder Dienophiles.

I.E.1-34 D. Ranganathan et al., J. Chem. Res. (S), 78 (1983); K. Uneyama, K. Takano and S. Torii, Bull. Chem. Soc. Jpn., 56, 2867 (1983).

O
‖
PhSCH$_2$CH$_2$NO$_2$ ArSe\\C=C/CHO

Nitroethylene Equivalent
Diels-Alder Dienophiles.

I.E.1-35 H. Schuster and J. Sauer, Tetrahedron Lett., 24, 4087 (1983); P. A. Harland and P. Hodge, Synthesis, 419 (1983).

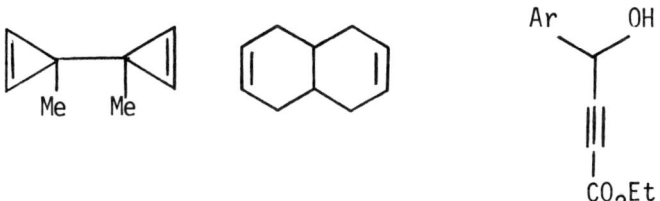

Diels-Alder Dienophiles.

I.E.1-36 J.B. Hendrickson and V. Singh, Chem. Commun., 837 (1983); F. Fringuelli, A. Taticchi, E. Wenkert et al., J. Org. Chem., 48, 1810 (1983); F. Fringuelli, A. Taticchi, E. Wenkert et al., ibid, 48, 2802 (1983); D. Caine, C. R. Harrison and D. G. Van Derveer, Tetrahedron Lett., 24, 1353 (1983).

R^1, R^2 = Me, Me; MeO, Me; MeO, MeO.

n = 1-3

Diels-Alder Dienophiles.

I.E.1-37 R. Block and J. Abecassis, Tetrahedron Lett., 24, 1247 (1983); R. Bloch, Tetrahedron, 39, 639 (1983); D. W. Landry, ibid, 39, 2761 (1983); T. V. RajanBabu, D. F. Eaton and T. Fukunaga, J. Org. Chem., 48, 652 (1983); W. B. Whalley et al., J. Chem. Res. (S), 273 (1983).

1) nBuLi
2) R^2X
3) 650°C

37-47%

I.E.1-38 S. F. Martin et al., J. Org. Chem., 48, 5170 (1983);
R. K. Boeckman, Jr. et al., ibid, 48, 4152 (1983); Tetrahedron
Lett., 24, 5035 (1983).

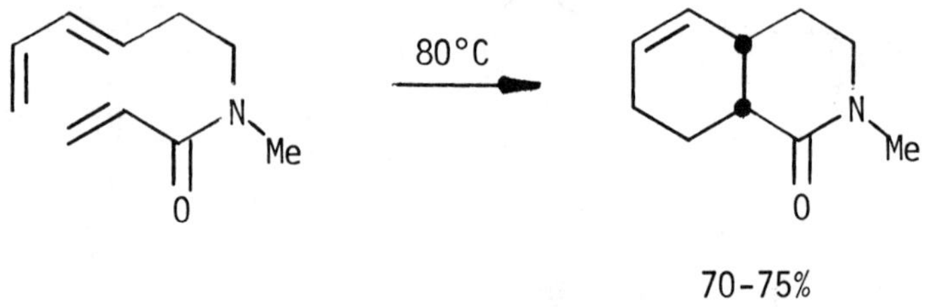

70-75%

I.E.1-39 B. B. Snider et al., J. Org. Chem., 48, 4370 (1983);
Tetrahedron Lett., 24, 3701 (1983).

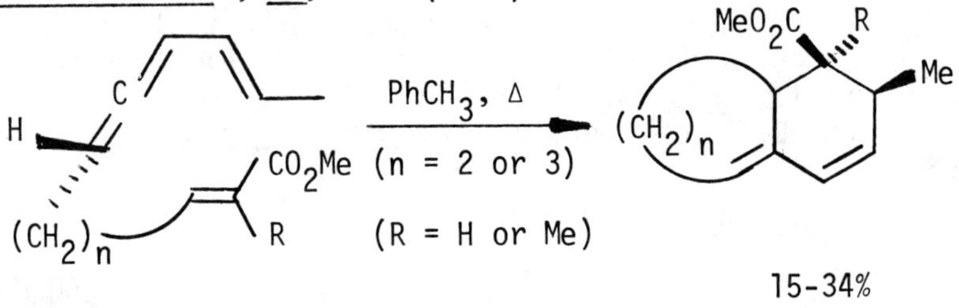

15-34%

I.E.1-40 S. D. Burke, T. H. Powner and M. Kageyama, Tetrahedron Lett., 24, 4529 (1983); K. J. Shea and J. W. Gilman, ibid, 24, 657 (1983); J. M. Hornback and R. D. Barrows, J. Org. Chem., 48, 90 (1983).

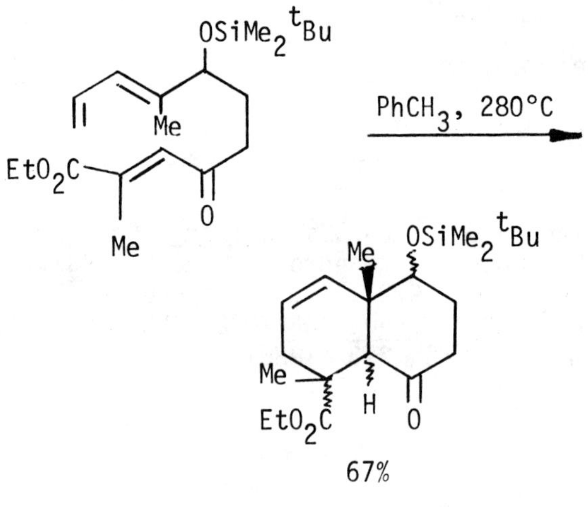

67%

I.E.1-41 Y. Ito, T. Saegusa et al., J. Am. Chem. Soc., 105, 1586 (1983); M. B. Glinski and T. Durst, Can. J. Chem., 61, 573 (1983); G. A. Kraus and M. D. Hagen, J. Org. Chem., 48, 3265 (1983).

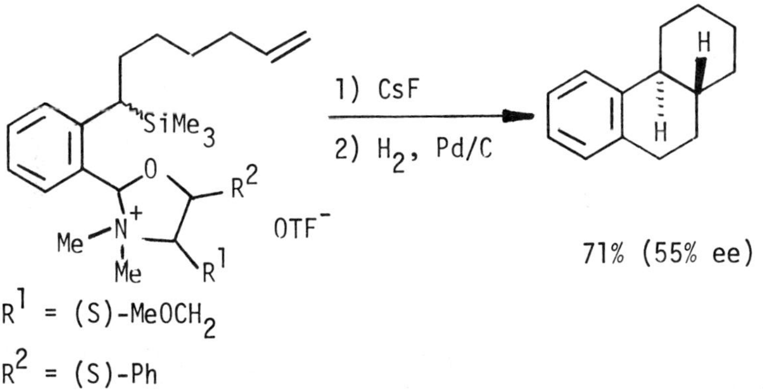
SiMe3
O
R2
N+
Me
Me
R1
OTF-
1) CsF
2) H2, Pd/C
H
H
71% (55% ee)
R1 = (S)-MeOCH2
R2 = (S)-Ph

I.E.1-42 T. Kametani et al., Chem. Commun., 852 (1983); J. Chem. Soc., Perkin I, 2569 (1983); J. Org. Chem., 48, 31 (1983).

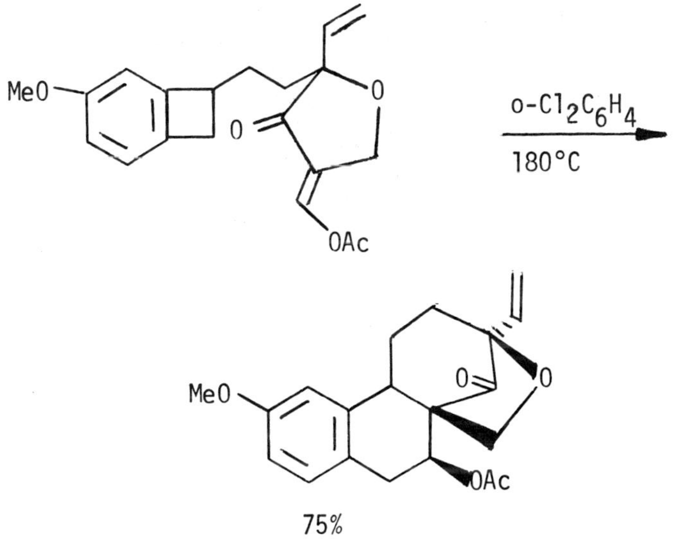
MeO
O
O
OAc
o-Cl2C6H4
180°C
MeO
O
O
OAc
75%

I.E.1-43 S. D. Burke, S. M. S. Strickland and T. H. Powner, J. Org. Chem., 48, 454 (1983); E.J. Thomas et al., J. Chem. Soc., Perkin I, 851 (1983); Tetrahedron Lett., 24, 5535 (1983).

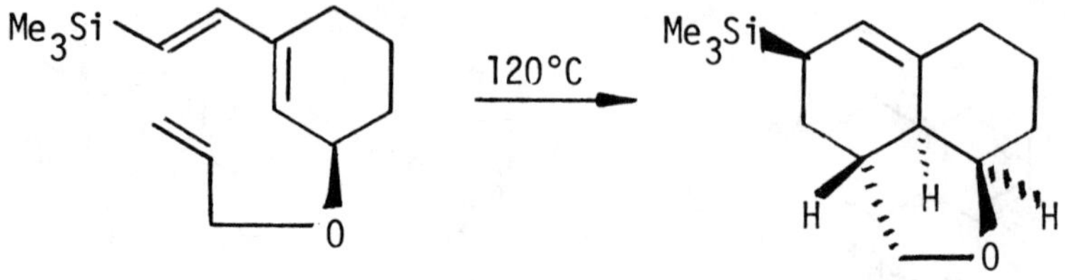

86%

I.E.1-44 R. H. Schlessinger et al., J.Org. Chem., 48, 1146 (1983); D. S. Watt et al., ibid, 48, 470 (1983); T. H. Kim, Y. Hayase and S. Isoe, Chem. Lett., 651 (1983); K. Sakan and B. M. Craven, J. Am. Chem. Soc., 105, 3732 (1983); K. J. Shea and P. D. Davis, Angew. Chem., Int. Ed. Engl., 22, 419 (1983).

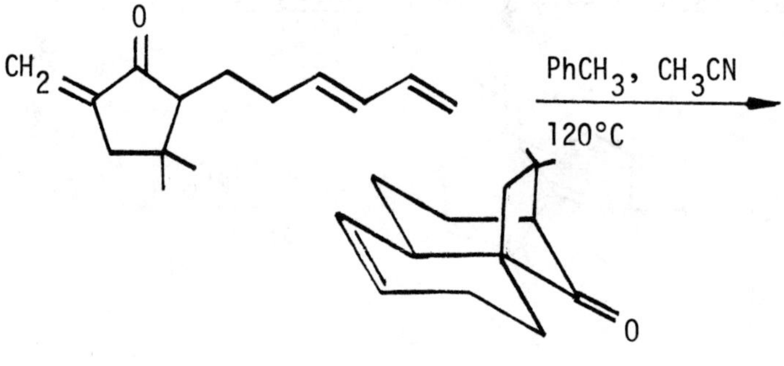

48%

I.E.1-45 D. D. Sternbach et al., <u>Tetrahedron Lett</u>., 24, 3295 (1983); T. S. Macas and P. Yates, <u>ibid</u>, 24, 147 (1983); H. D. Becker and K. Andersson, <u>ibid</u>, 24, 3273 (1983); O. Wallquist, M. Rey and A. S. Dreiding, <u>Helv. Chim. Acta</u>, 66, 1891 (1983).

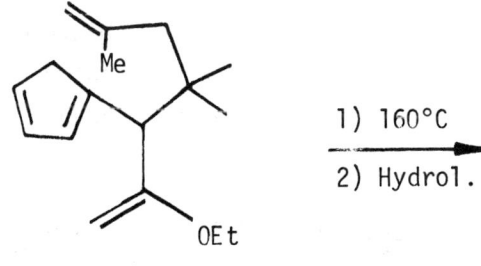

74%

I.E.1-46 T. Mukaiyama et al., <u>Bull. Chem. Soc. Jpn</u>., 56, 1107, 1669 (1983); B. J. Brisdon et al., <u>Tetrahedron Lett</u>., 24, 3037 (1983); L. A. Van Royen, R. Mijngheer and P. J. De Clercq, <u>ibid</u>, 24, 3145 (1983).

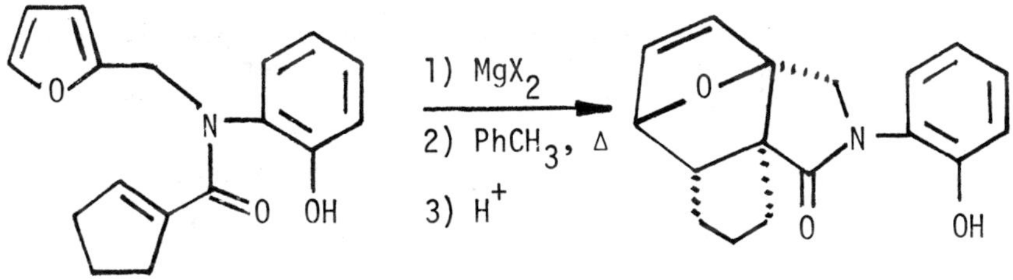

18-85%

I.E.1-47 W. H. Laarhoven, Rec. Trav. Chim., 102, 185, 241 (1983).

Review: "Photochemical Cyclizations and Intramolecular Cycloadditions of Conjugated Arylolefins.
Part 1: Photocyclization with Dehydrogenation.
Part 2: Photocyclizations without Dehydrogenation and Photocycloadditions."

I.E.1-48 H. Klein, G. Freyberger and H. Mayr, Angew. Chem., Int. Ed. Engl., 22, 49 (1983); N. L. Bauld and R. Pabon, J. Am. Chem. Soc., 105, 633 (1983); A. Mackor et al., Tetrahedron Lett., 24, 1419 (1983); L. Fitjer and S. Modaressi, ibid, 24, 5495 (1983).

$Me_2C=CH-CH\begin{subarray}{c}Me\\Cl\end{subarray}$ $\xrightarrow[-78°C]{Me_2C=CMe_2 \atop ZnCl_2, Et_2O}$ [cyclobutane with Me$_2$C-Cl, Me, Me, Me, Me, Me substituents]

72%

I.E.1-49 L. Ghosez et al., Tetrahedron Lett., 24, 2251 (1983); D. A. Jackson, M. Rey and A. S. Dreiding, ibid, 24, 4817 (1983).

$Me_2C=C\begin{subarray}{c}Cl\\NR_2\end{subarray}$ $\xrightarrow{\begin{array}{l}1)\ AgBF_4\ \text{or}\ ZnCl_2\\ 2)\ \diagup=\diagdown\\ 3)\ H_3O^+\end{array}}$ [cyclobutanone with Me, Me, Me, Me substituents]

Stereospecificity depends on nitrogen substituents.

I.E.1-50 G. Rosini, R. Ballini and V. Zanotti, Tetrahedron, 39, 1085 (1983); I. Erden and A. de Meijere, Tetrahedron Lett., 24, 3811 (1983).

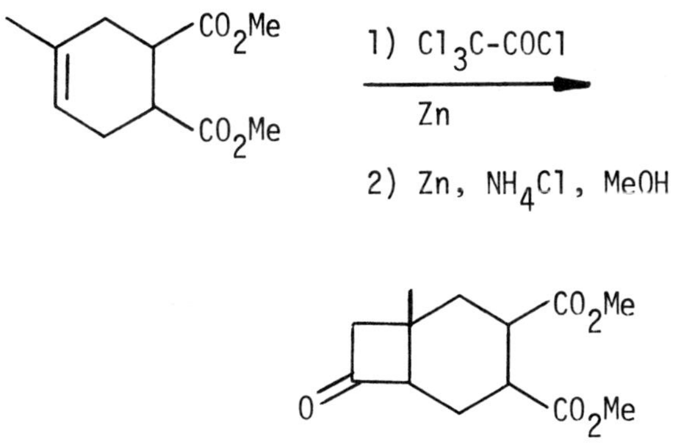

[structure] + 1) Cl$_3$C-COCl / Zn
2) Zn, NH$_4$Cl, MeOH
→ [bicyclic product] 65%

I.E.1-51 A. E. Greene, M. J. Luche and J. P. Depres, J. Am. Chem. Soc., 105, 2435 (1983); D. A. Jackson, M. Rey and A. S. Dreiding, Helv. Chim. Acta, 66, 2330 (1983); H. Jendralla, Synthesis, 111 (1983).

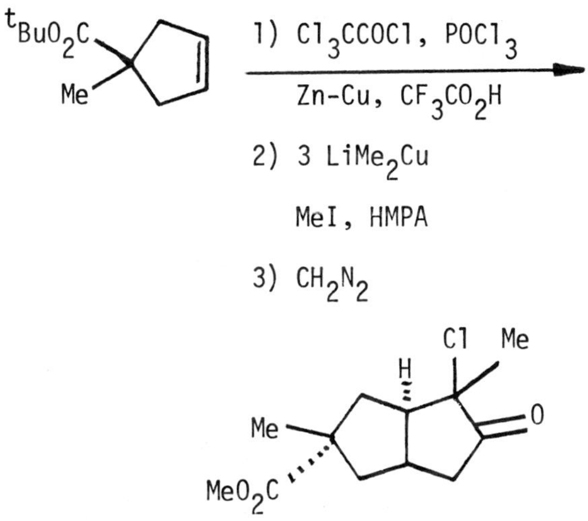

[cyclopentene structure with ^{t}BuO$_2$C and Me]

1) Cl$_3$CCOCl, POCl$_3$
   Zn-Cu, CF$_3$CO$_2$H
2) 3 LiMe$_2$Cu
   MeI, HMPA
3) CH$_2$N$_2$

→ [bicyclic product] 51%

I.E.1-52 K. Mizuno, T. Hashizume and Y. Otsuji, Chem. Commun., 772 (1983); W. Ried, O. Bellinger and J. W. Bats, Chem. Ber., 116, 3794 (1983).

ROCH=CH$_2$ + [2,3-dicyanonaphthalene] $\xrightarrow{h\nu, PhH}$ [cyclobutane with RO groups] 34-54%

cis:trans = 1:1

R = Alkyl, PhCH$_2$, Me$_3$Si.

I.E.1-53 H. W. Scheeren and A. E. Frissen, Synthesis, 794 (1983); H. W. Scheeren et al., Tetrahedron, 39, 1345 (1983); A. B. Smith, III and R. E. Richmond, J. Am. Chem. Soc., 105, 575 (1983); H. W. Scheeren et al., Rec. Trav. Chim., 102, 96, 130 (1983).

[Me/OMe alkene] + [NC/H alkene] $\xrightarrow{ZnCl_2}$ [cyclobutane product] 80%

I.E.1-54 M. Ikeda et al., J. Org. Chem., 48, 4241 (1983); G. Pattenden et al., J. Chem. Soc., Perkin I, 1885, 1893, 1901, 1905, 1913, 1919 (1983); T. Ogino et al., Tetrahedron Lett., 24, 2781 (1983).

[enone with allyl ether] $\xrightarrow[\text{Acetone}]{h\nu}$ [bicyclic product] 53-63%

I.E.1-55 K. Mizuno, H. Kagano and Y. Otsuji, Tetrahedron Lett., 24, 3849 (1983); M. Kuzuya, M. Tanaka and T. Okuda, ibid, 24, 4237 (1983); H. Mayer and J. Sauer, ibid, 24, 4091, 4095 (1983).

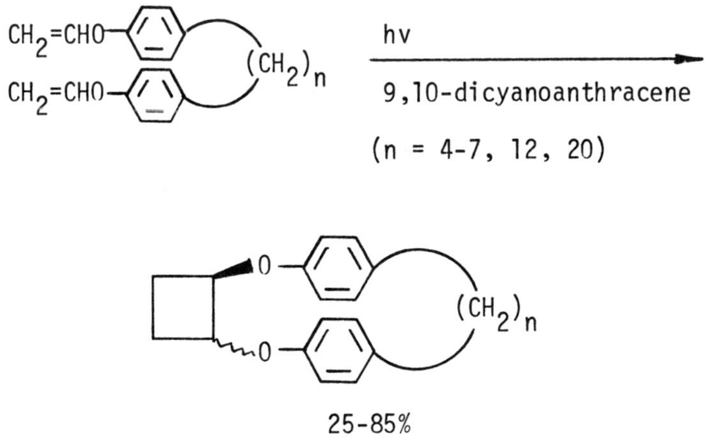

(n = 4-7, 12, 20)

25-85%

I.E.1-56 D. Becker et al., J. Org. Chem., 48, 2584 (1983); Z. Komiya and S. Nishida, ibid, 48, 1500 (1983).

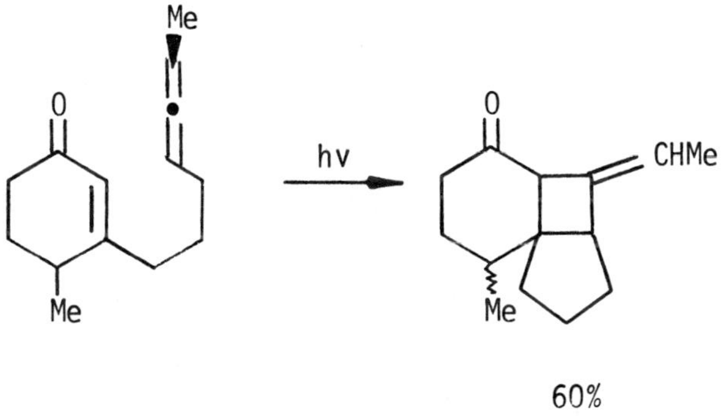

60%

I.E.1-57 R. L. Danheiser and H. Sard, Tetrahedron Lett., 24, 23 (1983); A. Hassner and J. L. Dillon, Jr., J. Org. Chem., 48, 3382 (1983); A. Bouvy, Z. Janousek and H. G. Viehe, Synthesis, 718 (1983); S. A. Shama and C. C. Wamser, Org. Syn., 61, 62 (1983).

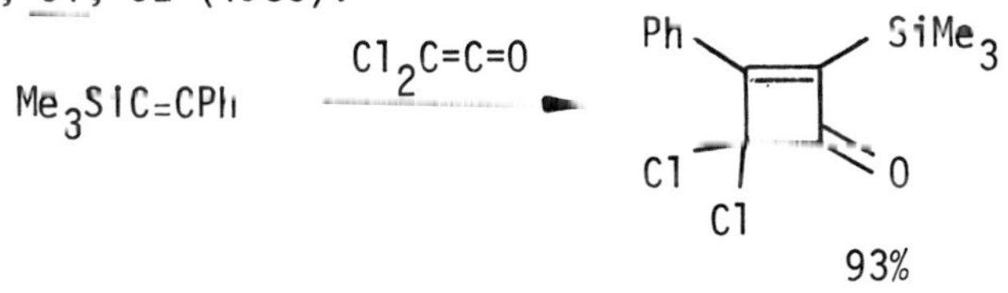

93%

I.E.1-58 A. Fadel, J. Salaun and J. M. Conia, Tetrahedron, 39, 1567 (1983); J. Ficini, D. Desmaele and A. M. Touzin, Tetrahedron Lett., 24, 1025 (1983); J. N. Vishwakarma, H. Ila and H. Junjappa, J. Chem. Soc., Perkin I, 1099 (1983); K. Tanaka and F. Toda, Chem. Commun., 593 (1983).

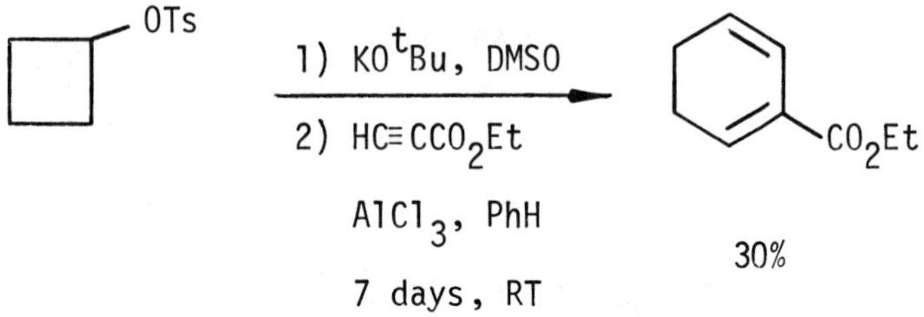
OTs
1) KO^{t}Bu, DMSO
2) HC≡CCO$_2$Et
AlCl$_3$, PhH
7 days, RT
CO_2Et
30%

I.E.1-59 T. Kametani et al., Tetrahedron, 39, 1123 (1983); E. Lee-Ruff, A. C. Hopkinson and H. Kazarians-Moghaddam, Tetrahedron Lett., 24, 2067 (1983).

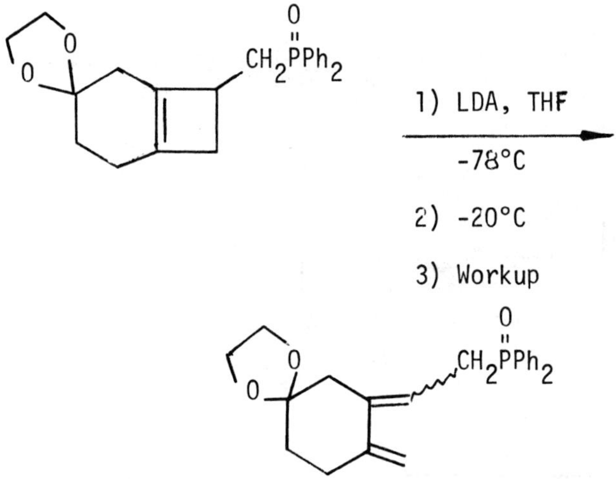
O
O
O
CH_2PPh_2
1) LDA, THF
-78°C
2) -20°C
3) Workup
O
O
O
CH_2PPh_2

13-90%

I.E.1-60 B. M. Trost and D. M. T. Chan, J. Am. Chem. Soc., 105, 2315, 2326 (1983); M. Calligaris, G. Carturan, A. Wojcicki et al., Organometallics, 2, 865 (1983).

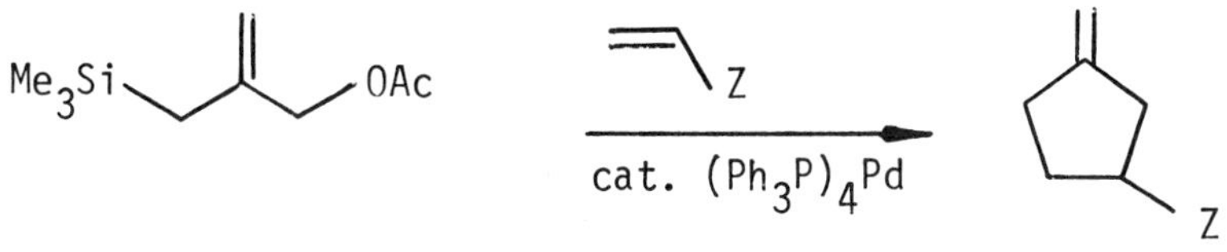

$$Me_3Si\diagdown\!\!\!\diagup\!\!\!\diagdown OAc \;+\; \diagup\!\!=\!\!Z \xrightarrow{cat.\;(Ph_3P)_4Pd} \text{[methylenecyclopentane-Z]}$$

10-85%

I.E.1-61 P. Binger, et al., Tetrahedron Lett., 24, 3599, 5847 (1983); Chem. Ber., 116, 2920 (1983).

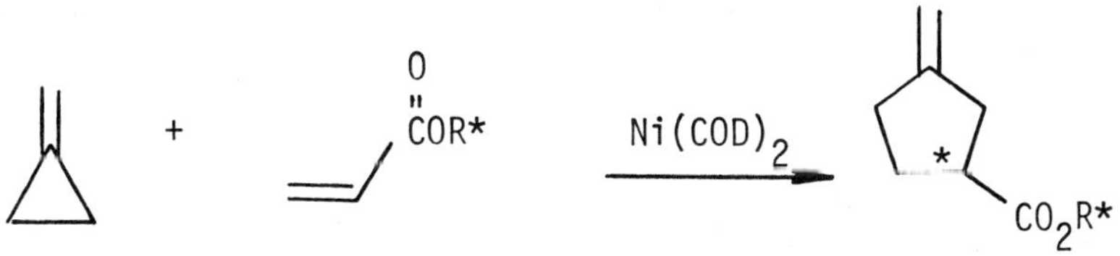

methylenecyclopropane + CH$_2$=CH−COR* $\xrightarrow{Ni(COD)_2}$ 3-methylene-cyclopentane-CO$_2$R*

90% (64% de)

I.E.1-62 A. P. Kozikowski et al., Tetrahedron Lett., 24, 3705 (1983); J. Org. Chem., 48, 366 (1983).

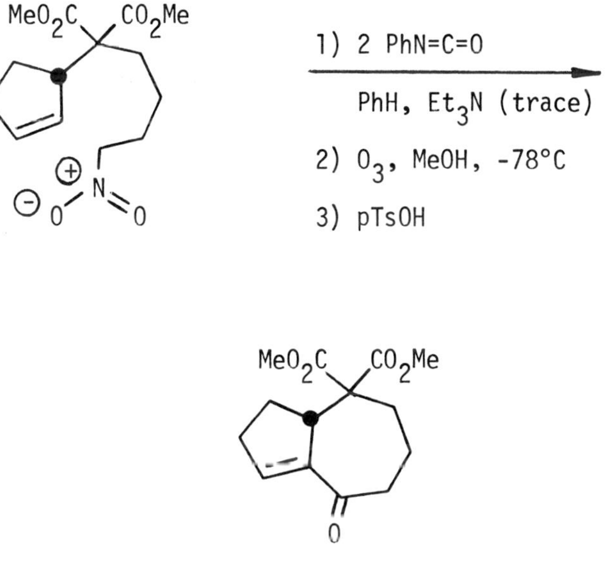

1) 2 PhN=C=O, PhH, Et$_3$N (trace)

2) O$_3$, MeOH, −78°C

3) pTsOH

66%

I.E.1-63 E. Turecek et al., <u>Chem. Commun.</u>, 805 (1983); C. Blackburn, R. F. Childs and R. A. Kennedy, <u>Can. J. Chem.</u>, <u>61</u>, 1981 (1983).

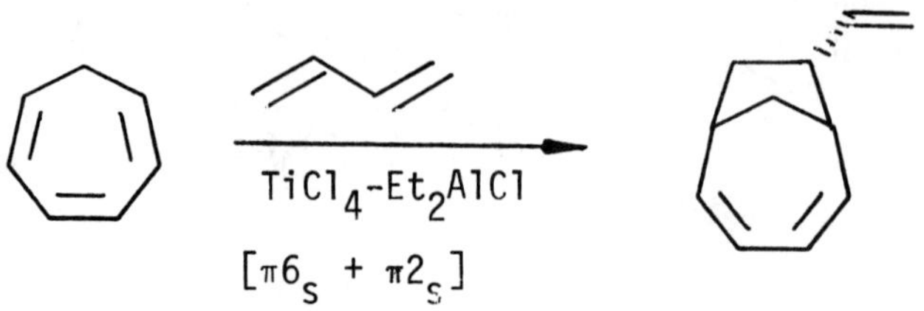

[cycloheptatriene] + [butadiene] → TiCl$_4$-Et$_2$AlCl [π6$_s$ + π2$_s$] → [bicyclic product] 78%

I.E.1-64 K. N. Houk et al, <u>J. Am. Chem. Soc.</u>, <u>105</u>, 6996 (1983); <u>J. Org. Chem.</u>, <u>48</u>, 403 (1983); M. E. Garst, V. A. Roberts and C. Prussin, <u>Tetrahedron, 39</u>, 581 (1983).

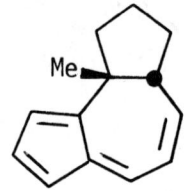

1) Et$_2$NH, K$_2$CO$_3$
   PhH, 5A Mol Sieve
2) 40°C

46-55%

I.D.1-65 K. N. Houk et al., J. Am. Chem. Soc., 105, 6714 (1983); J. Daub et al., Chem. Ber., 116, 2408 (1983).

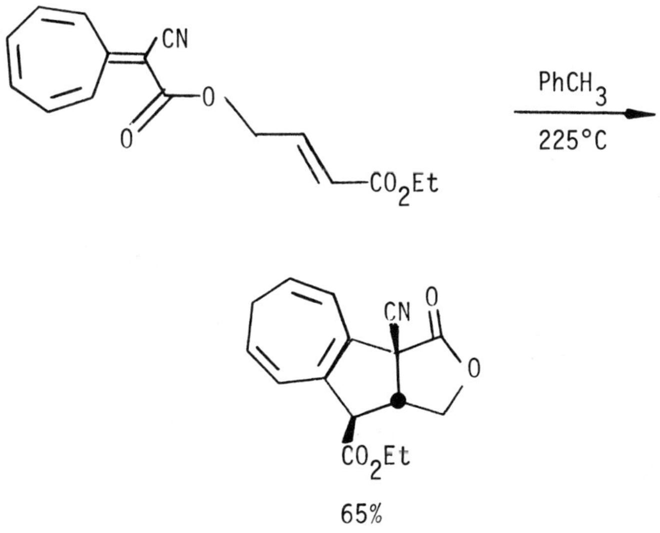

65%

I.E.1-66 H. M. R. Hoffmann and R. Henning, Helv. Chim. Acta, 66, 828 (1983).

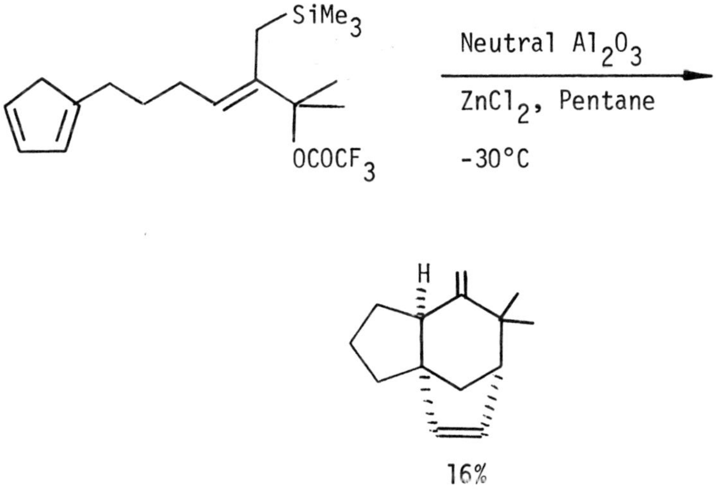

16%

I.E.1-67 K. Fujimori et al., Tetrahedron Lett., 24, 781 (1983); T. Toda et al., Chem. Lett., 523 (1983).

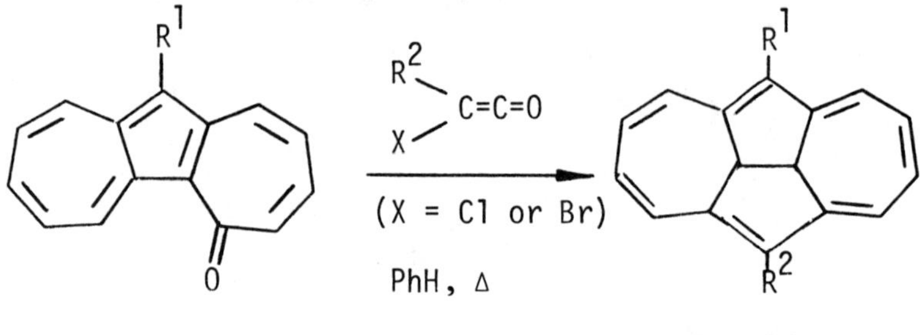

19-78%

I.E.1-68 D. T. Glatzhofer and D. T. Longone, Tetrahedron Lett., 24, 4413 (1983).

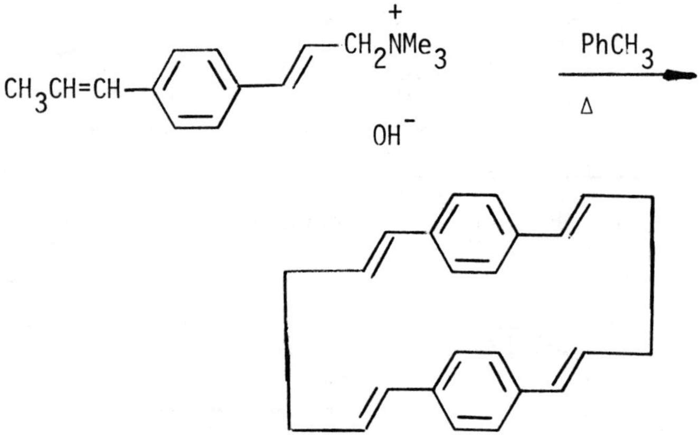

30%

I.E.2. Other Thermal Reactions

I.E.2-1 B. B. Snider and G. B. Phillips, J. Org. Chem., 48, 464, 3685 (1983).

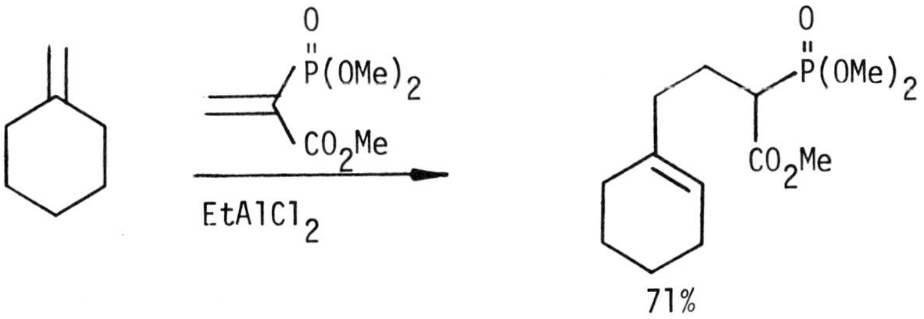

I.E.2-2 L. Moore, D. Gooding and J. Wolinsky, J. Org. Chem., 48, 3750 (1983); B. B. Snider and E. A. Deutsch, ibid, 48, 1822 (1983).

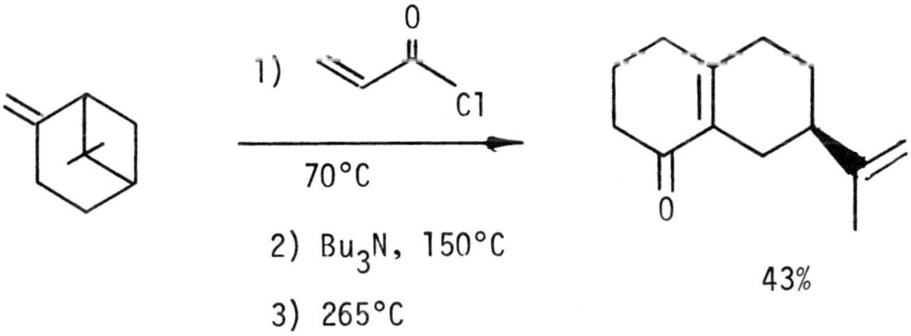

Ene-ene-retroene

I.E.2-3 J. K. Whitesell, D. Deyo and A. Bhattacharya, Chem. Commun., 802 (1983); L. Skattebol and Y. Stenstrom, Tetrahedron Lett., 24, 3021 (1983).

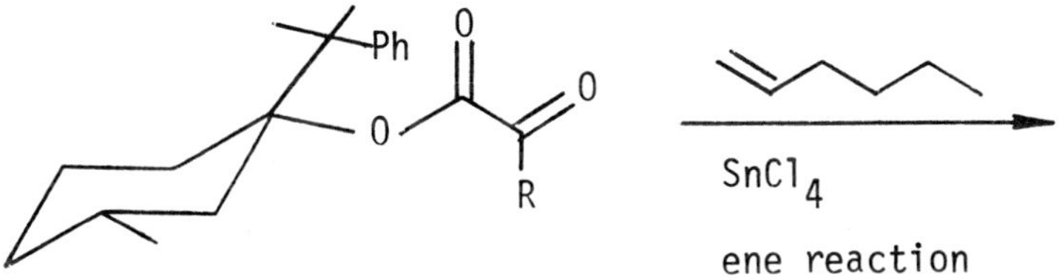

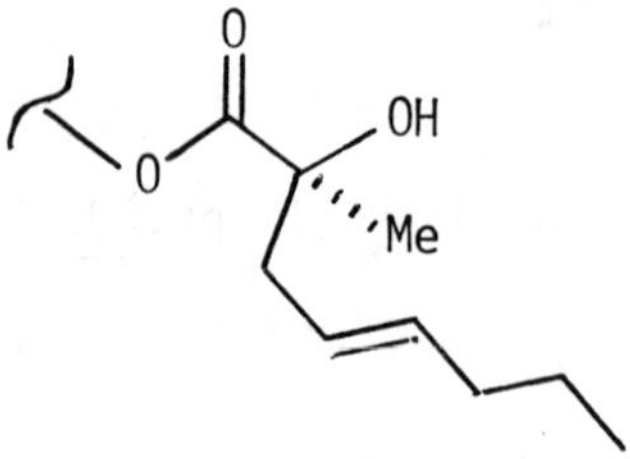

> 90% de

I.E.2-4 N. Bluthe, M. Malacria and J. Gore, Tetrahedron Lett., 24, 1157 (1983); L. E. Overman and A. F. Renaldo, ibid, 24, 3757 (1983); H. Urabe and I. Kuwajima, ibid, 24, 4241 (1983); P. A. Grieco and T. R. Vedananda, J. Org. Chem., 48, 3497 (1983).

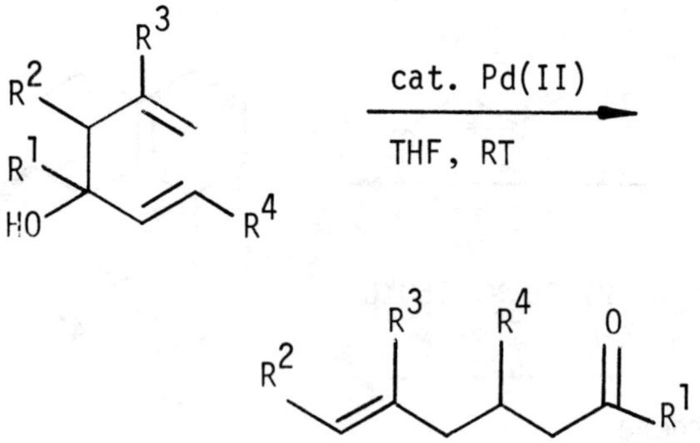

55-100%

(> 90% Stereoselectivity)

I.E.2-5 M. E. Jung and G. L. Hatfield, Tetrahedron Lett., 24, 2931 (1983); L. A. Paquette, D. R. Andrews and J. P. Springer, J. Org. Chem., 48, 1147 (1983); W. C. Still et al., J. Am. Chem. Soc., 105, 625 (1983).

Tandem anionic [1,3]-[3,3] sigmatropic rearrangement

I.E.2-6 G. W. Daub and S. R. Lunt, Tetrahedron Lett., 24, 4397 (1983).

Acid-Catalyzed Claissen.

I.E.2-7 R. K. Brunner and H. J. Borschberg, Helv. Chim. Acta, 66, 2608 (1983).

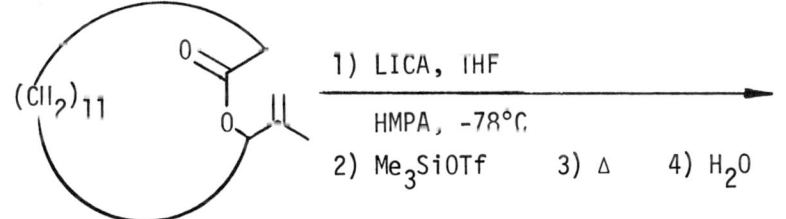

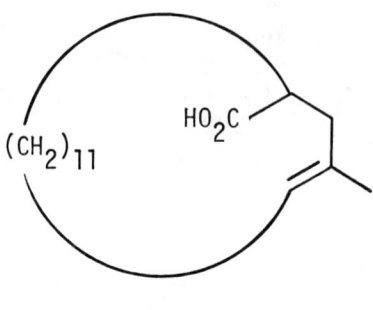

76%

I.E.2-8 G. Buchi and D. E. Vogel, J. Org. Chem., 48, 5406
(1983); J. Rodriquez, J. P. Dulcere and M. Bertrand, Tetrahedron Lett., 24, 4423 (1983); S. Suzuki, Y. Fujita and T. Nishida, ibid, 24, 5737 (1983).

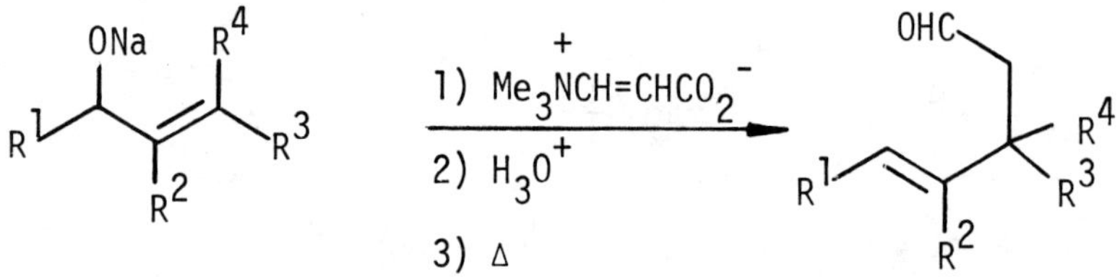

1) $Me_3NCH=CHCO_2^-$
2) H_3O^+
3) Δ

68-100%

Claissen Sequence that Avoids Mercury Catalysis.

I.E.2-9 S. E. Denmark and M. A. Harmata, J. Org. Chem., 48, 3369 (1983); G. W. Daub et al., ibid, 48, 3876 (1983); T. L. Ho and T. W. Hall, Chem. Ind. (London), 862 (1983).

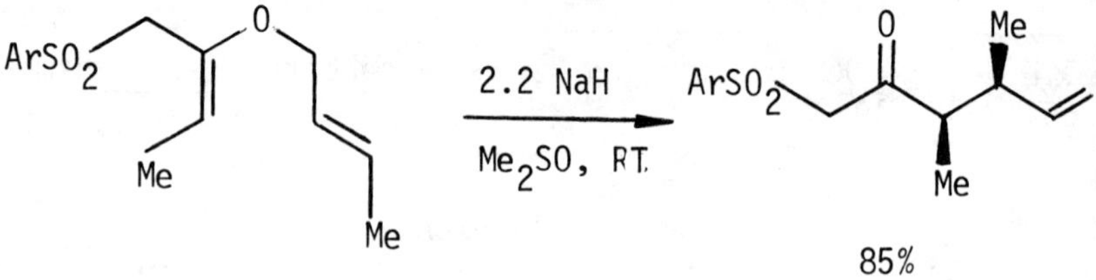

2.2 NaH
Me_2SO, RT.

85%

(98% Stereosel.)

I.E.2-10 T. Sato, K. Tajima and T. Fujisawa, Tetrahedron Lett., 24, 729 (1983); J. Kallmerten and T. J. Gould, ibid, 24, 5177 (1983); J. M. Vatele, ibid, 24, 1239 (1983); R. Malherbe, G. Rist and D. Bellus, J. Org. Chem., 48, 860 (1983).

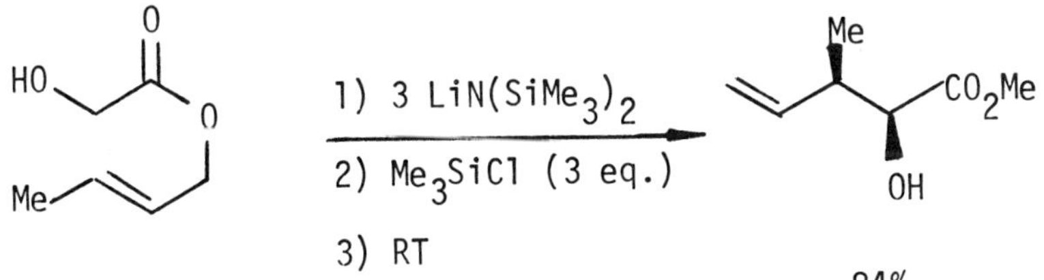

84%

(92% Erythro)

I.E.2-11 P. Beslin et al., Tetrahedron Lett., 24, 3617 (1983); E. Schaumann and J. Lindstaedt, Chem. Ber., 116, 1728 (1983).

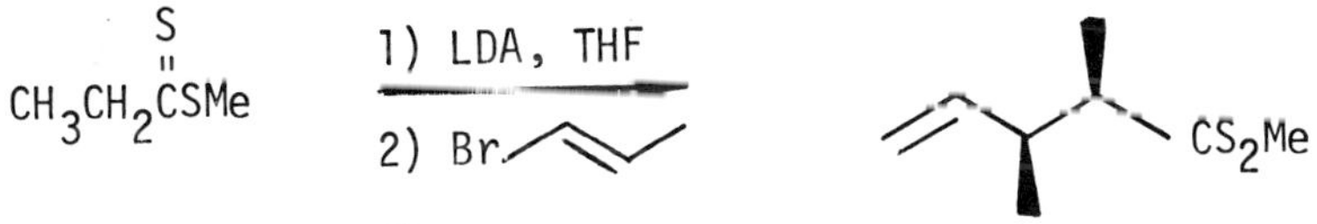

59%

Alkylation via cis lithium thioenolate then thio-Claissen.

I.E.2-12 E. H. Smith and N. D. Tyrrell, Chem. Commun., 285 (1983); M. J. Kurth and O. H. W. Decker, Tetrahedron Lett., 24, 4535 (1983).

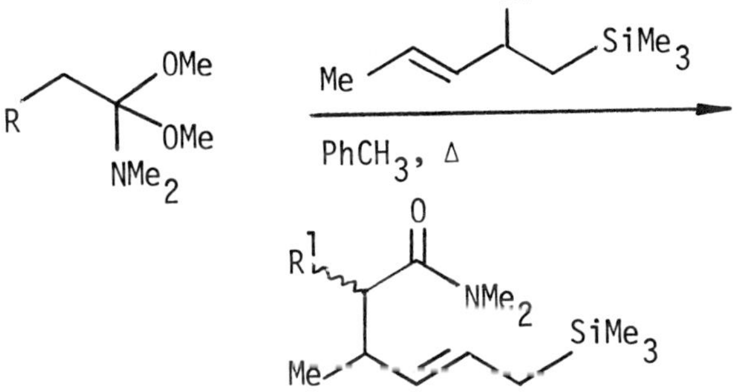

46-55% (threo:erythro = 1.2:1)

I.E.2-13 S. D. Burke, W. F. Fobare and G. J. Pacofsky, J. Org. Chem., 48, 5221 (1983).

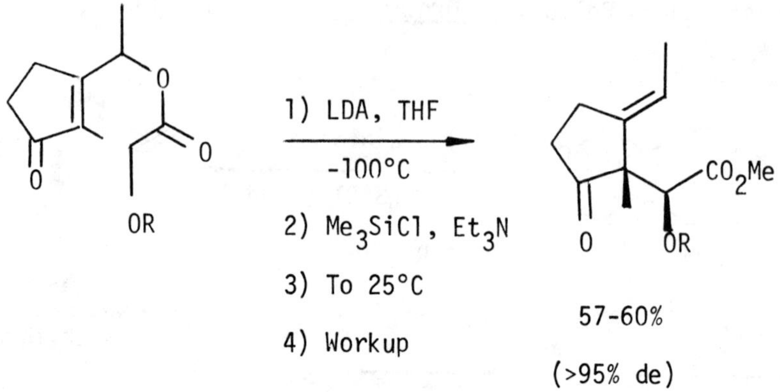

1) LDA, THF, -100°C
2) Me$_3$SiCl, Et$_3$N
3) To 25°C
4) Workup

57-60%
(>95% de)

I.E.2-14 J. H. Hoare, P. P. Policastro and G. A. Berchtold, J. Am. Chem. Soc., 105, 6264 (1983); R. E. Ireland and M. D. Varney, J. Org. Chem., 48, 1829 (1983); M. Koreada and L. Brown, ibid, 48, 2122 (1983); A. A. Ponaras, ibid, 48, 3866 (1983); Tetrahedron Lett., 24, 3 (1983).

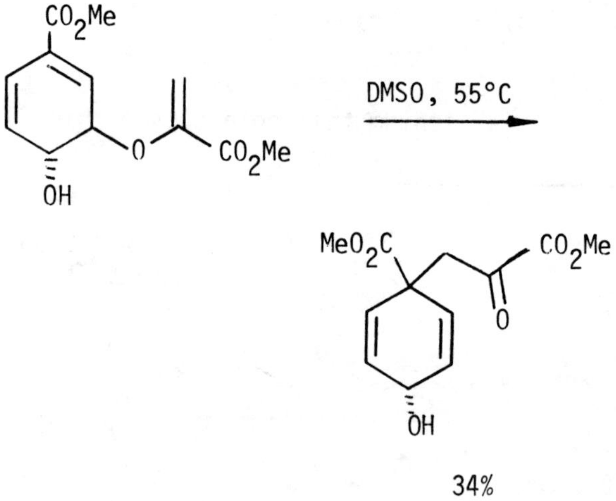

DMSO, 55°C

34%

I.E.2-15 W. D. Ollis et al., J. Chem. Soc., Perkin I, 1009, 1029, 1041, 1049 (1983); E. Morera and G. Ortar, J. Org. Chem., 48, 119 (1983).

Base Catalyzed Rearrangements Involving Ylide Intermediates.

I.E.2-16 T. Nakai et al., J. Org. Chem., 48, 279 (1983); K. Mikami, K. I. Azuma and T. Nakai, Chem. Lett., 1379 (1983); K. Mikami, K. Fujimoto and T. Nakai, Tetrahedron Lett., 24, 513 (1983).

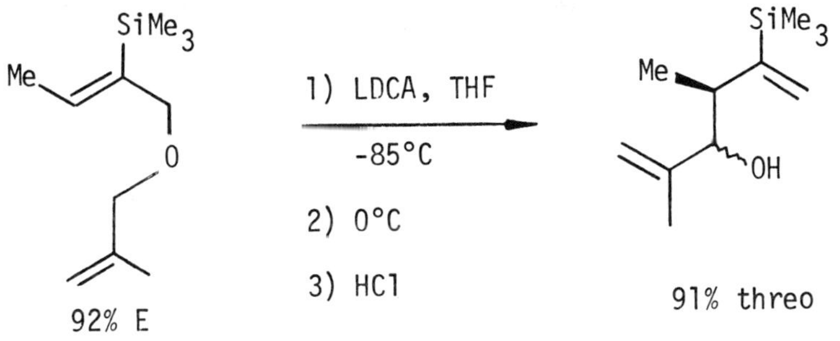

I.E.2-17 W. H. Okamura et al., J. Am. Chem. Soc., 105, 1626, 3588 (1983); J. Org. Chem., 48, 4030 (1983); H. J. Martens, G. J. Hoornaert et al., ibid, 48, 2188 (1983).

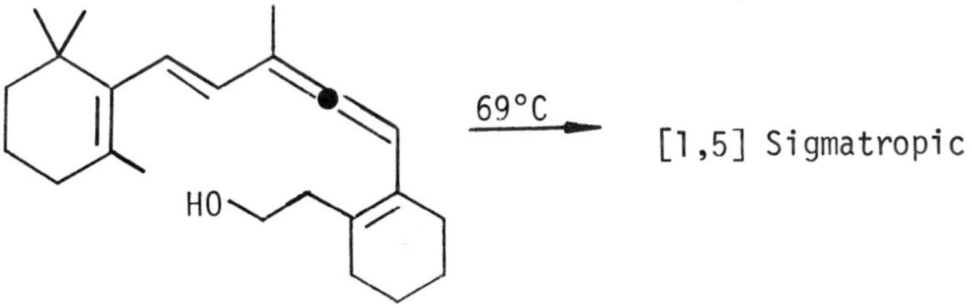

I.E.2-18 E. N. Marvell, et al., J. Org. Chem., 48, 4272, 5379 (1983); D. Tanner, O. Wennerstrom and T. Olsson, Tetrahedron Lett., 24, 5407 (1983); Y. Fujise et al., ibid, 24, 4261 (1983).

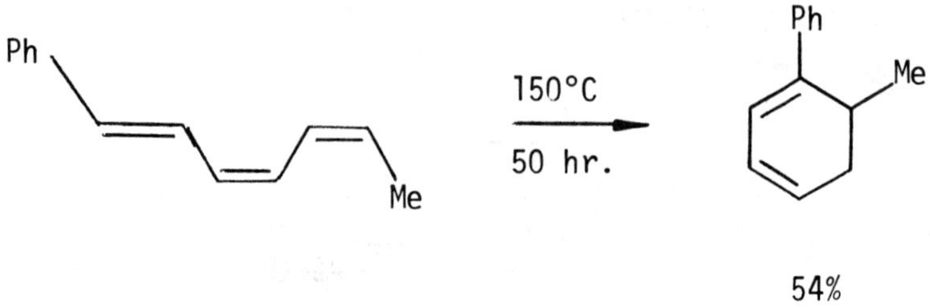

$$\xrightarrow[50 \text{ hr.}]{150°C}$$

54%

I.E.2-19 S. Nishida et al., Chem. Commun., 1191 (1983).

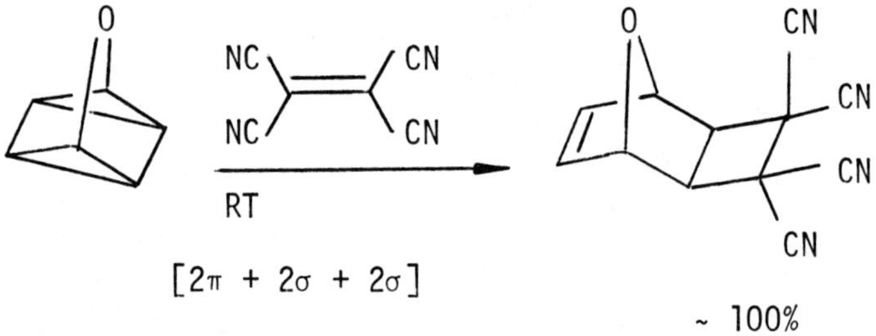

$[2\pi + 2\sigma + 2\sigma]$

~100%

I.E.2-20 R. M. Coates and L. A. Last, J. Am. Chem. Soc., 105, 7322 (1983); J. R. Williams, J. F. Callahan and C. Lin, J. Org. Chem., 48, 3162 (1983); P. Warner, I. S. Chu and W. Boulanger, Tetrahedron Lett., 24, 4165 (1983); K. J. Shea et al., ibid, 24, 4173 (1983).

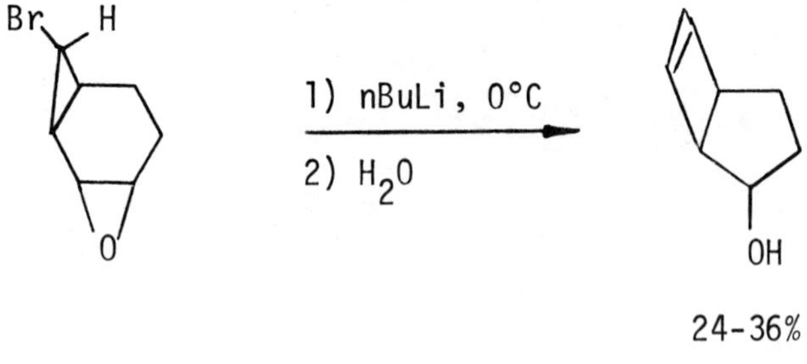

1) nBuLi, 0°C
2) H$_2$O

24-36%

Conrot. Ring-Opening then Conrot. Ring-Closure.

I.E.2-21 J. Huguet, M. Karpf and A. S. Dreiding, Tetrahedron Lett., 24, 4177 (1983); G. G. G. Manzardo, M. Karpf and A. S. Dreiding, Helv. Chim. Acta, 66, 627 (1983); E. Piero, C. K. Lau and I. Nagakura, Can. J. Chem., 61, 288 (1983).

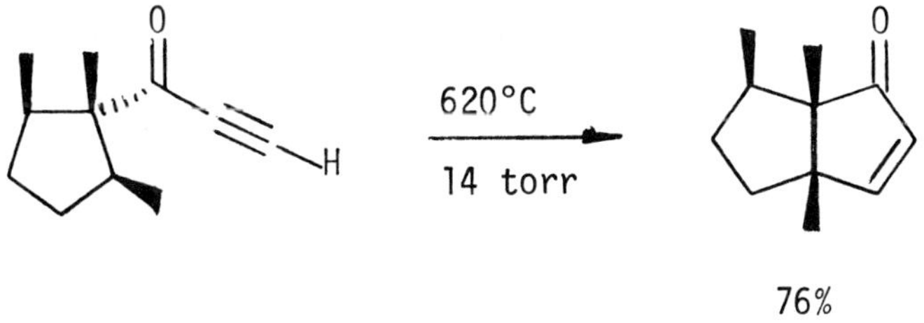
O
H
620°C
14 torr
O
76%

I.F. Aromatic Substitutions Forming a New Carbon-Carbon Bond

I.F.1. Friedel-Crafts Type Aromatic Substitution Reactions

I.F.1-1 S. Masuda, T. Nakajima and S. Suga, Bull. Chem. Soc. Jpn., 56, 1089 (1983).

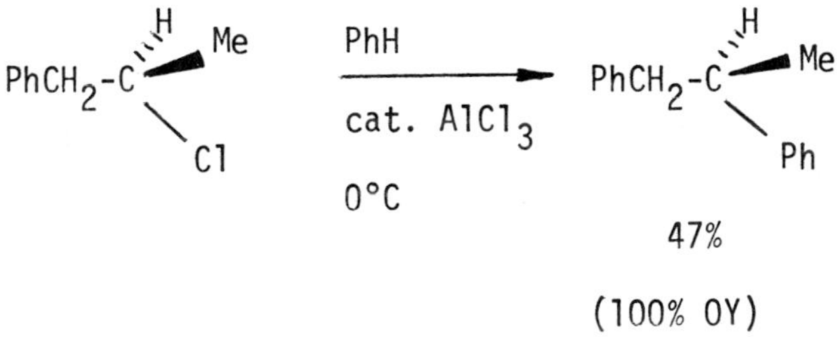
H
Me
PhCH2-C
Cl
PhH
cat. AlCl3
0°C
H
Me
PhCH2-C
Ph
47%
(100% OY)

1,1-Diphenylpropane is other product.

I.F.1-2 N. B. Nevrekar et al., Chem. Ind. (London), 206 (1983); Y. V. Pozdnyakovich et al., J. Org. Chem. (USSR), 19, 353 (1983)..

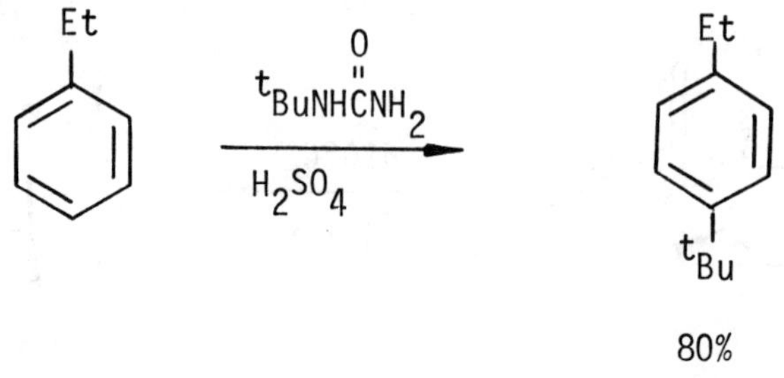

$$\text{Et-C}_6\text{H}_5 \xrightarrow[\text{H}_2\text{SO}_4]{^t\text{BuNHCNH}_2, \; O} \text{4-}^t\text{Bu-Et-C}_6\text{H}_4$$

80%

Also carried out using ^{t}BuOH, Urea and H_2SO_4.

I.F.1-3 E. Fujita et al., Chem. Pharm. Bull., 31, 86 (1983); A. Panunzi et al., J. Chem. Soc., Perkin II, 993 (1983).

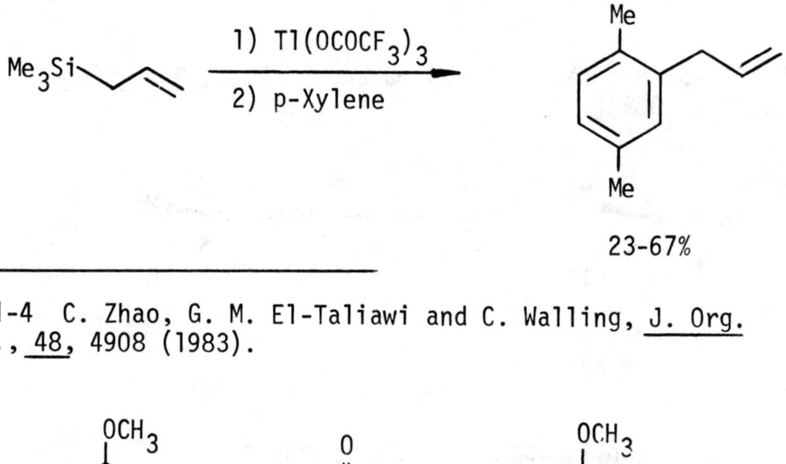

$$\text{Me}_3\text{Si-CH}_2\text{-CH=CH}_2 \xrightarrow[\text{2) p-Xylene}]{\text{1) Tl(OCOCF}_3)_3} \text{2,5-Me}_2\text{-C}_6\text{H}_3\text{-CH}_2\text{CH=CH}_2$$

23-67%

I.F.1-4 C. Zhao, G. M. El-Taliawi and C. Walling, J. Org. Chem., 48, 4908 (1983).

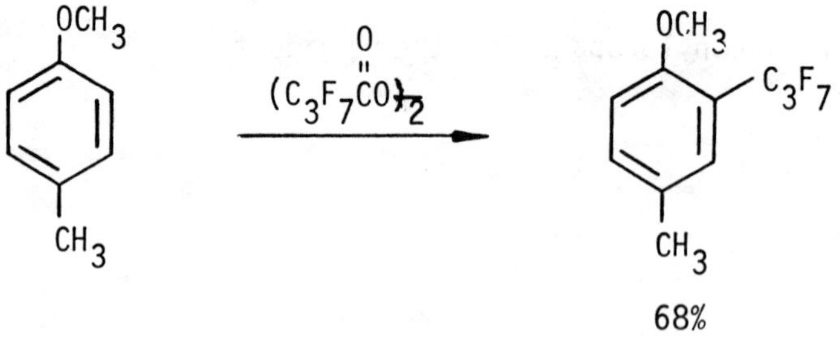

$$\text{4-CH}_3\text{-C}_6\text{H}_4\text{-OCH}_3 \xrightarrow{(C_3F_7CO)_2O} \text{product}$$

68%

I.F.1-5 S. K. Taylor, D. W. Brooks et al., J. Het. Chem., 20, 1745 (1983); J. Org. Chem., 48, 592 (1983).

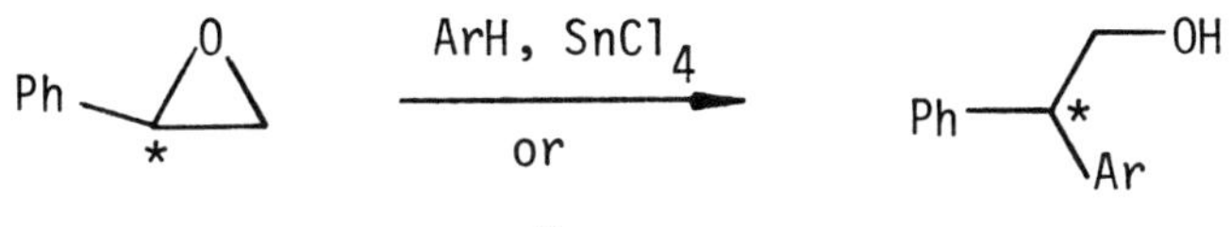

Ph—epoxide* → (ArH, SnCl$_4$ or ArMgBr) → Ph—*CH(Ar)—CH$_2$OH

20-34% ee

I.F.1-6 G. Sartori et al., Tetrahedron, 39, 1761 (1983).

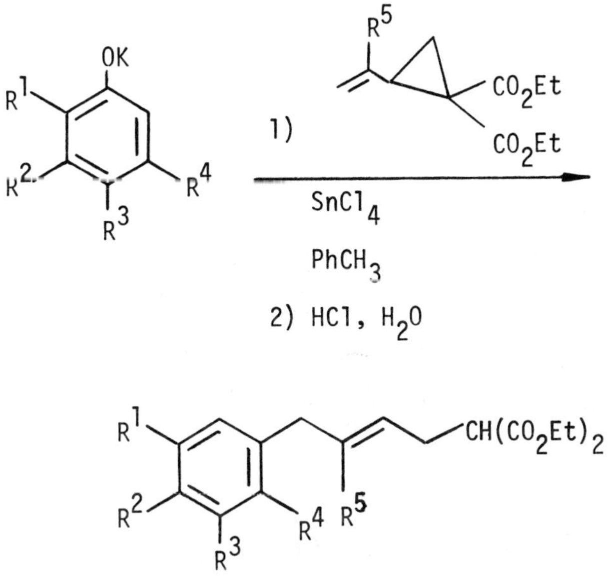

Aryl-OK (with R^1, R^2, R^3, R^4)

1) methylenecyclopropane with R^5, C(CO$_2$Et)$_2$ / SnCl$_4$ / PhCH$_3$

2) HCl, H$_2$O

→ Ar(R^1,R^2,R^3,R^4)-CH$_2$-C(R^5)=CH-CH$_2$-CH(CO$_2$Et)$_2$

25-50%

I.F.1-7 G. Casiraghi et al., Chem. Commun., 1210 (1983); G. Sartori et al., J. Chem. Soc., Perkin I, 1649 (1983).

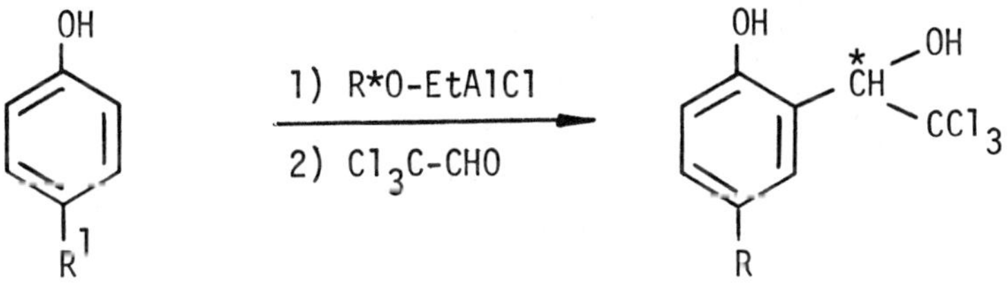

4-R^1-phenol → 1) R*O-EtAlCl   2) Cl$_3$C-CHO → 2-(*CH(OH)CCl$_3$)-4-R-phenol

51-98% (0-49% ee)

I.F.1-8 R. M. Williams and A. O. Stewart, Tetrahedron Lett., 24, 2715 (1983); M. A. McKervey and P. Ratananukul, ibid, 24, 117 (1983); S. Czernecki and V. Dechavanne, Can. J. Chem., 61, 533 (1983).

[Reaction: thiopyridyl glycoside + ArOTf, ArH (Active) → C-aryl glycoside, 50-57%]

Also, reaction with TMS enol ethers.

I.F.1-9 F. Effenberger et al., Chem. Ber., 116, 1183, 1195 (1983); G. P. Rizzi, Synth. Commun., 13, 1173 (1983).

$$R-\overset{O}{\underset{\|}{C}}-Cl \xrightarrow[\text{CH}_2\text{Cl}_2,\ -50°\text{C}]{\text{1) AgOSO}_2\text{CF}_3} Ar-\overset{O}{\underset{\|}{C}}-R$$

2) ArH

No Friedel-Crafts catalyst needed, even with deactivated aromatics.

I.F.1-10 J. K. Ruminski, Chem. Ber., 116, 970 (1983); M. Thyes et al., J. Med. Chem., 26, 800 (1983).

[Reaction: 2,6-dimethylphenol + phthalic anhydride, AlCl$_3$, Cl$_2$CHCHCl$_2$, 47°C → aroyl benzoic acid product, 74%]

I.F.1-11 M. R. Stillings et al., J. Med. Chem., 26, 1353 (1983).

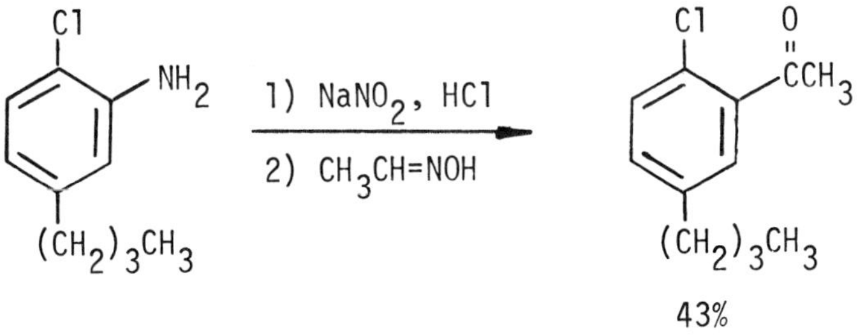

[2-chloro-5-butylaniline] → (1) $NaNO_2$, HCl; 2) $CH_3CH=NOH$) → 2'-chloro-5'-butylacetophenone

43%

I.F.1-12 S. F. Dyke et al., J. Organometal. Chem., 253, 399 (1983).

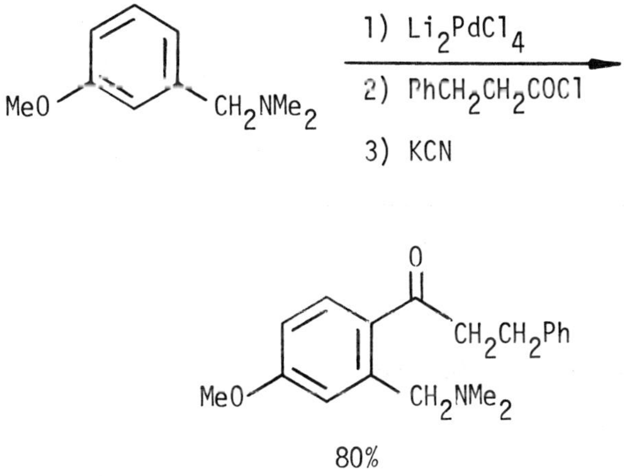

[3-methoxybenzyl dimethylamine] → 1) Li_2PdCl_4; 2) $PhCH_2CH_2COCl$; 3) KCN → 4-methoxy-2-(dimethylaminomethyl)phenyl 2-phenylethyl ketone

80%

I.F.1-13 M. E. Neubert and D. L. Fishel, Org. Syn., 61, 8 (1983).

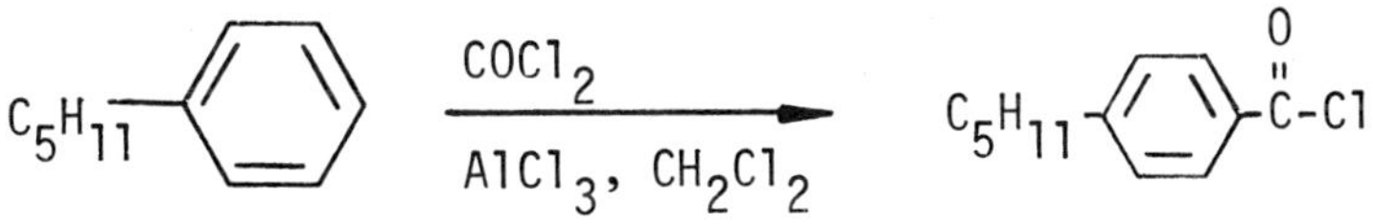

C_5H_{11}–C$_6$H$_5$ → ($COCl_2$, $AlCl_3$, CH_2Cl_2) → C_5H_{11}–C$_6$H$_4$–C(O)–Cl

Phosgene Prep: $(COCl)_2$ + $AlCl_3/CH_2Cl_2$.

I.F.1-14 G. A. Olah, M. Arvanaghi and V. V. Krishnamurthy, J. Org. Chem., 48, 3359 (1983).

PhO-C(=O)-R →[Nafion-H][PhNO$_2$, Δ] 2-hydroxyphenyl C(=O)-R

63-75%

Nafion-H = Solid superacid resin sulfonic acid catalyst.

I.F.1-15 P. Jacob, III and A. T. Shulgin, J. Med. Chem., 26, 746 (1983); F. W. Ullrich and E. Breitmaier, Synthesis, 641 (1983); M. Komiyama and H. Hirai, J. Am. Chem. Soc., 105, 2018 (1983); Y. Suzuki and H. Takahashi, Chem. Pharm. Bull, 31, 1751 (1983); J. Liebscher and U. Bechstein, Z. Chem., 23, 214 (1983).

4-MeO, 3-Me, 1-SMe-benzene →[CH$_3$OCHCl$_2$][AlCl$_3$] 4-MeO, 3-Me, 6-SMe, 1-CHO-benzene

11%

Vilsmeier-Haack reaction was ineffective.

I.F.1-16 A. McKillop, F. A. Madjdabadi and D. A. Long, Tetrahedron Lett., 24, 1933 (1983); D. R. Maulding, K. D. Lotts and S. A. Robinson, J. Org. Chem., 48, 2938 (1983).

4-R, 1-OMe-benzene →[MeOCH$_2$COCl][AlCl$_3$, CH$_3$NO$_2$ or CS$_2$] 4-R, 1-OMe, 2-CH$_2$Cl-benzene

70-99%

I.F.1-17 A. Pochini, G. Puglia and R. Ungaro, Synthesis, 906 (1983); A. K. Sinhababu and R. T. Borchardt, Synth. Commun., 13, 677 (1983).

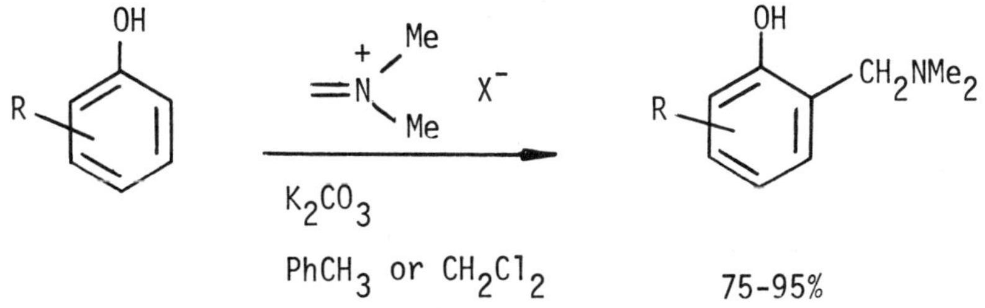

I.F.1-18 V. V. Mezheritskii, V. V. Tkachenko and O. N. Zhukovskaya, J. Org. Chem., 19, 360 (1983); K. Arai et al., Tetrahedron Lett., 24, 1531 (1983); A. M. Gorelik et al., J. Org. Chem. (USSR), 19, 183 (1983); S. Kurokawa and A. G. Anderson, Jr., Bull. Chem. Soc. Jpn., 56, 2059 (1983); P. L. Joshi, V. V. Dhekne and A. S. Rao, Ind. J. Chem., 22B, 23 (1983); S. Kumar and A. P. Bhaduri, ibid, 22B, 17 (1983).

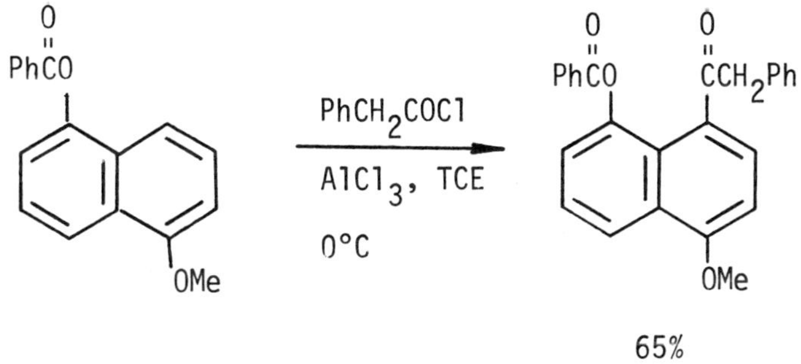

I.F.1-19 C. M. Wong, H. Y. Lam and W. Haque, Can. J. Chem., 61, 562 (1983); J. R. Merchant and R. B. Upasani, Chem. Ind. (London), 929 (1983); J. M. Aubry et al., Tetrahedron, 39, 623 (1983).

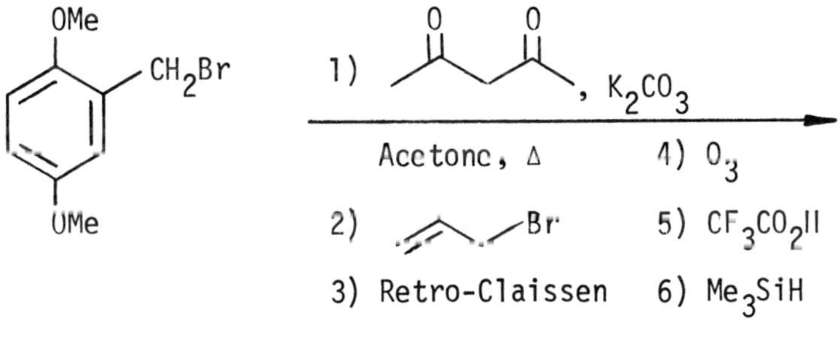

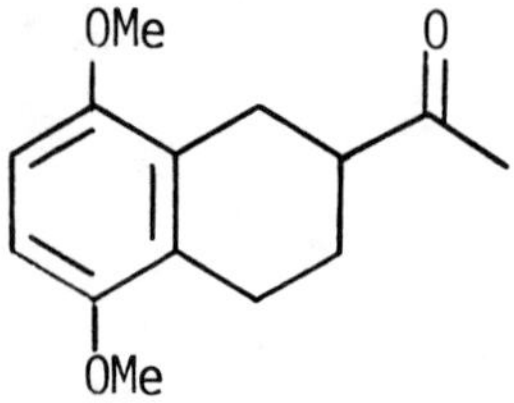

38 g Scale

I.F.1-20 A. M. Caporusso and L. Lardicci, J. Chem. Res. (S), 194 (1983); S. K. Taylor et al., J. Org. Chem., 48, 2449 (1983); J. H. Rigby, A. Kotnis and J. Kramer, Tetrahedron Lett., 24, 2939 (1983).

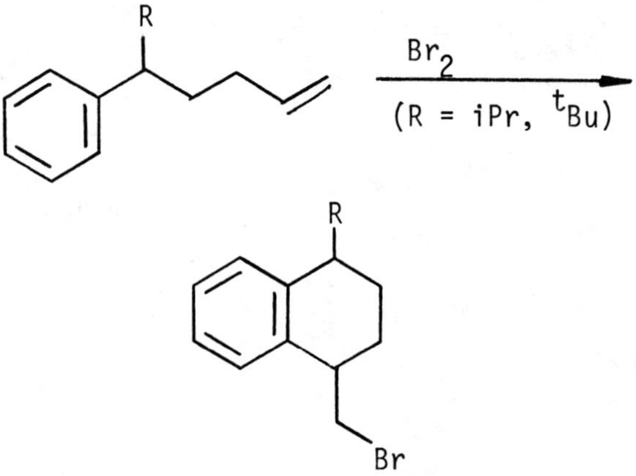

46-55%

I.F.1-21 E. E. Van Tamelen et al., J. Am. Chem. Soc., 105, 142 (1983); A. V. R. Rao, K. B. Reddy and A. R. Mehendale, Chem. Commun., 564 (1983).

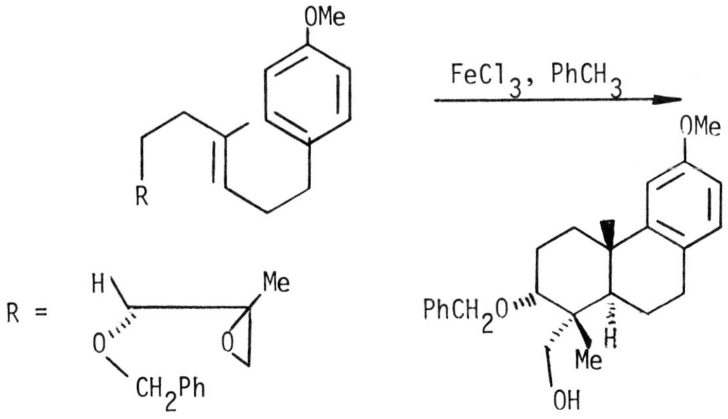

>55%

I.F.1-22 M. Shibuya, Tetrahedron Lett., 24, 1175 (1983); S. Fung et al., Can. J. Chem., 61, 368 (1983); L. M. Deck and G. H. Daub, J. Org. Chem., 48, 3577 (1983); M. S. Newman and S. Veeraraghavan, ibid, 48, 3246, 3249 (1983).

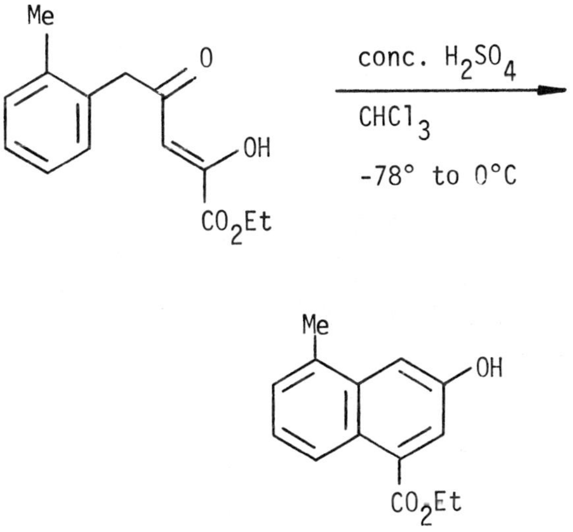

70%

I.F.1-23 B. M. Trost and J. I. Yoshida, Tetrahedron Lett., 24, 4895 (1983).

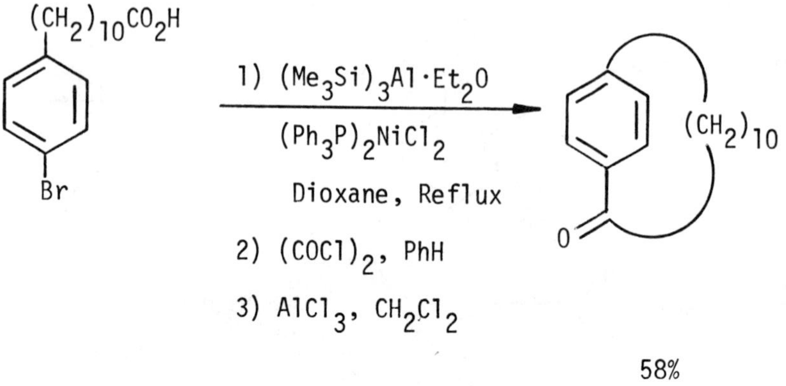

$(CH_2)_{10}CO_2H$
Br
1) $(Me_3Si)_3Al \cdot Et_2O$
$(Ph_3P)_2NiCl_2$
Dioxane, Reflux
2) $(COCl)_2$, PhH
3) $AlCl_3$, CH_2Cl_2
$(CH_2)_{10}$
O
58%

I.F.1-24 G. Casnati et al., Pure Appl. Chem., 55, 1677 (1983).

Review: "Template Catalysis via Non-Transition Metal Complexes. New Highly Selective Syntheses on Phenol Systems."

I.F.1-25 W. S. Murphy and S. Wattanasin, Chem. Soc. Rev., 12, 213 (1983).

Review: "Anionic Cyclization of Phenols."

I.F.1-26 K. Krohn and B. Sarstedt, Angew. Chem., Int. Ed. Engl., 22, 875 (1983); L. A. Mitscher, T. S. Wu and I. Khanna, Tetrahedron Lett., 24, 4809 (1983); A. V. R. Rao, B. Chanda and H. B. Borate, Ind. J. Chem., 22B, 521 (1983); A. Pochini et al., J. Org. Chem., 48, 3783 (1983).

O
O
OH
O
OH
CHO
Alkaline
Dithionite

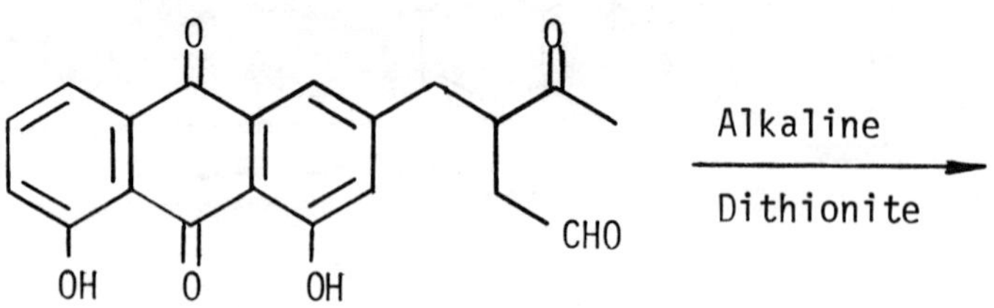

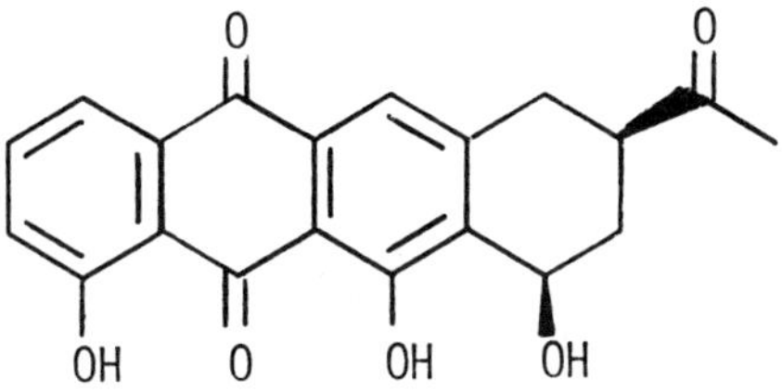

60%

I.F.2. Coupling Reactions to Form an Aromatic Carbon-Carbon Bond

I.F.2-1 S. Ozasa et al., Chem. Pharm. Bull., 31, 1572 (1983); E. Brown et al., Tetrahedron, 39, 2707, 2795 (1983); F. S. Tanaka, R. G. Wien and B. L. Hoffer, Synth. Commun., 13, 951 (1983).

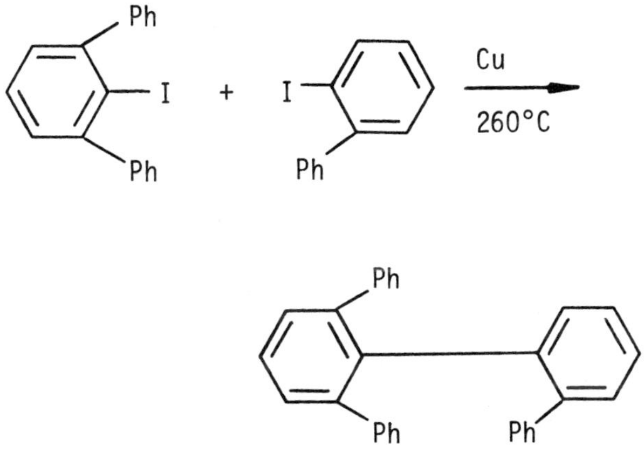

39%

I.F.2-2 M. Uchiyama, T. Suzuki and Y. Yamazaki, Chem. Lett., 1165, 1201 (1983); R. Nakajima et al., Bull. Chem. Soc. Jpn., 56, 1113 (1983).

$$ArI\text{-}Ar \;\; \overset{+}{X^-} \xrightarrow[\text{THF, RT}]{\text{Zn}, \text{ Pd-Cat.}} Ar\text{-}Ar$$

80-93% (GC)

I.F.2-3 C. S. Chao, C. H. Cheng and C. T. Chang, J. Org. Chem., 48, 4904 (1983); H. Matsumoto, S. I. Inaba and R. D. Rieke, ibid, 48, 840 (1983).

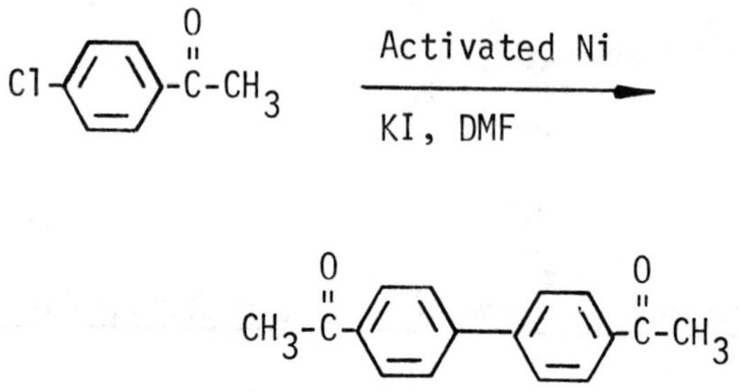

$$Cl-C_6H_4-C(O)-CH_3 \xrightarrow{\text{Activated Ni}, \text{KI, DMF}} CH_3-C(O)-C_6H_4-C_6H_4-C(O)-CH_3$$

72-90%

Unprecidented homo coupling of aryl chlorides to biphenyls.

I.F.2-4 R. A. Kjonaas and D. C. Shubert, J. Org. Chem., 48, 1924 (1983); A. K. Yatsimirsky, S. A. Deiko and A. D. Ryabov, Tetrahedron, 39, 2381 (1983).

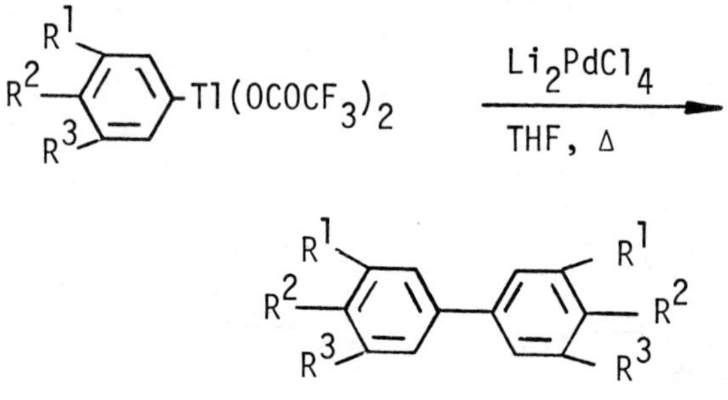

$$R^1R^2R^3C_6H_2-Tl(OCOCF_3)_2 \xrightarrow{Li_2PdCl_4, \text{THF}, \Delta} \text{biphenyl}(R^1,R^2,R^3)_2$$

20-97%

I.F.2-5 J. D. White, W. K. M. Chong and K. Thirring, J. Org. Chem., 48, 2300 (1983); L. Castedo et al., Tetrahedron Lett., 24, 5419 (1983); M. Sainsbury et al., J. Chem. Soc., Perkin I, 2053, 2059 (1983); M. Tashiro and T. Yamato, Chem. Commun., 617 (1983).

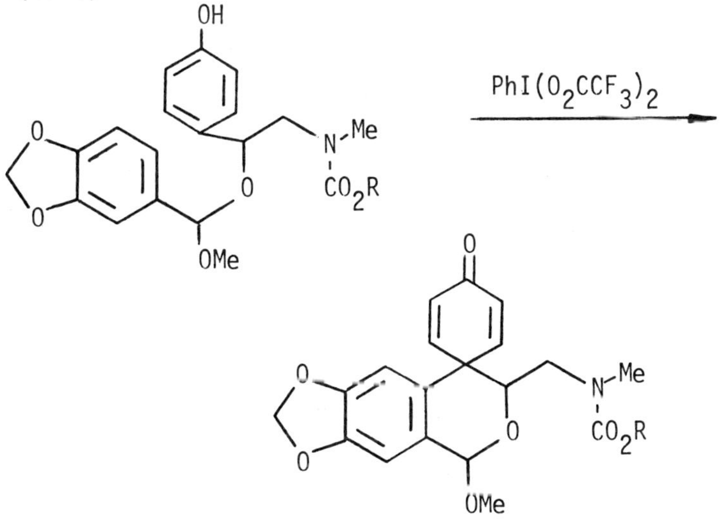

Further Reactions

I.F.2-6 J. Perichon et al., Chem. Commun., 793 (1983); Y. Xu et al., Synthesis, 556 (1983); A. Spencer, J. Organometal. Chem., 247, 117 (1983); ibid, 258, 101 (1983).

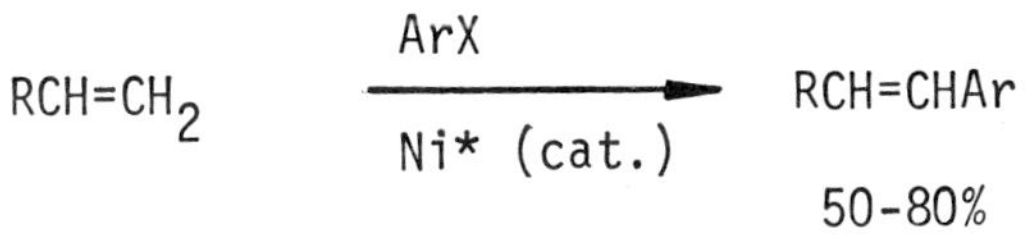

$$RCH=CH_2 \xrightarrow[\text{Ni* (cat.)}]{\text{ArX}} RCH=CHAr$$

50-80%

Ni* = Electrochemical generated organonickel-phosphine complex.

I.F.2-7 F. Naso et al., Tetrahedron Lett., 24, 4603 (1983); G. D. Hartman, R. D. Hartman and D. W. Cochran, J. Org. Chem., 48, 4119 (1983).

$$CH_2=C\begin{smallmatrix}CO_2H\\NHAc\end{smallmatrix} \quad \xrightarrow[\substack{Pd(OAc)_2 \text{ (cat.)}\\Ph_3P,\ 110°C\\Et_3N \text{ or TMEDA}}]{ArBr} \quad Ar-CH=C\begin{smallmatrix}CO_2H\\NHAc\end{smallmatrix} \quad 25-75\%$$

I.F.2-8 R. C. Larock, K. Narayanan and S. S. Hershberger, J. Org. Chem., 48, 4377 (1983); G. A. Molander et al., Tetrahedron Lett., 24, 5449 (1983).

$$ArHgCl \quad \xrightarrow[\substack{(X = Br \text{ or } I)\\ \text{cat. } (Ph_3P)_3RhCl\\HMPA}]{RCH=CHX} \quad RCH=CHAr \quad 31-70\%$$

I.F.2-9 K. Kikukawa et al., Chem. Lett., 1337 (1983); N. D. Obushak et al., J. Org. Chem. (USSR), 19, 1391 (1983).

$$\underset{H}{\overset{Ph}{\diagdown}}C=C\underset{H}{\overset{SiMe_3}{\diagup}} \quad \xrightarrow[\substack{Pd(dba)_2 \text{ (cat.)}\\CH_3CN,\ RT}]{ArN_2BF_4} \quad \underset{H}{\overset{Ph}{\diagdown}}C=C\underset{Ar}{\overset{H}{\diagup}} \quad 49-80\%$$

I.F.2-10 T. Yamamoto and Y. Kurata, Can. J. Chem., 61, 86 (1983); T. K. Dougherty, K. S. Y. Lau and F. L. Hedberg, J. Org. Chem., 48, 5273 (1983); E. T. Sabourin and A. Onopchenko, ibid, 48, 5135 (1983); K. Kikukawa et al., Bull. Chem. Soc. Jpn., 56, 961 (1983); H. Yamanaka et al., Synthesis, 312 (1983).

$$PhC\equiv CH \quad \xrightarrow[Cu,\ 155°C]{PhI} \quad PhC\equiv C-Ph \quad 66\%$$

I.F.3. Other Aromatic Substitutions

I.F.3-1 J. T. Pinhey et al., Aust. J. Chem., 36, 311, 789 (1983); Tetrahedron Lett., 24, 1301 (1983).

$$\text{Ph-C(O)-[furanone]} \xrightarrow[\text{CHCl}_3,\ C_5H_5N]{\text{ArPb(OAc)}_3} \text{Ph-C(O)-C(Ar)=CH}_2$$

62-93%

I.F.3-2 S. Cacchi and A. Arcadi, J. Org. Chem., 48, 4236 (1983); K. Kikukawa et al., ibid, 48, 1333 (1983); S. A. Buntin and R. F. Heck, Org. Syn., 61, 82 (1983).

$$\text{ArCH=CHCR(O)} \xrightarrow[\text{HCO}_2\text{H, Et}_3\text{N}]{\text{PhI, Pd(0) (cat.)}} \underset{\text{Ph}}{\overset{\text{Ar}}{>}}\text{CH-CH}_2\text{-C(O)-R}$$

21-61%

I.F.3-3 C. E. Russell and L. S. Hegedus, J. Am. Chem. Soc., 105, 943 (1983); E. I. Negishi et al., Tetrahedron Lett., 24, 3823 (1983).

$$\text{CH}_2\text{=C(OEt)(H)} \xrightarrow[\text{THF, 25°C}]{\substack{1)\ ^t\text{BuLi, -70°C} \\ 2)\ \text{ZnCl}_2 \\ 3)\ \text{ArX, Cat. Pd}^\circ}} \text{Ar-C(O)-CH}_3$$

33-84%

Also, alkenyl halides. Allenic ethers give α,β-unsaturated ketones.

I.F.3-4 T. Migita et al., Chem. Commun., 344 (1983); A. Osuka, T. Kobayashi and H. Suzuki, Synthesis, 67 (1983); Chem. Lett., 589 (1983).

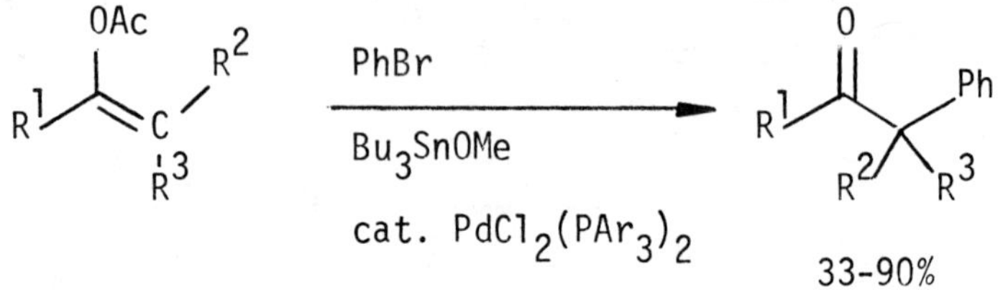

R^2, R^3 = H,H; H,Me; Me,Me.

I.F.3-5 M. F. Semmelhack et al., J. Am. Chem. Soc., 105, 2034 (1983); Organometallics, 2, 467 (1983); E. P. Kundig, V. Desobry and D. P. Simmons, J. Am. Chem. Soc., 105, 6962 (1983).

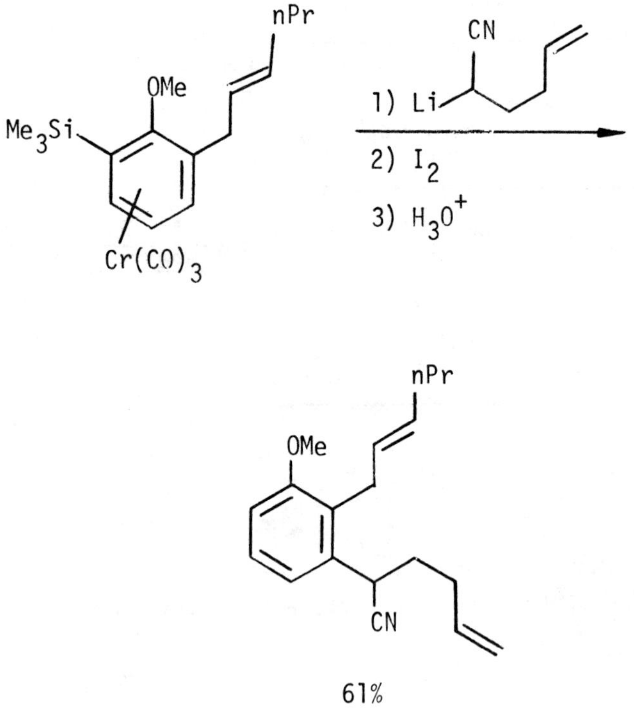

I.F.3-6 H. Yamamoto et al., J. Am. Chem. Soc., 105, 7177 (1983).

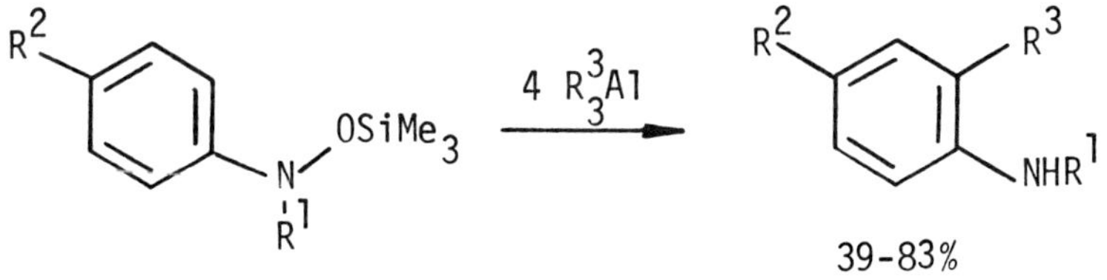

R^2-C₆H₃(NR¹-OSiMe₃) →[4 R³₃Al] R^2,R^3-C₆H₃-NHR¹

39-83%

I.F.3-7 J. E. McMurry and S. Mohanraj, Tetrahedron Lett., 24, 2723 (1983).

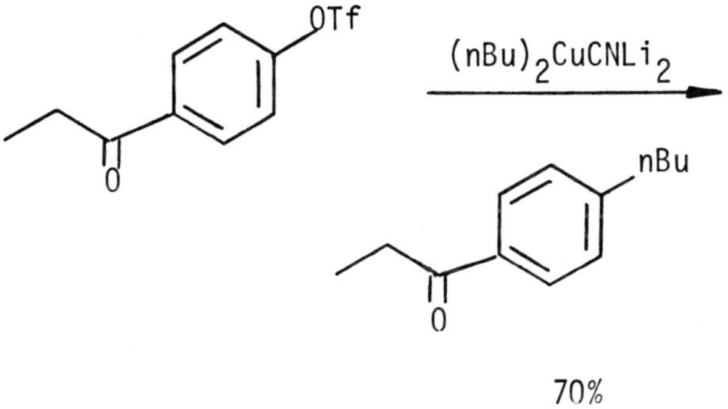

aryl-OTf propiophenone →[(nBu)₂CuCNLi₂] 4-nBu-propiophenone

70%

I.F.3-8 H. C. Brown, S. C. Kim and S. Krishnamurthy, Organometallics, 2, 779 (1983).

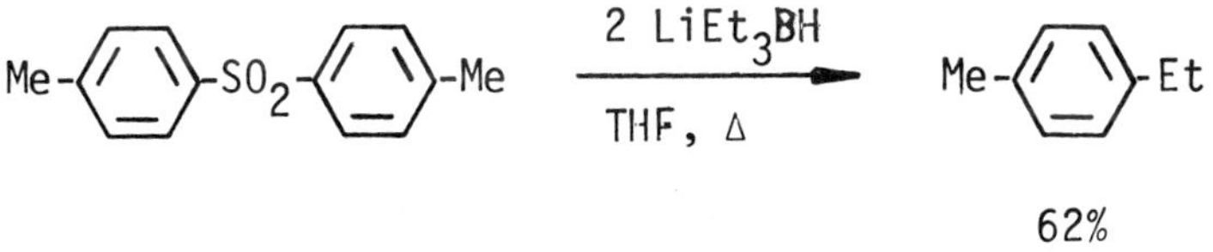

Me-C₆H₄-SO₂-C₆H₄-Me →[2 LiEt₃BH / THF, Δ] Me-C₆H₄-Et

62%

I.F.3-9 J. F. Wolfe et al., J. Org. Chem., 48, 1180, 2392 (1983).

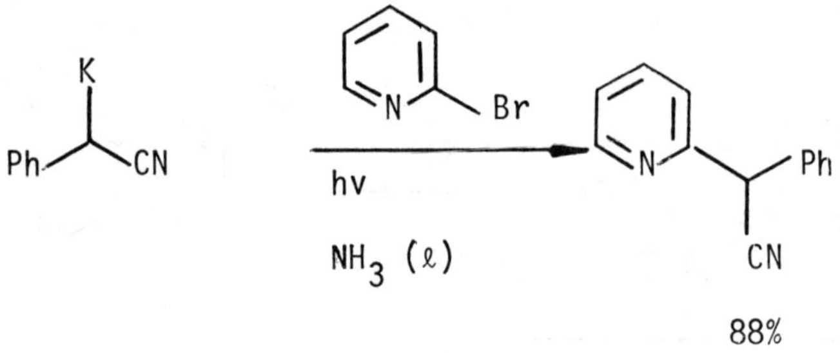

Ph-CH(K)-CN + 2-Br-pyridine $\xrightarrow[NH_3 (\ell)]{h\nu}$ 2-pyridyl-CH(Ph)(CN) 88%

I.F.3-10 T. Klingstedt and T. Frejd, Organometallics, 2, 598 (1983); E. Wenkert et al., Synthesis, 701 (1983); M. Tiecco, E. Wenkert et al., Tetrahedron, 39, 2289 (1983).

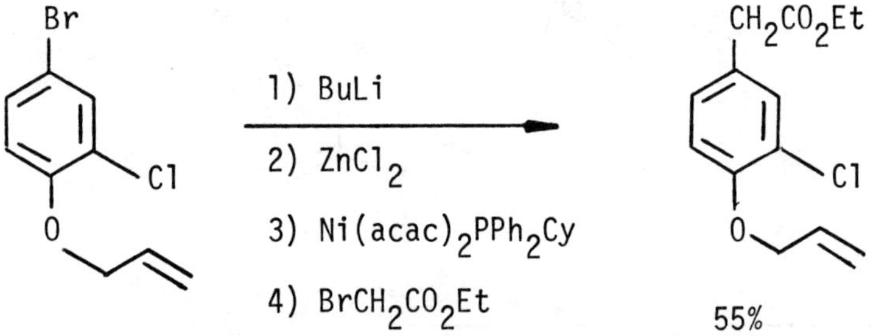

4-Br-2-Cl-1-(allyloxy)benzene
1) BuLi
2) $ZnCl_2$
3) $Ni(acac)_2 PPh_2Cy$
4) $BrCH_2CO_2Et$
→ 4-(CH_2CO_2Et)-2-Cl-1-(allyloxy)benzene   55%

I.F.3-11 S. D. Lee, M. A. Brook and T. H. Chan, Tetrahedron Lett., 24, 1569 (1983).

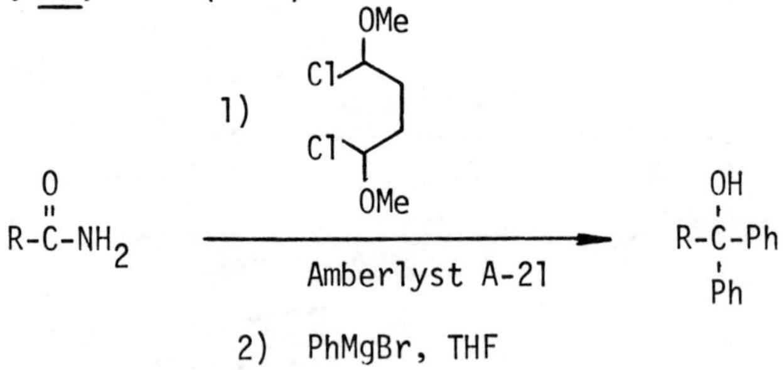

$R-\underset{\underset{}{\|}}{C}(=O)-NH_2$
1) Cl-CH(OMe)-CH_2-CH_2-CH(OMe)-Cl, Amberlyst A-21
2) PhMgBr, THF
→ R-C(OH)(Ph)(Ph)

Via acylpyrroles

I.F.3-12 J. L. Luche et al., <u>J. Org. Chem.</u>, <u>48</u>, 3837 (1983);
D. E. Bergstrom, M. W. Ng and J. J. Wong, <u>ibid</u>, <u>48</u>, 1902
(1983); I. W. Lawston and T. D. Inch, <u>J. Chem. Soc., Perkin
I</u>, 2629 (1983).

Ar-Br $\xrightarrow[\text{Ultrasound}]{\text{1) Li, ZnBr}_2 \text{ THF}}$

2) $R^1\diagdown_{R^2}\diagup^{R^3}\diagdown_{R^4}=O$

$\underset{Ar}{\overset{R^1\; R^2}{\diagdown\diagup}}\underset{R^3}{\diagdown}\overset{O}{\diagup}R^4$

70-98%

I.F.3-13 R. M. Carlson et al., <u>Synth. Commun.</u>, <u>13</u>, 21 (1983);
C. Tintel, et al., <u>Rec. Trav. Chim.</u>, <u>102</u>, 14, 224, 228 (1983).

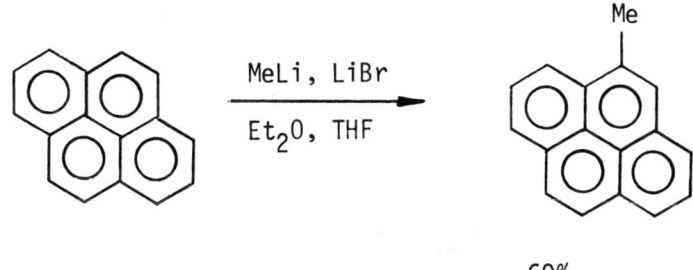

60%

I.F.3-14 M. Makosza et al., <u>Synthesis</u>, 40, 1023 (1983);
<u>Tetrahedron Lett.</u>, <u>24</u>, 3277, 3279 (1983); P. J. Atkins and
V. Gold, <u>Chem. Commun.</u>, 140 (1983); L. M. Gnanadoss and N.
Radha, <u>J. Org. Chem.</u>, <u>48</u>, 570 (1983).

Ar-NO$_2$ $\xrightarrow[\text{NaOH, DMSO}]{\overset{R^2}{\underset{|}{Cl-CH-SO_3R^1}}}$ $O_2N-\underset{}{\bigcirc}-\underset{R^2}{\overset{R^3}{CH-SO_3R^1}}$

(R_1 = neo-C_5H_{11})

42-83%

(ortho + para)

I.F.3-15   N. S. Narasimhan and R. S. Mali, Synthesis, 957 (1983).

Review: "Synthesis of Heterocyclic Compounds Involving Aromatic Lithiation Reactions in the Key Step."

I.F.3-16   Tetrahedron, 39, Issue 12 (1983).

Reviews: "Heteroatom-Directed Metallations in Heterocyclic Synthesis."
(Tetrahedron Report - 14 Papers.)

I.F.3-17   A. I. Meyers and P. D. Pansegrau, Tetrahedron Lett., 24, 4935 (1983).

[Scheme: 3-chlorophenyl oxazoline → 1) 3 nBuLi, -78° to RT; 2) E$^+$; 3) H$^+$ → 2-nBu-3-E-benzoic acid (CO_2H), 53-70%]

I.F.3-18   L. S. Chen, G. J. Chen and C. Tamborski, J. Organometal. Chem., 251, 139, 149 (1983); H. Hart and G. C. Nwokogu, Tetrahedron Lett., 24, 5721, 5725 (1983); A. K. Sinhababu and R. T. Borchardt, J. Org. Chem., 48, 2356 (1983); W. Neugebauer, T. Clark and P. v. R. Schleyer, Chem. Ber., 116, 3283 (1983); J. Novak and C. A. Salemink, Synthesis, 597 (1983); L. A. Cate, ibid, 385 (1983).

[Scheme: 1,4-dibromobenzene → 1) nBuLi, -78°C; 2) CF_3CO_2Me; 3) nBuLi; 4) E$^+$; 5) H$^+$ → 4-E-C$_6$H$_4$-C(O)-CF$_3$, 65-86%]

E^+ = MeI, CO_2, CF_3CO_2Me, MeCONMe$_2$, HCONMe$_2$, S, H_2O.

I.F.3-19   D. L. Comins and J. D. Brown, Tetrahedron Lett., 24, 5465 (1983);   L. Dashan and S. Trippett, ibid, 24, 2039 (1983); A. I. Meyers and K. A. Lutomski, Synthesis, 105 (1983).

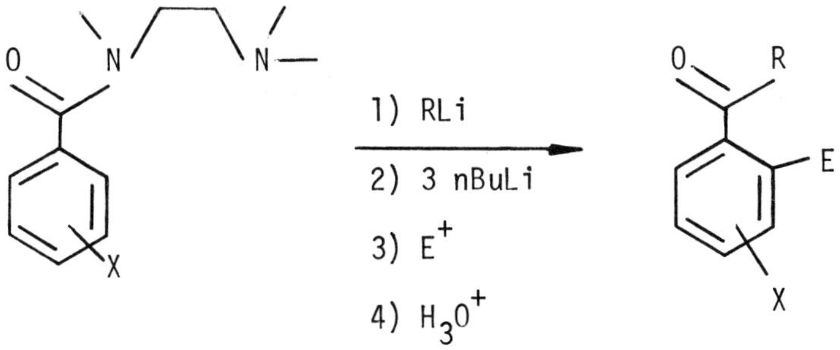

E^+ = MeI, MeSSMe, Me$_3$SiCl

33-87%

I.F.3-20   V. Snieckus et al., J. Org. Chem., 48, 1565, 1935 (1983);   E. Napolitano et al., ibid, 48, 3653 (1983).

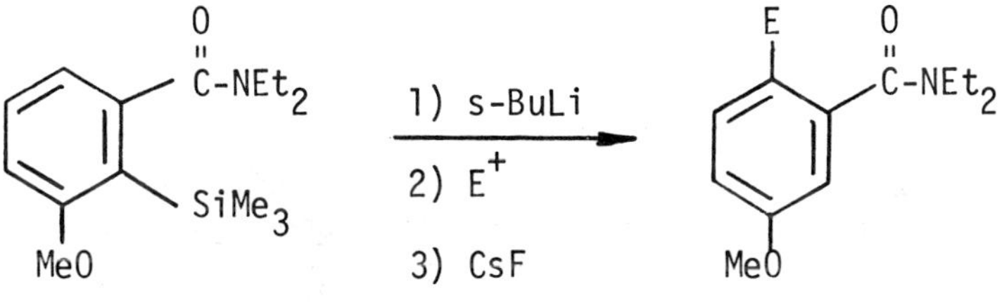

E^+ = MeI, DMF, ClCONEt$_2$.

63-75%

I.F.3-21 M. Watanabe, H. Maenosono and S. Furukawa, Chem. Pharm. Bull, 31, 2662 (1983); S. Penco et al., J. Org. Chem., 48, 405 (1983); J. K. Ray and R. G. Harvey, ibid, 48, 1352 (1983); J. S. Swenton et al., Tetrahedron Lett., 24, 1329 (1983).

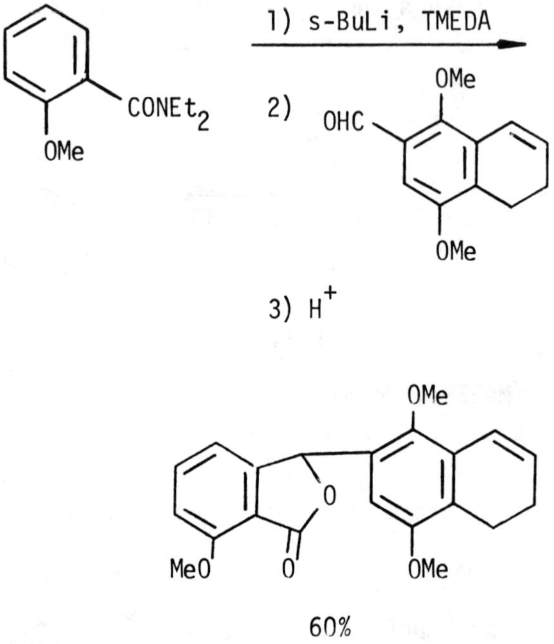

60%

I.F.3-22 M. Uemura et al., J. Org. Chem., 48, 2349 (1983); M. Uemura, K. Take and Y. Hayashi, Chem. Commun., 858 (1983); M. Fukui, T. Ikeda and T. Oishi, Chem. Pharm. Bull., 31, 466 (1983); N. F. Masters and D. A. Widdowson, Chem. Commun., 955 (1983).

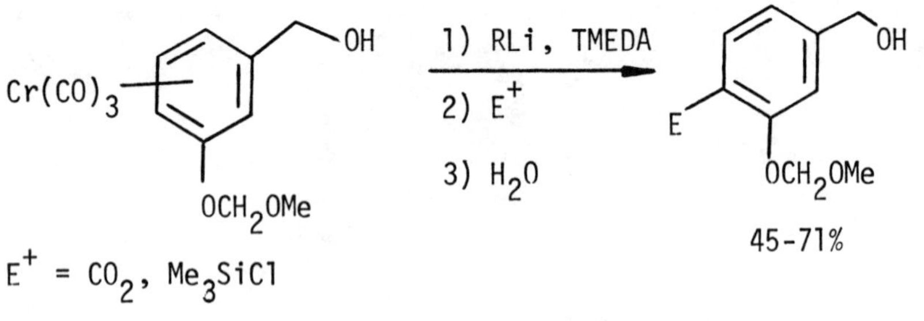

E^+ = CO_2, Me_3SiCl

45-71%

I.F.3-23 J. A. Turner, J. Org. Chem., 48, 3401 (1983); S. L. Taylor, D. Y. Lee and J. C. Martin, ibid, 48, 4156 (1983); M. Iwao and T. Kuraishi, Tetrahedron Lett., 24, 2649 (1983); D. J. Ager, ibid, 24, 5441 (1983); J. Epsztajn et al., ibid, 24, 4735 (1983); P. Breant, F. Marsais and G. Queguiner, Synthesis, 822 (1983).

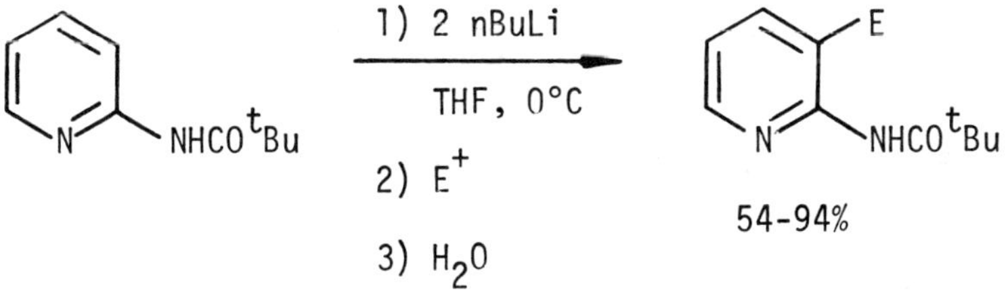

E^+ = D_2O, Me_3SiCl, MeI, PhCHO, $ClCO_2Et$, $HCONMe_2$, MeSSMe.

I.F.3-24 T. H. Chan and G. J. Kang, Tetrahedron Lett., 24, 3051 (1983); T. H. Chan et al., Can. J. Chem., 61, 688 (1983).

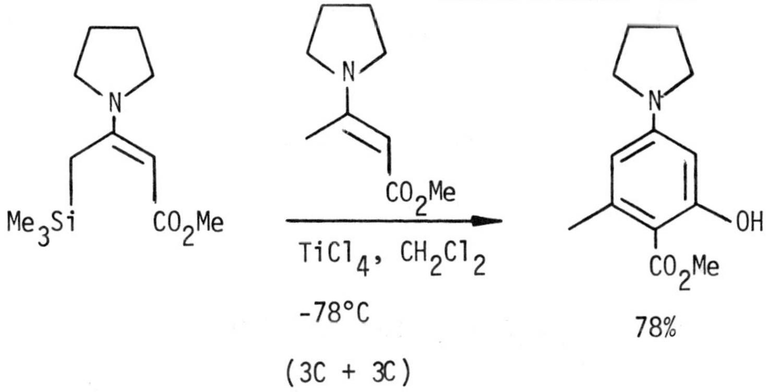

(3C + 3C)

Enamines from cyclic ketones give aromatic compounds by 4C + 2C annulation.

I.F.3-25 M. A. Tius et al., Synthesis, 467 (1983); J. Org. Chem., 48, 3839 (1983).

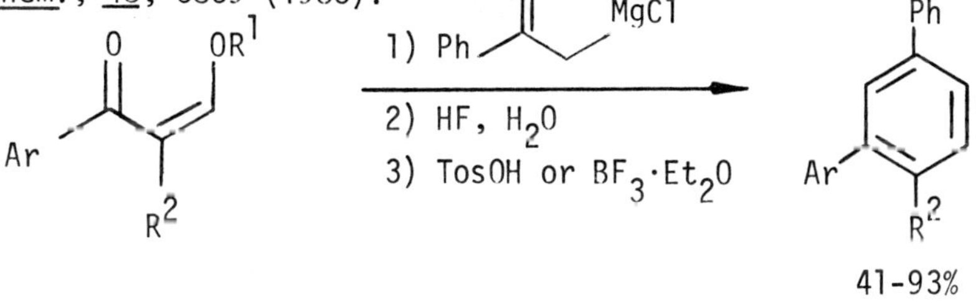

I.F.3-26 R. D. Sands, J. Org. Chem., **48**, 3362 (1983); R. O. Duthaler, Helv. Chim. Acta, **66**, 1475, 2543 (1983); S. Tobinaga et al., Chem. Pharm. Bull., **31**, 4355, 4360 (1983); Y. M. Dangyan et al., J. Gen. Chem. (USSR), **53**, 819 (1983).

$$E-CH_2-\overset{O}{\underset{\|}{C}}-CH_2-E \quad + \quad R^1-\overset{O}{\underset{\|}{C}}-CH_2-\overset{O}{\underset{\|}{C}}-R^2 \xrightarrow{KOH, MeOH} \text{HO}-\underset{E}{\overset{E}{\bigcirc}}\begin{matrix}R^1\\R^2\end{matrix}$$

Bicyclo[3.3.1] nonanes formed in two cases.

I.F.3-27 J. B. Dickenson and W. Reusch, Synth. Commun., **13**, 303 (1983); S. Trippett et al., J. Chem. Res. (S), 14 (1983).

Reagents:
1) PhSOCH$_2$Li
2) H$_3$O$^+$
3) LDA
4) CH$_2$=CHCO$_2$Me
5) THF, Δ

67%

I.F.3-28 J. Sepiol, Synthesis, **504**, 559 (1983).

Reagents:
1) CH$_2$(CN)$_2$, NH$_4$OAc, AcOH
2) H$_2$SO$_4$

76%

I.F.3-29 S. Auricchio, A. Ricca and O. V. de Pava, J. Org. Chem., 48, 602 (1983); L. Anandan, H. H. Mathur and G. K. Trivedi, Ind. J. Chem., 22B, 542 (1983).

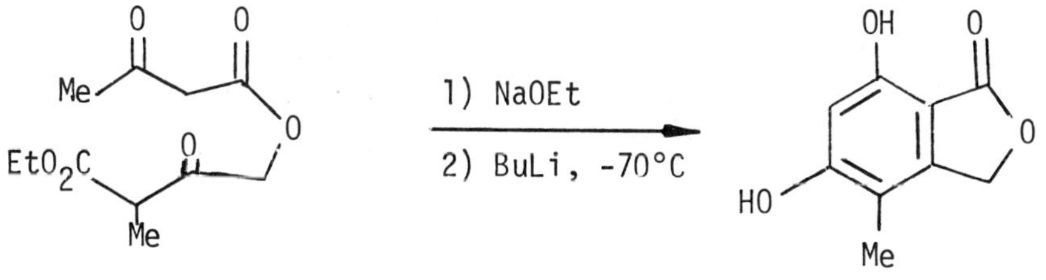

[Reaction: β-ketoester substrate treated with 1) NaOEt, 2) BuLi, −70°C yields a hydroxy-methyl isobenzofuranone, 23%]

I.F.3-30 P. v. R. Schleyer et al., Chem. Ber., 116, 751 (1983).

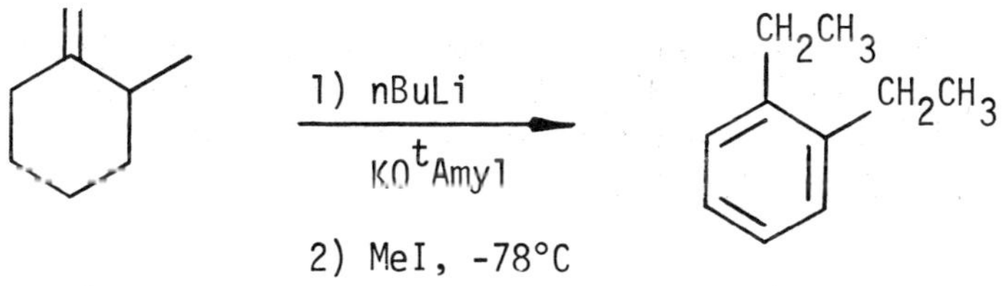

[Reaction: methylenemethylcyclohexane treated with 1) nBuLi / KO^{t}Amyl, 2) MeI, −78°C gives 1,2-diethylbenzene]

I.F.3-31 B. C. Berris, G. H. Hovakeemian and K. P. C. Vollhardt, Chem. Commun., 502 (1983).

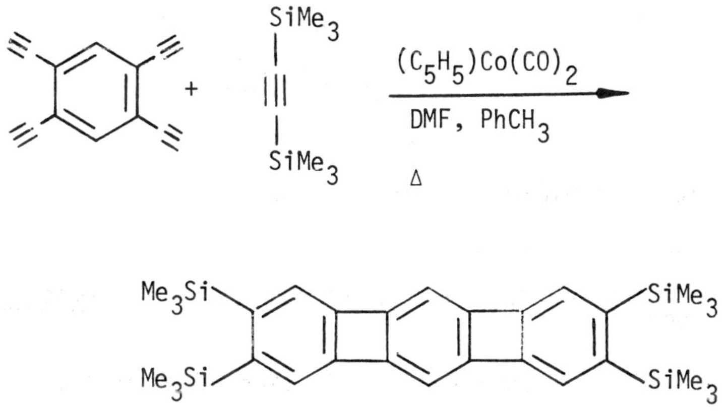

[Reaction: 1,2,4,5-tetraethynylbenzene + bis(trimethylsilyl)acetylene with $(C_5H_5)Co(CO)_2$ in DMF, PhCH$_3$, Δ gives tetrakis(trimethylsilyl)-substituted linear [3]phenylene, 71%]

I.F.3-32 D. N. Nicolaides and K. E. Litinas, J. Chem. Res. (S), 57 (1983).

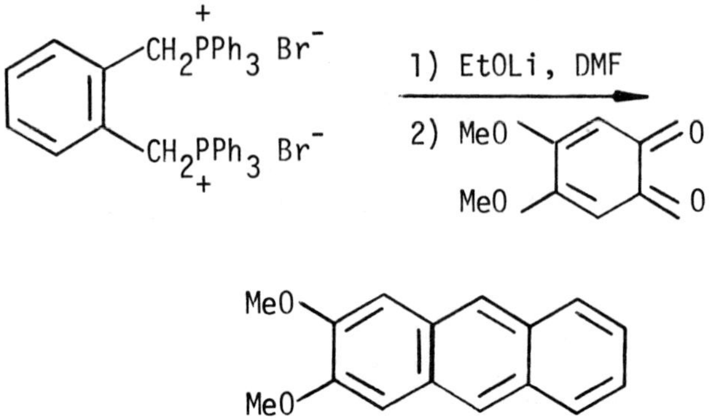

[o-xylylene bis(triphenylphosphonium) dibromide]

1) EtOLi, DMF
2) MeO-substituted bis-methylene cyclohexadienedione

→ 2,3-dimethoxyanthracene

8-20%

I.F.3-33 M. Azadi-Ardakani and T. W. Wallace, Tetrahedron Lett., 24, 1829 (1983); A. Hussain and J. Parrick, ibid, 24, 609 (1983); O. Abou-Teim, M. C. Goodland and J. F. W. McOmie, J. Chem. Soc., Perkin I, 2659 (1983).

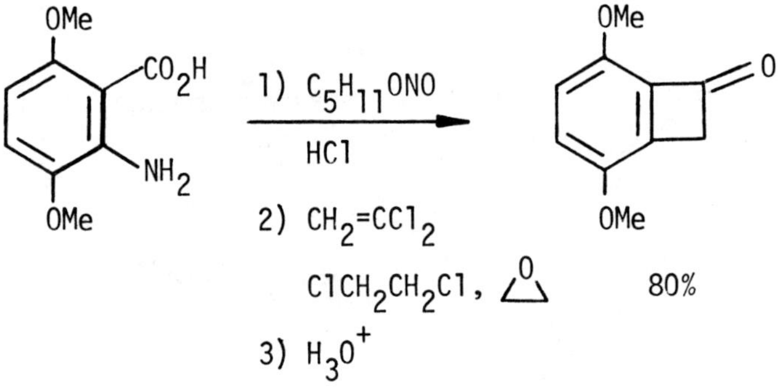

3,6-dimethoxyanthranilic acid

1) $C_5H_{11}ONO$, HCl
2) $CH_2=CCl_2$, $ClCH_2CH_2Cl$, Δ
3) H_3O^+

→ 3,6-dimethoxybenzocyclobutenone   80%

Product a versatile quinone precursor.

I.F.3-34 H. N. C. Wong, T. K. Ng and T. Y. Wong, Heterocycles, 20, 1815 (1983).

Review: "Arene Synthesis by Extrusion of Heteroatoms from 7-Hetero-Bicyclo[2.2.1]Heptene Systems."

I.F.3-35 L. Boisvert and P. Brassard, Tetrahedron Lett., 24, 2453 (1983); T. H. Kim and S. Isoe, Chem. Lett., 539 (1983).

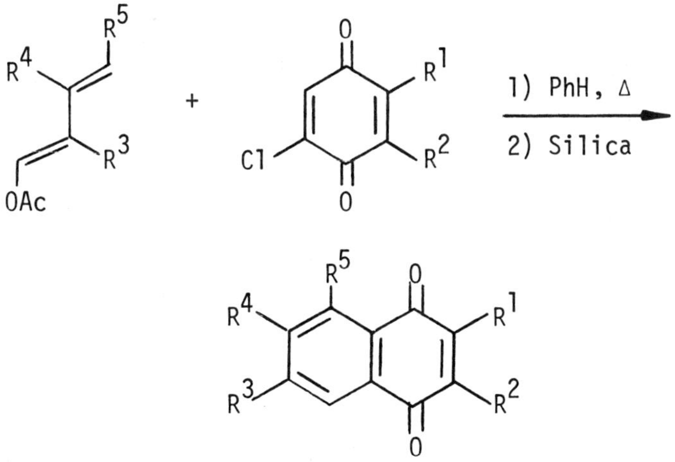

65-96%

I.F.3-36 O. Tsuge, E. Wada and S. Kanemasa, Chem. Lett., 239 (1983); J. P. Gesson, J. C. Jacquesy and M. Mondon, Nouv. J. Chem., 7, 205 (1983); T. R. Kelly et al., Tetrahedron Lett., 24, 2331 (1983).

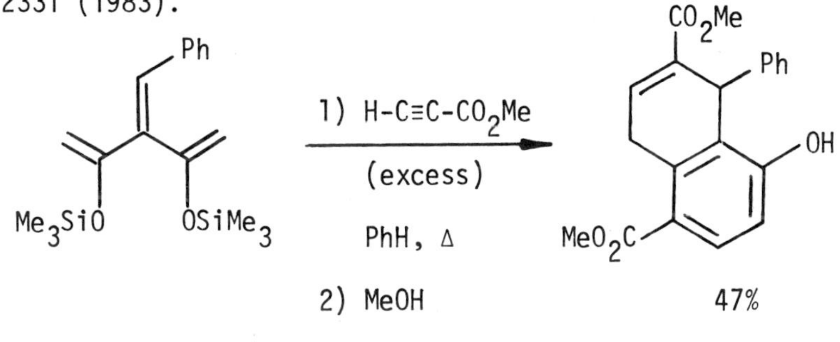

47%

I.F.3-37 A. J. Guilford and R. W. Turner, Chem. Commun., 466 (1983); L. A. Levy and S. Kumar V. P., Tetrahedron Lett., 24, 1221 (1983); G. W. Gribble et al., Synthesis, 502 (1983); C. S. LeHoullier and G. W. Gribble, J. Org. Chem., 48, 2364 (1983); S. Mondal, T. K. Bandyopadhyay and A. J. Bhattacharya, Ind. J. Chem., 22B, 225, 448 (1983); L. A. Levy, Synth. Commun., 13, 639 (1983); R. K. Boeckman, Jr. and S. H. Cheon, J. Am. Chem. Soc., 105, 4112 (1983);

furan + 1) PhSO$_2$CH=C=CH$_2$ 2) nBuLi 3) H$_2$O → 2-methyl-3-(phenylsulfonyl)phenol 24%

I.F.3-38 C. S. Le Houllier and G. W. Gribble, J. Org. Chem., 48, 1682 (1983); Y. Himeshima, T. Sonoda and H. Kobayashi, Chem. Lett., 1211 (1983).

1,5-bis(tosyloxy)-2,6-dibromonaphthalene + 1) PhLi, 1,2,5-trimethylpyrrole; 2) MCPBA, CHCl$_3$ → 1,4,7,10-tetramethylchrysene 52%

I.F.3-39 R. E. Ireland et al., J. Am. Chem. Soc., 105, 1988
(1983); Y. Tamura et al., Chem. Pharm. Bull., 31, 2691 (1983);
D. L. Boger and M. D. Mullican, Tetrahedron Lett., 24, 4939
(1983); L. Birkofer and B. Wahle, Chem. Ber., 116, 3309
(1983); F. Farina et al., J. Org. Chem., 48, 5373 (1983); H.
L. Gingrich, D. M. Roush and W. A. Van Saun, ibid, 48, 4869
(1983).

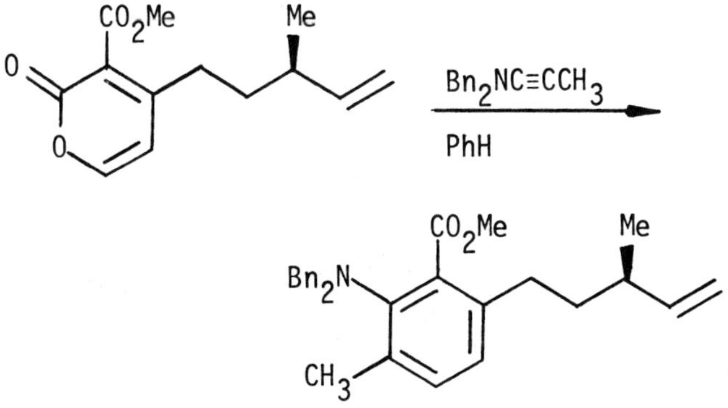

$$\text{pyranone} \xrightarrow[\text{PhH}]{Bn_2NC\equiv CCH_3} \text{aniline product}$$

60-80%

I.F.3-40 G. A. Kraus et al., J. Org. Chem., 48, 3439 (1983);
F. M. Hauser, S. Prasanna and D. W. Combs, ibid, 48, 1328
(1983).

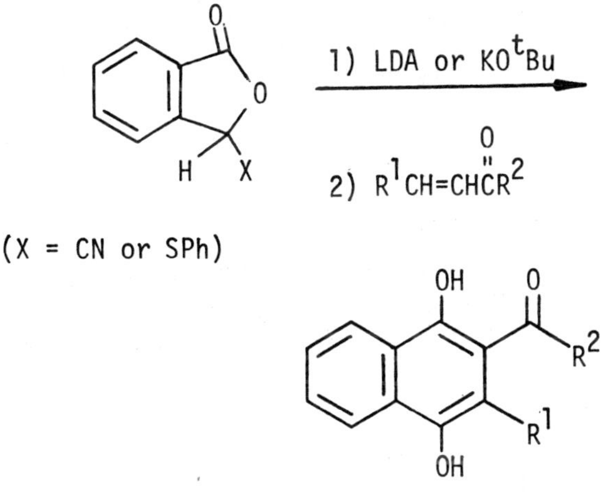

(X = CN or SPh)

1) LDA or KO^{t}Bu
2) R^{1}CH=CHĊR^2 (with C=O)

→ 1,4-dihydroxynaphthalene with R^1, R^2

48-85%

I.F.3-41 T. Kometani, Y. Takeuchi and E. Yoshii, J. Org. Chem., 48, 2630 (1983); T. Kumamoto et al., Bull. Chem. Soc. Jpn., 56, 1665 (1983); U. M. Dzhemilev et al., J. Org. Chem. (USSR), 19, 920 (1983).

[Naphthyl-O-CH(CO$_2$Me)-CH$_2$CH$_2$-SPh]

1) MCPBA, 0°C
2) 200°C

[1-hydroxy-2-(CH$_2$CH=CHCO$_2$Me)-naphthalene]

59%

I.F.3-42 T. J. Lee and W. J. Holtz, Tetrahedron Lett., 24, 2071 (1983).

[2,6-dichlorobenzyl-SMe$_2^+$ BF$_4^-$]

NaH, HMPA

NaCH(CO$_2$Me)$_2$

[2-Cl-6-(CH$_2$SMe)-C$_6$H$_3$-CH$_2$CH(CO$_2$Me)$_2$]

39%

I.F.3-43 J. J. Brunet, C. Sidot and P. Caubere, J. Org. Chem., 48, 1166 (1983); Y. Fujiwara et al., J. Organometal. Chem., 256, C35 (1983); Z. I. Yoshida et al., Tetrahedron Lett., 24, 3869 (1983).

$$\text{Ar-X} \xrightarrow[\substack{\text{PhH, 5N NaOH} \\ \text{Bu}_4\text{N}^+ \text{Br}^-, 65°C, h\nu}]{\text{Co}_2(\text{CO})_8, \text{CO}} \text{Ar-CO}_2\text{Na}$$

17-97%

I.F.3-44 K. Suzuki, E. Katayama and G. I. Tsuchihashi, Tetrahedron Lett., 24, 4997 (1983); K. Fujii, K. Nakao and T. Yamauchi, Synthesis, 444 (1983).

[Reaction: diol with Me, Ar, R substituents → 1) MsCl, Et₃N; 2) Et₃Al, -78°C → ketone product, 75-96%]

I.F.3-45 B. Kumar and N. Kaur, J. Org. Chem., 48, 2281 (1983); S. C. Roy and U. R. Ghatak, J. Chem. Res. (S), 138 (1983); K. Tsujimoto, K. Abe and M. Ohashi, Chem. Commun., 984 (1983).

[Reaction: benzophenone oxime → hv, CH₃OH → hydroxyfluorene, 34%]

I.F.3-46 D. Armesto et al., Tetrahedron Lett., 24, 1089, 1197 (1983); A. Padwa, M. J. Pulwer and R. J. Rosenthal, ibid, 24, 983 (1983); J. Fuhrmann, T. Dietzsch and H. G. Henning, Z. Chem., 23, 52 (1983).

[Reaction: Ph-C(=O)-C(Ph)(Ph)-N=C(Ph)Ph → 1) hv, tBuOH; 2) Silica Gel → ortho-dibenzoyl product]

1,3 or 1,5-Benzoyl Migration

I.F.3-47 R. C. Larock and K. Takagi, Tetrahedron Lett., 24, 3457 (1983); S. Cacchi and G. Palmieri, Tetrahedron, 39, 3373 (1983); T. D. Lee and G. D. Davies, Jr., J. Org. Chem., 48, 399 (1983).

$$\text{CH}_2=\text{CHCH}_2\text{CH}_2\text{CH}=\text{CH}_2 \xrightarrow[\text{Li}_2\text{PdCl}_4]{\text{PhHgCl}} \text{Ph(CH}_2)_2\text{CH} \cdots \text{CH}_2$$
$$\underset{\text{Cl/2}}{\overset{|}{\text{Pd}}}$$

61%

I.F.3-48 M. Catellani and G. P. Chiusoli, et al., J. Organometal. Chem., 247, C59 (1983); Tetrahedron Lett., 24, 813 (1983); G. Consiglio, F. Morandini and O. Piccolo, Tetrahedron, 39, 2699 (1983); T. Y. Luh, K. S. Lee and S. W. Tam, J. Organometal. Chem., 248, 221 (1983).

norbornene $\xrightarrow[\text{HC(OEt)}_3]{\text{PhBr, Pd cat.}}$ product

34%

I.F.3-49 T. Funabiki, H. Nakamura and S. Yoshida, J. Organometal. Chem., 243, 95 (1983); H. Vorbruggen and K. Krolikiewicz, Synthesis, 316 (1983); W. K. Fife, J. Org. Chem., 48, 1375 (1983); N. Hata and N. Nishida, Chem. Lett., 1043 (1983); J. M. Adam and T. Winkler, Helv. Chim. Acta, 66, 411 (1983).

1-bromonaphthalene $\xrightarrow[\text{KOH, H}_2\text{O, H}_2]{\text{KCN, CoCl}_2 \text{ (cat.)}}$ 1-cyanonaphthalene

78%

I.G. Synthesis via Organometallics

I.G.1 Synthesis via Organoboranes

I.G.1-1 A. Suzuki, Top. Curr. Chem., 112, 67 (1983).
Review: "Some Aspects of Organic Synthesis Using Organoborates."

I.G.1-2 S. Hara, H. Dojo and A. Suzuki et al., Chem. Lett., 285, 933, 1125 (1983); Synth. Commun., 13, 367, 1149 (1983).

$$R_3^{\bar{}}B-C\equiv C-R^2 \quad \xrightarrow[\text{2) [O]}]{\text{1) } R^3C(OR^4)_3, \text{ TiCl}_4} \quad \begin{array}{c} R^1 \\ R^1 \end{array}\!\!C=C\!\!\begin{array}{c} R^2 \\ \underset{\|}{C}-R^3 \\ O \end{array}$$

33-100% (GC)

I.G.1-3 A. Pelter et al., Tetrahedron Lett., 24, 623, 627, 631, 635, 637 (1983); A. Pelter, S. Singaram and H. C. Brown, ibid, 24, 1433 (1983).

Mes$_2$B-CH$_2$-G   Dimesitylboron Group in Organic Synthesis: Formation of carbanion α to boron and "Boron Wittig" (G = H, R, CH=CH$_2$, MR$_n$ where M is Si, Sn, Pb, S, Hg).

I.G.1-4 H. J. Bestmann and T. Roder, Angew. Chem., Int. Ed. Engl., 22, 782 (1983).

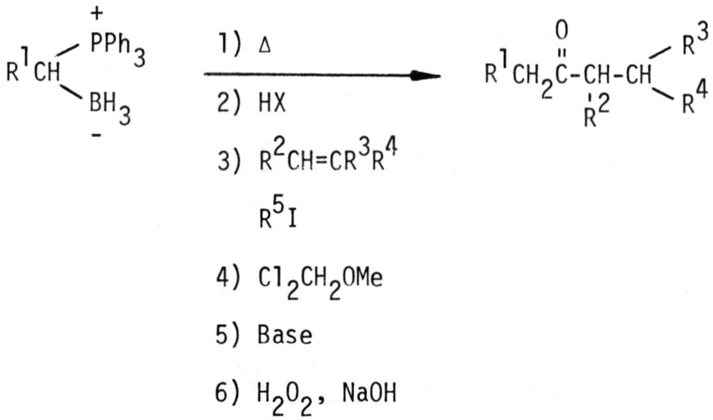

$$R^1CH \begin{matrix} +PPh_3 \\ BH_3^- \end{matrix} \quad \begin{matrix} 1)\ \Delta \\ 2)\ HX \\ 3)\ R^2CH=CR^3R^4 \\ \quad R^5I \\ 4)\ Cl_2CH_2OMe \\ 5)\ Base \\ 6)\ H_2O_2,\ NaOH \end{matrix} \quad R^1CH_2\overset{O}{\underset{}{C}}-\underset{R^2}{CH}-CH\begin{matrix}R^3\\R^4\end{matrix}$$

I.G.1-5 E. I. Negishi and H. C. Brown, Org. Syn., 61, 103 (1983).

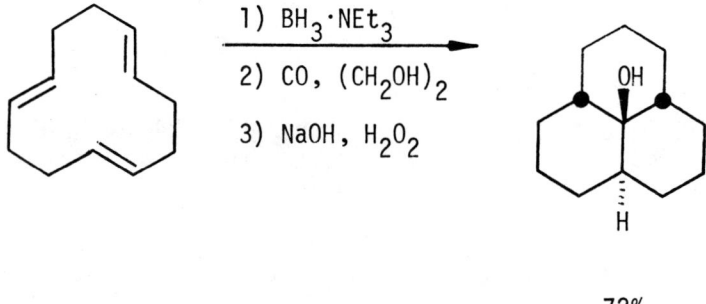

72%

I.G.1-6 D. S. Matteson et al., J. Am. Chem. Soc., 105, 2077 (1983); Organometallics, 2, 230, 236, 1083, 1529, 1536, 1543 (1983).

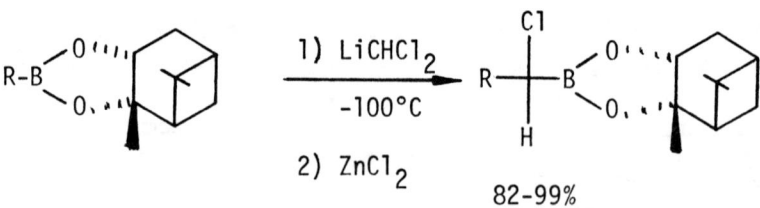

82-99%

(96-99% Diasterioselectivity)

I.G.1-7 H. C. Brown et al., J. Am. Chem. Soc., 105, 6285 (1983); H. C. Brown and D. Basavaiah, Synthesis, 283 (1983); H.C. Brown et al., ibid, 885, 886 (1983).

$$R-B\langle\!\!\begin{array}{c}O\\O\end{array}\!\!\rangle \xrightarrow[\substack{\text{THF}\\ \text{2) HgCl}_2 \\ \text{3) H}_2O_2,\ pH\ 8}]{\text{1) PhS(MeO)CHLi}} R-CHO \quad 60\text{-}78\%$$

I.G.2. Carbonylation Reactions

I.G.2-1 V. P. Baillargeon and J. K. Stille, J. Am. Chem. Soc., 105, 7175 (1983); N. Chatani, S. Murai and N. Sonoda, ibid, 105, 1370 (1983); A. Kasahara, T. Izumi and H. Yanai, Chem. Ind. (London), 898 (1983); S. Murai et al., Organometallics, 2, 1883 (1983).

$$RX \xrightarrow[\substack{\text{cat. Pd(0)}\\ 50°C}]{\text{CO, nBu}_3\text{SnH}} RCHO \quad 53\text{-}77\%$$

RX = Aryl, Vinyl, Benzyl and Allyl Halides.

I.G.2-2 Y. Matsui and M. Orchin, J. Organometal. Chem., 246, 57 (1983); F. Piacenti et al., ibid, 247, 89 (1983); P. W. N. M. van Leeuwen and C. F. Roobeek, ibid, 258, 343 (1983); R. Lazzaroni et al., ibid, 258, 351 (1983); S. Pucci et al., ibid, 247, C56 (1983); G. Cavinato and L. Toniolo, ibid, 241, 275 (1983); M. J. Mirbach et al., Chem. Ber., 116, 1422 (1983).

$$\begin{array}{c}Ph\\Ph\end{array}\!\!\!C=CH_2 \xrightarrow[\text{[RhCl(CO)}_2]_2]{\text{CO, H}_2} Ph_2CHCH_2\text{-CHO} \quad 85\%$$

Cobalt-catalysis yields mostly hydrogenation.

I.G.2-3 R. Grigg, G. J. Reimer and A. R. Wade, J. Chem. Soc., Perkin I, 1929 (1983).

$$\text{CH}_2=\text{CHCH}_2\text{CH(CO}_2\text{Et)}_2 \xrightarrow[\text{RT}]{\substack{1\% \text{ Rh(H)CO(PPh}_3)_3 \\ \text{CO, PhH}}}$$

[cyclopentane with OH and two CO_2Et groups]

57% (GC)

I.G.2-4 M. Foa et al., J. Organometal. Chem., 243, 87 (1983); M. Foa, A. Umani-Ronchi et al., ibid, 248, 225 (1983); G. Tanguy, B. Weinberger and H. des Abbayes, Tetrahedron Lett., 24, 4005 (1983); R. A. Sawicki, J. Org. Chem., 48, 5382 (1983); J. J. Brunet, C. Sidot and P. Caubere, ibid, 48, 1919 (1983).

$$\underset{\substack{|\\ \text{H}}}{\overset{\substack{\text{Ph}\\|}}{\text{X}-\text{C}-\text{Me}}} \xrightarrow[\substack{\text{Aq. NaOH}\\ {}^t\text{BuOMe}}]{\text{Co}_2(\text{CO})_8, \text{ CO}} \underset{\substack{|\\ \text{H}}}{\overset{\substack{\text{Ph}\\|}}{\text{Me}-\text{C}-\text{CO}_2\text{H}}}$$

73%

(45% op)

I.G.2-5 H. Alper et al., Chem. Commun., 1270 (1983); T. Fuchikami, K. Ohishi and I. Ojima, J. Org. Chem., 48, 3803 (1983); S. Gladiali et al., J. Organometal. Chem., 244, 289 (1983).

$$\text{RCH=CH}_2 \xrightarrow[\substack{\text{PdCl}_2, \text{ CuCl}_2, \text{ HCl}\\ \text{O}_2, \text{ RT}}]{\text{CO + H}_2\text{O}} \underset{\substack{|\\ \text{CO}_2\text{H}}}{\text{RCHCH}_3}$$

30-100%

I.G.2-6 H. Alper and J. F. Petrignani, <u>Chem. Commun.</u>, 1154 (1983).

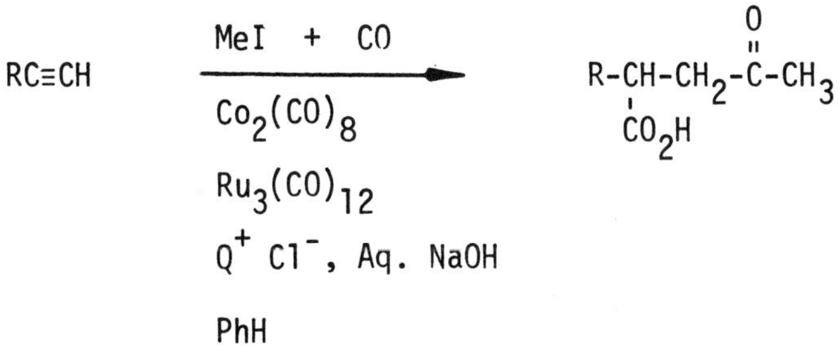

$$RC\equiv CH \xrightarrow[\substack{Co_2(CO)_8 \\ Ru_3(CO)_{12} \\ Q^+ Cl^-, \text{ Aq. NaOH} \\ PhH}]{MeI + CO} R-CH-CH_2-\overset{O}{\underset{\displaystyle CO_2H}{C}}-CH_3$$

I.G.2-7 E. P. Kundig and D. P. Simmons, <u>Chem. Commun.</u>, 1320 (1983).

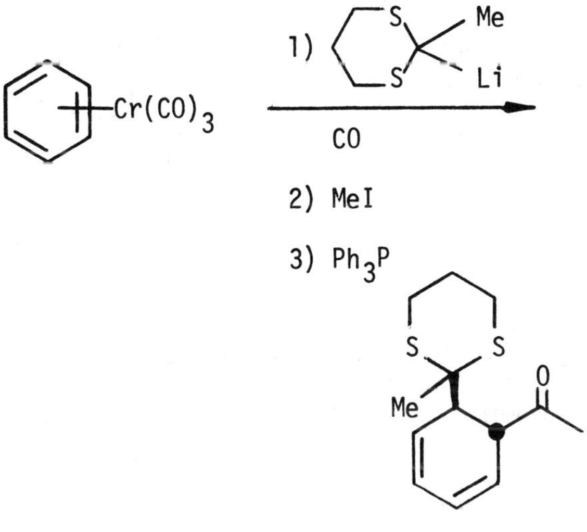

[arene-Cr(CO)₃] →
1) 2-lithio-2-methyl-1,3-dithiane, CO
2) MeI
3) Ph₃P

→ product (89%)

I.G.2-8 D. Seyferth et al., <u>Tetrahedron Lett.</u>, <u>24</u>, 4907 (1983); J. Org. Chem., <u>48</u>, 1144 (1983).

cyclohexanone $\xrightarrow[\substack{THF, \text{ MeOMe} \\ -135°C}]{nBuLi, CO (1:1)}$ 1-hydroxy-1-(pentanoyl)cyclohexane (73%)

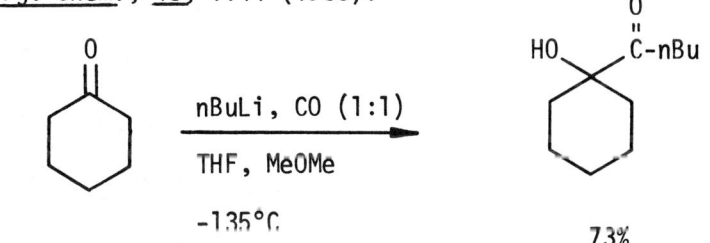

I.G.2-9 S. P. Current, J. Org. Chem., 48, 1779 (1983); H. Hoberg, F. J. Fananas and H. J. Riegel, J. Organometal. Chem., 254, 267 (1983).

$$2 \text{ ROH} \xrightarrow[\substack{\text{Pd(OAc)}_2 \\ \text{Co(OAc)}_2 \\ \text{Ph}_3\text{P, Benzoquinone}}]{2 \text{ CO, 0.5 O}_2} \underset{60-95\%}{\text{RO-C(=O)-C(=O)-OR}}$$

I.G.2-10 J. J. Eisch, A. A. Aradi and K. I. Han, Tetrahedron Lett., 24, 2073 (1983); M. F. Semmelhack, J. W. Herndon and J. K. Liu, Organometallics, 2, 1885 (1983); P. Eilbracht, M. Acker and W. Totzauer, Chem. Ber., 116, 238 (1983).

$$\text{Ph-C} \equiv \text{C-Ph} \xrightarrow[\substack{\text{Me}_3\text{Si-NC} \\ 2) \text{ HCl}}]{1) \text{ Ni(COD)}_2} \text{tetraphenylcyclopentadienone}$$

50%

I.G.2-11 J. C. Clinet, E. Dunach and K. P. C. Vollhardt, J. Am. Chem. Soc., 105, 6710 (1983); C. Exon and P. Magnus, ibid, 105, 2477 (1983); D. C. Billington, Tetrahedron Lett., 24, 2905 (1983); T. Kajimoto and J. Tsuji, J. Org. Chem., 48, 1685 (1983).

[Enyne substrate with SiMe₃ alkyne, MeO-aryl, and RO-methylenic side chain] $\xrightarrow[\text{Xylene, }\Delta]{\text{CpCo(CO)}_2}$ [steroidal product with Me₃Si, Me, OR, MeO, CoCp substituents]

72%

I.G.3. Other Syntheses via Organometallics

I.G.3-1 E. Dinjus et al., Z. Chem., 23, 303 (1983).

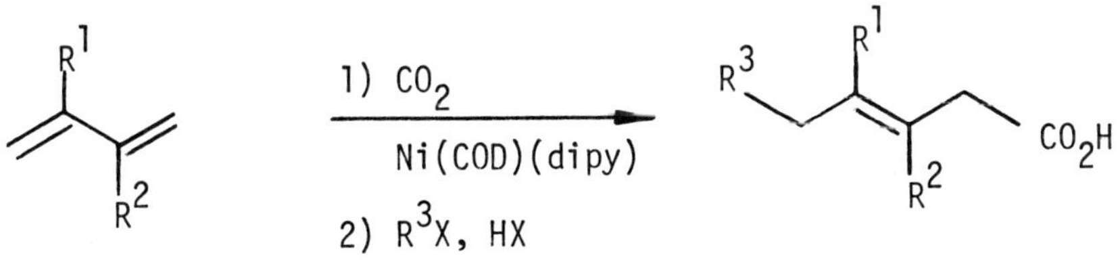

Reagents/conditions: 1) CO_2, Ni(COD)(dipy); 2) R^3X, HX

Starting material: diene with R^1, R^2 substituents → product with R^1, R^2, R^3 and CO_2H group.

I.G.3-2 J. W. Faller and K. H. Chao, J. Am. Chem. Soc., 105, 3893 (1983).

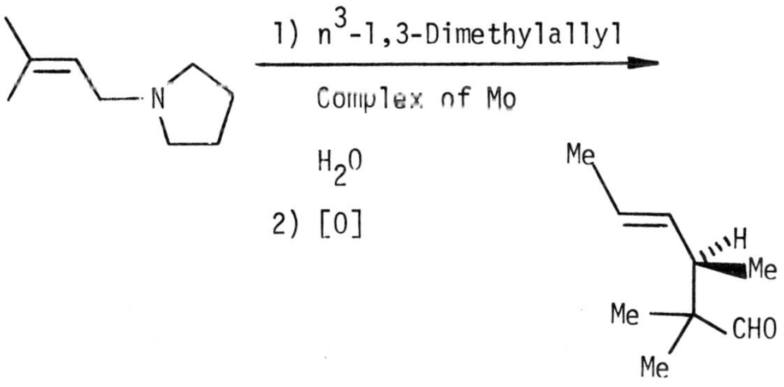

Reagents/conditions: 1) η^3-1,3-Dimethylallyl Complex of Mo, H_2O; 2) [O]

60%
(>96% ee)

I.G.3-3 K. P. Vora, Synth. Commun., 13, 99 (1983).

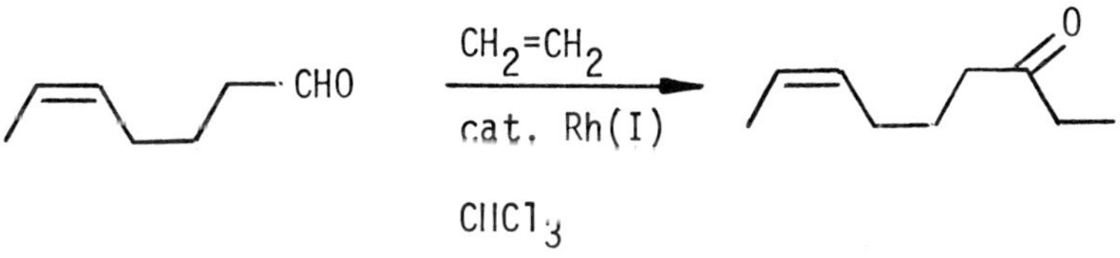

Reagents/conditions: $CH_2=CH_2$, cat. Rh(I), $CHCl_3$

26%

I.G.3-4   J. Sabadie and G. Descotes, <u>Bull. Soc. Chim. Fr. II</u>, 253 (1983).

$$CH_3(CH_2)_8CH_2CH_2OH \xrightarrow[\substack{\text{Ni or Pd Cat.} \\ 160\text{-}200°C}]{\text{MeOH, NaOMe}} CH_3(CH_2)_8\overset{\overset{\displaystyle CH_3}{|}}{C}HCH_2OH$$

70-80% Selectivity

Improved Guerbet reaction

I.G.3-5   P. Brun, A. Tenaglia and B. Waegell, <u>Tetrahedron Lett.</u>, <u>24</u>, 385 (1983);   Y. Yamada et al., <u>ibid</u>, <u>24</u>, 921 (1983);   M. Iyoda, M. Oda et al., <u>Chem. Commun.</u>, 1058 (1983);   K. Ito, Y. Otsuji et al., <u>Chem. Lett.</u>, 657 (1983).

$$2 \diagup\!\!\diagdown\!\!\diagup\!\!\diagdown_{OSiMe_3} \xrightarrow[PhH]{Ni(0)} \text{cyclooctadiene with } OSiMe_3 \text{ groups}$$

90%

I.G.3-6   H. Lehmkuhl and Y. L. Tsien, <u>Chem. Ber.</u>, <u>116</u>, 2437 (1983);   B. R. James and C. G. Young, <u>Chem. Commun.</u>, 1215 (1983);   L. E. Overman and A. F. Renaldo, <u>Tetrahedron Lett.</u>, <u>24</u>, 2235 (1983);   A. Doutheau, J. Sartoretti and J. Gore, <u>Tetrahedron</u>, <u>39</u>, 3059 (1983).

$$\text{1,5-hexadiene} \xrightarrow[\substack{\text{iPrMgBr} \\ \text{THF}}]{Cp_2TiCl, Cp_2TiCl_2} \text{methylenecyclopentane}$$

76%

I.G.3-7   J. C. Mol, <u>CHEMTECH</u>, 250 (1983).

Review: "Metathesis of Functionalized Olefins."

I.G.3-8 G. C. N. van den Aardweg, R. H. A. Bosma and J. C. Mol, Chem. Commun., 262 (1983); D. Villemin, Tetrahedron Lett., 24, 2855 (1983).

$$CH_2=CH(CH_2)_nCN \xrightarrow[Et_4Sn]{Re_2O_7-Al_2O_3} NC-(CH_2)_nCH=CH(CH_2)_n-CN$$

(n>1)   86-98%

I.G.4. Organometallic Reviews

I.G.4-1 D. Walther, E. Dinjus and J. Sieler, Z. Chem., 23, 237 (1983).

Review: "Activation of Carbon Dioxide on Transition Metal Centers: New Routes for Organic and Metalorganic Synthesis."

I.G.4-2 I. P. Beletskaya, J. Organometal. Chem., 250, 551 (1983).

Review: "The Cross Coupling Reactions of Organic Halides with Organic Derivatives of Tin, Mercury and Copper Catalyzed by Palladium."

I.G.4-3 N. R. Natale, Org. Prep. Proced. Int., 15, 387 (1983).

Review: "Lanthanides in Organic Synthesis."

I.G.4-4 J. V. N. V. Prasad and C. N. Pillai, J. Organometal. Chem., 259, 1 (1983).

Review: "Carbometallation: Addition of Organometallic Compounds to Isolated Multiple Bonds in Functionally Substituted Compounds."

I.G.4-5 K. H. Dotz, Pure Appl. Chem., 55, 1689 (1983).

Review: "Carbon-Carbon Bond Formation via Carbonyl-Carbene Complexes."

I.G.4-6 M. Franck-Neumann, Pure Appl. Chem., 55, 1715 (1983).

Review: "Synthetic Applications of Some Metal Carbonyl Complexes."

I.G.4-7 R. Scheffold et al., Pure Appl. Chem., 55, 1791 (1983).

Review: "Formation of (C-C) Bonds Catalyzed by Vitamin B_{12}."

I.G.4-8 P. Pino and G. Consiglio, Pure Appl. Chem., 55, 1781 (1983).

Review: "Organometallic Catalysis in Asymmetric Synthesis."

I.G.4-9 R. G. Salomon, Tetrahedron, 39, 485 (1983).

Review: "Homogeneous Metal-Catalysis in Organic Photochemistry."

I.G.4-10 H. K. Hall, Jr., Angew. Chem., Int. Ed. Engl., 22, 440 (1983).

Review: "Bond-Forming Initiation in Spontaneous Addition and Polymerization Reactions of Alkenes."

I.G.4-11 I. S. Akhrem, N. M. Chistovalova and M. E. Vol'pin, Russ. Chem. Rev., 52, 542 (1983).

Review: "The Splitting of the Silicon-Carbon Bond by Compounds of Platinum and Palladium."

I.G.4-12 J. Organometal. Chem., 242 (1983).

Annual Surveys: π Complexes.
Heteronuclear Metal-Metal Bond Complexes.
Magnesium, Technetium and Rhenium.
Cobalt, Rhodium and Iridium.

I.G.4-13 J. Organometal. Chem., 245 (1983).

Annual Surveys: Boron.
Bismuth.
Antimony.
Thallium.
Transition Metals.
Ruthenium and Osmium.

I.G.4-14 J. Organometal. Chem., 257 (1983).

Annual Surveys: Berylium.
Aluminum.
Arsenic.
Chromium, Molybdenum and Tungsten.
Ferrocene.
π-Complexes.
Heteronuclear Metal-Metal Bond Complexes.

II
OXIDATIONS

II.A. C—O Oxidations

1. Alcohol → Ketone, Aldehyde

II.A.1-1 R. Baker, V. B. Rao, P. D. Ravenscroft, et al., Synthesis, 572 (1983).

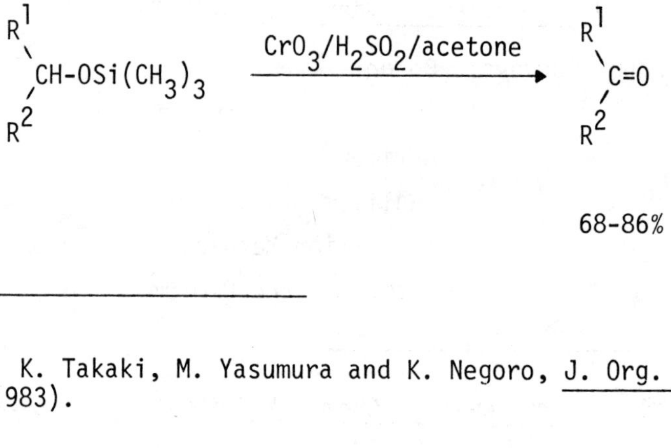

$$\begin{array}{c} R^1 \\ \diagdown \\ CH\text{-}OSi(CH_3)_3 \\ \diagup \\ R^2 \end{array} \xrightarrow{CrO_3/H_2SO_2/acetone} \begin{array}{c} R^1 \\ \diagdown \\ C=O \\ \diagup \\ R^2 \end{array}$$

68-86%

II.A.1-2 K. Takaki, M. Yasumura and K. Negoro, J. Org. Chem., 48, 54 (1983).

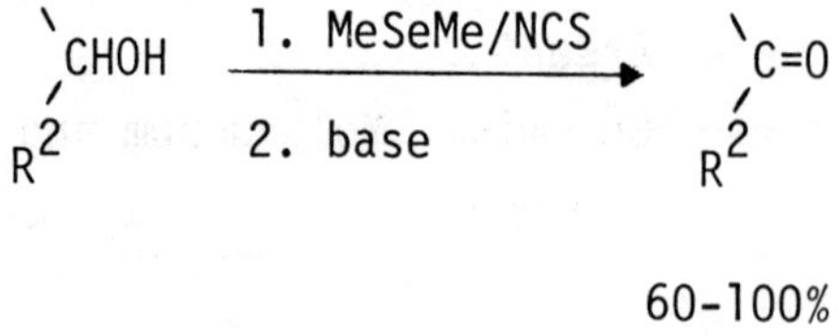

$$\begin{array}{c} R^1 \\ \diagdown \\ CHOH \\ \diagup \\ R^2 \end{array} \xrightarrow[\text{2. base}]{\text{1. MeSeMe/NCS}} \begin{array}{c} R^1 \\ \diagdown \\ C=O \\ \diagup \\ R^2 \end{array}$$

60-100%

OXIDATIONS

II.A.1-3 M. F. Semmelhack, C. S. Chou and D. A. Cortes, J. Am. Chem. Soc., 105, 4492 (1983).

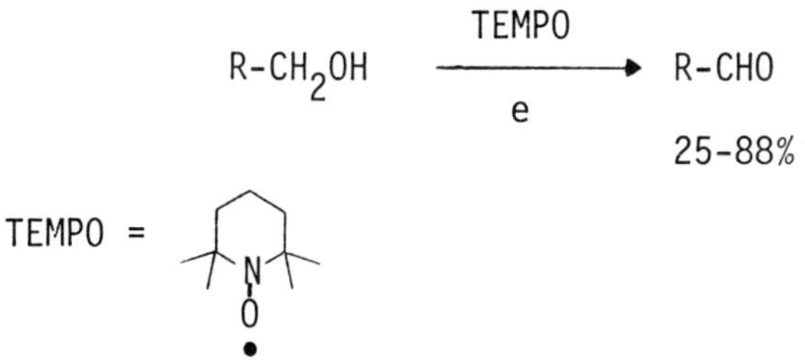

$$R-CH_2OH \xrightarrow[e]{TEMPO} R-CHO$$

25-88%

TEMPO =

II.A.1-4 T. Imamoto, Y. Hatanaka and M. Tokoyama, Tetrahedron Lett., 24, 2399 (1983)

$$ArRCHOH \xrightarrow[(NH_4)_2Ce(NO_3)_6\text{-Charcoal}]{Air} ArRC=O$$

68-92%

R = H, Ar'CO

II.A.1-5 K. Oshima, S. Kanemoto, S. Matsubara et al., Tetrahedron Lett., 24, 2185 (1983).

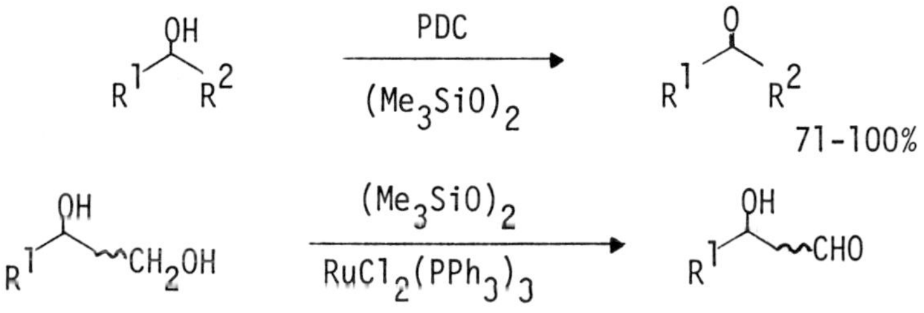

$$R^1\underset{R^2}{\overset{OH}{\diagup}} \xrightarrow[(Me_3SiO)_2]{PDC} R^1\underset{R^2}{\overset{O}{\diagup}}$$

71-100%

$$R^1-CH(OH)-CH_2OH \xrightarrow[RuCl_2(PPh_3)_3]{(Me_3SiO)_2} R^1-CH(OH)-CHO$$

II.A.1-6 G. L. Gard, H. B. Davis, R. M. Sheets et al., Heterocycles, 20, 2029 (1983).

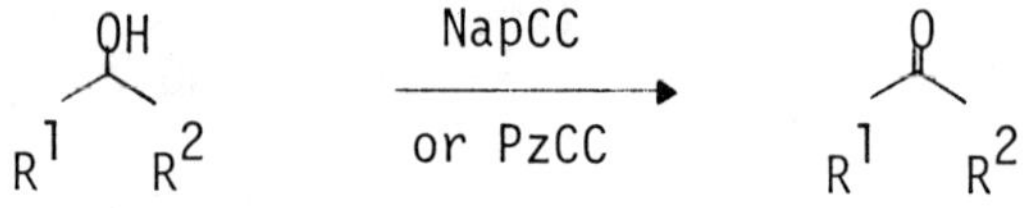

$$\underset{R^1 \quad R^2}{\overset{OH}{>\!\!-\!\!<}} \xrightarrow{\text{NapCC or PzCC}} \underset{R^1 \quad R^2}{\overset{O}{>\!\!=}}$$

82-100%

NapCC = [1,8-naphthyridinium]$^+$ CrO$_3$Cl$^-$

PzCC = [pyrazinium]$^+$ CrO$_3$Cl$^-$

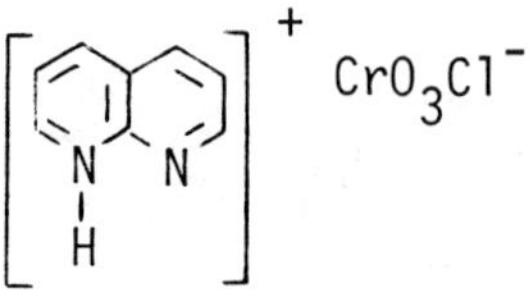

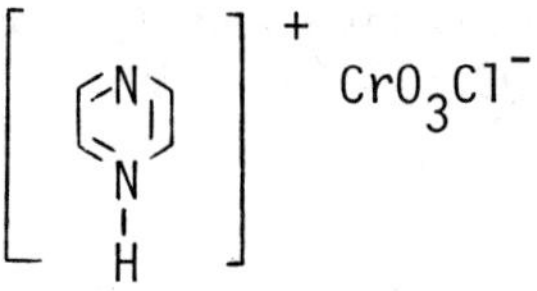

II.A.1-7 Y. Ueno, T. Kageyama and M. Okawara, Synthesis, 815, (1983).

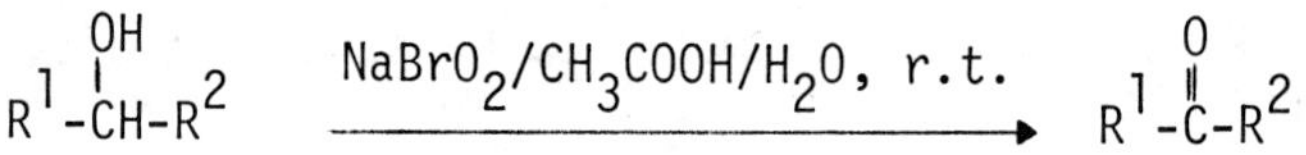

$$R^1\text{-CH(OH)-}R^2 \xrightarrow{\text{NaBrO}_2/\text{CH}_3\text{COOH}/\text{H}_2\text{O, r.t.}} R^1\text{-C(O)-}R^2$$

82-100%

II.A.1-8 E. Santaniello, F. Milani and R. Casati, <u>Synthesis</u>, 749 (1983).

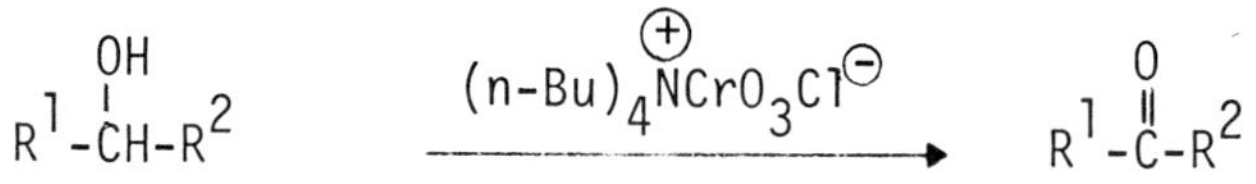

$$R^1\text{-CH(OH)-}R^2 \xrightarrow{(n\text{-Bu})_4\overset{\oplus}{N}CrO_3Cl^{\ominus}} R^1\text{-C(O)-}R^2$$

50-88%

II.A.1-9 C. Palomo and J. Aizpurua, <u>Tetrahedron Lett.</u>, <u>40</u>, 4367 (1983).

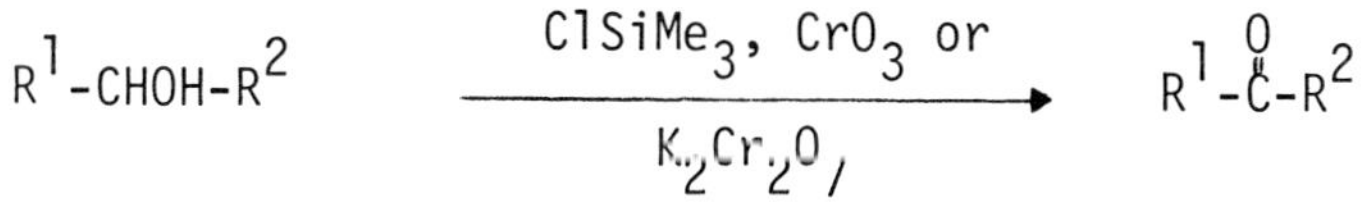

$$R^1\text{-CHOH-}R^2 \xrightarrow[K_2Cr_2O_7]{ClSiMe_3,\ CrO_3\ \text{or}} R^1\text{-C(O)-}R^2$$

81-98%

II.A.1-10 G. Rosini and R. Ballini, <u>Synthesis</u>, 543 (1983).

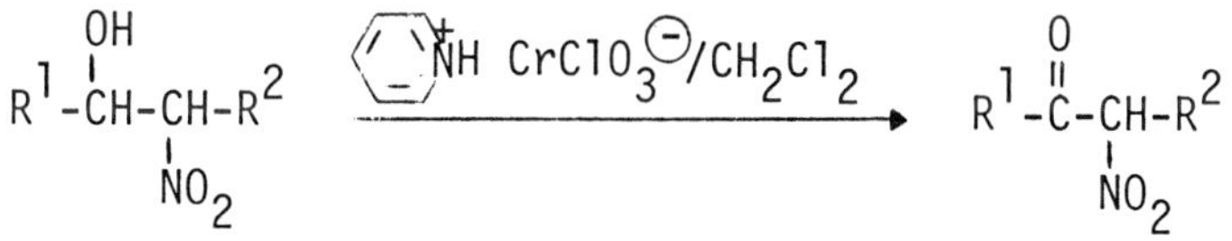

$$R^1\text{-CH(OH)-CH(NO}_2\text{)-}R^2 \xrightarrow{C_5H_5\overset{+}{N}H\ CrClO_3^{\ominus}/CH_2Cl_2} R^1\text{-C(O)-CH(NO}_2\text{)-}R^2$$

R^1, R^2 = alkyl

61-87%

II.A.1-11/II.A.2-1 S. Ohta, T. Tachi and M. Okamoto, Synthesis, 291 (1983).

$$Ar\text{-}CH=O + 0.5\ O_2 + NaH \xrightarrow{\text{pyrazole}} Ar\text{-}COONa$$

$$Ar\text{-}CH_2\text{-}OH + O_2 + NaH \xrightarrow{\text{pyrazole}} Ar\text{-}COONa$$

66-100%

II.A.2. Alcohol, Aldehyde → Acid, Acid Derivative

II.A.2-2 Z. Yoshida, Y. Tamaru and Y. Amada, Synthesis, 474 (1983).

$$R\text{-}CH=O + NH\underset{\smile}{\frown}O \xrightarrow[\text{PhBr, } K_2CO_3]{Pd(OAc)_2} R\text{-}\underset{\parallel}{\overset{O}{C}}\text{-}N\underset{\smile}{\frown}O$$

25-93%

II.A.2-3 F. Yuste, A. E. Origel and L. J. Brena, Synthesis, 109 (1983).

$$\underset{R^1}{\underset{R^2}{\text{Ar}}}\text{-}CH=O \xrightarrow[\substack{2.\ NaOH/H_2O/DMSO \\ BzNEt_3^{\oplus}\ Cl^{\ominus} \\ 3.\ O_2}]{1.\ (R^3)_2NH_2Cl/NaCN} \underset{R^1}{\underset{R^2}{\text{Ar}}}\text{-}\overset{O}{\underset{\parallel}{C}}\text{-}N(R^3)_2$$

50-66%

II.A.2-4   N. Arumugam, S. Sivasubramanian and P. Manisankar, Indian J. Chem. Sect. B., 21, 454 (1982).

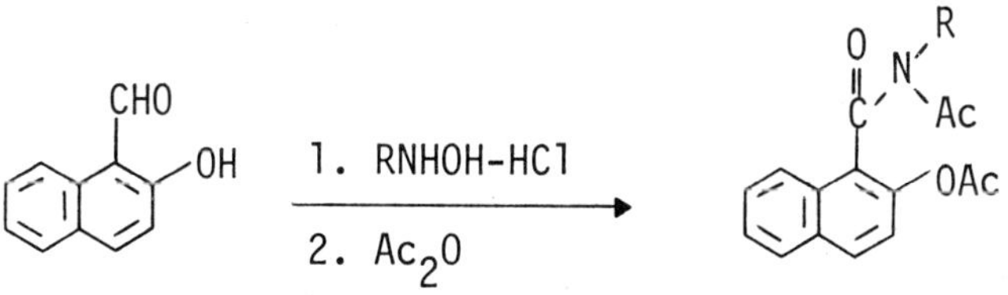

[Reaction: 2-hydroxy-1-naphthaldehyde → (1) RNHOH·HCl, (2) Ac_2O → 1-(N-acetyl-N-R-carbamoyl)-2-acetoxynaphthalene]

II.A.2-5   J. O. Amupitan, Synthesis, 730 (1983).

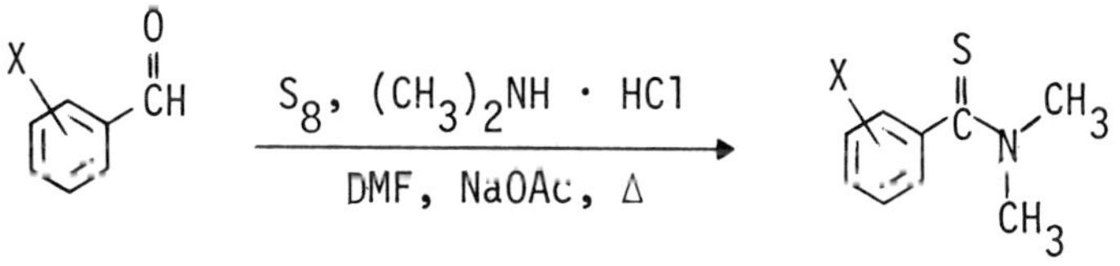

[Reaction: X-substituted benzaldehyde → S_8, $(CH_3)_2NH\cdot HCl$, DMF, NaOAc, Δ → X-substituted N,N-dimethylbenzothioamide]

72-92%

II.B. C—H Oxidations

1. C—H → C-O

II.B.1-1 R. M. Coates and C. H. Cummins, J. Org. Chem., 48, 2070 (1983).

$$\underset{R}{\overset{R'}{>}}CH-C\underset{H}{\overset{\nearrow O}{\searrow}} \quad \xrightarrow[\substack{\text{2. R"COCl, Et}_3\text{N} \\ \text{3. H}_2\text{O}}]{\text{1. t-BuNHOH, TsOH,} \atop \text{Na}_2\text{SO}_4,\text{CH}_2\text{Cl}_2} \quad \underset{R}{\overset{R'}{>}}\overset{\overset{\text{O} \atop \|}{\text{OCR"}}}{\underset{|}{C}}-C\underset{H}{\overset{\nearrow O}{\searrow}}$$

22-95%

<cyclohexanone> $\xrightarrow{\text{similar conditions}}$ <2-acyloxycyclohexanone, OCR>

26-59%

II.B.1-2 B. Ganem and A. Biloski, J. Org. Chem., 48, 3118 (1983).

$$R^1-CH_2C\underset{OCH_3}{\overset{N-R^2}{\diagup}} \xrightarrow[\substack{\text{4-vinylpyriline} \\ \text{polymer}}]{\text{Pb(OAc)}_4} R^1-CHC\underset{OCH_3}{\overset{N-R^2}{\diagup}} \atop \underset{\text{OAc}}{|}$$

~75%

II.B.1-3 A. N. Kashin, M. L. Tulchinskii, N. A. Bumagin et al., J. Org. Chem. (USSR), 18, 1390 (1982).

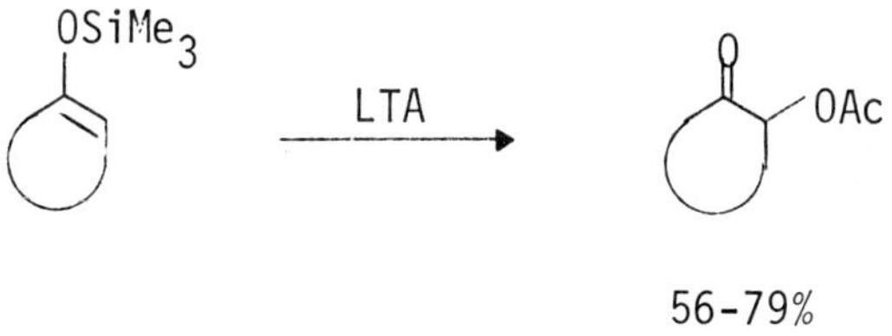

[Cyclohexene with OSiMe$_3$] $\xrightarrow{\text{LTA}}$ [Cyclohexanone with OAc at α position]

56-79%

II.B.1-4 G. M. Rubottom, R. Marrero and J. M. Gruber, Tetrahedron, 39, 861 (1983).

$$R^1R^2C\text{-CHOTMS} \xrightarrow{\text{LTA}} R^1R^2\underset{|}{\overset{OAc}{C}}\text{-CHO}$$

II.B.1-5 K. Takai, S. Matsubara and H. Nozaki, Bull. Chem. Soc. Jap., 56, 2029 (1983).

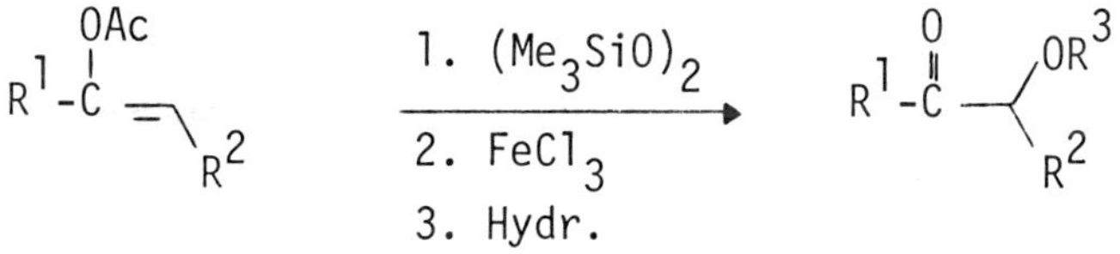

$$R^1\underset{|}{\overset{OAc}{C}}=\underset{R^2}{} \xrightarrow[\substack{\text{2. FeCl}_3 \\ \text{3. Hydr.}}]{\text{1. (Me}_3\text{SiO)}_2} R^1\overset{O}{\underset{\|}{C}}-\underset{R^2}{\overset{OR^3}{C}}$$

R^3 = H, Ac

II.B.1-6 T. Shono, Y. Matsumura and K. Inoue, J. Org. Chem., 48, 1388 (1983).

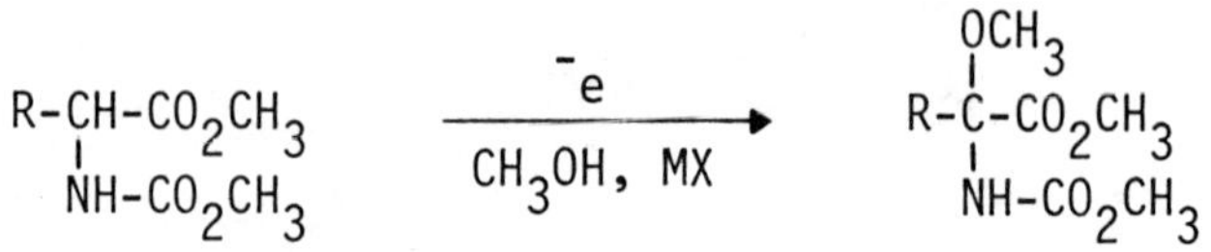

$$\underset{\underset{NH-CO_2CH_3}{|}}{R-CH-CO_2CH_3} \xrightarrow[CH_3OH,\ MX]{-e} \underset{\underset{NH-CO_2CH_3}{|}}{\overset{\overset{OCH_3}{|}}{R-C-CO_2CH_3}}$$

62-90%

[β-lactam substrate] $\xrightarrow[NaCl,\ MeOH]{-e}$ [methoxylated β-lactam product]

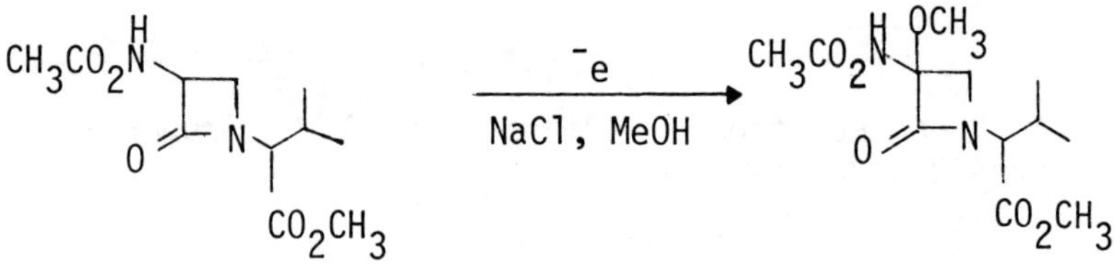

92%

II.B.1-7 A. DallaCort, A. LaBarbera and L. Mandolini, J. Chem. Res. Synop., 44 (1983).

$$ArCH_3 + 2Ce^{IV} \xrightarrow{ROH} ArCH_2OR + 2Ce^{III} + 2H^+$$

46-53%

II.B.1-8 R. G. Harvey and H. Lee, J. Org. Chem., 48, 749, (1983).

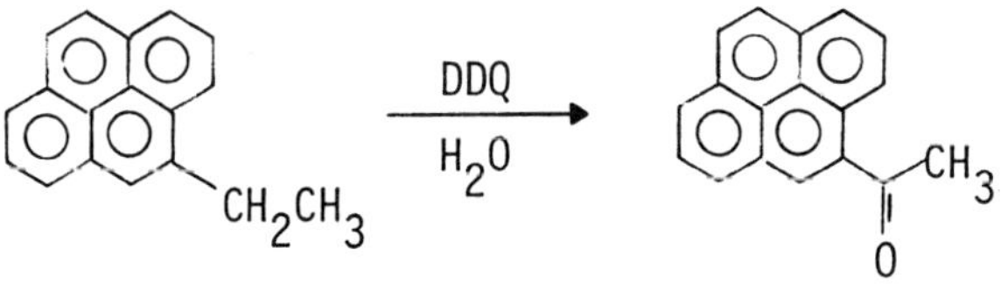

pyrene-CH$_2$CH$_3$ $\xrightarrow{\text{DDQ} \atop \text{H}_2\text{O}}$ pyrene-C(=O)CH$_3$

48%

benz[a]anthracene-CH$_3$ $\xrightarrow{\text{DDQ} \atop \text{H}_2\text{O}}$ benz[a]anthracene-CHO

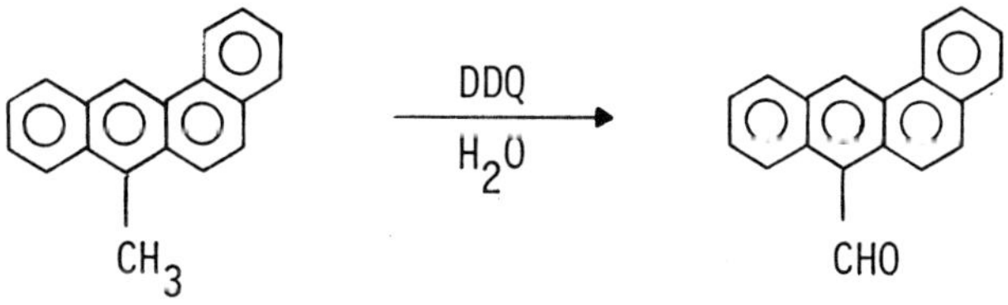

63%

II.B.1-9 M. Hirama and M. Shimizu, Synth. Commun., 13, 781 (1983).

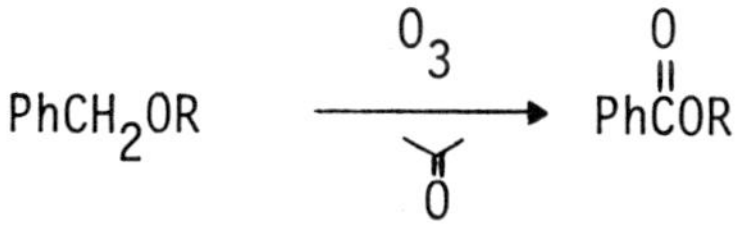

$$\text{PhCH}_2\text{OR} \xrightarrow[\text{acetone}]{O_3} \text{PhCOR}$$

51-87%

II.B.1-10 P. F. Schuda, M. B. Cichowicz and M. R. Heimann, Tetrahedron Lett. 24, 3829 (1983).

$$R-O-CH_2Ph \xrightarrow[CH_3CN/H_2O]{RuO_2/NaIO_4/CCl_4} R-O-\overset{O}{\overset{\|}{C}}Ph$$

54-96%

II.B.1-11 K. Schank and C. Lick, Synthesis, 392 (1983).

$$\begin{array}{c} R^1-\overset{O}{\overset{\|}{C}} \\ \diagdown \\ CH_2 \\ \diagup \\ R^2-\underset{\|}{\underset{O}{C}} \end{array} \xrightarrow[2.\ O_3/CH_2Cl_2,\ -40°C]{1.\ PhI(OAc)_2} \begin{array}{c} R^1-\overset{O}{\overset{\|}{C}} \\ \diagdown \\ C=O \\ \diagup \\ R^2-\underset{\|}{\underset{O}{C}} \end{array}$$

65-92%

II.B.1-12 T. R. Beebe, et al., J. Org. Chem., 48, 3126 (1983).

$$R^1\underset{\underset{OH}{}}{\overset{R^2}{\diagup\diagdown\diagup}} \xrightarrow{NIS} \underset{O}{\overset{R^2}{\square}}-R^1$$

52-94%

II.B.1-13 U. K. Banerjee and R. V. Venkateswaran, Tetrahedron Lett., 24, 4625 (1983).

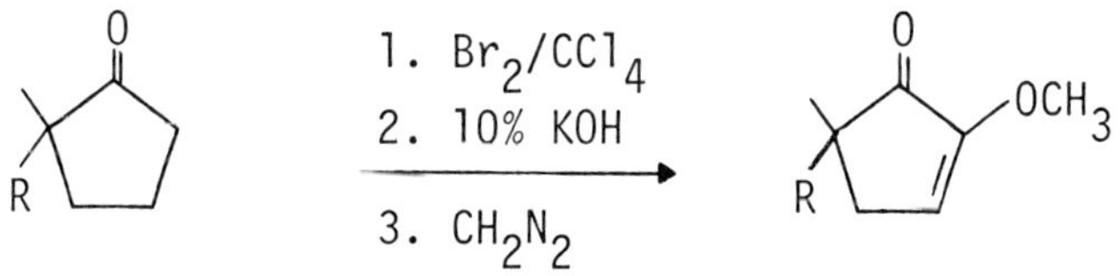

$$\text{cyclopentanone with } R, CH_3 \xrightarrow[\text{3. } CH_2N_2]{\substack{\text{1. } Br_2/CCl_4 \\ \text{2. } 10\% \text{ KOH}}} \text{2-methoxy cyclopentenone with } R, CH_3, OCH_3$$

50-58%

II.D.2.   C—H → C—Hal

II.B.2-1 M. Bellassoued, F. Habbachi and M. Gaudemar, Synthesis, 745 (1983).

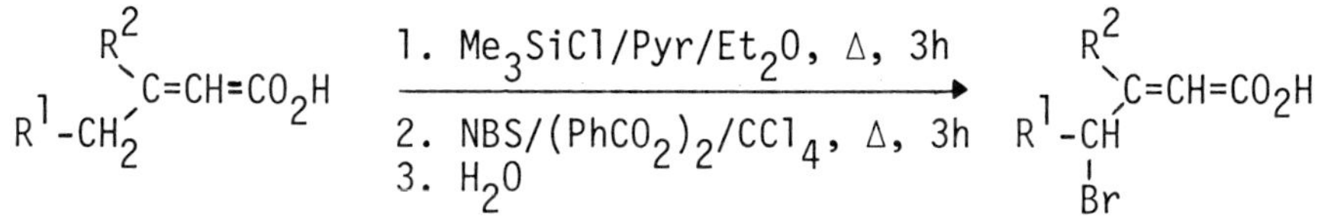

$$R^1\text{-}CH_2\underset{R^2}{\diagup}C=CH=CO_2H \xrightarrow[\substack{\text{2. NBS/(PhCO}_2)_2/CCl_4, \Delta, 3h \\ \text{3. } H_2O}]{\text{1. Me}_3SiCl/Pyr/Et_2O, \Delta, 3h} R^1\text{-}\underset{Br}{\overset{R^2}{CH}}C=CH=CO_2H$$

66-73%

II.B.2-2 R. J. Crawford, J. Org. Chem., 48, 1364 (1983).

$$RCH_2CO_2H + Cl_2 \xrightarrow[150°C]{ClSO_3H,\ TCNQ} RCHClCO_2H + HCl$$

80-90%

II.B.2-3 A. Guy, M. Lemaire and J.-P. Guette, Synthesis, 1018 (1982).

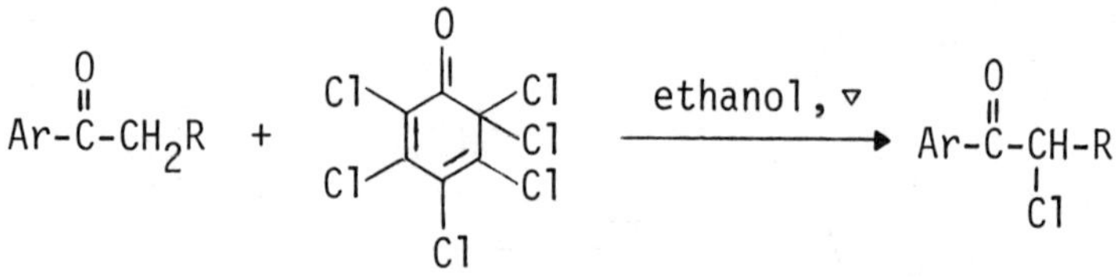

$$Ar-\overset{O}{\underset{\|}{C}}-CH_2R + \text{(hexachlorocyclohexadienone)} \xrightarrow{\text{ethanol}, \triangle} Ar-\overset{O}{\underset{\|}{C}}-\underset{\underset{Cl}{|}}{CH}-R$$

50-100%

II.B.2-4 G. Piancatelli et al., Synth. Commun., 12, 1127 (1982).

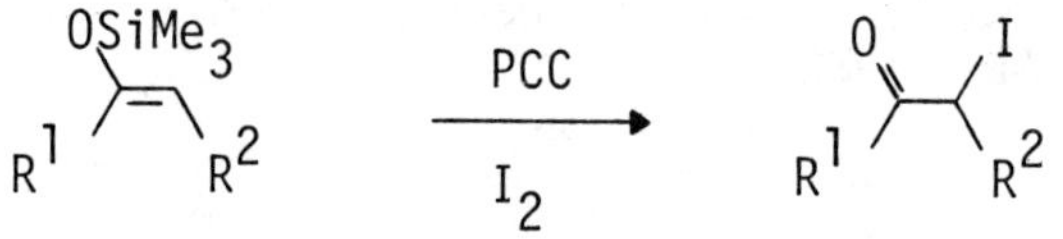

$$\underset{R^1}{\overset{OSiMe_3}{\diagup}}\!\!=\!\!\underset{R^2}{\diagdown} \xrightarrow[I_2]{PCC} \underset{R^1}{\overset{O}{\diagup}}\!\!\overset{I}{\underset{R^2}{\diagdown}}$$

R^1 = H, alkyl, aryl   76-100%
R^2 = alkyl

II.B.2-5 S. Motohashi and M. Satomi, Synthesis, 1021 (1982).

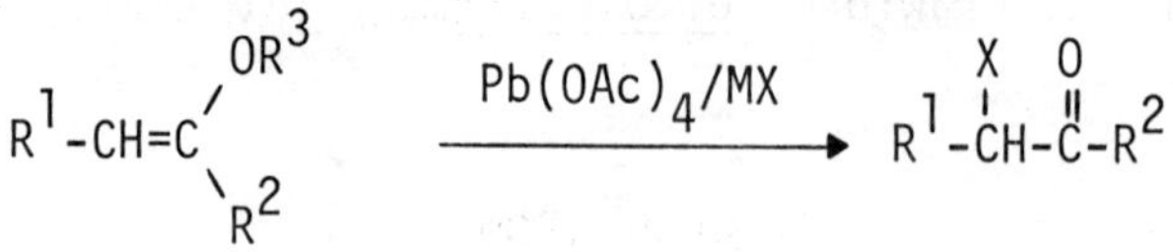

$$R^1-CH=C\underset{R^2}{\overset{OR^3}{\diagup}} \xrightarrow{Pb(OAc)_4/MX} R^1-\underset{\underset{X}{|}}{CH}-\overset{O}{\underset{\|}{C}}-R^2$$

X = Br, Cl, I   58-99%

II.B.2-6 E. Laurent, R. Tardivel and H. Thiebault, Tetrahedron Lett., 24, 903 (1983).

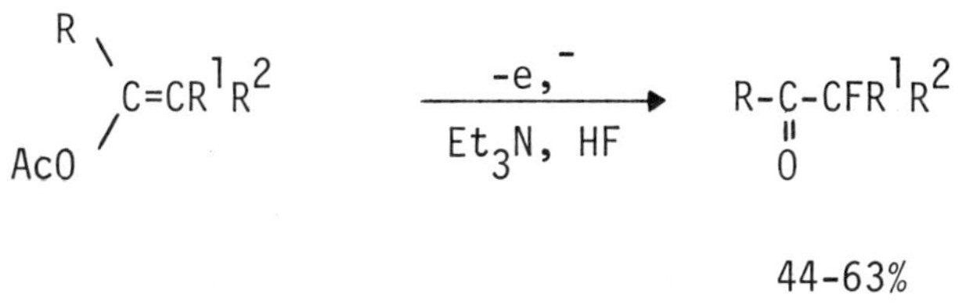

$$\underset{AcO}{\overset{R}{\diagdown}}C=CR^1R^2 \xrightarrow[Et_3N, HF]{-e, -} R-\underset{\underset{O}{\|}}{C}-CFR^1R^2$$

44-63%

II.B.2-7 O. Lerman and S. Rozen, J. Org. Chem., 48, 724 (1983).

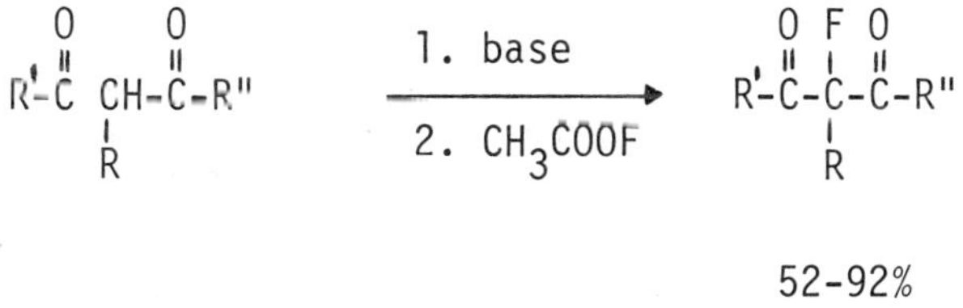

$$R'-\overset{O}{\underset{\|}{C}}-\underset{\underset{R}{|}}{CH}-\overset{O}{\underset{\|}{C}}-R'' \xrightarrow[2.\ CH_3COOF]{1.\ base} R'-\overset{O}{\underset{\|}{C}}-\overset{F}{\underset{\underset{R}{|}}{C}}-\overset{O}{\underset{\|}{C}}-R''$$

52-92%

II.B.2-8 S. T. Purrington and W. A. Jones, J. Org. Chem., 48, 761 (1983).

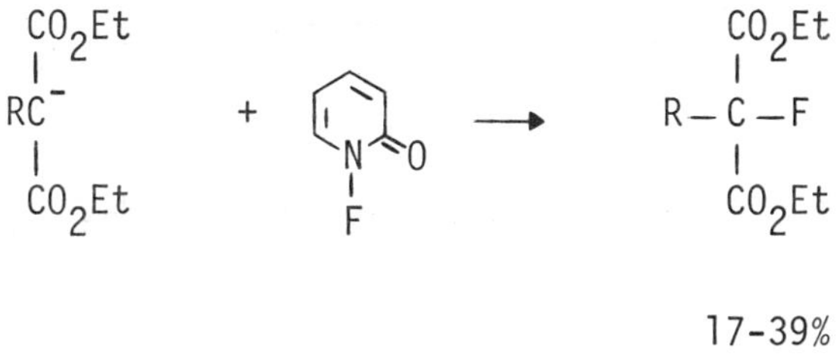

$$\underset{CO_2Et}{\overset{CO_2Et}{\underset{|}{|}}}RC^- \;+\; \underset{F}{\overset{}{\bigcirc}}\!\!\!N\text{-}O \longrightarrow \underset{CO_2Et}{\overset{CO_2Et}{\underset{|}{|}}}R-C-F$$

17-39%

II.B.2-9 F. Bellesia et al., J. Chem. Res. Synop., 16 (1983).

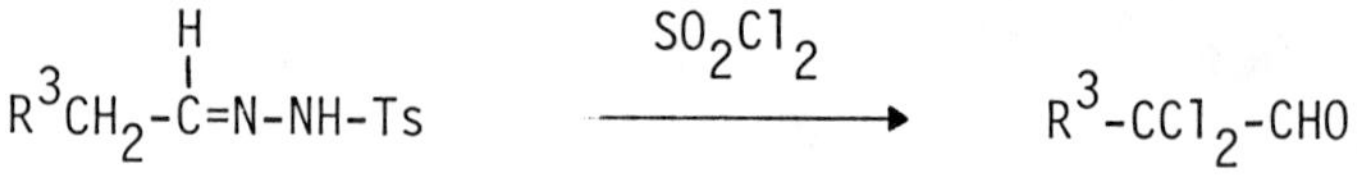

$$R^3CH_2-\overset{H}{\underset{|}{C}}=N-NH-Ts \xrightarrow{SO_2Cl_2} R^3-CCl_2-CHO$$

59-75%

II.B.2-10 M. Ouertani, P. Girard and H. B. Kagan, Bull. Soc. Chim. Fr., II-327 (1982).

$$ArCH_3 \xrightarrow[Br_2, \; CCl_4, \; h\nu]{La(OAc)_3} ArCH_2Br$$

50-90%

II.B.2-11 G. A. Olah et al., J. Org. Chem., 48, 3356 (1983).

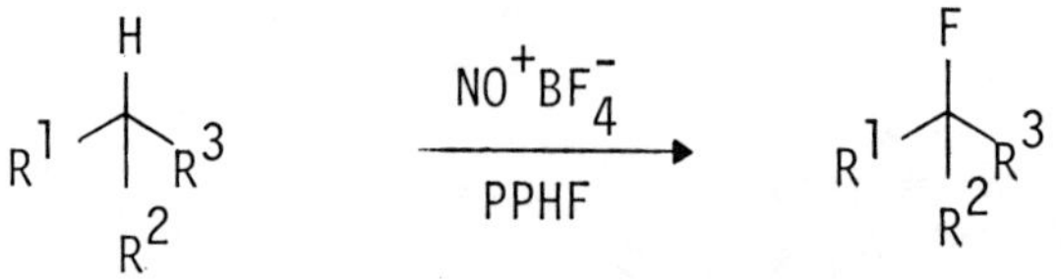

$$\underset{R^2}{\overset{H}{R^1 {\textstyle\diagup\!\!\!\diagdown} R^3}} \xrightarrow[PPHF]{NO^+BF_4^-} \underset{R^2}{\overset{F}{R^1 {\textstyle\diagup\!\!\!\diagdown} R^3}}$$

90-95%

PPHF = Pyridine Polyhydrogen Fluoride

II.B.2-12 J. R. L. Smith and L. C. McKeer, <u>Tetrahedron Lett.</u>, <u>24</u>, 3117 (1983).

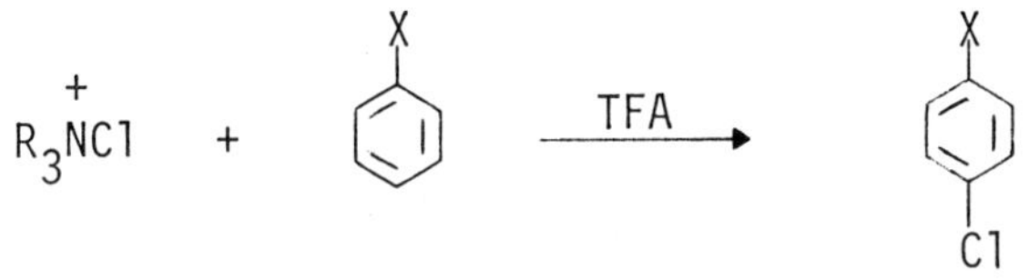

$R_3\overset{+}{N}Cl$ + PhX $\xrightarrow{TFA}$ 4-Cl-C$_6$H$_4$-X

X = electron donator

II.B.2-13 F. O. Ayorinde, <u>Tetrahedron Lett.</u>, <u>24</u>, 2077 (1983).

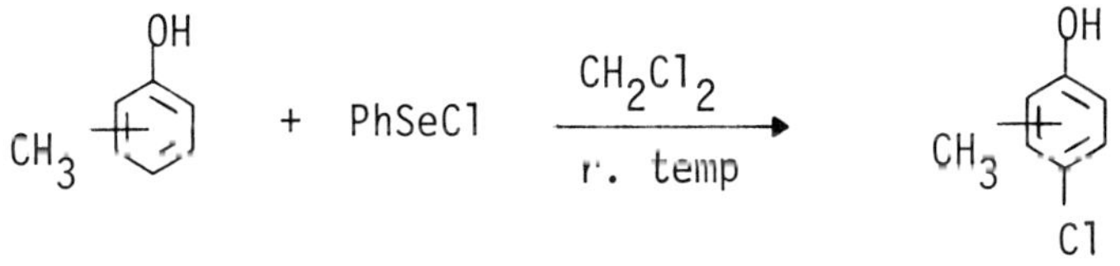

3-CH$_3$-C$_6$H$_4$-OH + PhSeCl $\xrightarrow[\text{r. temp}]{CH_2Cl_2}$ 3-CH$_3$-4-Cl-C$_6$H$_3$-OH

II.B.2-14 S. Stavber and M. Zupan, <u>J. Chem. Soc. Chem. Commun.</u>, 563 (1983).

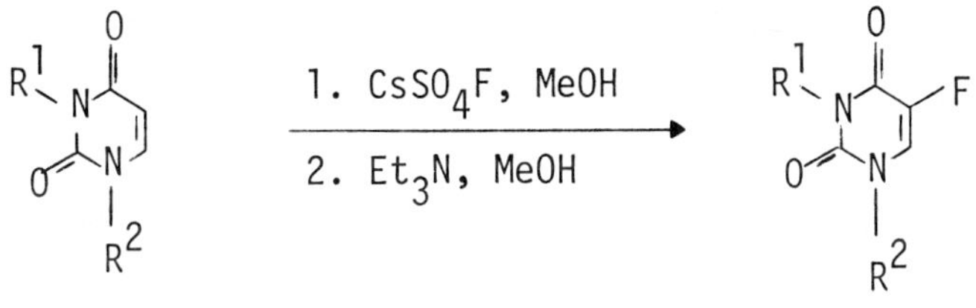

uracil (N1-R^1, N3-R^2) $\xrightarrow[\text{2. Et}_3\text{N, MeOH}]{\text{1. CsSO}_4\text{F, MeOH}}$ 5-fluoro uracil (N1-R^1, N3-R^2)

79-89%

II.B.2-15 T. P. Mohandas, A. S. Mamman and P. M. Nair, Tetrahedron, 39, 1187 (1983).

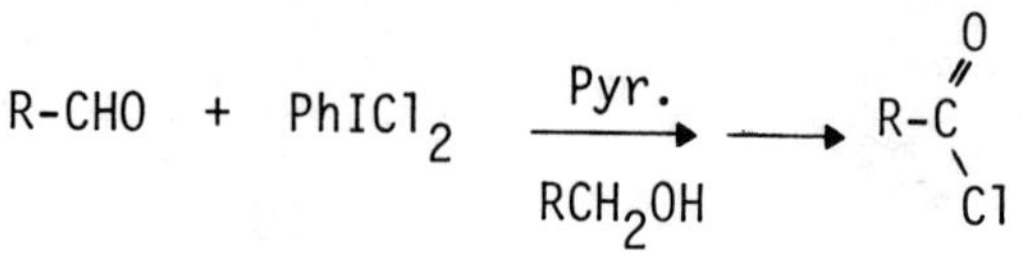

$$R\text{-CHO} + PhICl_2 \xrightarrow[RCH_2OH]{Pyr.} \longrightarrow R\text{-C}(=O)Cl$$

X = Cl; R = alkyl   56-83%

II.B.2-16 N. Kornblum, H. K. Singh and W. J. Kelly, J. Org. Chem., 48, 332 (1983).

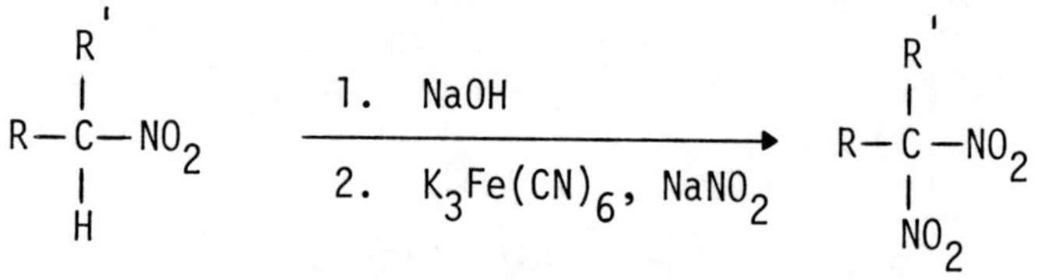

$$\underset{H}{\underset{|}{R-\overset{R'}{\overset{|}{C}}-NO_2}} \xrightarrow[2.\ K_3Fe(CN)_6,\ NaNO_2]{1.\ NaOH} \underset{NO_2}{\underset{|}{R-\overset{R'}{\overset{|}{C}}-NO_2}}$$

64-90%

$$\underset{NO_2}{\underset{|}{R-\overset{R'}{\overset{|}{C}}-H}} \xrightarrow[I_2]{base} \underset{NO_2}{\underset{|}{R-\overset{R'}{\overset{|}{C}}-I}} \xrightarrow{R''SO_2^-} \underset{NO_2}{\underset{|}{R-\overset{R'}{\overset{|}{C}}-\overset{O}{\overset{\|}{\underset{\|}{\underset{O}{S}}}}-R''}}$$

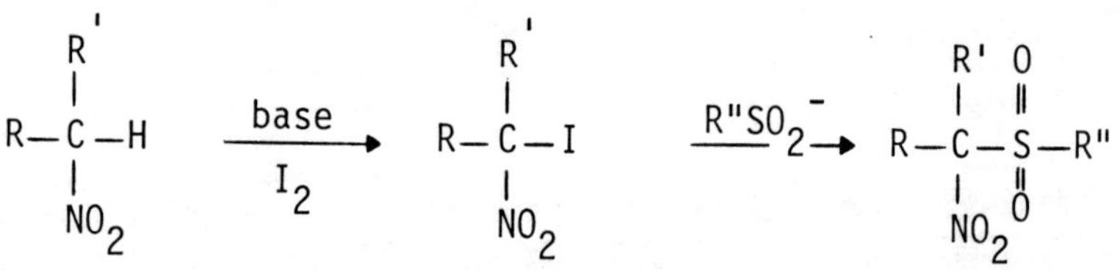

60-86%

II.B.3. Other C—H Oxidations

II.B.3-1 H. Feuer et al., Synthesis, 187 (1983).

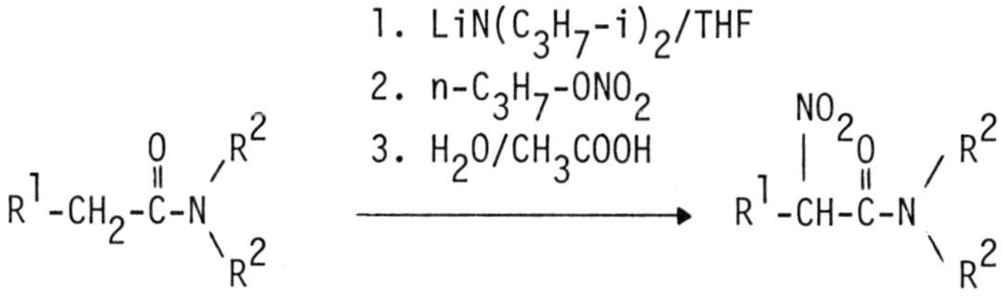

$$R^1-CH_2-\overset{O}{\overset{\|}{C}}-N\begin{matrix}R^2\\R^2\end{matrix} \xrightarrow[\begin{array}{l}1.\ \text{LiN}(C_3H_7\text{-}i)_2/\text{THF}\\2.\ n\text{-}C_3H_7\text{-ONO}_2\\3.\ H_2O/CH_3COOH\end{array}]{} R^1-\overset{NO_2}{\underset{|}{CH}}-\overset{O}{\overset{\|}{C}}-N\begin{matrix}R^2\\R^2\end{matrix}$$

67-77%

II.B.3-2 P. Dampawan and W. W. Zajac, Jr., Synthesis, 545 (1983).

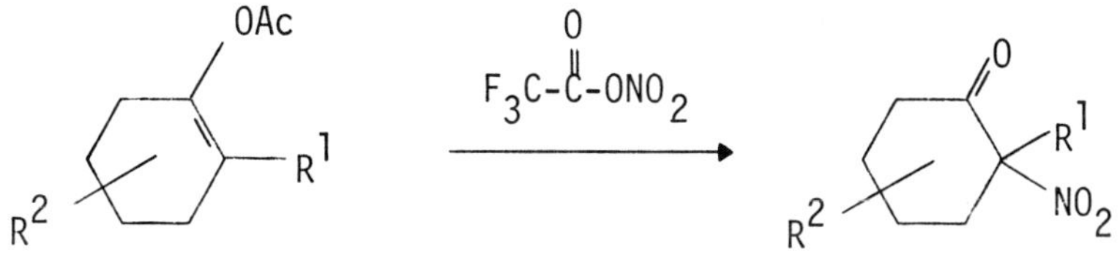

R^1 = prim-alkyl

41-100%

II.B.3-3 S. Lociuro, L. Pellacani and P. A. Tardella, Tetrahedron Lett., 24, 593 (1983).

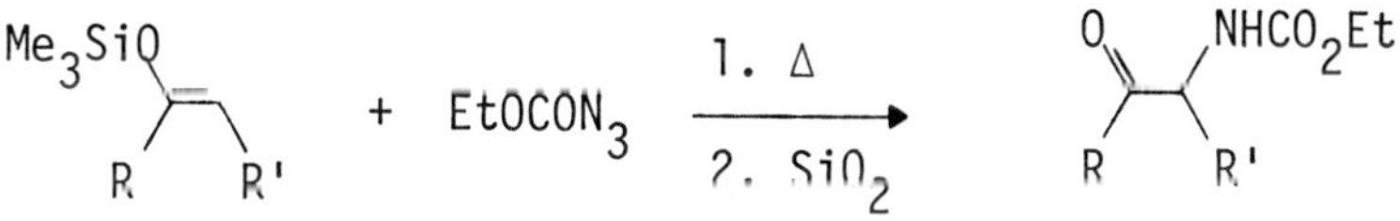

35-65%

II.B.3-4 G. Kresze and H. Muensterer, J. Org. Chem., 48, 3561 (1983).

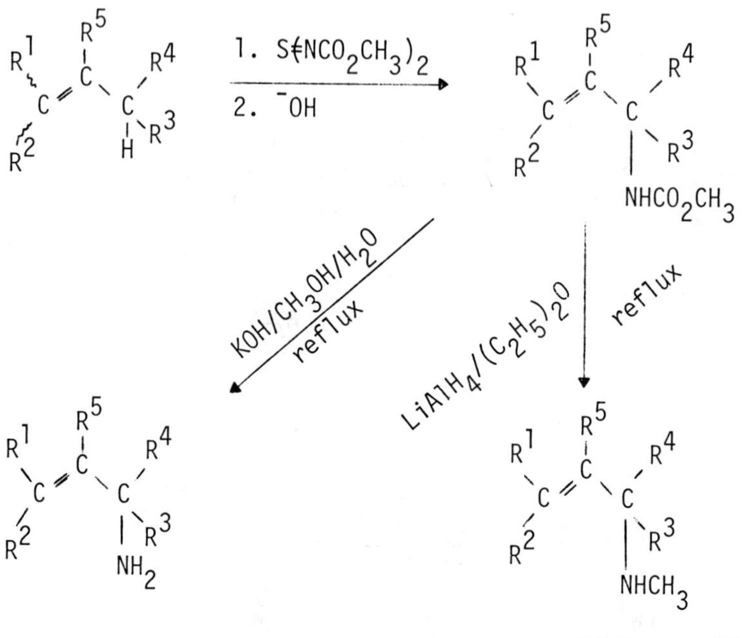

Reagents: 1. S(NCO$_2$CH$_3$)$_2$; 2. $^-$OH

Products via LiAlH$_4$/(C$_2$H$_5$)$_2$O, reflux → NHCH$_3$ product (~40% overall)

Via KOH/CH$_3$OH/H$_2$O, reflux → NH$_2$ product

II.B.3-5 J. N. Reed and V. Snieckus, Tetrahedron Lett., 24, 3795 (1983).

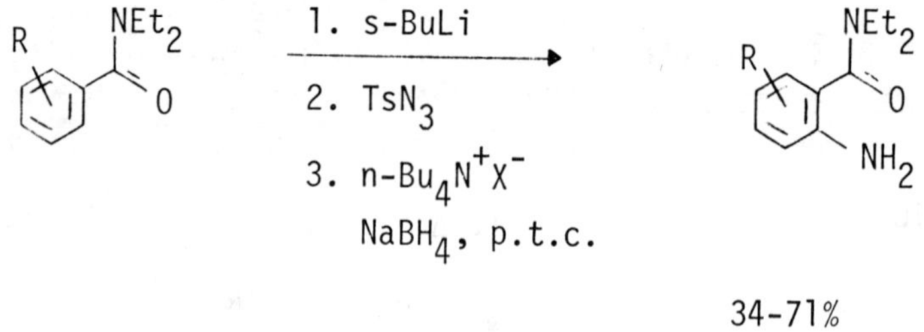

Ar–C(=O)NEt$_2$ →
1. s-BuLi
2. TsN$_3$
3. n-Bu$_4$N$^+$X$^-$, NaBH$_4$, p.t.c.

→ ortho-NH$_2$ substituted Ar–C(=O)NEt$_2$

34–71%

II.B.3-6 Y. Endo, K. Shudo and T. Okamoto, <u>Synthesis</u>, 471 (1983).

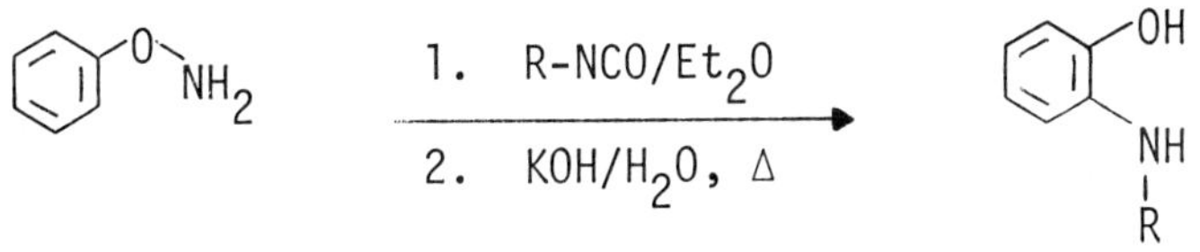

PhO-NH$_2$ $\xrightarrow{\text{1. R-NCO/Et}_2\text{O}}_{\text{2. KOH/H}_2\text{O, }\Delta}$ 2-(NHR)-C$_6$H$_4$-OH

40-75%

II.B.3-7 C. Caristi et al., <u>Tetrahedron Lett.</u>, <u>25</u>, 2685 (1983).

R-C(O)-N(CH$_3$)$_2$ + 2NBS → R-C(O)-N(CH$_3$)-CH$_2$-(succinimide)

almost quantitative

II.B.3-8 N. Miyoshi et al., <u>Tetrahedron Lett.</u>, <u>23</u>, 4813 (1982).

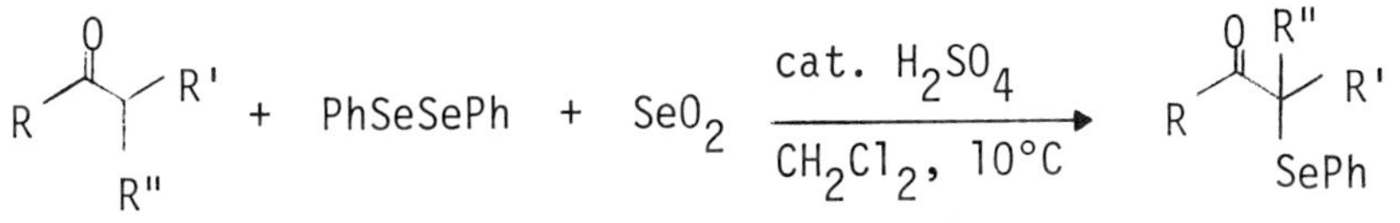

R-C(O)-CHR'R" + PhSeSePh + SeO$_2$ $\xrightarrow[\text{CH}_2\text{Cl}_2,\ 10°\text{C}]{\text{cat. H}_2\text{SO}_4}$ R-C(O)-C(R')(R")(SePh)

R = Alkyl, aryl, H   38-87%

II.B.3-9 H. Emde and G. Simchen, Liebigs Ann. Chem., 816 (1983).

$$R^3-CH_2-C{\overset{O}{\underset{OR^4}{\nwarrow}}} \quad \xrightarrow[\text{Ether, 0-25°C}]{F_3CSO_2OSi-R^1{\atop |}R^2\atop |R^2} \quad {R^3\atop H}C=C{R^2\atop \underset{OR^4}{|}}{OSi-R^1\atop |R^2} \quad + \quad {R^3\atop |}H-C-C{\overset{O}{\nwarrow}\atop OR^4}\atop \underset{R^1\underset{R^2}{\diagup}{Si}\diagdown R^2}{|}$$

II.B.3-10 J. L. Soto et al., Org. Prep. Proced. Int., 15, 41 (1983).

[Reaction scheme: aryl with R_1, R_2, R_3 substituents bearing a $-C(O)-CH_2-CN$ group, treated with 1. NOCl, 2. TosCl, giving the corresponding $-C(O)-C(=NOTs)-CN$ product]

55-85%

II.B.3-11 R. J. Billedau, M. P. Sibi and V. Snieckus, Tetrahedron Lett., 24, 4515 (1983).

[Reaction scheme: R_3SiO-substituted aryl $C(O)NEt_2$ → s-BuLi / TMEDA / THF/-78° → ortho-lithiated intermediate → 1. → RT, 2. H_2O → HO-substituted aryl with ortho-SiR_3 and $C(O)NEt_2$]

47-78%

II.B.3-12 N. S. Narasimhan and R. Ammanamanchi, Tetrahedron Lett., 24, 4733 (1983).

$$\text{ArLi} \xrightarrow[\text{2. Ni-Al}]{\text{1. TosN}_3} \text{ArNH}_2$$

for introducing ortho amino group

34-85%

II.C. C—N Oxidations

II.C-1 A. R. Katritzky et al., Recl. J. R. Neth. Chem. Soc., 102, 51 (1983).

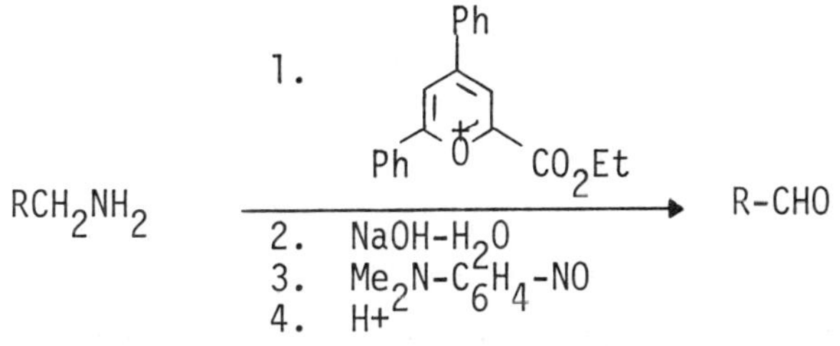

$$\text{RCH}_2\text{NH}_2 \xrightarrow[\substack{\text{2. NaOH-H}_2\text{O} \\ \text{3. Me}_2\text{N-C}_6\text{H}_4\text{-NO} \\ \text{4. H+}}]{\text{1. } \text{Ph-pyrylium-CO}_2\text{Et}} \text{R-CHO}$$

II.C-2 K. Takabe and T. Yamada, Chem. Ind. (London), 959 (1982).

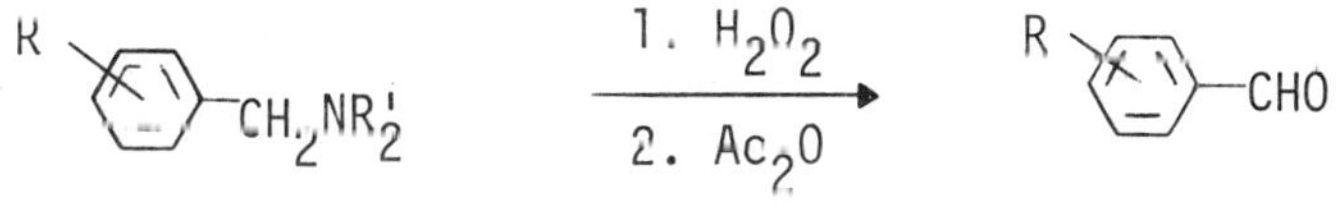

$$\text{R-C}_6\text{H}_4\text{-CH}_2\text{NR}'_2 \xrightarrow[\text{2. Ac}_2\text{O}]{\text{1. H}_2\text{O}_2} \text{R-C}_6\text{H}_4\text{-CHO}$$

38-60%

II.C-3 Y. Ohshiro et al., Tetrahedron Lett., 24, 3465 (1983).

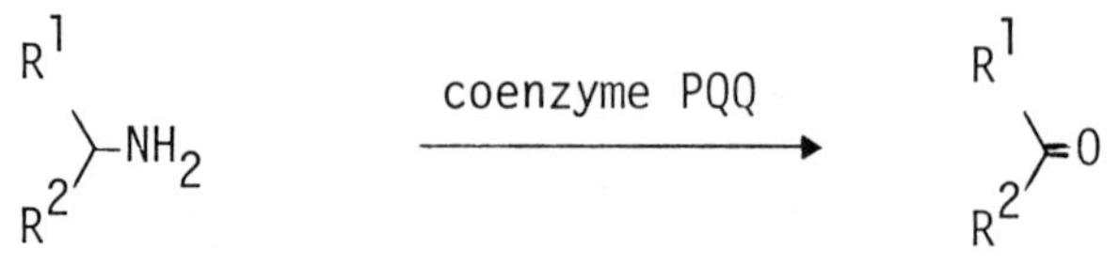

R1
NH2
R2
coenzyme PQQ
R1
O
R2

II.C-4 C. Klein, G. Schulz and W. Steglich, Liebigs Ann. Chem. 1623 (1983).

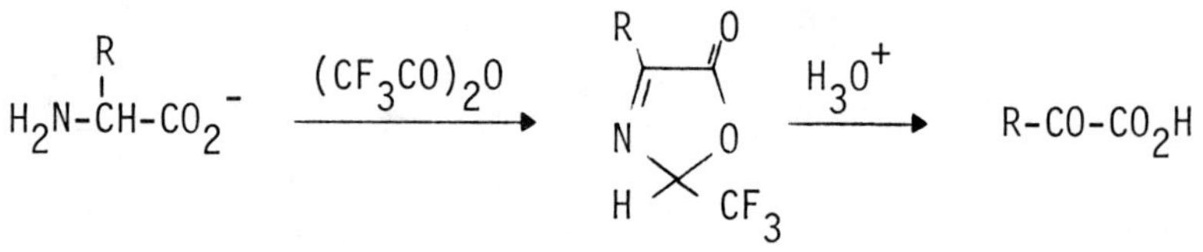

R
H2N-CH-CO2-
(CF3CO)2O
R
O
N
O
H
CF3
H3O+
R-CO-CO2H

R = ω-guanidinoalky or ω-ureidoalkyl

II.C-5 C. Klein, G. Schulz and W. Steglich, Liebigs Ann. Chem., 1638 (1983).

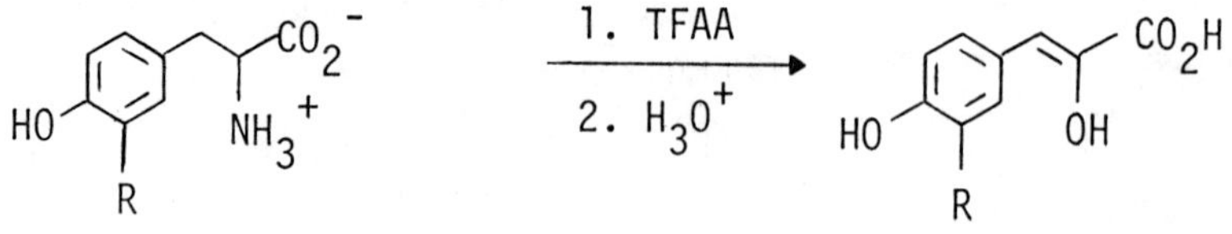

CO2-
HO
NH3+
R
1. TFAA
2. H3O+
CO2H
HO
OH
R

61-84%

II.C-6 T. Shono, Y. Matsumura and S. Kashimura, J. Org. Chem., 48, 3338 (1983).

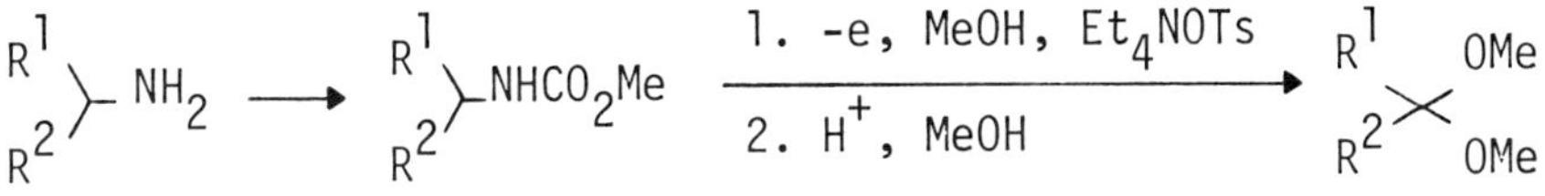

$$R^1R^2C(NH_2) \longrightarrow R^1R^2C(NHCO_2Me) \xrightarrow{\text{1. -e, MeOH, Et}_4\text{NOTs}}_{\text{2. H}^+\text{, MeOH}} R^1R^2C(OMe)_2$$

60-90%

II.C-7 K. Takabe, T. Yamada and T. Katagiri, Chem. Lett., 1987 (1982).

$$R^1R^2C=CR^3CH_2NR_2 \xrightarrow[\text{2. }(CH_3CO)_2O]{\text{1. }H_2O_2} R^1R^2C=CR^3CHO$$

56-76%

II.C-8 J. Nokami, T. Sonoda and S. Wakabayashi, Synthesis, 763 (1983).

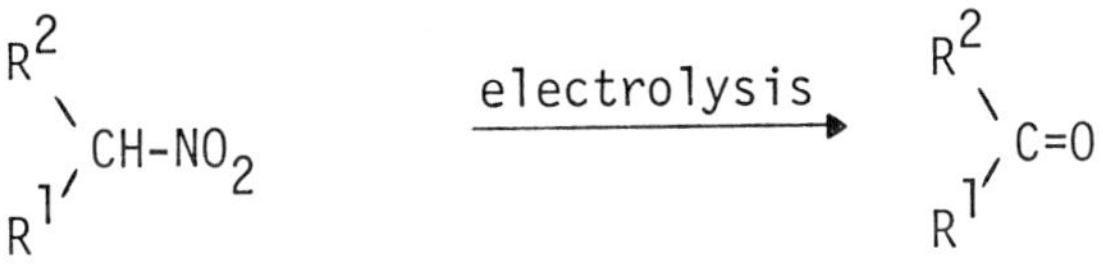

$$R^1R^2CH-NO_2 \xrightarrow{\text{electrolysis}} R^1R^2C=O$$

43-90%

II.C-9 W. T. Monte, M. M. Baizer and R. D. Little, J. Org. Chem., 48, 803 (1983)

$$RR'CHNO_2 \xrightarrow{O_2,\ e^-} RR'C=O$$

68-86%

II.C-10 M. R. Semmelhack and C. R. Schmid, J. Am. Chem. Soc., 105, 6732 (1983).

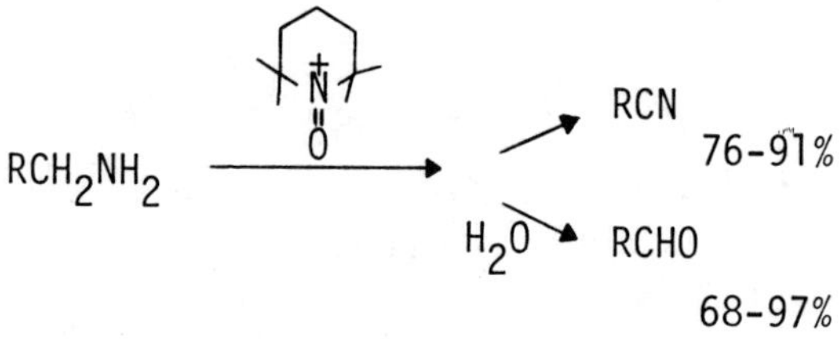

$$RCH_2NH_2 \xrightarrow{\text{TEMPO}} \begin{array}{l} RCN \quad 76\text{-}91\% \\ \xrightarrow{H_2O} RCHO \quad 68\text{-}97\% \end{array}$$

II.C-11 H. Singh, S. K. Aggarwal and N. Malhotra, Synthesis, 791 (1983).

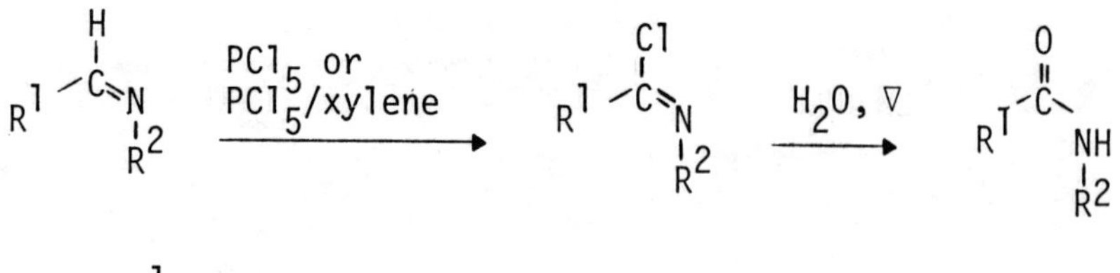

$$R^1-\underset{R^2}{\underset{|}{C}}=\underset{H}{\overset{}{N}} \xrightarrow[\text{PCl}_5/\text{xylene}]{\text{PCl}_5 \text{ or}} R^1-\underset{R^2}{\underset{|}{C}}=\underset{Cl}{\overset{}{N}} \xrightarrow{H_2O,\ \nabla} R^1-\underset{\underset{R^2}{|}}{\overset{O}{\overset{\|}{C}}}-NH$$

R^1 = aryl   30-60%

II.D. Amine Oxidations

II.D-1 A. J. Biloski and B. Ganem, Synthesis, 537 (1983).

$$R^1NHR^2 \xrightarrow[\substack{\text{or } Na_2HPO_4 \\ 2.\ (PhCO_2)_2 \\ 3.\ KOMe}]{1.\ \text{poly-4-vinylpyridine}} R^1N(OH)R^2$$

II.D-2 R. V. Hoffman and E. L. Belfoure, Synthesis, 34 (1983).

$$2\ R\text{-}NH_2 + [Ar\text{-}SO_2\text{-}O]_2 \longrightarrow R\text{-}NH\text{-}O\text{-}SO_2Ar$$

63-96%

II.D-3 L. Castedo, R. Riguera and M. P. Vazquez, J. Chem. Soc. Chem. Commun., 301 (1983).

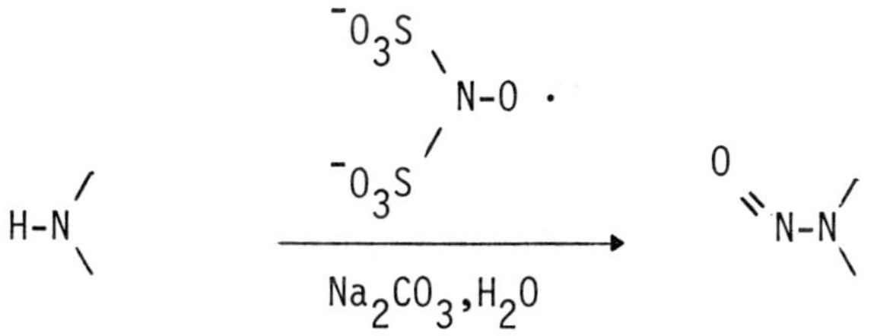

$$H\text{-}N\diagdown \xrightarrow[Na_2CO_3, H_2O]{\substack{^-O_3S \\ \diagdown N\text{-}O\cdot \\ ^-O_3S\diagup}} \overset{O}{\underset{}{\diagdown}}N\text{-}N\diagdown$$

up to 40%

II.D-4   R. G. Srivastava, R. L. Pandey and P. S. Venkataramani, Indian J. Chem. Sect. B 20B, 995 (1981).

$$2 ArNH_2 \xrightarrow{BaMnO_4} ArN=NAr$$

30-77%

II.D-5   G. D. Hartman, J. E. Schwering and R. D. Hartman, Tetrahedron Lett., 24, 1011 (1983).

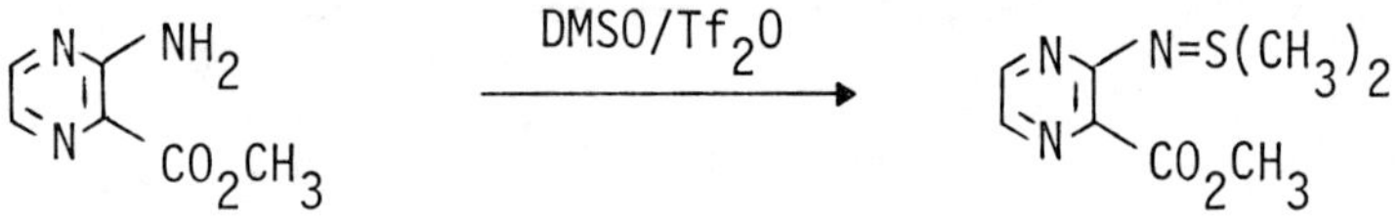

[pyrazine with NH_2 and CO_2CH_3 substituents] $\xrightarrow{DMSO/Tf_2O}$ [pyrazine with $N=S(CH_3)_2$ and CO_2CH_3 substituents]

85%

II.D-6/II.E-1   A. McKillop and J. A. Tarbin, Tetrahedron Lett., 24, 1505 (1983).

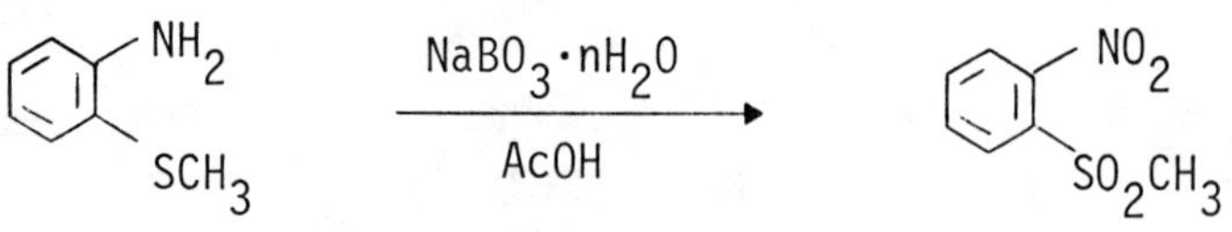

[benzene with NH_2 and SCH_3 substituents] $\xrightarrow[AcOH]{NaBO_3 \cdot nH_2O}$ [benzene with NO_2 and SO_2CH_3 substituents]

81%

II.E-2 Sulfur Oxidations

II.E-2 T. Takata and W. Ando, Tetrahedron Lett., 24, 3631 (1983).

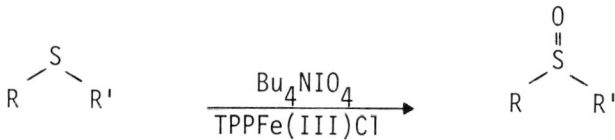

72-96%

TPPFe(III)Cl = 5,10,15,20-tetraphenylporphinato iron(III)chloride

---

II.E-3 J. Yoshida, H. Sofuku and N. Kawabata, Bull. Chem. Soc. Jpn. 56, 1243 (1983).

$$\text{RSR'} \xrightarrow[\substack{\text{Poly(4-vinylpyridine)·HBr} \\ \text{CH}_3\text{CN, H}_2\text{O}}]{\text{electric current}} \text{R}\overset{\overset{\text{O}}{\|}}{\text{S}}\text{R'}$$

72-95%

II.E-4 J. M. Aizpurua and C. Palomo, Tetrahedron Lett., 24, 4367 (1983).

$$2 \text{ R-SH} \xrightarrow[\text{or CrO}_3, \text{ClSiMe}_3]{\text{K}_2\text{Cr}_2\text{O}_7} (\text{R-S})_2$$

44-96%

II.E-5 A. Cornelis et al., Tetrahedron Lett., 24, 3103 (1983).

$$2\,RSH \xrightarrow{Fe(NO_3)_3 \cdot 9\,H_2O} RSSR$$

R = aryl, prim- or sec-alkyl

39-97%

II.E-6 E. Santaniello, F. Milani and R. Casati, Synthesis, 749 (1983).

$$2\,R\text{-}SH \xrightarrow[CHCl_3]{n\text{-}Bu_4N^{\oplus}CrO_3Cl^{\ominus}} R\text{-}S\text{-}S\text{-}R$$

80-87%

II.E-7 H. Firouzabadi et al., Synth. Commun. 13, 611 (1983).

$$2\,R\text{-}SH \xrightarrow{BPCP} RS\text{-}SR$$

60-100%

BPCP = (
[2,2'-bipyridine])$_2$Cu(MnO$_4$)$_2$

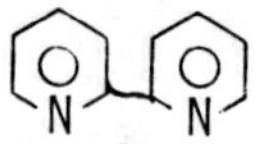

II.E-8 G. A. Olah and B. G. B. Grupta, J. Org. Chem., 48, 3585 (1983).

$$RSOR' \xrightarrow{NO_2BF_4} RSO_2R'$$

65-92%

II.E-9 T. Kageyama, Y. Ueno and M. Okawara, Synthesis, 815 (1983).

$$R^1\text{-}S\text{-}R^2 \xrightarrow[\text{r.t.}]{NaBrO_2} R^1\text{-}\overset{\overset{O}{\|}}{S}\text{-}R^2$$

18-97%

II.E-10 F. Gasparrini et al., Tetrahedron, 39, 3181 (1983).

$$(Ar)R\text{-}S\text{-}R' \xrightarrow[CH_3NO_2/HNO_3\ 10\%,\ r.t.]{Au(III)catalyst} (Ar)R\text{-}\overset{\overset{O}{\|}}{S}\text{-}R'$$

76-97%

II.E.-11 F. A. Davis, O. D. Stringer and J. M. Billmers, Tetrahedron Lett., 24, 1213 (1983).

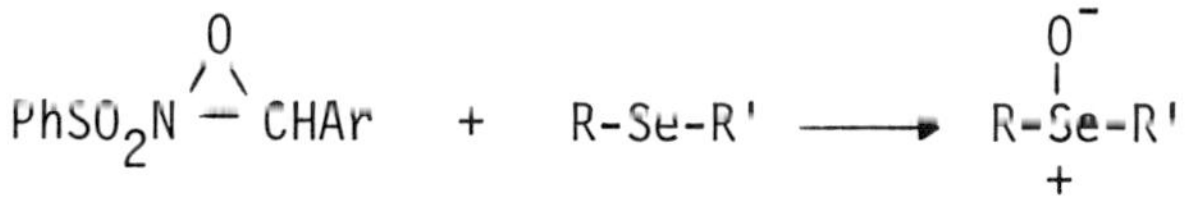

$$PhSO_2N\overset{O}{\overset{\diagup\diagdown}{-}}CHAr\ +\ R\text{-}Se\text{-}R' \longrightarrow \overset{+}{R}=\overset{O^-}{\underset{|}{Se}}\text{-}R'$$

II.F. Oxidative Additions to C—C Multiple Bonds

1. Epoxidations

II.F.1-1 R. Antonioletti et al., Tetrahedron, 39, 1765 (1983).

$$\underset{R^2\ \ \ \ H}{\overset{R^1\ \ \ \ R^3}{\diagup\!\!=\!\!\diagdown}} \quad \xrightarrow[\text{2. } Al_2O_3]{\text{1. } I_2,\ PDC} \quad \underset{R^2\ \ \ \ H}{\overset{R^1\ \ O\ \ R^3}{\triangle}}$$

30-65%

II.F.1-2 F. Camps et al., Chem. Lett., 971 (1983).

$$\underset{R^2\ \ \ R^3}{\overset{R^1\ \ \ \ CH(OCH_3)_2}{\diagup\!\!=\!\!\diagdown}} \quad \xrightarrow{KF/NaF/m\text{-}CPBA} \quad \underset{R^2\ \ \ R^3}{\overset{R^1\ \ O\ \ CH(OCH_3)_2}{\triangle}}$$

70-96%

II.F.1-3 G. Sturtz and Pondaven-Raphalen, Bull. Soc. Chim. Fr., II-125 (1983).

$$(R^1O)_2\underset{O}{\overset{\|}{P}}-(CH_2)_n-\underset{R^4}{\overset{R^2\ \ \ \ R^3}{C\!\!=\!\!C}} \quad \xrightarrow[\substack{\text{or}\\ t\text{-BuOOH/Mo(CO)}_6}]{H_2O\ /Na_2WO_4} \quad (R^1O)_2\underset{O}{\overset{\|}{P}}-(CH_2)_n-\underset{O}{\overset{R^2\ \ \ \ R^3}{C\!\!-\!\!C}}\diagdown R^4$$

70-80%

II.F.1-4   L. A. Arias et al., J. Org. Chem., 48, 888 (1983).

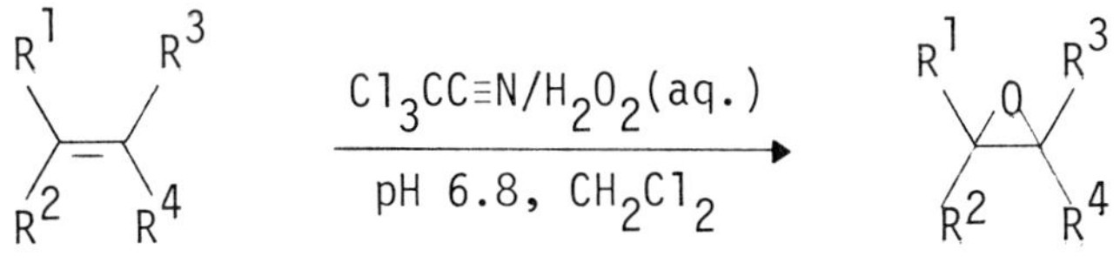

$$\underset{R^2R^4}{\overset{R^1R^3}{\diagdown\mkern-6mu/}}=\diagup\mkern-14mu\diagdown \quad\xrightarrow[\text{pH 6.8, CH}_2\text{Cl}_2]{\text{Cl}_3\text{CC}\equiv\text{N/H}_2\text{O}_2\text{(aq.)}}\quad \underset{R^2R^4}{\overset{R^1R^3}{\triangle}}$$

57-90%

II.F.1-5   T. Hiyama and M. Obayashi, Tetrahedron Lett., 24, 395 (1983).

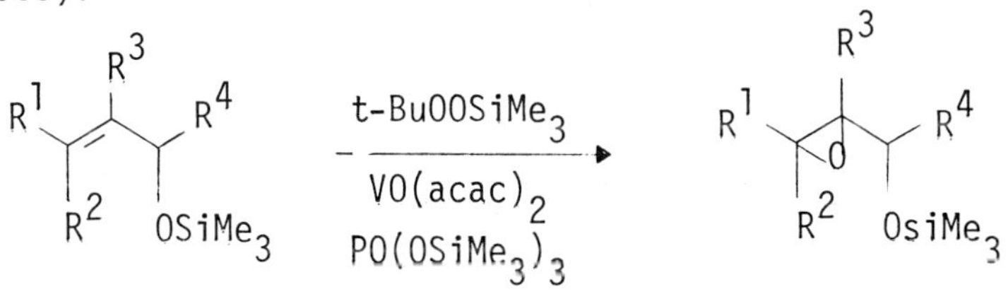

$$\xrightarrow[\substack{\text{VO(acac)}_2 \\ \text{PO(OSiMe}_3)_3}]{\text{t-BuOOSiMe}_3}$$

21-85%

same conditions

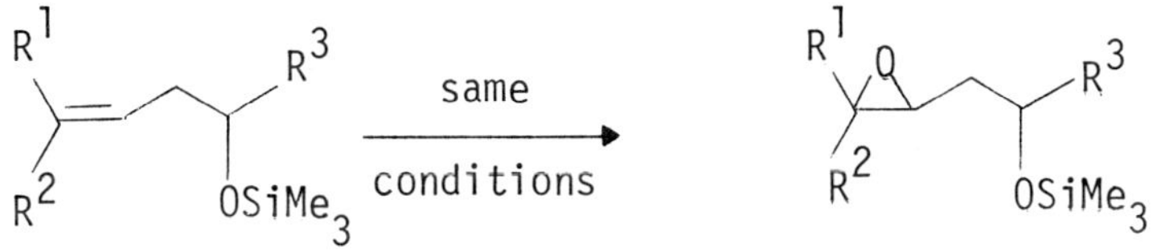

60%

II.F.1-6/II.F.2-1 A. Bongini et al., J. Org. Chem., 47, 4626 (1982).

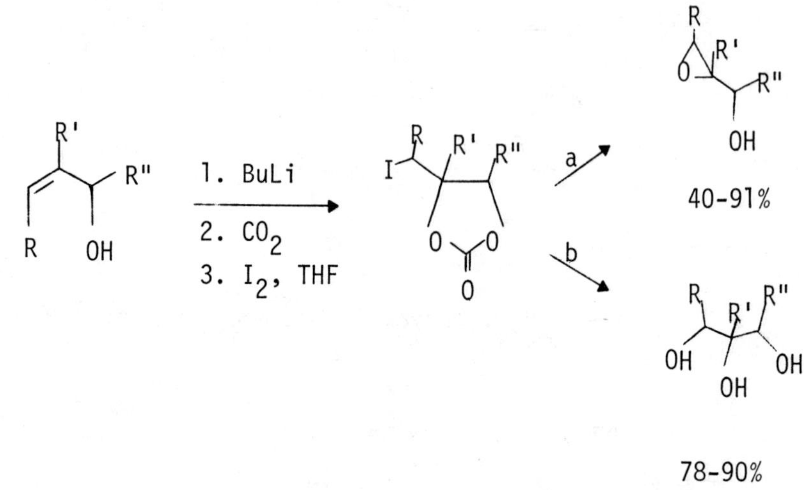

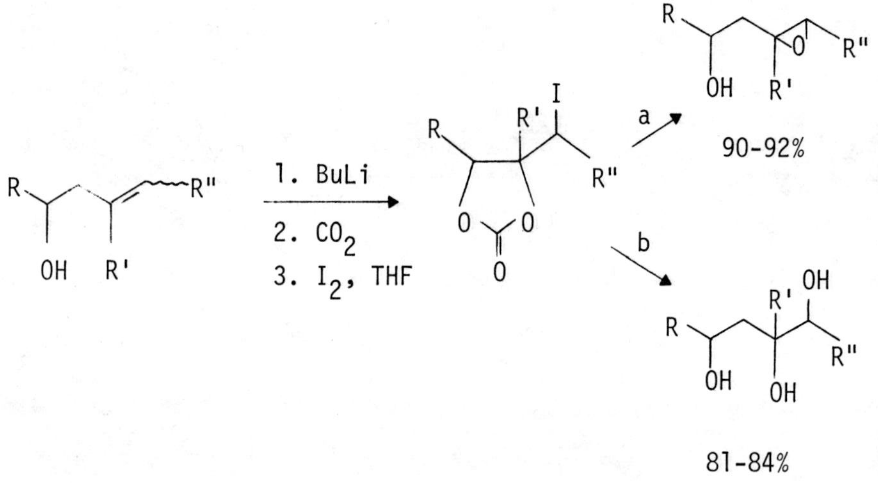

a: Amberlyst A26 OH⁻ form, MeOH
b: Amberlyst A26 $CO_3^=$ form, C_6H_6

II.F.2. Hydroxylation

II.F.2-2 T. Mukaiyama, F. Tabusa and Suzuki, K., *Chem. Lett.*, 173 (1983).

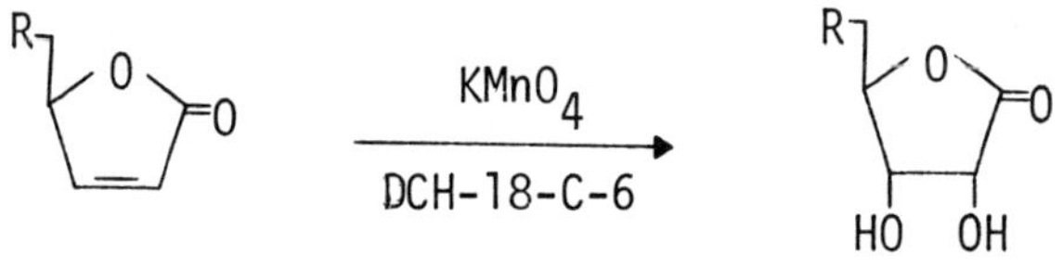

$$\text{butenolide with R group} \xrightarrow[\text{DCH-18-C-6}]{\text{KMnO}_4} \text{dihydroxy lactone}$$

27-66%

II.F.2-3 A. Citterio and C. Arnoldi, *J. Chem. Soc. Perkin Trans. 1*, 801 (1983).

$$X\text{-Ar-}\underset{R^1}{\overset{R^3}{C}}=\underset{R^2}{C} \xrightarrow[\text{AcOH}]{S_2O_8^{2-},\ Cu^{2+}} X\text{-Ar-}\underset{R^1}{\overset{R^3}{C(OAc)}}-\underset{R^2}{COAc}$$

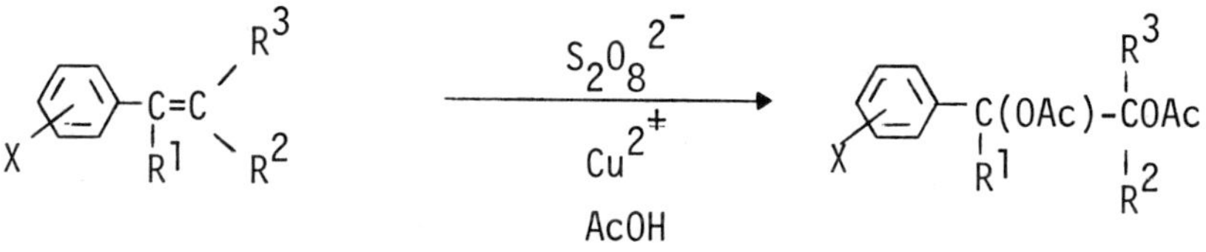

II.F.2-4 W. E. Fristad and J. R. Peterson, *Tetrahedron Lett.*, **24**, 4547 (1983).

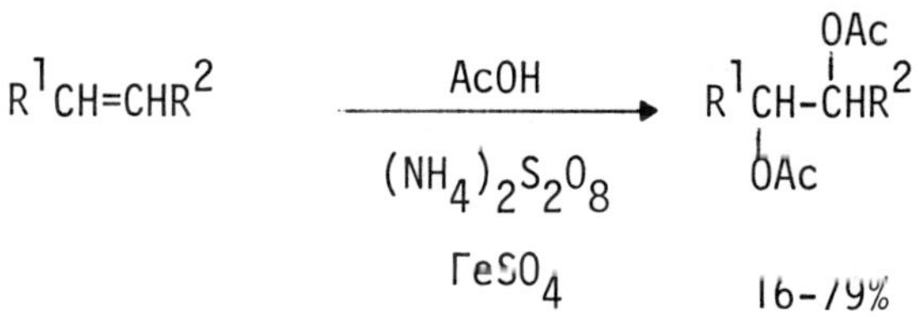

$$R^1CH=CHR^2 \xrightarrow[\substack{(NH_4)_2S_2O_8 \\ FeSO_4}]{AcOH} R^1\underset{\ }{\overset{OAc}{CH}}-\underset{OAc}{CHR^2}$$

16-79%

II.F.3. Other Oxidative Additions to C—C Multiple Bonds

II.F.3-1 N. Afza, A. Malik and W. Voelter, *J. Chem. Soc. Perkin Trans. 1*, 1349 (1983).

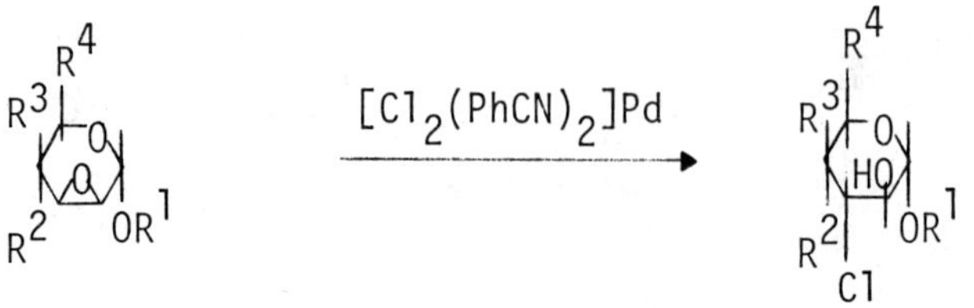

[pyranose with R^{1}O, R^2, R^3, R^4 substituents and internal epoxide] $\xrightarrow{[Cl_2(PhCN)_2]Pd}$ [pyranose with R^{1}O, R^2, R^3, R^4, HO, Cl substituents]

II.F.3-2 G. Palumbo, C. Ferreri and R. Caputo, *Tetrahedron Lett.*, 24, 1307 (1983).

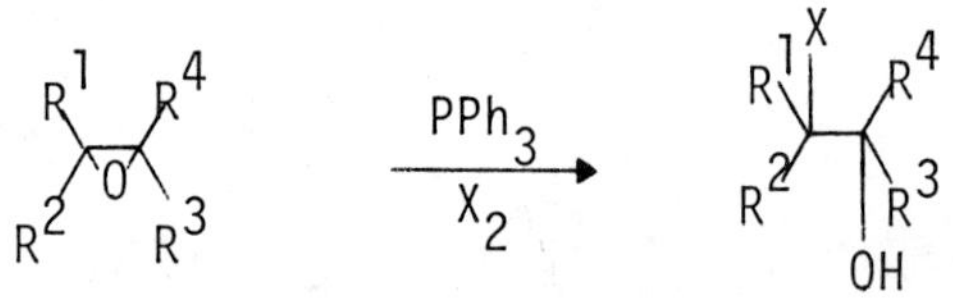

[epoxide with R^1, R^2, R^3, R^4] $\xrightarrow[X_2]{PPh_3}$ [product with X, R^1, R^2, R^3, R^4, OH]

X = Cl, Br, I

50-97%

II.F.3-3 M. Fiorenza et al., *Synthesis*, 640 (1983).

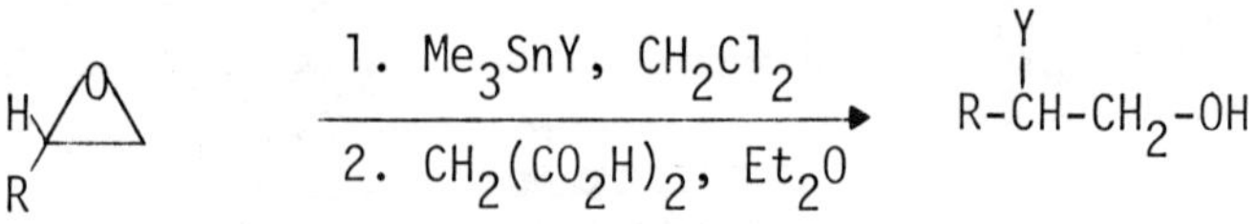

[epoxide: H, R on CH-CH$_2$-O] $\xrightarrow[\text{2. } CH_2(CO_2H)_2, Et_2O]{\text{1. } Me_3SnY, CH_2Cl_2}$ R-CH(Y)-CH$_2$-OH

Y = N$\langle$, OPh, Cl, I

~80%

II.F.3-4 I. W. J. Still and F. J. Ablenas, Synth. Commun., 12, 1103 (1982).

$$R\text{–}\triangleleft^O + ArSO_2Na \xrightarrow[\text{(cat.)}]{Mg(NO_3)_2} ArSO_2CH_2CH(OH)R$$

R = alkyl   23-83%

II.F.3-5 Y.-H. Kang and J. L. Kice, Tetrahedron Lett., 23, 5373 (1982).

$$R^1CH=C=C\begin{smallmatrix}R^2\\ \\R^3\end{smallmatrix} \xrightarrow[\substack{2.\ H_2O_2\\3.\ H_2O}]{1.\ ArSO_2SePh,\ h\nu} R^1CH=C\text{–}\underset{\underset{ArSO_2}{|}}{\overset{\overset{OH}{|}}{C}}\text{–}R^2 \;\; |\;R^3$$

70-98%

II.F.3-6 V. L. Heasley et al., J. Org. Chem., 48, 1377 (1983).

$$\underset{R_2}{\overset{R_1}{>}}C=C\underset{R_3}{\overset{\overset{O}{\|}}{<}}R_4 + NBS \xrightarrow[H_2SO_4]{CH_3OH} R_1\text{–}\underset{OCH_3}{\overset{R_2}{|}}\text{–}\underset{Br}{\overset{R_3}{|}}\text{–}\overset{O}{\|}R_4$$

12-99%

II.F.3-7 C. Blandy, R. Choukroun and D. Gervais, Tetrahedron Lett., 24, 4189 (1983).

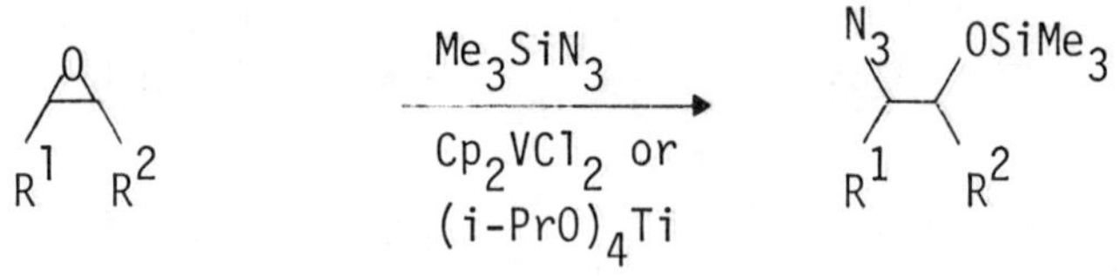

$$\underset{R^1 \quad R^2}{\triangle^O} \xrightarrow[\text{(i-PrO)}_4\text{Ti}]{\text{Me}_3\text{SiN}_3, \text{ Cp}_2\text{VCl}_2 \text{ or}} \underset{R^1 \quad R^2}{\overset{N_3 \quad OSiMe_3}{\diagdown\diagup}}$$

II.F.3-8 H. Kohn and S.-H. Jung, J. Am. Chem. Soc., 105, 4106, (1983).

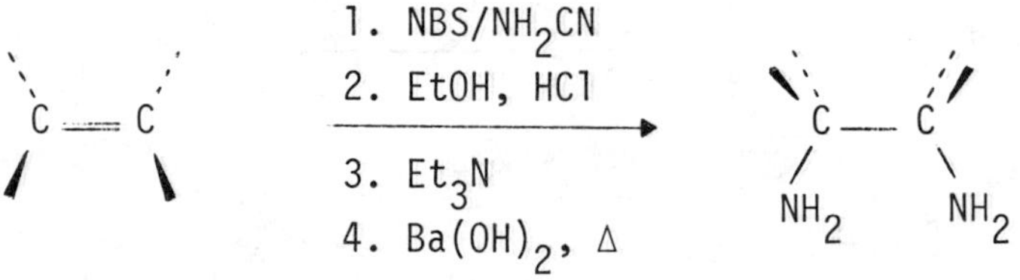

1. NBS/NH$_2$CN
2. EtOH, HCl
3. Et$_3$N
4. Ba(OH)$_2$, Δ

II.F.3-9 P. N. Becker and R. G. Bergman, Organometallics, 2, 787 (1983).

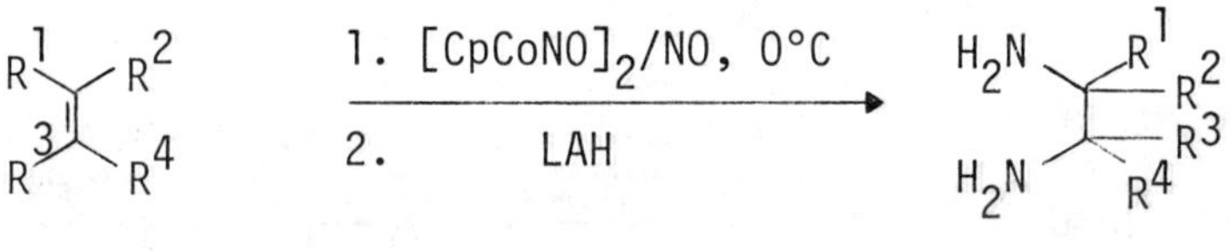

1. [CpCoNO]$_2$/NO, 0°C
2. LAH

43-90%

II.F.3-10 A. P. Croft and R. A. Bartsch, *J. Org. Chem.*, <u>48</u>, 3353 (1983).

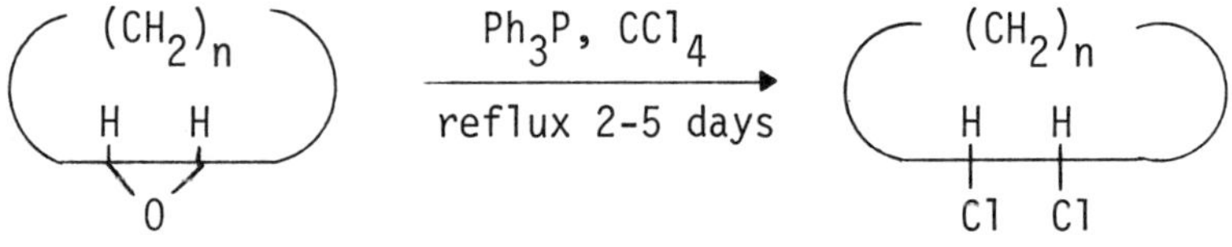

$$\text{cyclo-}(CH_2)_n\text{-CH(H)-CH(H)-O} \xrightarrow[\text{reflux 2-5 days}]{Ph_3P,\ CCl_4} \text{cyclo-}(CH_2)_n\text{-CH(H)(Cl)-CH(H)(Cl)}$$

70-80%

II.F.3-11 V. L. Heasley et al., *J. Org. Chem.*, <u>48</u>, 3195 (1983).

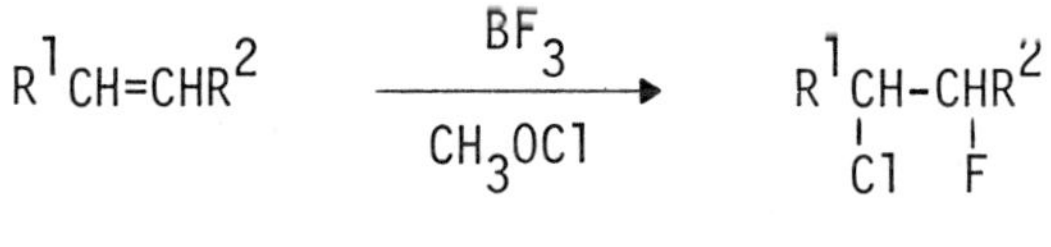

$$R^1CH=CHR^2 \xrightarrow[CH_3OCl]{BF_3} R^1\underset{Cl}{C}H-\underset{F}{C}HR^2$$

8-75%

II.F.3-12 G. Cardillo et al., *J. Chem. Soc. Chem. Commun.*, 1308 (1982).

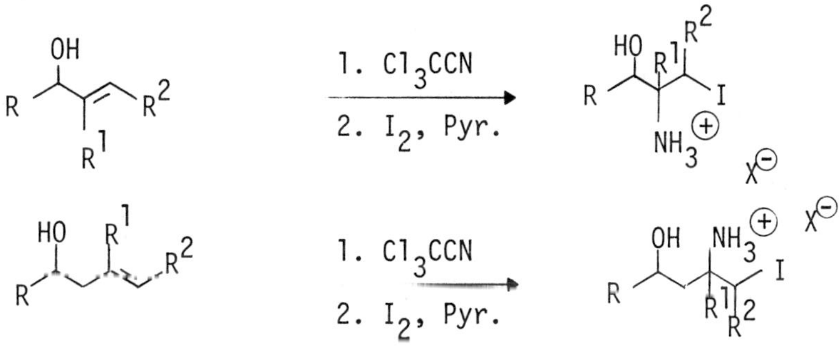

$$R\underset{\underset{R^1}{\|}}{CH(OH)}-C=CHR^2 \xrightarrow[2.\ I_2,\ Pyr.]{1.\ Cl_3CCN} R-CH(OH)-\underset{NH_3^+\ X^-}{C}R^1(R^2)-CHI$$

$$R-CH(OH)-CH(R^1)-CH=CHR^2 \xrightarrow[2.\ I_2,\ Pyr.]{1.\ Cl_3CCN} R-CH(OH)-CH_2-\underset{R^1,R^2}{C}(NH_3^+\ X^-)-CHI$$

~50-80%

II.F.3-13 M. Taddei et al., Tetrahedron Lett., 24, 2311 (1983).

$$\underset{R^1 \quad R^2}{\overset{S}{\triangle}} \xrightarrow[2.\ CH_2(CO_2H)_2]{1.\ Me_3SiY} \underset{R^1 \quad R^2}{SH \quad Y}$$

Y = NR_2, CN   ~70%

II.F.3-14 T. G. Back, S. Collins and R. G. Kerr, J. Org. Chem., 48, 3077 (1983).

$$RC{\equiv}CR' \xrightarrow[AIBN,\ \Delta]{PhSeSO_2Ar} \underset{PhSe\quad R^1}{\overset{R\quad SO_2Ar}{>=<}}$$

1. m-CPBA → dioxolane product with R, R^1, SO_2Ar substituents   74-90%

2. KOH, HOCH_2CH_2OH

$HClO_4$ → $R\text{-CO-CHR}^1\text{-SO}_2Ar$ 60-98%

II.F.3-15 A. Ourari, R. Comdom and R. Guedj, Can. J. Chem., 60, (1982).

$$\underset{R^2\quad\overset{\diagdown}{O}\diagup}{\overset{R^1\quad CO_2R^3}{\diagup\!\!\!\diagdown}} \xrightarrow[2.\ CrO_3]{1.\ HF/Pyr.} \underset{R^2\ F\quad O}{R^1\quad CO_2R^3}$$

58-87%

II.F.3-16 A. Zombeck, D. E. Hamilton and R. S. Drago, J. Am. Chem. Soc., 104, 6782 (1982).

$$R-CH=CH_2 + O_2 \xrightarrow{CoSalMDPT} R-CO-CH_3 + R-CH(OH)-CH_3$$

CoSalMDPT = cobalt(II) bis(salicylidene-γ-iminopropyl) methylamine.

II.F.3-17 T. Hosokawa, T. Ohta and S. Murahashi, J. Chem. Soc. Chem. Commun., 848 (1983).

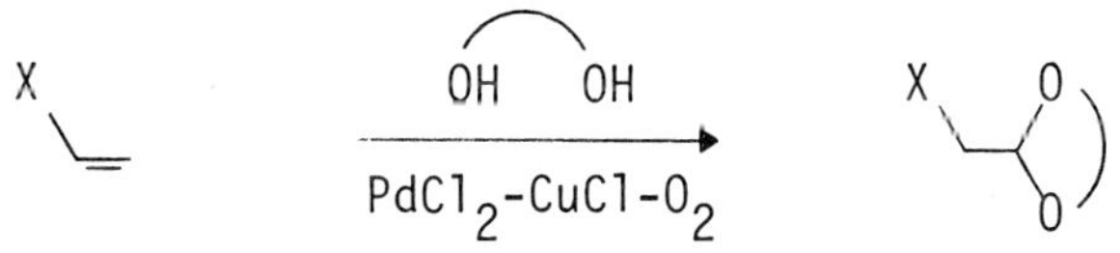

$$X-CH_2-CH=CH_2 \xrightarrow[PdCl_2-CuCl-O_2]{HO-CH_2-CH_2-OH} \text{(cyclic acetal)}$$

X = electron-withdrawing group

II.H. Oxidative Cleavages

II.H-1 J. A. Cella, Synth. Commun., 13, 93 (1983).

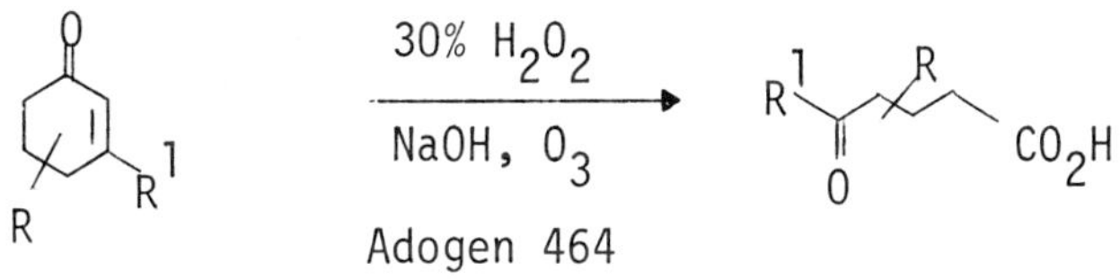

$$\text{cyclohexenone} \xrightarrow[\substack{NaOH, O_3 \\ \text{Adogen 464}}]{30\% \; H_2O_2} R^1-CO-CHR-CH_2-CO_2H$$

84-95%

II.H-2   I. Saito et al., Tetrahedron Lett., 24, 4439 (1983).

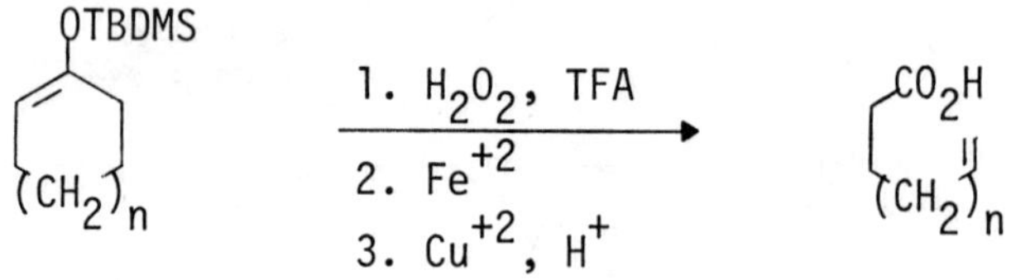

$$\text{cyclohexene-OTBDMS, (CH}_2)_n \xrightarrow[\text{2. Fe}^{+2}]{\text{1. H}_2\text{O}_2, \text{TFA}} \text{CO}_2\text{H, (CH}_2)_n\text{-CHO}$$

1. H_2O_2, TFA
2. Fe^{+2}
3. Cu^{+2}, H^+

44-77%

II.H-3   S. Ito and M. Matsumoto, J. Org. Chem., 48, 1133 (1983).

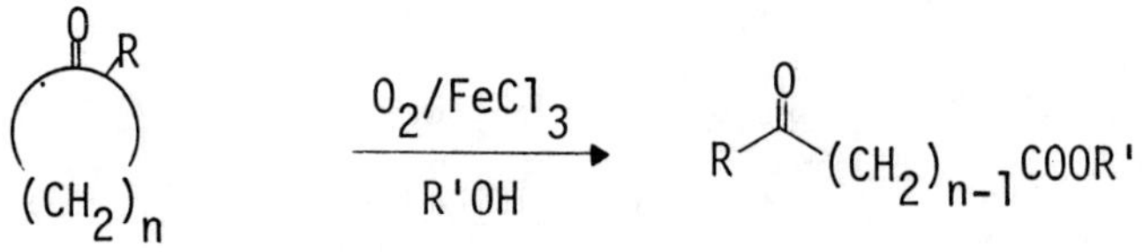

$$\text{cyclic ketone (CH}_2)_n \xrightarrow[\text{R'OH}]{\text{O}_2/\text{FeCl}_3} R-\overset{\text{O}}{\text{C}}-(CH_2)_{n-1}\text{COOR'}$$

R = alkyl

63-93%

II.H-4   S. Matsubara, K. Takai and H. Nozaki, Bull. Chem. Soc. Jpn., 56, 2029 (1983).

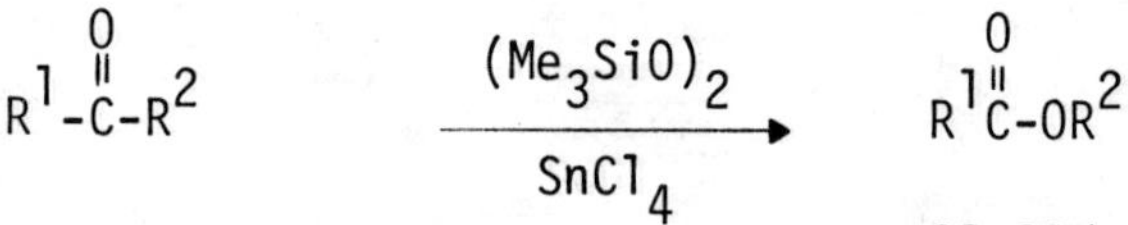

$$R^1-\overset{\text{O}}{\underset{\|}{\text{C}}}-R^2 \xrightarrow[\text{SnCl}_4]{(\text{Me}_3\text{SiO})_2} R^1-\overset{\text{O}}{\underset{\|}{\text{C}}}-OR^2$$

38-93%

II.H-5 S. L. Schreiber and W.-F. Liew, Tetrahedron Lett., 24, 2363 (1983).

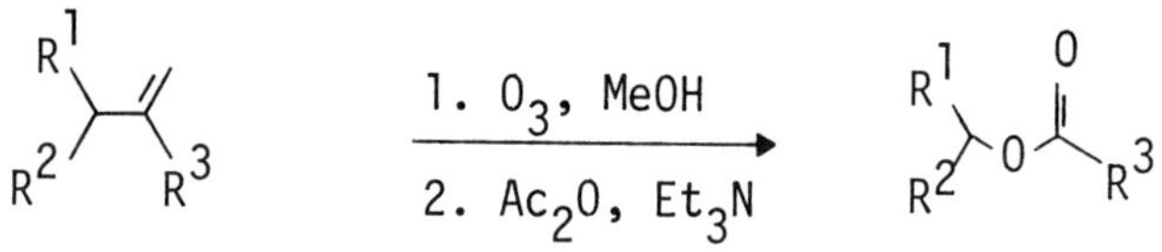

$$\underset{R^2}{\overset{R^1}{>}}\!\!=\!\!CH_2 \text{ (}R^3\text{)} \quad \xrightarrow[\text{2. } Ac_2O,\ Et_3N]{\text{1. } O_3,\ MeOH} \quad R^1R^2CH\text{-}O\text{-}C(O)\text{-}R^3$$

II.H-6 A. K. Chakraborti and U. R. Ghatak, Synthesis, 746 (1983).

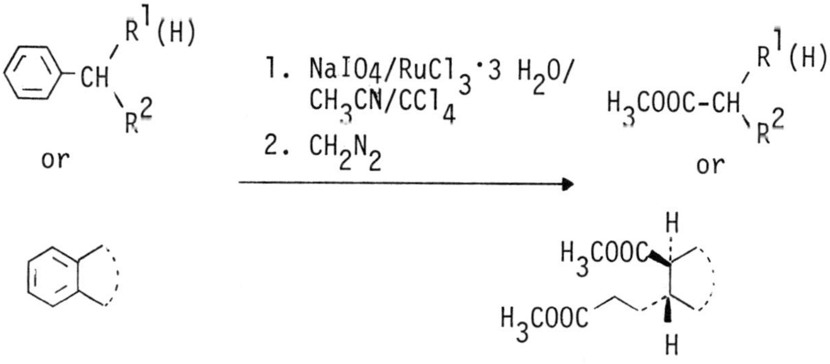

Ph-CH(R¹(H))(R²) or tetralin-type substrate

$\xrightarrow[\text{2. } CH_2N_2]{\text{1. } NaIO_4/RuCl_3 \cdot 3\,H_2O/\ CH_3CN/CCl_4}$

$H_3COOC\text{-}CH(R^1(H))(R^2)$ or cis-cyclohexane-1,2-diyl bearing H_3COOC and $H_3COOC\text{-}CH_2\text{-}$ substituents

80-90%

II.H-7 D. Wenkert, K. M. Eliasson and D. Rudisill, J. Chem. Soc. Chem. Commun., 392 (1983).

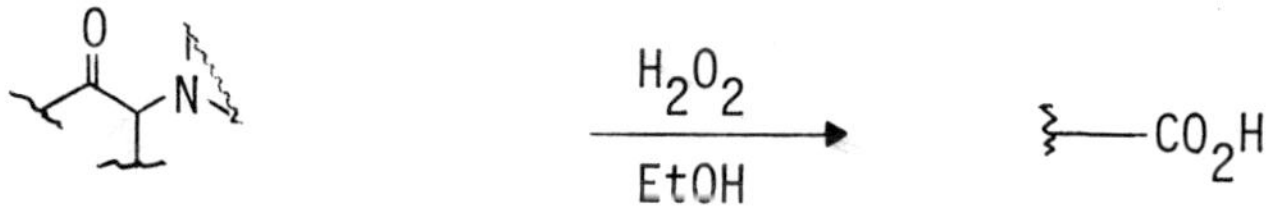

$$\text{acyl-N}_3 \xrightarrow[\text{EtOH}]{H_2O_2} \text{R-}CO_2H$$

72-96%

II.H-8 G. Weber et al., Synthesis, 191 (1983).

$$Ar\text{-}\overset{O}{\underset{\|}{C}}\text{-}CHBr_2 \xrightarrow{NaN_3} Ar\text{-}\overset{O}{\underset{\|}{C}}\text{-}N_3$$

45-92%

II.I. Photosensitized Oxygenations

II.I-1 T. R. Beebe et al., J. Org. Chem., 48, 3126 (1983)

$$ArCH_2OH \xrightarrow[h\nu]{NIS} ArCHO$$

66-93%

II.I-2 M. L. Graziano et al., Synthesis, 125 (1983).

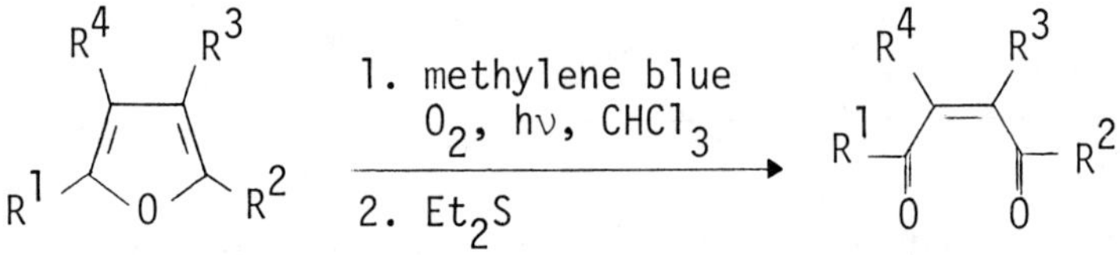

1. methylene blue O_2, hν, $CHCl_3$
2. Et_2S

70-98%

II.I-3 J. Muzart, P. Pale and J. P. Pete, Tetrahedron Lett., 24, 4567 (1983)

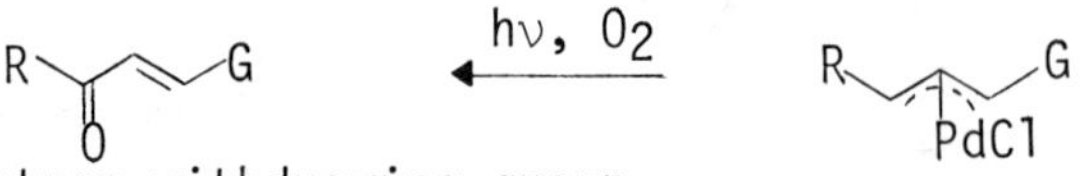

$$\xleftarrow{h\nu,\ O_2}$$

G = electron-withdrawing group

$$\underset{2}{\overset{PdCl}{}}$$

17-89%

II.J. Dehydrogenation

II.J-1 I. Shimizu, I. Minami and J. Tsuji, *Tetrahedron Lett.*, **24**, 1797 (1983).

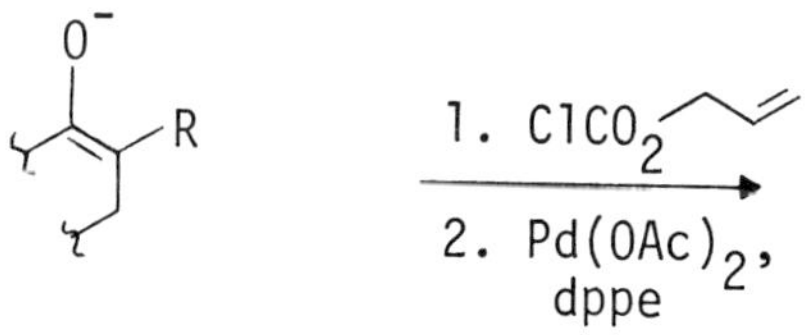

[enolate] → 1. ClCO$_2$CH$_2$CH=CH$_2$ 2. Pd(OAc)$_2$, dppe → [α,β-unsaturated ketone with R]

II.J-2 D. H. R. Barton, X. Lusinchi and P. Milliet, *Tetrahedron Lett.*, **23**, 4949 (1982).

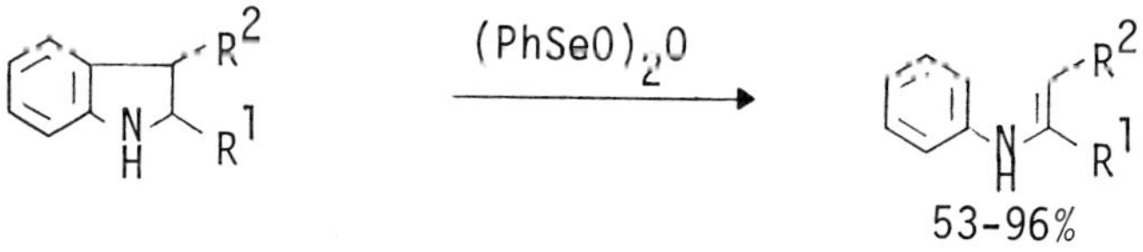

[indoline with R^2, R^1] → (PhSeO)$_2$O → [indole with R^2, R^1] 53-96%

II.J-3 T. Shono et al., *J. Am. Chem. Soc.*, **104**, 6697 (1982).

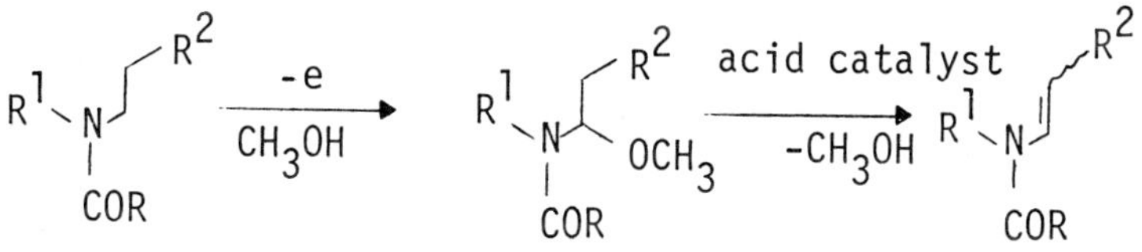

R^1-N(COR)-CH$_2$-R^2 $\xrightarrow{-e, CH_3OH}$ R^1-N(COR)-CH(OCH$_3$)-R^2 $\xrightarrow[\text{acid catalyst}]{-CH_3OH}$ R^1-N(COR)-CH=R^2

38-96%

II.J-4 T. Mukaiyama, M. Ohshima and T. Nakatsuka, Chem. Lett., 1207 (1983).

R–CH$_2$–CHO $\xrightarrow[\text{O-NMe (morpholine)}]{\text{PdCl}_2(\text{PhCN})_2\text{-AgOTf}}$ R–CH=CH–CHO   38-88%

R–CH$_2$–C(O)–R' $\xrightarrow[\text{NEt (pyrrolidine)}]{\text{Sn(OTf)}_2}$ $\xrightarrow[\text{iPr}_2\text{NEt}]{\text{PdCl}_2(\text{CH}_3\text{CN})_2}$ R–CH=CH–C(O)–R'   37-78%

II.K. Other Oxidations and Reviews

II.K-1 M. Ochiai et al., Tetrahedron Lett., 24, 777 (1983).

$\text{CH}_2=\text{C}(R)\text{CH}_2\text{SiMe}_3$ $\xrightarrow[\substack{\text{BF}_3\text{-Et}_2\text{O} \\ \text{dioxane}}]{\text{C}_6\text{H}_5\text{I=O}}$ $\text{CH}_2=\text{C}(R)\text{CHO}$ 63-72%

II.K-2 D. T. W. Chu, J. Org. Chem., 48, 3571 (1983).

RCH$_2$SPh $\xrightarrow[\text{2. H}_2\text{O, SiO}_2]{\text{1. SO}_2\text{Cl}_2}$ RCHO   58-72%

II.K-3 S. M. Paraskewas and A. A. Danopoulos, Synthesis, 638 (1983).

$$R-NCS \xrightarrow[O_2]{PdCl_2} R-NCO$$

49-96%

II.K-4 J. S. Swenton, Acc. Chem. Res. 16, 74 (1983).

Review: "Quinone Bis- and Monoketals via Electrochemical Oxidation. Versatile Intermediates for Organic Synthesis."

II.K-5 J.-E. Backvall, Acc. Chem. Res., 16, 335 (1983).

Review: "Palladium in Some Selective Oxidation Reactions."

III
REDUCTIONS

III.A. C=O Reductions   (see also: III.F.1)

II.A-1 C. F. Nutaitis and G. W. Gribble, Tetrahedron Lett., 24, 4287 (1983).

$$RCHO \xrightarrow[C_6H_6]{n\text{-}Bu_4NBH(OAc)_3} RCH_2OH$$

80-96%

III.A-2 C. Camacho, G. Uribe and R. Contreras, Synthesis, 1027 (1982).

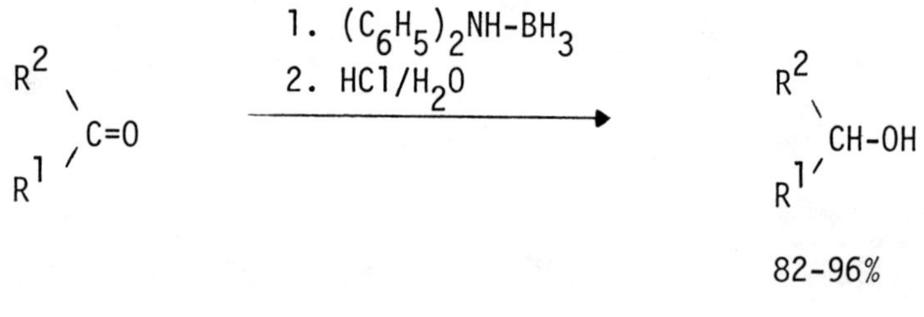

$$\underset{R^1}{\overset{R^2}{>}}C=O \xrightarrow[\text{2. HCl/H}_2O]{\text{1. }(C_6H_5)_2NH\text{-}BH_3} \underset{R^1}{\overset{R^2}{>}}CH\text{-}OH$$

82-96%

III.A-3 C. Chuit et al., Synthesis, 981 (1982);
F. J. P. Corriu, R. Perz and C. Reye, Tetrahedron, 39, 999 (1983).

$$\begin{array}{c} R \\ R^1 \end{array}\!\!C=O \quad \xrightarrow[\text{2. } H_3O^+]{\text{1. HSi}\!\!\leftarrow, \text{ salt}} \quad \begin{array}{c} R \\ R^1 \end{array}\!\!CHOH$$

72-99%

$$HSi\!\!\leftarrow \;=\; \begin{array}{c} Me \\ | \\ H-Si-OEt \\ | \\ OEt \end{array} \quad \text{or} \quad Me_3Si-O\left[\begin{array}{c} Me \\ | \\ -Si-O \\ | \\ H \end{array}\right]_n\!\!-SiMe_3$$

III.A-4 G. Dupas et al., Tetrahedron Lett., 23, 5141 (1982).

$$\begin{array}{c} R_1 \\ R_2 \end{array}\!\!C=O \;+\; \text{[dihydropyridine-CON}\!\!<\!\!\begin{array}{c}A\\A'\end{array}\text{, N-(P)]} \quad \xrightarrow[\text{2. } H_2O]{\text{1. } CH_3CN/C_6H_6,\; Mg^{2+}} \quad \begin{array}{c} R_1 \\ R_2 \end{array}\!\!C\!\!<\!\!\begin{array}{c}OH\\H\end{array}$$

III.A-5 Y. Kamitori et al., Synthesis, 387 (1983).

$$\begin{array}{c} R^1 \\ R^2 \end{array}\!\!C=O \quad \xrightarrow{LiAlH_4/SiO_2/n\text{-}C_6H_{14} \text{ or } C_6H_6} \quad \begin{array}{c} R^1 \\ R^2 \end{array}\!\!CH\text{-}OH$$

R^1, R^2 = alkyl, aryl, H   69-100%

III.A-6 Y. Kamitori et al., Tetrahedron Lett., 24, 2575 (1983).

$$X-\langle O \rangle-\overset{O}{\underset{\|}{C}}-R \xrightarrow{LAH-SiO_2} X-\langle O \rangle-CHOH-R$$

$X = NO_2, CN$

82 & 77%

II.A-7 D. Wang and T.H. Chan, Tetrahedron Lett., 24, 1573 (1983).

$$R^1\underset{R^2}{\rangle}=O + \text{(pinanyl)}SiMeH_2 \xrightarrow{(Ph_3P)_3RhCl} \xrightarrow{CH_3OK-CH_3OH} R^1\underset{R^2}{\rangle}\overset{*}{C}H-OH$$

79-96%   8.8-25.7% e.e.

III.A-8 A. Mori et al., Tetrahedron Lett., 24, 4581 (1983).

$$R^1\underset{R^2}{\rangle}=O \xrightarrow[H^+]{\underset{OH\ OH}{R\diagup\diagdown R}} \underset{R^1\diagdown R^2}{\overset{O\diagup\diagdown O}{\bigcirc}} \xrightarrow[\substack{1.\ R_2AlH \\ 2.\ [O] \\ 3.\ Base}]{} R^1\underset{S}{\overset{H\ \ OH}{\diagdown\!\!\diagup}}R^2$$

R = Br, Cl, i-Bu

64-99%, 78-96% e.e.

III.A-9 H. Suda et al., Tetrahedron Lett., 24, 1513 (1983).

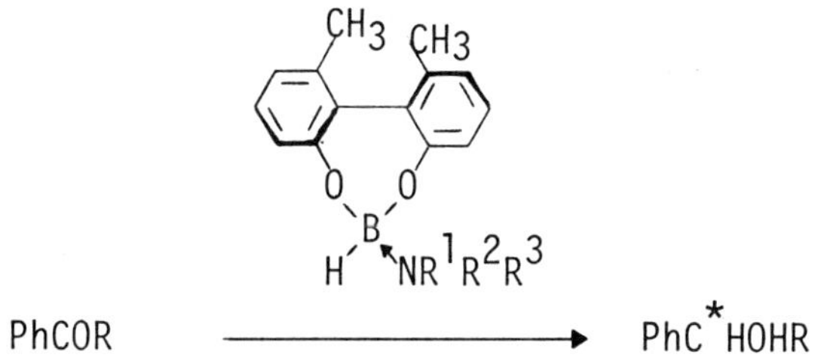

PhCOR → PhC*HOHR

(reagent: biphenyl-bridged dioxaborine with H and NR^{1}R^{2}R^3 substituents on B; ortho-CH$_3$ groups on aryl rings)

III.A-10 S. Itsuno et al., J. Chem. Soc. Perkin Trans. 1, 1673 (1983).

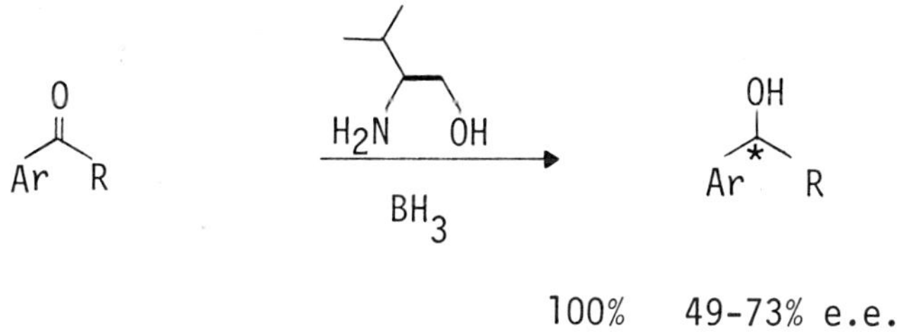

Ar-CO-R →[H$_2$N-CH(iPr)-CH$_2$OH / BH$_3$]→ Ar-C*H(OH)-R

100% 49-73% e.e.

III.A-11 S. Itsuno et al., J. Chem. Soc. Chem. Commun., 469 (1983).

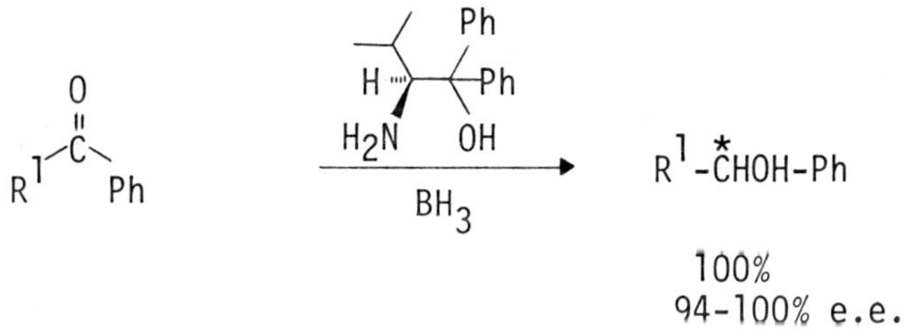

R^1-CO-Ph →[amino alcohol (H, iPr, H$_2$N, OH, CPh$_2$) / BH$_3$]→ R^1-C*HOH-Ph

100%
94-100% e.e.

III.A-12 H. C. Brown and G. G. Pai, J. Org. Chem., 48, 1784 (1983).

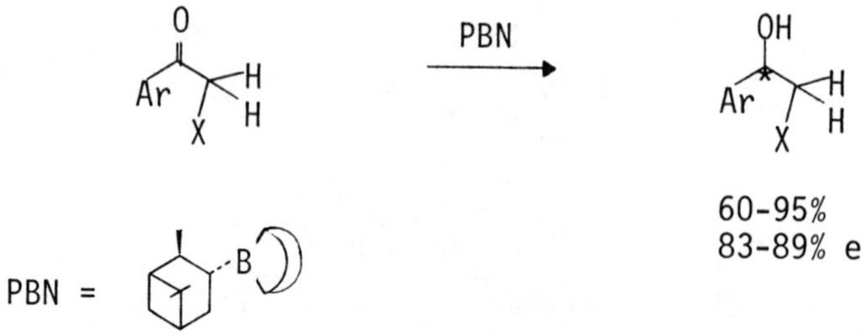

Ar-CO-CHXH →(PBN) Ar-C*(OH)H-CHXH

60-95%
83-89% e.e.

PBN = [pinanyl-B structure]

III.A-13 T. Nakata, T. Tanaka and T. Oishi, Tetrahedron Lett., 24, 2653 (1983).

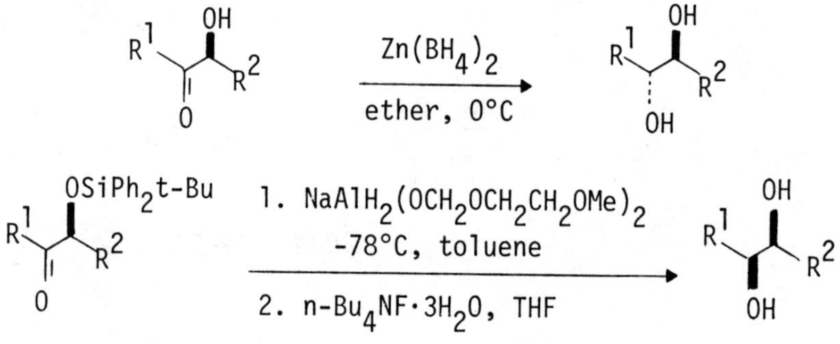

R¹-C(=O)-CH(OH)-R² →[Zn(BH$_4$)$_2$, ether, 0°C] R¹-CH(OH)-CH(OH)-R²

R¹-C(=O)-CH(OSiPh$_2$t-Bu)-R² →[1. NaAlH$_2$(OCH$_2$OCH$_2$CH$_2$OMe)$_2$, -78°C, toluene; 2. n-Bu$_4$NF·3H$_2$O, THF] R¹-CH(OH)-CH(OH)-R²

III.A-14 A. K. Samaddar, S. K. Konar and D. Nasipuri, J. Chem. Soc. Perkin Trans. 1, 1449 (1983).

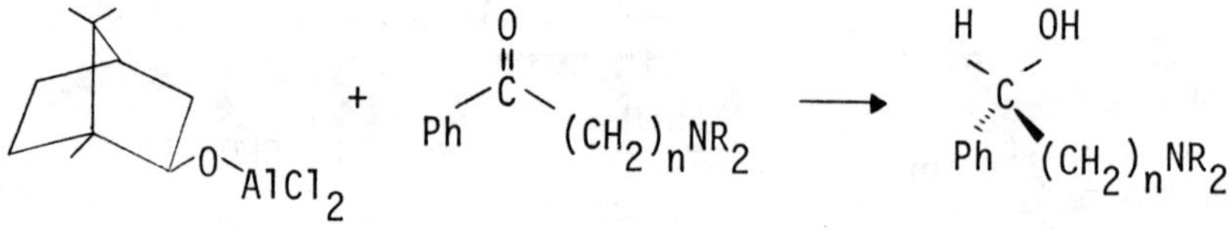

[bornyloxy-AlCl$_2$] + Ph-C(=O)-(CH$_2$)$_n$NR$_2$ → Ph-CH(OH)-(CH$_2$)$_n$NR$_2$

n = 1,2

75-90%
58-92% e.e.

III.A-15 A. M. Caporusso, G. Giacomelli and L. Lardicci, J. Org. Chem., 47, 4640 (1982).

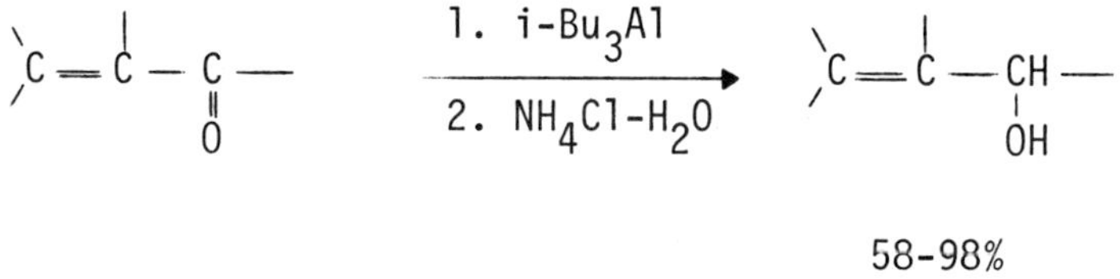

$$\diagup C=C-C(=O)- \xrightarrow[\text{2. NH}_4\text{Cl-H}_2\text{O}]{\text{1. i-Bu}_3\text{Al}} \diagup C=C-CH(OH)-$$

58-98%

III.A-16 T. Sato et al., Tetrahedron Lett., 24, 4123 (1983).

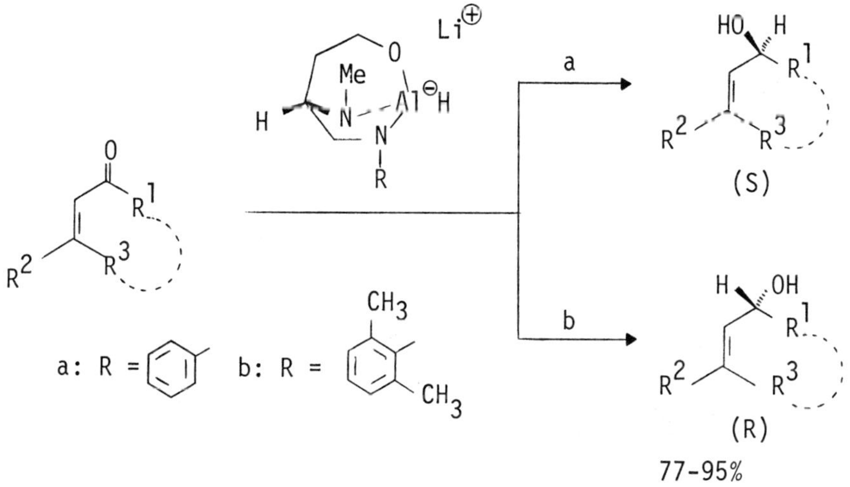

a: R = phenyl   b: R = 2,6-dimethylphenyl

77-95%

III.A-17 J. K. Rasmussen et al., Synthesis, 457 (1983).

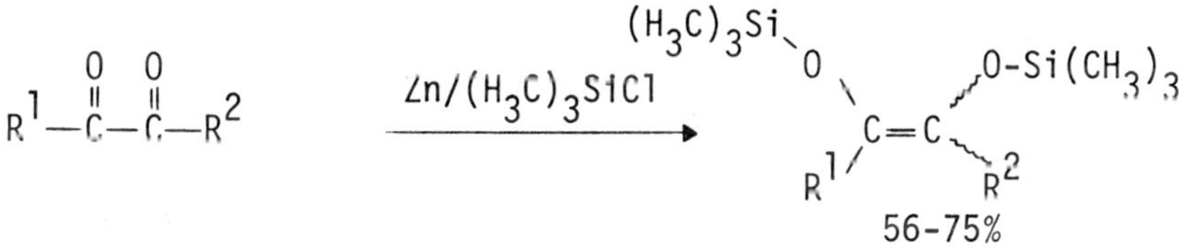

$$R^1-\underset{O}{\overset{O}{C}}-\underset{O}{\overset{O}{C}}-R^2 \xrightarrow{Zn/(H_3C)_3SiCl} \underset{R^1}{\overset{(H_3C)_3Si-O}{\diagdown}}C=C\underset{R^2}{\overset{O-Si(CH_3)_3}{\diagup}}$$

56-75%

III.A-18 M. Hirama, M. Shimizu and M. Iwashita, J. Chem. Soc. Chem. Commun., 599 (1983).

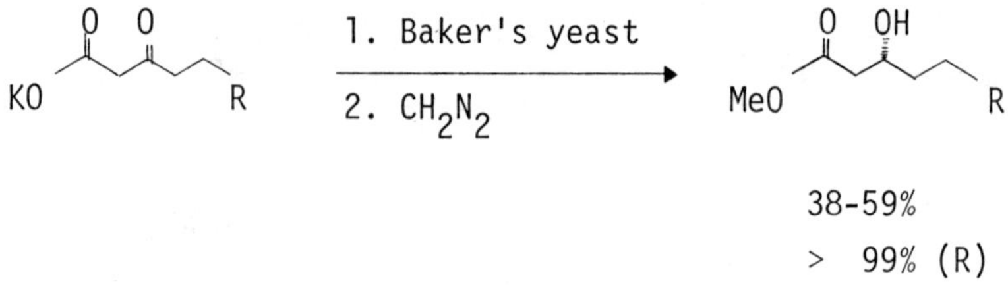

[Reaction: KO-CO-CH₂-CO-CH₂-R → (1. Baker's yeast, 2. CH₂N₂) → MeO-CO-CH₂-CH(OH)-CH₂-R]

38-59%
> 99% (R)

III.A-19 S. Cortes and H. Kohn, J. Org. Chem., 48, 2246 (1983).

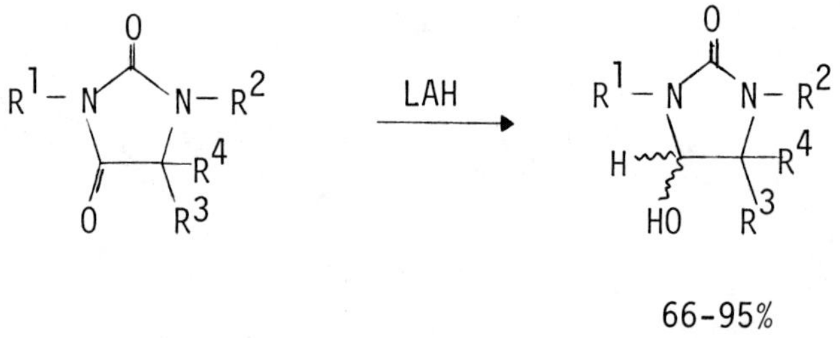

[Reaction: hydantoin-type R¹-N-C(O)-N(R²)-C(R³)(R⁴)-C(O) → LAH → R¹-N-C(O)-N(R²)-C(R³)(R⁴)-CH(OH)]

66-95%

III.B. C-N Multiple Bond Reductions

2. Imine Reductions

III.B.2-1 K. Yamada, M. Takeda and T. Iwakuma, J. Chem. Soc. Perkin Trans. 1, 265 (1983).

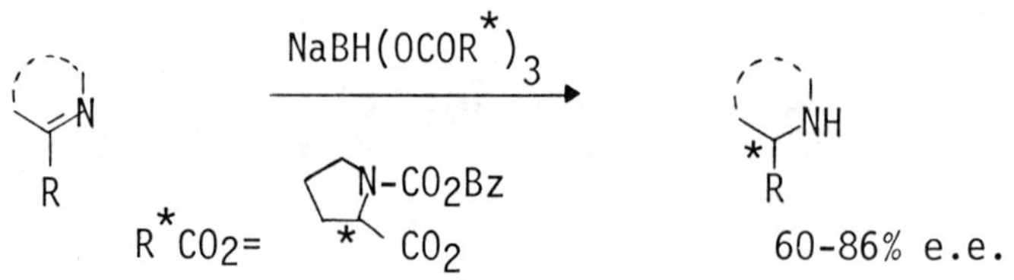

[Reaction: cyclic imine with R → NaBH(OCOR*)₃ → cyclic amine NH with R, *]

$R^*CO_2 =$ proline derivative N-CO₂Bz with *CO₂

60-86% e.e.

III.B.2-2 J. Barluenga, B. Olano and S. Fustero, J. Org. Chem., 48, 2255 (1983).

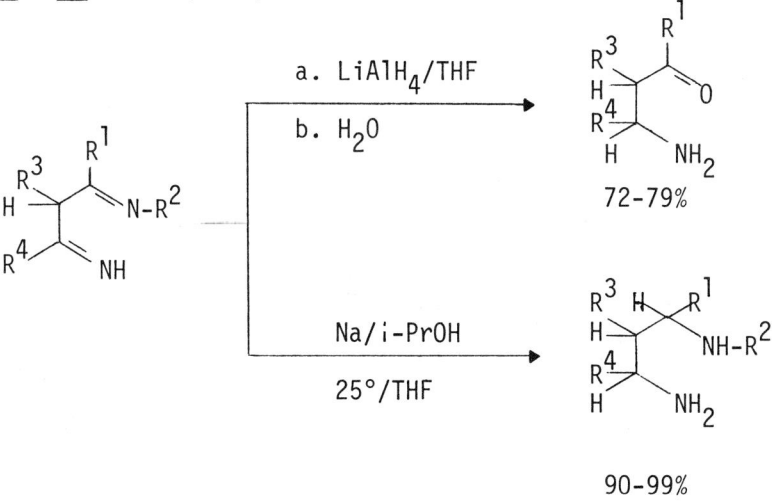

III.B.2-3 B. P. Branchaud, J. Org. Chem., 48, 3531 (1983).

III.B.2-4 H. G. Howell, Synth. Commun., 13, 635 (1983).

III.C. Reduction of Sulfur Compounds

III.C-1 J. R. Irving, E. Perrone and R. J. Stoodley, Tetrahedron Lett., 24, 1429 (1983).

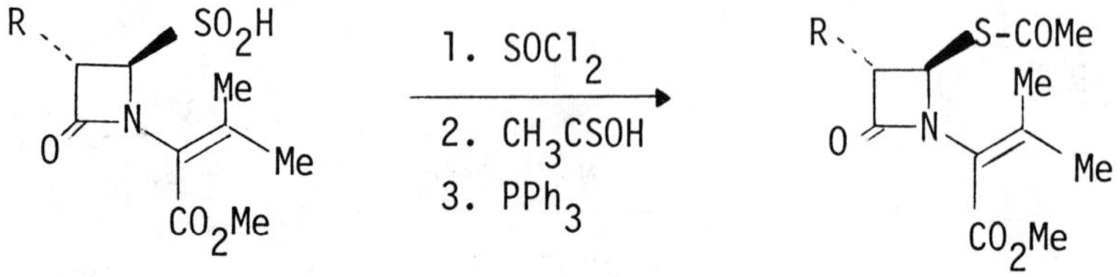

[Reaction: β-lactam with SO$_2$H, Me substituents and N-C(CO$_2$Me)=C(Me)Me side chain → 1. SOCl$_2$, 2. CH$_3$CSOH, 3. PPh$_3$ → β-lactam with S-COMe, Me substituents and same side chain]

30-78%

III.D. N—O Reductions

III.D-1 J. George and S. Chandrasekaran, Synth. Commun., 13, 495 (1983).

$$R\text{-}NO_2 \xrightarrow[\text{TiCl}_4]{\text{HgCl}_2\text{-Mg}} R\text{-}NH_2$$

87-98%

III.D-2 A. Ono, H. Sasaki and F. Yaginuma, Chem. Ind. (London), 480 (1983).

$$Ar\text{-}NO_2 \xrightarrow[\text{EtOH}]{\text{NaBH}_4\text{-FeCl}_2} Ar\text{-}NH_2$$

75-90%

III.D-3 Y. Watanabe et al., Tetrahedron Lett., 24, 4121 (1983).

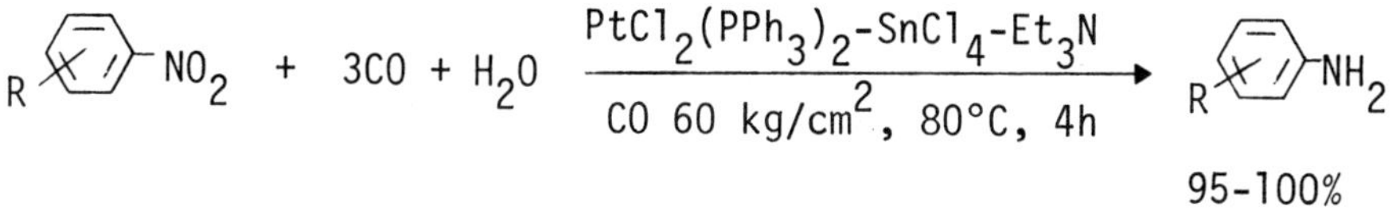

$$R\text{-}C_6H_4\text{-}NO_2 + 3CO + H_2O \xrightarrow[\text{CO 60 kg/cm}^2,\ 80°C,\ 4h]{PtCl_2(PPh_3)_2\text{-}SnCl_4\text{-}Et_3N} R\text{-}C_6H_4\text{-}NH_2$$

95-100%

III.D-4 N. R. Natale, Tetrahedron Lett., 23, 5009 (1982).

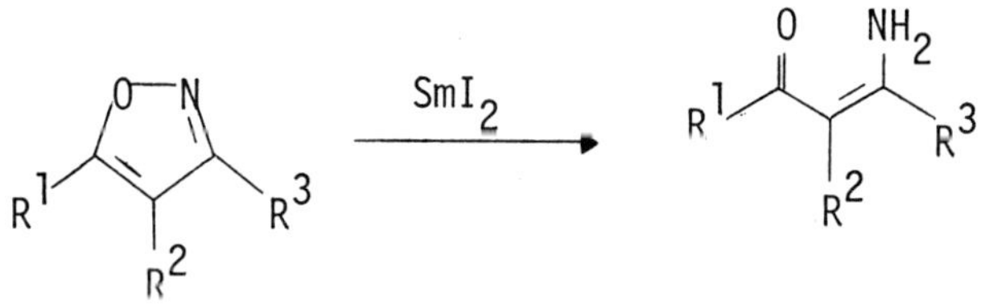

$$\underset{R^2}{\underset{|}{R^1\text{-isoxazole-}R^3}} \xrightarrow{SmI_2} R^1\text{-}\underset{R^2}{\underset{|}{C(=O)\text{-}C}}\text{=}C(NH_2)\text{-}R^3$$

III.D-5 D. Monti et al., Tetrahedron Lett., 24, 417 (1983).

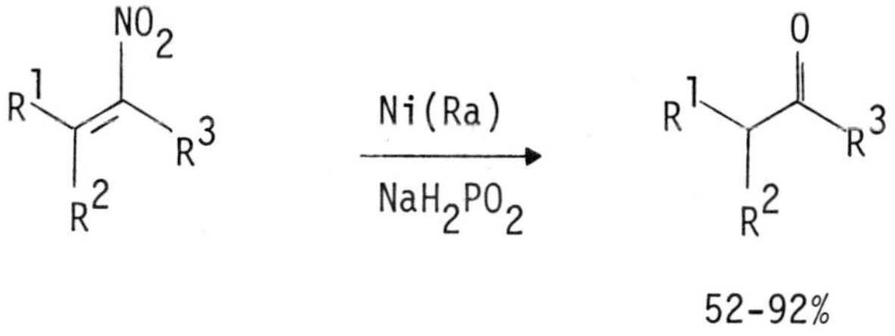

$$\underset{R^2}{\underset{|}{R^1\text{-}C}}\text{=}\underset{}{C(NO_2)\text{-}R^3} \xrightarrow[NaH_2PO_2]{Ni(Ra)} R^1\text{-}\underset{R^2}{\underset{|}{CH}}\text{-}C(=O)\text{-}R^3$$

52-92%

III.D-6 S. Torii, H. Tanaka and T. Katoh, Chem. Lett., 607 (1983).

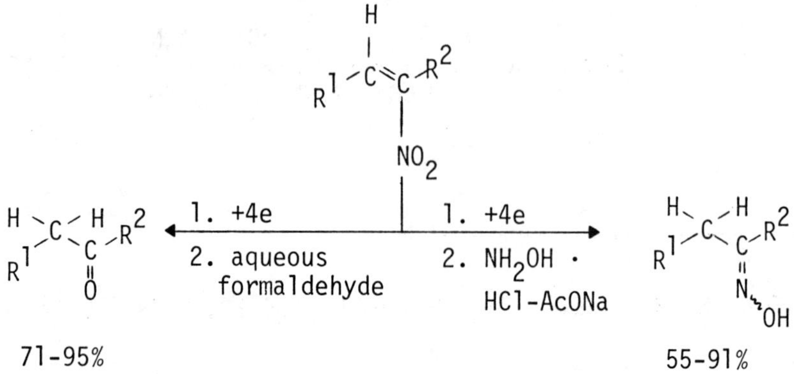

Starting material: $R^1-CH=C(R^2)-NO_2$

Left pathway: 1. +4e; 2. aqueous formaldehyde → $R^1-CH(R^2)-C(=O)H$ type product, 71-95%

Right pathway: 1. +4e; 2. $NH_2OH \cdot HCl$-AcONa → oxime product, 55-91%

III.E. C-C Multiple Bond Reductions

1. C=C Reductions

III.E.1-1 Y. Kumar and L. Florvall, Synth. Commun., 13, 489 (1983).

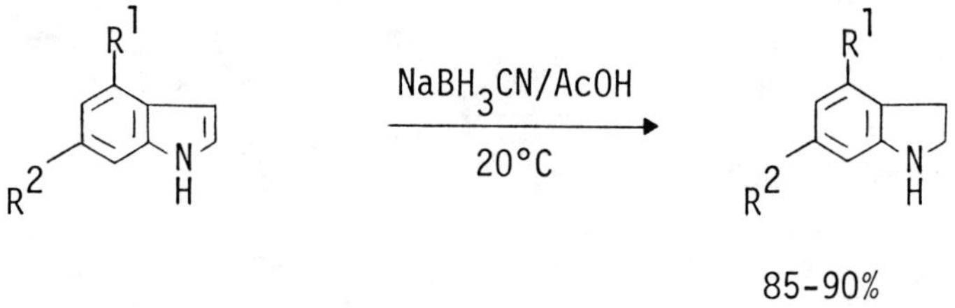

Indole (with R^1 at 4-position, R^2 at 6-position) $\xrightarrow{NaBH_3CN/AcOH,\ 20°C}$ indoline, 85-90%

III.E.1-2   A. Alberola et al., Synthesis, 413 (1983).

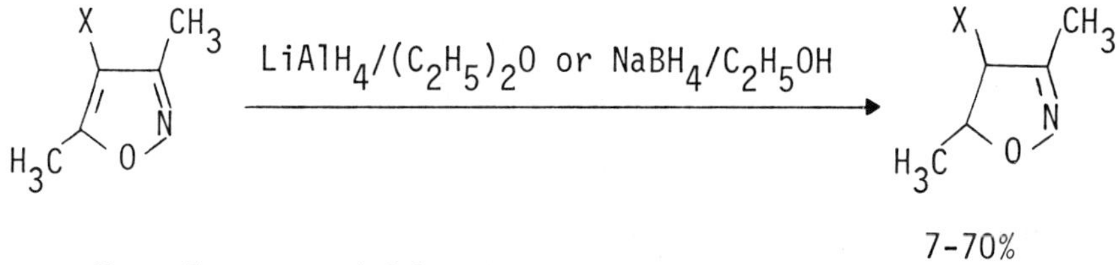

$$\text{isoxazole (X, CH}_3\text{, H}_3\text{C)} \xrightarrow{\text{LiAlH}_4/(C_2H_5)_2O \text{ or NaBH}_4/C_2H_5OH} \text{dihydroisoxazole}$$

7-70%

X = electron withdrawing group

III.E.1-3   A. K. Sinhababu and R. T. Borchardt, Tetrahedron Lett., 24, 227 (1983).

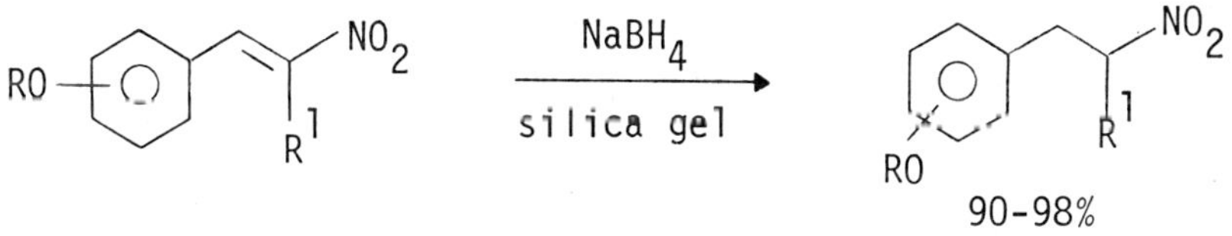

$$\text{RO-Ar-CH=C(R}^1\text{)NO}_2 \xrightarrow[\text{silica gel}]{\text{NaBH}_4} \text{RO-Ar-CH}_2\text{-CH(R}^1\text{)NO}_2$$

90-98%

III.E.1-4   H. Cikashita et al., Synth. Commun., 13, 1033 (1983).

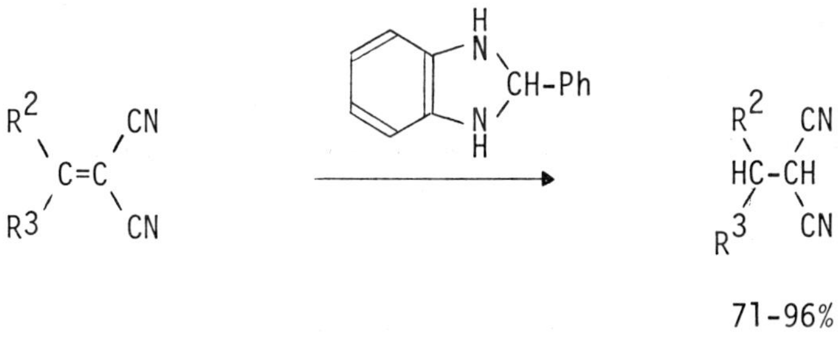

$$\underset{R^3}{\overset{R^2}{>}}C=C\underset{CN}{\overset{CN}{<}} \xrightarrow{\text{benzimidazoline-CH-Ph}} \underset{R^3}{\overset{R^2}{>}}HC-CH\underset{CN}{\overset{CN}{<}}$$

71-96%

III.E.1-5 A. M. Caporusso, G. Giacomelli and L. Lardicci, J. Org. Chem., 47, 4640 (1982).

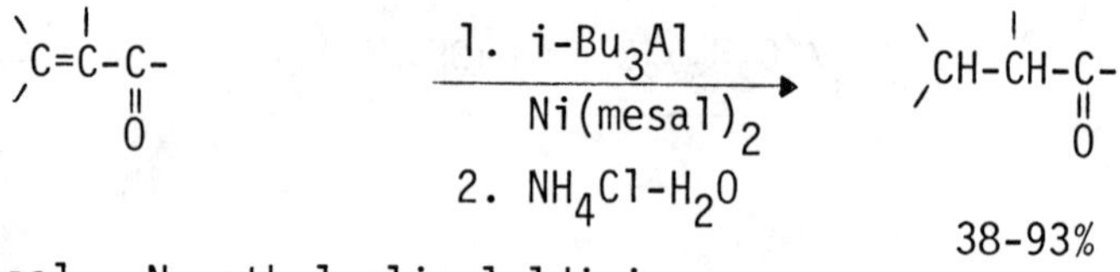

$$\diagdown C=\underset{\underset{O}{\|}}{C}-C- \quad \xrightarrow[\text{2. } NH_4Cl-H_2O]{\text{1. } i-Bu_3Al \quad \text{Ni(mesal)}_2} \quad \diagdown CH-\underset{\underset{O}{\|}}{CH}-C-$$

38-93%

mesal = N-methylsalicylaldimine

III.E.1-6 A. Uehara, T. Kubota and R. Tsuchiya, Chem. Lett., 441 (1983).

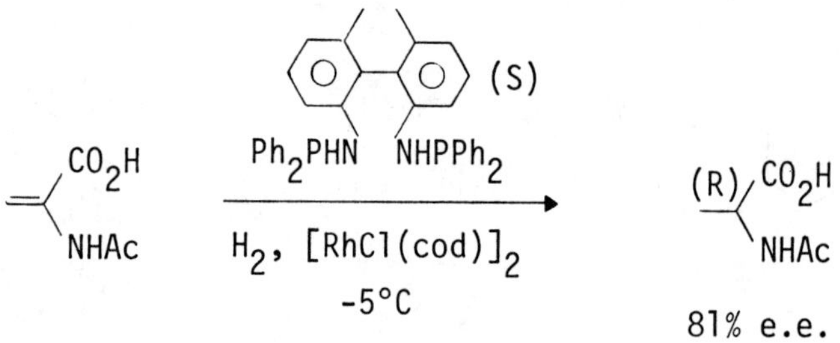

81% e.e.

III.E.1-7 B. L. Sondengam et al., J. Chem. Soc. Perkin Trans. 1, 1219 (1983).

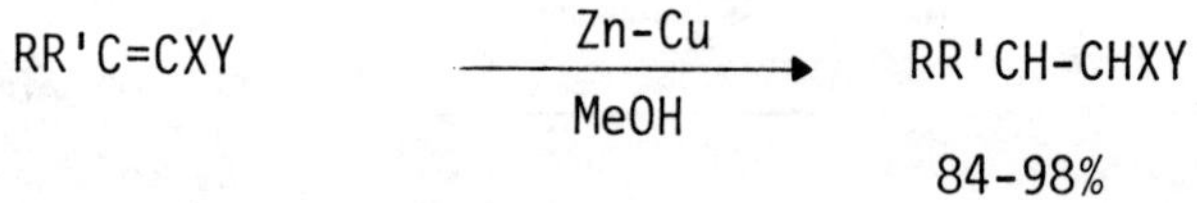

$$RR'C=CXY \quad \xrightarrow[\text{MeOH}]{\text{Zn-Cu}} \quad RR'CH-CHXY$$

84-98%

X,Y = electronegative substituents

III.E.2. C≡C Reductions

III.E.2-1 J. Rajaram et al., Tetrahedron, 39, 2315 (1983).

$$R-C\equiv C-R^1 \xrightarrow[\text{MnCl}_2]{\text{Pd-CaCO}_3\text{-PbO}} R-CH=CH-R^1$$

91-97%
selective

III.E.2-2 N. Suzuki et al., J. Chem. Soc. Chem. Commun., 515 (1983).

$$R^1C\equiv CR^2 \xrightarrow[\text{PEG-CH}_2\text{Cl}_2]{\text{NaBH}_4\text{-PdCl}_2} \underset{H}{\overset{R^1}{\diagdown}} C=C \underset{H}{\overset{R^2}{\diagup}}$$

77-91%

III.E.2-3 J. A. Miller and G. Zweifel, J. Am. Chem. Soc., 105, 1383 (1983).

$$R-C\equiv C-C\equiv CSiMe_3 \xrightarrow[2.\ H_3O^+]{1.\ Li[AlH(i\text{-Bu})_2\text{-}n\text{-Bu}]} \underset{H}{\overset{R}{\diagdown}}C=C\underset{\equiv\text{-SiMe}_3}{\overset{H}{\diagup}}$$

90-95%

$$\xrightarrow[2.\ \text{NaOH}]{1.\ i\text{-Bu}_2\text{AlH}} \text{(R,H)CH=CH-CH=CH(SiMe}_3\text{)}$$

91-97%

III.E.3. Reduction of Aromatic Rings

III.E.3-1 R. A. Benkeser, F. G. Belmonte and J. Kang, J. Org. Chem., 48, 2796 (1983).

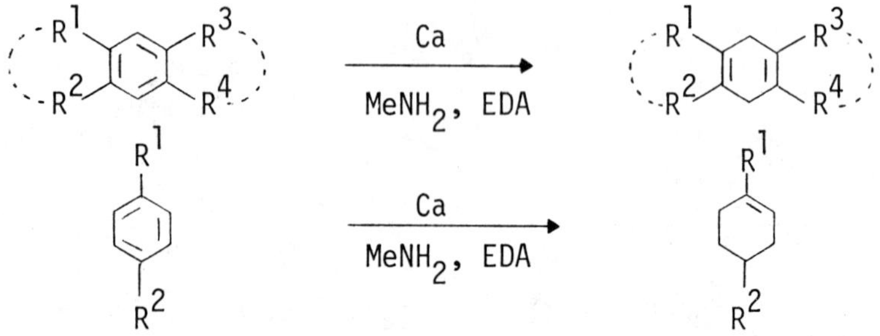

66-98%

III.E.3-2 K. R. Januszkiewicz and H. Alper, Organometallics, 2, 1055 (1983).

31-100%

III.F. Hydrogenolysis of Hetero Bonds

1. C—O → C—H

III.F.1-1 K. K. Ogilvie et al., Tetrahedron Lett., 24, 865 (1983).

$$\text{RO-furanose(B)-HO, OSiMe}_2(t\text{-Bu}) \xrightarrow[\text{2. }(n\text{-Bu})_3\text{SnH}]{\text{1. Im-C(=S)-Im}} \text{RO-furanose(B)-OSiMe}_2(t\text{-Bu})$$

R = methoxytrityl, dimethoxytrityl   35-55%

III.F.1-2 M. J. Robins et al., J. Am. Chem. Soc. 105, 4059 (1983).

$$\text{R(R')CH-OH} \xrightarrow[\text{2. Bu}_3\text{SnH, AIBN}]{\text{1. PTC-Cl}} \text{R-CH}_2\text{-R'}$$

PTC-Cl = PhO-C(=S)-Cl

57-85%

III.F.1-3 H. Suzuki et al., Chem. Lett., 247 (1983).

$$\text{Ar-C(R}^1\text{)(R}^2\text{)-OH} \xrightarrow[\text{C}_6\text{H}_6,\ \text{N}_2,\ \Delta]{\text{P}_2\text{I}_4} \text{Ar-C(R}^1\text{)(R}^2\text{)-H}$$

(Ar = R^3, R^4, R^5, R^6, R^7-substituted phenyl)

59-98%

III.F.1-4 Y. Ueno, C. Tanaka and M. Okawara, Chem. Lett., 795 (1983).

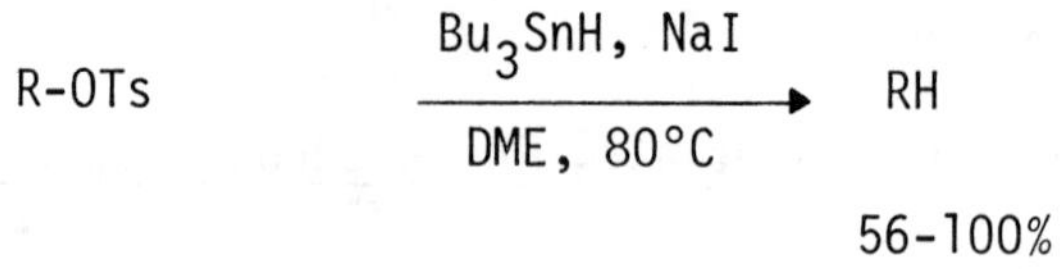

$$R\text{-OTs} \xrightarrow[\text{DME, }80°C]{\text{Bu}_3\text{SnH, NaI}} RH$$

56-100%

III.F.1-5 K. Nakamura et al., Tetrahedron Lett., 24, 2001 (1983)

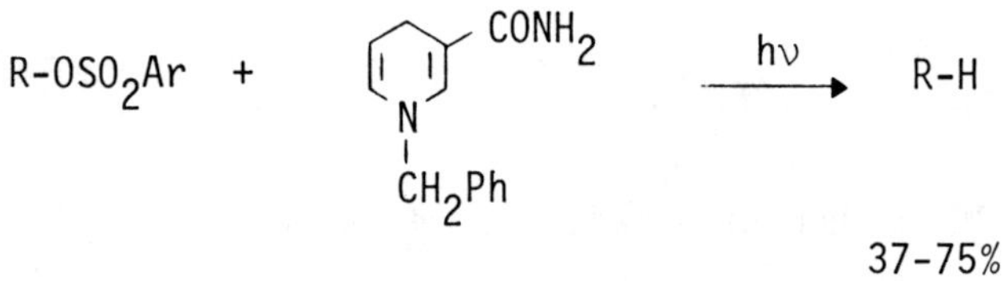

$$R\text{-OSO}_2Ar + \underset{\underset{CH_2Ph}{|}}{\text{[3-CONH}_2\text{-1-benzyl-dihydropyridine]}} \xrightarrow{h\nu} R\text{-H}$$

37-75%

III.F.1-6 Y. D. Vankar, P. S. Arya and C. T. Rao, Synth. Commun., 13, 869 (1983).

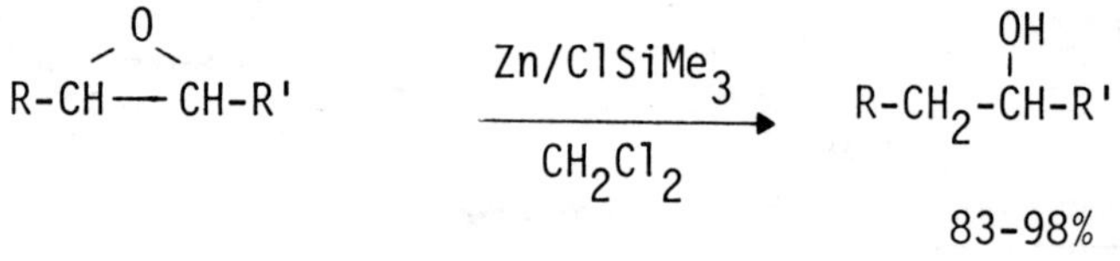

$$R\text{-CH}\overset{O}{-}\text{CH-R'} \xrightarrow[CH_2Cl_2]{\text{Zn/ClSiMe}_3} R\text{-CH}_2\text{-}\underset{}{\overset{OH}{\text{CH}}}\text{-R'}$$

83-98%

III.F.1-7 S. Fernandez, R. Quintanilla and J. E. Hernandez, Synth. Commun., 13, 621 (1983).

$$\text{tetrahydrofuran-2-yl-C(=O)-C}_{10}\text{H}_{21} \xrightarrow[-78°]{\text{HMPA-Li}} \text{HO-CH}_2\text{CH}_2\text{CH}_2\text{CH}_2\text{-C(=O)-C}_{10}\text{H}_{21}$$

66%

III.F.1-8 M. Hojo, et al., Synthesis, 387 (1983).

$$R-C(=O)-OR \xrightarrow[\text{n-hexane}]{\text{LAH, SiO}_2} R-CH_2OH$$

62-100%

III.F.1-9 F. J. P. Coriu, et al., Synthesis, 981 (1983); F. J. P. Corriu, R. Perz and C. Rey, Tetrahedron, 39, 999 (1983).

$$R^1-C(=O)(OR^2) + H-Si\underset{1 \text{ or } 2}{\diagdown} \xrightarrow{\text{KF/DMSO}} \xrightarrow{H^{\oplus} \text{ or } CH_3OH} R-CH_2-OH$$

$$\underset{1}{\overset{CH_3}{\underset{OC_2H_5}{H-Si-OC_2H_5}}} \qquad \underset{2}{(H_3C)_3Si-O{\left[-\underset{CH_3}{\overset{H}{Si}}-O\right]}_n-Si(CH_3)_3}$$

III.F.1-10 M. Hojo et al., Tetrahedron Lett., 24, 2575 (1983).

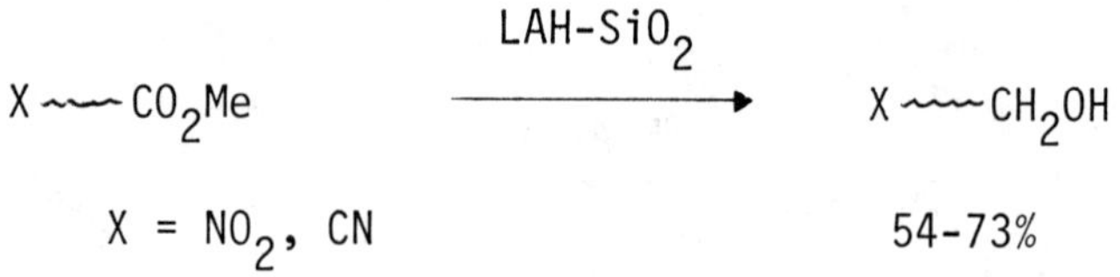

$$X \sim CO_2Me \xrightarrow{\text{LAH-SiO}_2} X \sim CH_2OH$$

$X = NO_2$, CN   54-73%

III.F.1-11 C. Camacho, G. Uribe and R. Contreras, Synthesis, 1027 (1982).

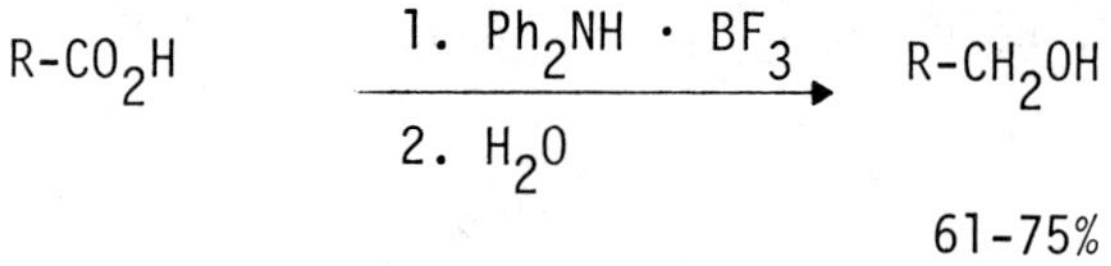

$$R-CO_2H \xrightarrow[\text{2. } H_2O]{\text{1. } Ph_2NH \cdot BF_3} R-CH_2OH$$

61-75%

III.F.1-12 T. Fujisawa, T. Mori and T. Sato, Chem. Lett., 835 (1983).

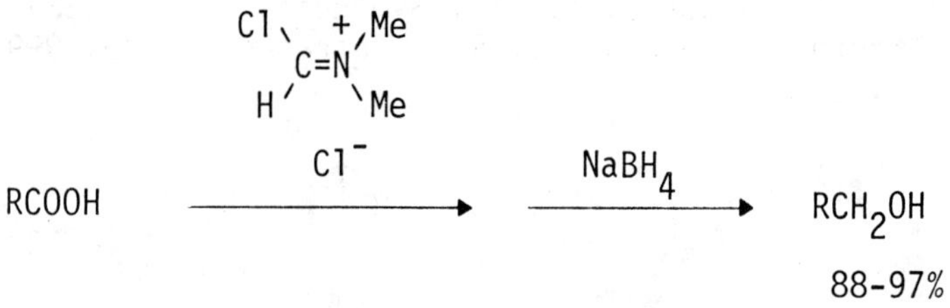

$$RCOOH \xrightarrow{\underset{Cl^-}{\overset{Cl}{\underset{H}{>}}C=N\overset{+}{\underset{Me}{<}}Me}} \xrightarrow{NaBH_4} RCH_2OH$$

88-97%

III.F.1-13 T. Fujisawa et al., Tetrahedron Lett., 24, 1543 (1983).

RCO_2H $\xrightarrow[\text{2. LiAlH(O-t-Bu)}_3, \text{ CuI, } -78°C]{\text{1. } \underset{\text{Pyr., } -30°C}{\overset{Cl^{\ominus}}{\underset{H}{\overset{Cl}{\diagdown}}\underset{}{C=\overset{\oplus}{N}}\underset{Me}{\overset{Me}{\diagup}}}}}$ $RCHO$

55-90%

III.F.1-14 R. Knorr, P. Hassel and P. Loew, Synthesis, 785 (1983).

$\underset{R^2}{\overset{R^1}{\diagdown}}CH-CO_2H$ $\xrightarrow{H_3CO-CH=\overset{\oplus}{N}\underset{R^3}{\overset{R^3}{\diagup}} \quad H_3CO-SO_3^-}$ $\underset{R^2}{\overset{R^1}{\diagdown}}C=C\underset{N\underset{R^3}{\diagdown}\overset{R^3}{\diagup}}{\overset{H}{\diagup}}$

36-89%

III.F.1-15 H. Suzuki et al., Chem. Lett., 909 (1983).

Ar-C(=O)-R^6 $\xrightarrow[C_6H_6, N_2, \Delta]{LiAlH_4/P_2I_4}$ Ar-CH_2R^6

(Ar = substituted phenyl with R^1, R^2, R^3, R^4, R^5)

III.F.1-16 M. Yamashita and I. Ojima, J. Am. Chem. Soc., 105, 6339 (1983).

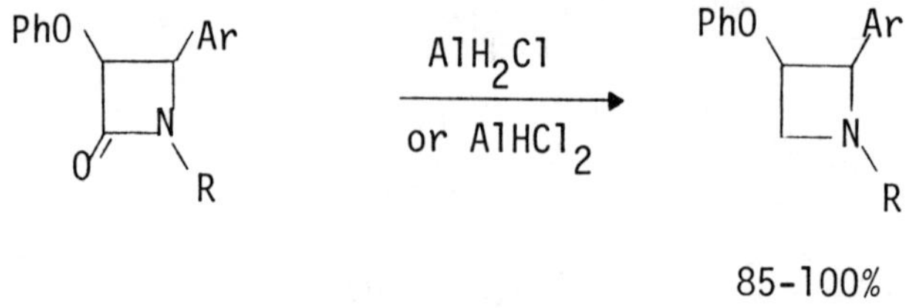

PhO-[azetidinone with Ar, N-R, C=O] → (AlH$_2$Cl or AlHCl$_2$) → PhO-[azetidine with Ar, N-R]

85-100%

III.F.1-17 K. Soai, A. Ookawa and H. Hayashi, J. Chem. Soc. Chem. Commun., 668 (1983).

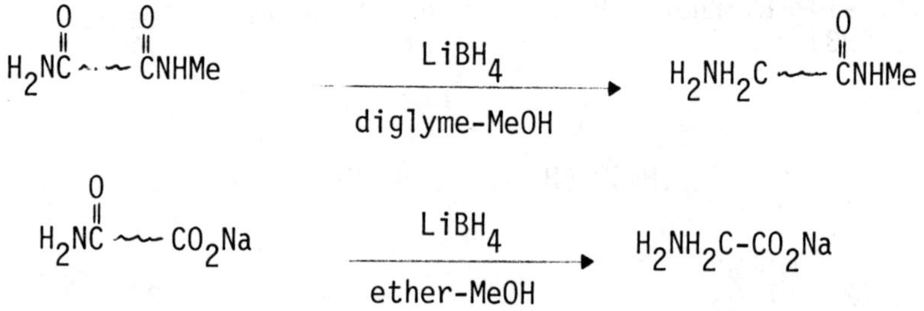

$H_2NC(O)\sim\sim C(O)NHMe$ → (LiBH$_4$, diglyme-MeOH) → $H_2NH_2C\sim\sim C(O)NHMe$

$H_2NC(O)\sim\sim CO_2Na$ → (LiBH$_4$, ether-MeOH) → $H_2NH_2C-CO_2Na$

III.F.1-18 S. Cortes and H. Kohn, J. Org. Chem., 48, 2246 (1983)

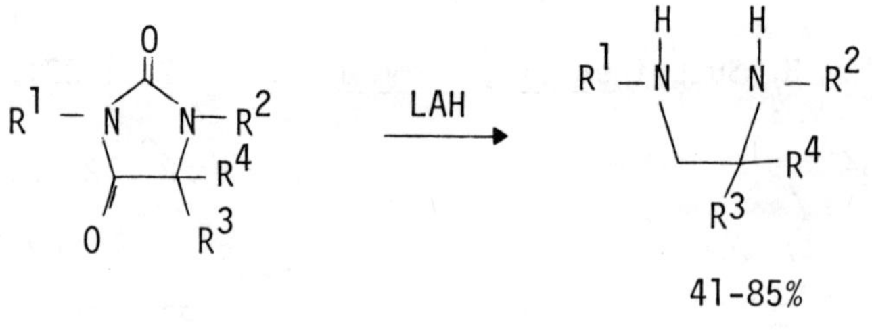

[hydantoin: R^1-N, C=O, N-R^2, C(R^3)(R^4), C=O] → LAH → R^1-NH-CH$_2$-C(R^3)(R^4)-NH-R^2

41-85%

III.F.2. C—Hal → C—H

III.F.2-1 A. L. J. Beckwith and S. H. Goh, J. Chem. Soc. Chem. Commun. 907 (1983).

$$R\text{-}X \xrightarrow[(t\text{-}BuO)_2,\ h\nu]{LAH,} RH$$

53-100%

III.F.2-2 S. Krishnamurthy and H. C. Brown, J. Org. Chem., 48, 3085 (1983).

$$R\text{-}X \xrightarrow{LiEt_3BH} R\text{-}H$$

X = Br, I
R = alkyl

88-100%

III.F.2-3 S. Kim, C. Y. Hong and S. Yang, Angew. Chem. Int. Ed. Engl. 22, 562 (1983).

$$RX + Zn(BH_4)_2 \xrightarrow{Et_2O} RH$$

R = t-alkyl, benzylic

81-99%

III.F.2-4 S. Kim, Y. J. Kim and K. H. Ahn, Tetrahedron Lett., 24, 3369 (1983).

$$RX \xrightarrow{NaBH_3CN\text{-}ZnCl_2} RH$$

R = tertiary, allyl, benzyl

47-96%

III.F.2-5 T. Hirao et al., Bull. Chem. Soc. Jpn., 56, 1881 (1983).

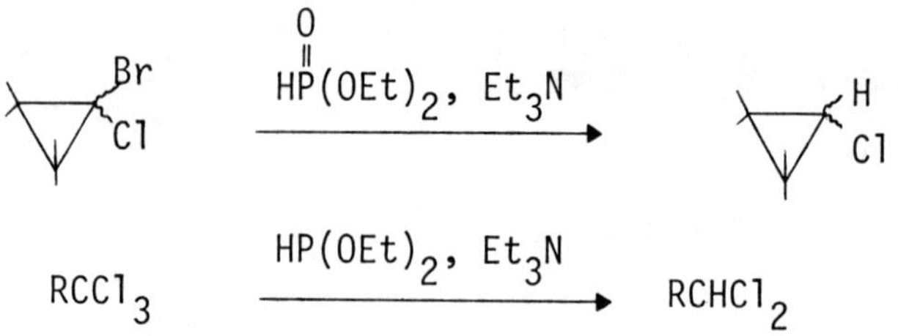

$$\text{cyclopropane(Br)(Cl)} \xrightarrow{\text{HP(OEt)}_2,\ \text{Et}_3\text{N}} \text{cyclopropane(H)(Cl)}$$

$$\text{RCCl}_3 \xrightarrow{\text{HP(OEt)}_2,\ \text{Et}_3\text{N}} \text{RCHCl}_2$$

III.F.2-6 M. Muehlbacher and K.-H. Ongania, Z. Naturforsch., 37b, 1352 (1982).

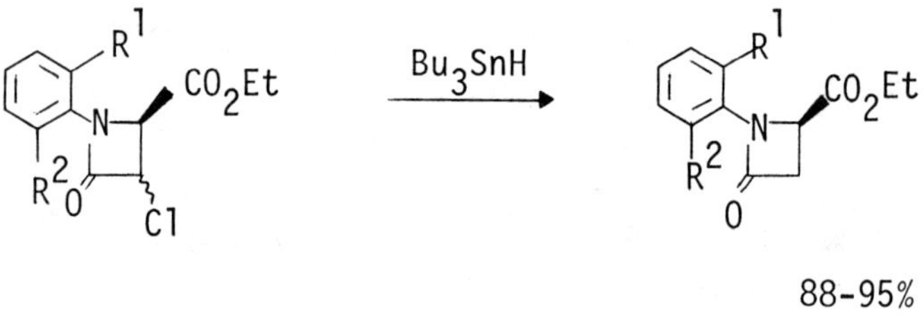

$$\text{β-lactam (Cl, CO}_2\text{Et)} \xrightarrow{\text{Bu}_3\text{SnH}} \text{β-lactam (H, CO}_2\text{Et)}$$

88-95%

III.F.2-7 A. Osuka and H. Suzuki, Chem. Lett., 119 (1983).

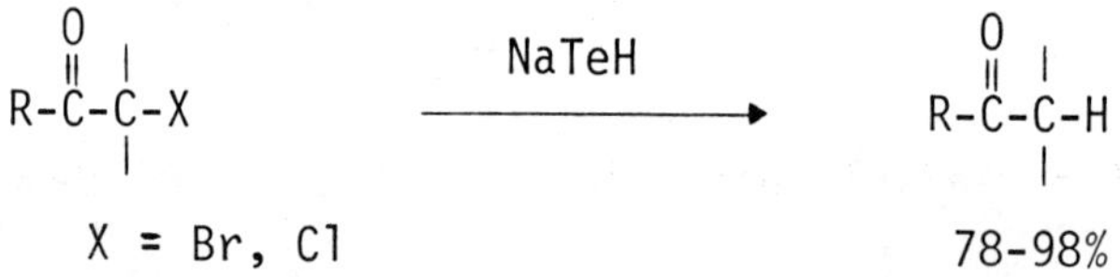

$$\underset{X\ =\ Br,\ Cl}{R-\overset{O}{\underset{\|}{C}}-\overset{|}{\underset{|}{C}}-X} \xrightarrow{\text{NaTeH}} R-\overset{O}{\underset{\|}{C}}-\overset{|}{\underset{|}{C}}-H$$

78-98%

III.F.2-8 J.-E. Dubois, C. Lion and J.-Y. Dugast, Tetrahedron Lett., 24, 4207 (1983).

$$\begin{array}{c}\diagdownOBr\\ -C-\overset{\|}{C}-C\diagup\\ \diagup\diagdown\end{array} \xrightarrow[H_2O]{\underline{iPr_2}NLi} \begin{array}{c}\diagdownOH\\ -C-\overset{\|}{C}-C\diagup\\ \diagup\diagdown\end{array} \underline{1}$$

$$\xrightarrow[SiMe_3Cl]{iPr_2NLi} \begin{array}{c}\diagdownOSiMe_3\\ -C-\overset{|}{C}=C\diagup\\ \diagup\diagdown\end{array} \underline{2}$$

1: 78-97%   2: 81-98%

III.F.2-9 M. Yamana, T. Ishihara and T. Ando, Tetrahedron Lett., 24, 507 (1983).

$$CF_2ClCR\!\!\overset{O}{\|}\ + Zn(0)\ +\ Me_3SiCl \xrightarrow[MeCN]{60°C} \begin{array}{c}F\diagdown\diagup OSiMe_3\\ C=C\\ F\diagup\diagdown R\end{array}$$

35-74%

III.F.2-10 T. Hirao et al., Tetrahedron Lett., 24, 399 (1983).

41-84%

III.F.3   C-S → C-H

III.F.3-1 H. Alper, S. Ripley and T. L. Prince, J. Org. Chem., 48, 250 (1983).

$$RSH \xrightarrow[FeCl_2, THF-C_6H_6]{Na(C_2H_5)_3BH/Al_2O_3} RH$$

R = t-alkyl, aryl   39-73%

III.F.3-2 V. S. Karavan, D. A. Simonov and T. I. Temnikova, J. Org. Chem. USSR 18, 1767 (1982).

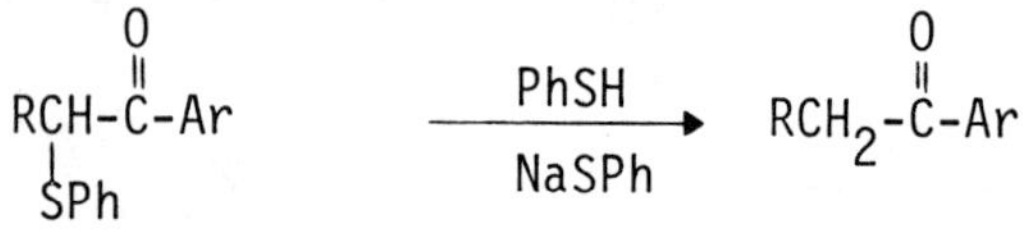

$$\underset{\underset{SPh}{|}}{RCH}-\overset{O}{\underset{||}{C}}-Ar \xrightarrow[NaSPh]{PhSH} RCH_2-\overset{O}{\underset{||}{C}}-Ar$$

III.F.3-3 C. Kashima et al., J. Chem. Soc. Perkin Trans. 1, 1799 (1983).

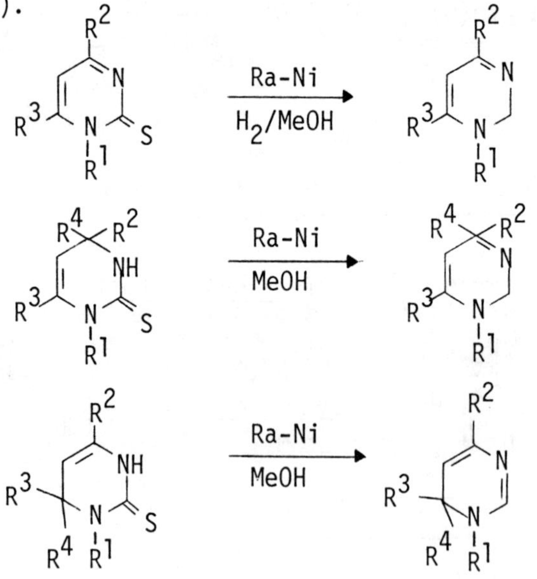

1: 35-66%   2: 54-84%   3: 27-84%

III.F.3-4 D. R. Williams and J. L. Moore, Tetrahedron Lett., 24, 339 (1983).

$$\underset{X,Y\ =\ S,O,NCOR}{\left[\begin{array}{c}X\\Y\end{array}\right\rangle=S}\quad\xrightarrow{Bu_3SnH}\quad\underset{50-94\%}{\left[\begin{array}{c}X\\Y\end{array}\right\rangle<\begin{array}{c}H\\H\end{array}}$$

III.F.3-5 T. Cuvigny et al., Tetrahedron Lett., 24, 4319 (1983).

$$\underset{\substack{M\ =\ Pd,\ Ni}}{\text{Et}\diagup\hspace{-0.3em}\diagdown\hspace{-0.3em}\underset{n\text{-Hex}}{\overset{SO_2Ph}{|}}}\quad\xrightarrow[\substack{ML_n,\ 2L'n\\ THF,\ 1hr.,\ r.t.}]{n\text{-BuMgCl}}\quad\underset{\sim 50\%}{\text{Et}\diagup\hspace{-0.3em}\diagdown\hspace{-0.3em}\diagup\hspace{-0.3em}\diagdown n\text{-Hex}}$$

III.F.3-6 M. Ochiai, T. Ukita and E. Fujita, J. Chem. Soc. Chem. Commun., 619 (1983).

$$\underset{R^1\quad R^2}{\overset{SO_2Ph}{\diagdown\hspace{-0.3em}=\hspace{-0.3em}\diagup}}\quad\xrightarrow[\text{2. SiO}_2]{\text{1. LiSnBu}_3}\quad\underset{64-98\%}{R^1\diagdown=\diagup R^2}$$

III.F.3-7 K. Ogura, K. Arai and G. Tsuchihashi, Bull. Chem. Soc. Jpn. 55, 3669 (1982).

$$\underset{H}{\overset{Ar}{>}}C=C\underset{SOCH_3}{\overset{SCH_3}{<}} \xrightarrow[2.\ H^+]{1.\ C_2H_5MgCl} \underset{H}{\overset{Ar}{>}}C=C\underset{H}{\overset{SCH_3}{<}}$$

41-82%

III.F.3-8 J.-L. Fabre and M. Julia, Tetrahedron Lett., 24, 4311 (1983).

$$\underset{R^4SO_2}{\overset{R^1}{>}}C=C\underset{R^3}{\overset{R^2}{<}} \xrightarrow[\substack{Pd(acac)_2 \\ P(nBu)_3}]{n\text{-BuMgCl}} \underset{}{\overset{R^1}{>}}C=C\underset{R^3}{\overset{R^2}{<}}$$

R^1 or R^2 = H   65-70%

III.F.4.   C—N → C—H

III.F.4-1 G. Rosini, R. Ballini and V. Zanotti, Synthesis, 137 (1983).

$$R^1-\underset{}{\overset{NH\text{-}Tos}{\underset{\|}{C}}}-\underset{NO_2}{\overset{R^2}{\underset{|}{C}}}-R^3 \xrightarrow[THF]{LAH} R^1-\underset{}{\overset{NH\text{-}Tos}{\underset{\|}{C}}}-\overset{R^2}{\underset{|}{CH}}-R^3$$

81-94%

III.F.4-2 N. Ono, R. Tamura and A. Kaji, J. Am. Chem. Soc., 105, 4017 (1983).

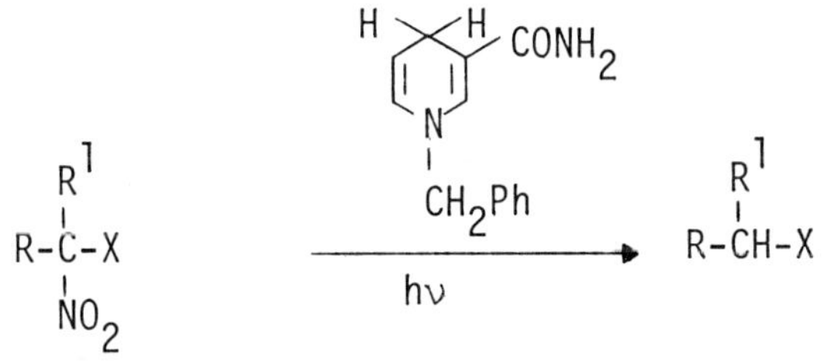

$$\underset{NO_2}{\overset{R^1}{R-C-X}} \xrightarrow[h\nu]{\text{1-benzyl-1,4-dihydronicotinamide}} \underset{}{\overset{R^1}{R-CH-X}}$$

X = CN, CO_2R', COR'   80-99%

III.F.4-3 E. Hofer and R. Keuper, Synthesis, 466 (1983).

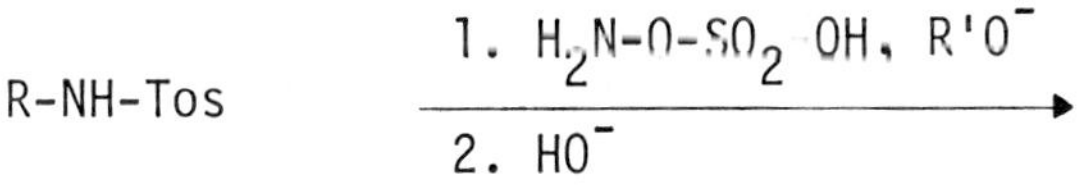

$$R-NH-Tos \xrightarrow[\text{2. } HO^-]{\text{1. } H_2N-O-SO_2-OH, \, R'O^-} RH$$

III.F.4-4 J.-A. Fehrentz and B. Castro, Synthesis, 676 (1983).

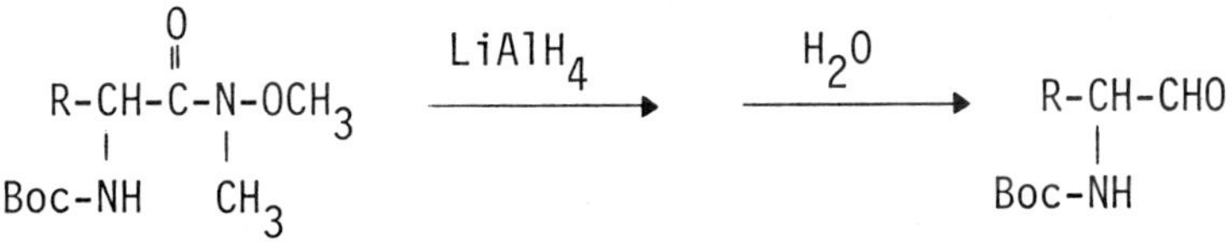

$$\underset{\underset{CH_3}{Boc-NH}}{R-CH-\overset{O}{\overset{\|}{C}}-N-OCH_3} \xrightarrow{LiAlH_4} \xrightarrow{H_2O} \underset{Boc-NH}{R-CH-CHO}$$

III.G. Reductive Eliminations

III.G-1 E. Keinan, M. Sahai and I. Kirson, <u>J. Org. Chem.</u>, <u>48</u>, 2550 (1983).

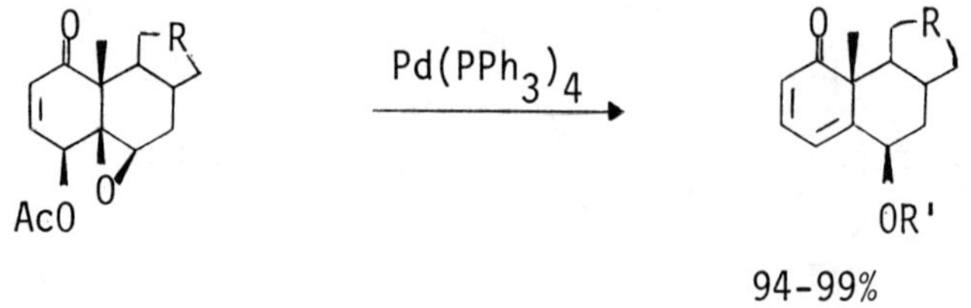

[Structure: bicyclic ketone with AcO and O (epoxide), R substituent] →[Pd(PPh$_3$)$_4$] [Structure: bicyclic dienone with OR' substituent]

94-99%

III.G-2 R. N. Baruah, R. P. Sharma and J. N. Baruah, <u>Chem. Ind.</u>, 524 (1983).

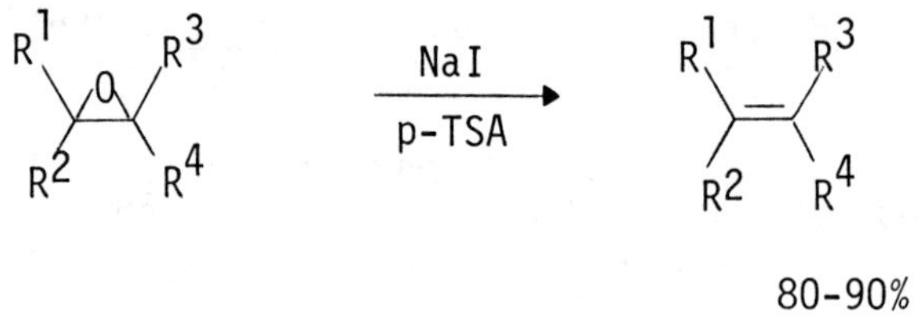

R^1, R^2, R^3, R^4 epoxide →[NaI / p-TSA] R^1, R^2, R^3, R^4 alkene

80-90%

III.G-3 J. R. Schauder, J. N. Denis and A. Krief, <u>Tetrahedron Lett.</u>, <u>24</u>, 1657 (1983).

[thiirane] ⟶ [alkene]

various reagents were tested

III.G-4 N. C. Barua <u>et al.</u>, <u>Chem. Ind.</u>, 956 (1982).

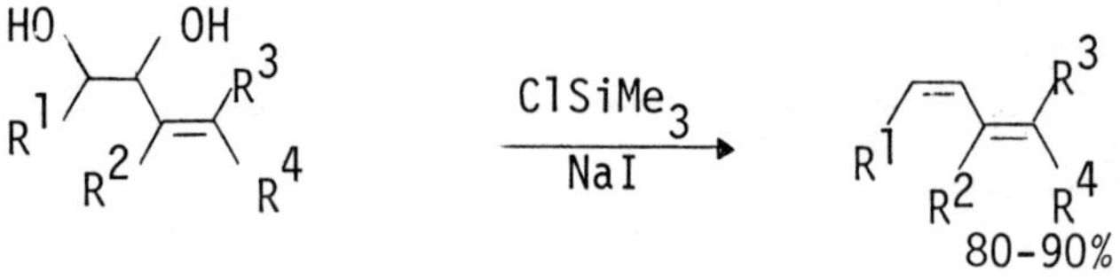

HO, OH diol with R^1, R^2, R^3, R^4 →[ClSiMe$_3$ / NaI] diene with R^1, R^2, R^3, R^4

80-90%

III.G-5 J. Nakayama, H. Machida and M. Hoshino, Tetrahedron Lett., 24, 3001 (1983).

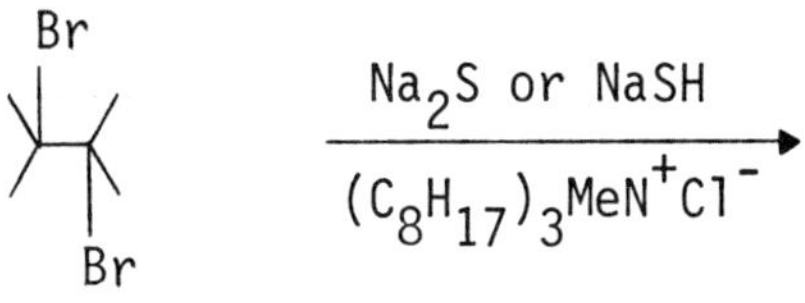

$$\text{(CH}_3\text{)}_2\text{C(Br)C(Br)(CH}_3\text{)}_2 \xrightarrow[\text{(C}_8\text{H}_{17}\text{)}_3\text{MeN}^+\text{Cl}^-]{\text{Na}_2\text{S or NaSH}} \text{(CH}_3\text{)}_2\text{C=C(CH}_3\text{)}_2$$

12-97%

III.G-6 V. Reutrakul and P. Poochaivatanon, Tetrahedron Lett., 24, 531 (1983).

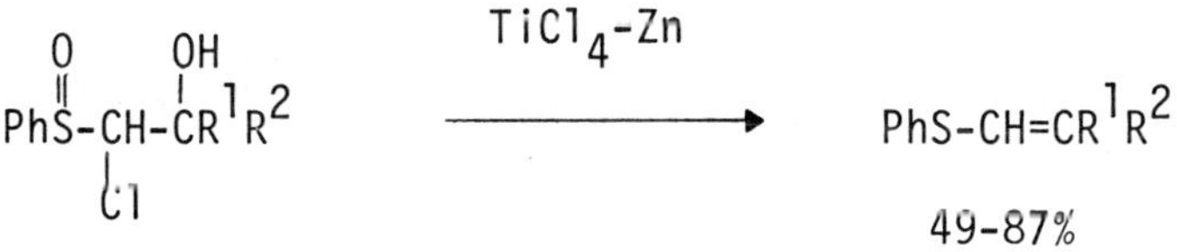

$$\underset{\underset{\text{Cl}}{|}}{\text{PhS(O)-CH-CR}^1\text{R}^2(\text{OH})} \xrightarrow{\text{TiCl}_4\text{-Zn}} \text{PhS-CH=CR}^1\text{R}^2$$

49-87%

III.G-7 Y. Izawa, M. Takeuchi and H. Tomioka, Chem. Lett., 1297 (1983).

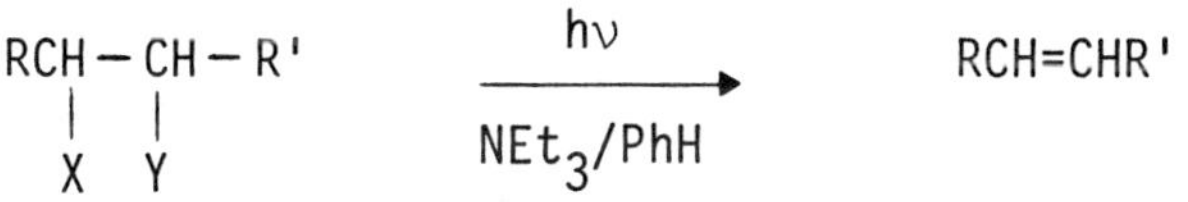

$$\underset{\underset{X \; Y}{|\;\;|}}{\text{RCH-CH-R}'} \xrightarrow[\text{NEt}_3/\text{PhH}]{h\nu} \text{RCH=CHR}'$$

35-100%

X,Y = halide

III.G-8 W. Wierenga, J. Griffin and M. A. Warpehoski, Tetrahedron Lett., 24, 2437 (1983).

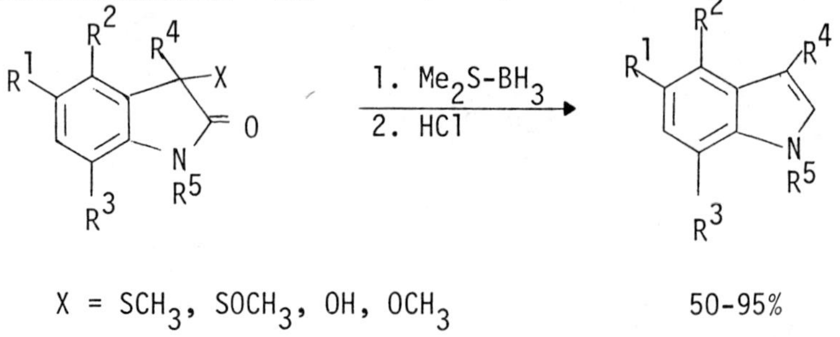

X = SCH_3, $SOCH_3$, OH, OCH_3

1. Me_2S-BH_3
2. HCl

50-95%

III.H. Reductive Cleavages

III.H-1 D. H. R. Barton and W. B. Motherwell, J. Chem. Soc. Chem. Commun., 939 (1983).

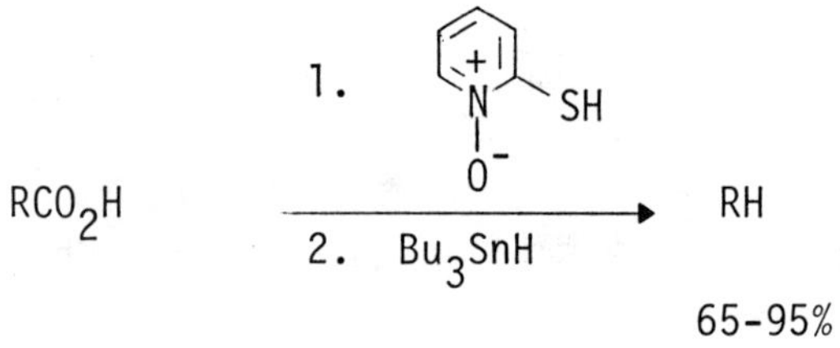

RCO_2H →(1. [pyridine-N-oxide-2-thiol] 2. Bu_3SnH)→ RH

65-95%

III.H-2 W. E. Fristad, M. A. Fry and J. A. Klang, J. Org. Chem., 48, 3575 (1983).

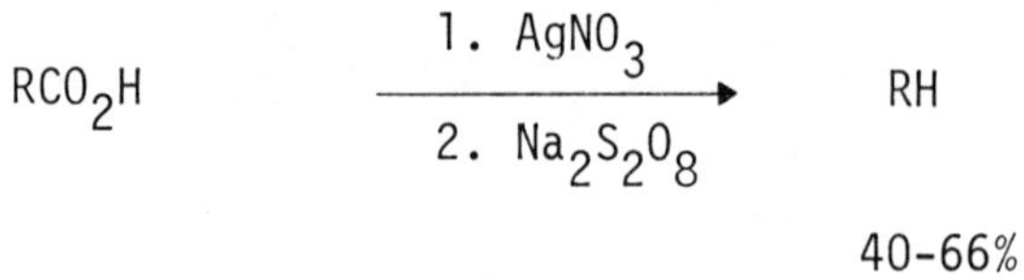

RCO_2H →(1. $AgNO_3$ 2. $Na_2S_2O_8$)→ RH

40-66%

III.H-3 R. Aneja et al., Tetrahedron Lett., 24, 4641 (1983).

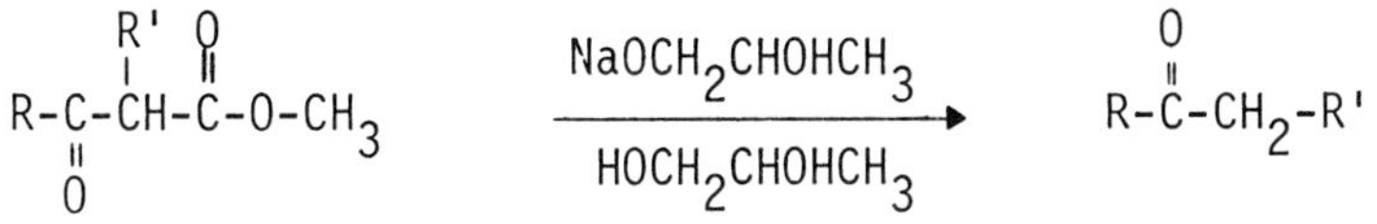

$$\underset{O}{\overset{R'}{R-C-CH-C-O-CH_3}} \xrightarrow[HOCH_2CHOHCH_3]{NaOCH_2CHOHCH_3} R-\overset{O}{\underset{\|}{C}}-CH_2-R'$$

65-98%

III.H-4 G. A. Olah, K. Laali and A. K. Mehrotra, J. Org. Chem., 48, 3360 (1983).

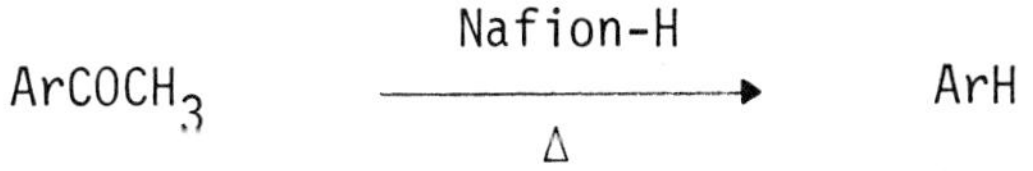

$$ArCOCH_3 \xrightarrow[\Delta]{Nafion-H} ArH$$

56-99%

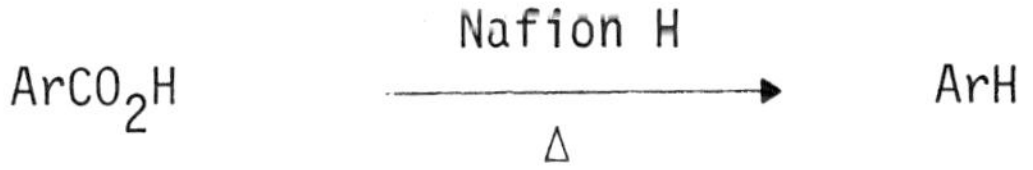

$$ArCO_2H \xrightarrow[\Delta]{Nafion\ H} ArH$$

5-80%

III.H-5 E. P. Goel and U. Krolls, Tetrahedron Lett., 24, 163 (1983).

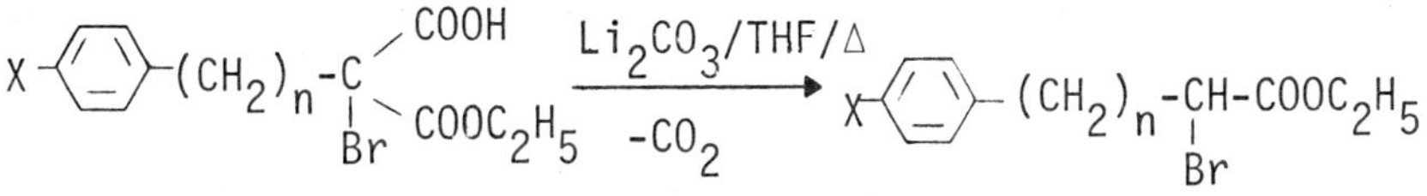

$$X-\langle\underline{}\rangle-(CH_2)_n-\underset{Br}{\overset{COOH}{\underset{|}{C}}}\underset{COOC_2H_5}{} \xrightarrow[-CO_2]{Li_2CO_3/THF/\Delta} X-\langle\underline{}\rangle-(CH_2)_n-\underset{Br}{\overset{|}{CH}}-COOC_2H_5$$

84-95%

III.H-6 U. Schmidt and A. Lieberknecht, Angew. Chem. Int. Ed. Engl., 22, 550 (1983).

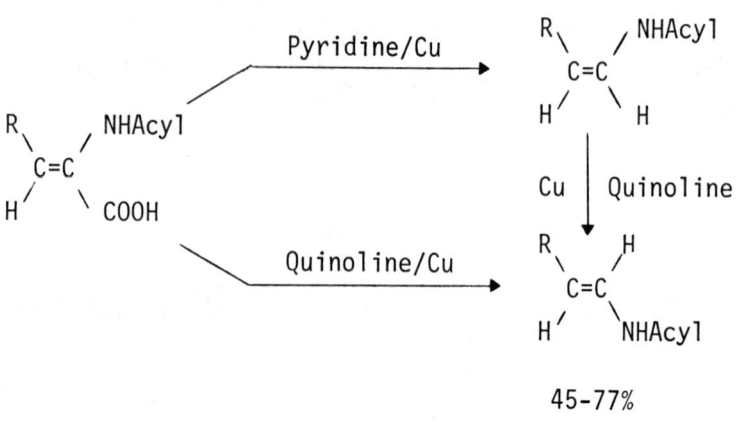

45-77%

III.H-7 J. L. Belletire and D. R. Walley, Tetrahedron Lett., 24, 1475 (1983).

70-95%

III.H-8 J. Mulzer and O. Lammer, Angew. Chem. Int. Ed. Engl., 22, 628 (1983).

60-96%

III.I. Hydroboration (Reduction only)

III.I-1 C. Camacho, G. Uribe and R. Contreras, <u>Synthesis</u>, 1027 (1982).

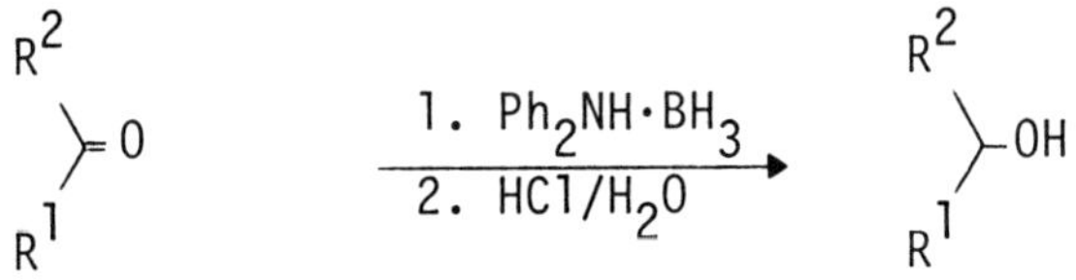

$$R^2R^1C=O \xrightarrow[\text{2. HCl/H}_2\text{O}]{\text{1. Ph}_2\text{NH·BH}_3} R^2R^1CH-OH$$

82-96%

III.I-2 H. C. Brown and G. G. Pai, <u>J. Org. Chem.</u>, <u>48</u>, 1784 (1983).

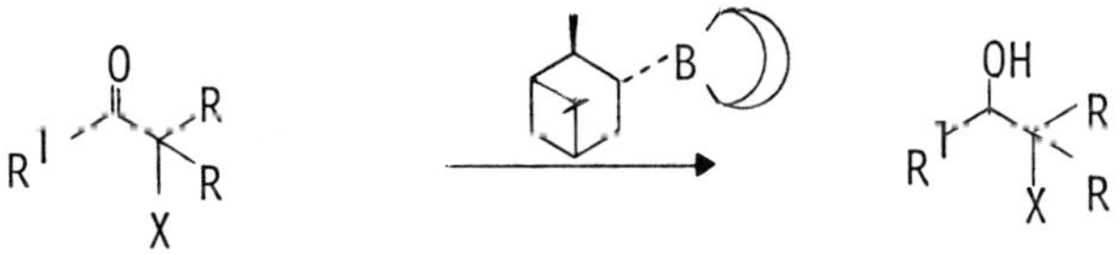

57-95%
32-88% e.e.

III.J. Other Reductions

III.J-1 T. Gartiser, C. Selve and J.-J. Delpuech, <u>Tetrahedron Lett.</u>, <u>24</u>, 1609 (1983).

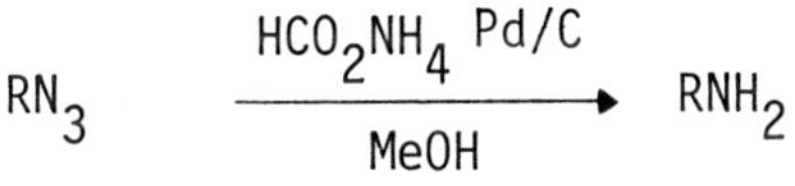

$$RN_3 \xrightarrow[\text{MeOH}]{\text{HCO}_2\text{NH}_4 \text{ Pd/C}} RNH_2$$

74-93%

III.J-2 M. Vaultier, N. Knouzi and R. Carrie, Tetrahedron Lett., 24, 763 (1983).

$$R-N_3 \xrightarrow[\text{2. }H_2O]{\text{1. }PPh_3/THF} R-NH_2$$

79-95%

III.J-3 H. Oda et al., Tetrahedron Lett., 24, 2877 (1983).

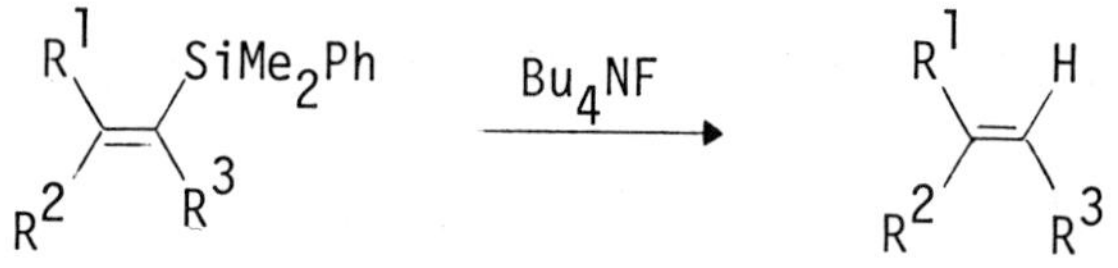

$$\underset{R^2\quad R^3}{\overset{R^1\quad SiMe_2Ph}{\diagup\!\!\!\diagdown}} \xrightarrow{Bu_4NF} \underset{R^2\quad R^3}{\overset{R^1\quad H}{\diagup\!\!\!\diagdown}}$$

60-100%

III.J-4 E. C. Chukovskaya, R. K. Freidlina and N. A. Kuzmina, Synthesis, 773 (1983).

Review: "Reduction of Trichloromethyl Compounds by Hydrogen Donors Induced by Transition Metal Carbonyls, Their Complexes, or Their Salts."

III.J-5 W. S. Knowles, Acc. Chem. Res., 16, 106 (1983).

Review: "Asymmetric Hydrogenation."

III.J-6 E. Block, Org. Reactions, 30, 457 (1984).

    Review: "Olefin Synthesis by Deoxygenation of Vicinal Diols."

III.J-7 R. Noyori and Y. Hayakawa, Org. Reactions, 29, 163 (1983).

    Review: "Reductive Dehalogenation of Polyhalo Ketones with Low-Valent Metals and Related Reducing Agents."

III.J-8 P. Caubere, Angew. Chem. Int. Ed. Engl., 22, 599 (1983).

    Review: "Complex Reducing Agents (CRA's) -- Versatile, Novel Ways of Using Sodium Hydride in Organic Synthesis."

III.J-9 J. W. Huffman, Acc. Chem. Res., 16, 399 (1983).

    Review: "Metal-Ammonia Reductions of Cyclic Aliphatic Ketones."

IV
SYNTHESIS OF HETEROCYCLES

IV.A. Aziridines

IV.A-1 R. S. Tewari, A. K. Awasthi and A. Awasthi, <u>Synthesis</u>, 330 (1983).

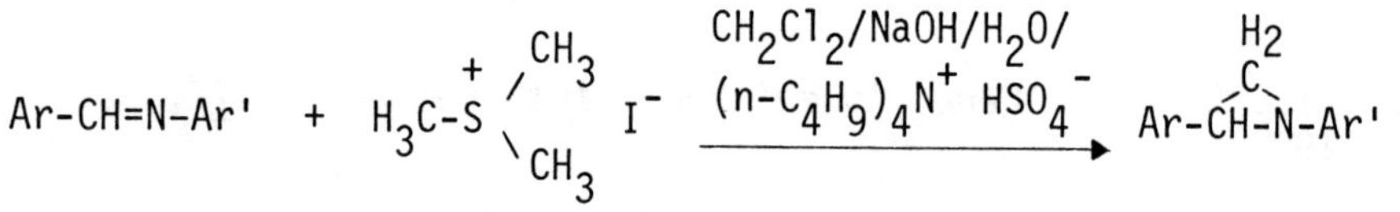

$$Ar-CH=N-Ar' + H_3C-\overset{+}{S}(CH_3)_2 \; I^- \xrightarrow[(n-C_4H_9)_4N^+ HSO_4^-]{CH_2Cl_2/NaOH/H_2O/} Ar-\underset{\underset{\displaystyle N}{|}}{\overset{\overset{\displaystyle CH_2}{\diagup \diagdown}}{CH}}-Ar'$$

80-94%

IV.A-2 G. Aumaitre et al., <u>Synthesis</u>, 816 (1983).

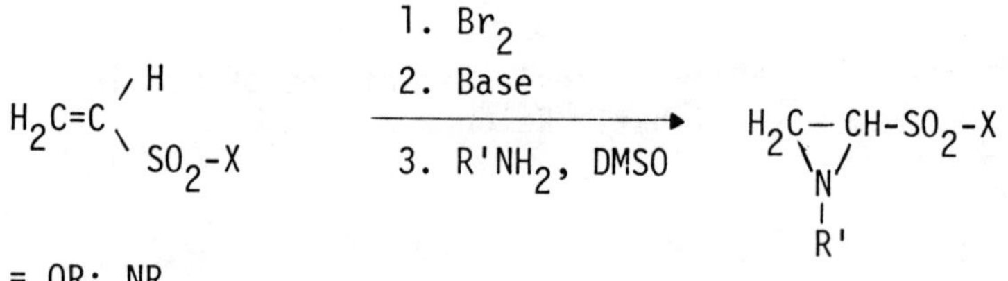

$$H_2C=C\begin{matrix}H\\SO_2-X\end{matrix} \xrightarrow[\text{3. R'NH}_2\text{, DMSO}]{\text{1. Br}_2 \quad \text{2. Base}} H_2C-CH-SO_2-X \atop \underset{\underset{\displaystyle R'}{|}}{\diagdown N \diagup}$$

X = OR; NR$_2$
R' = alkyl

17-75%

IV.A-3 G. Cardillo et al., J. Chem. Soc. Chem. Commun., 1309 (1982).

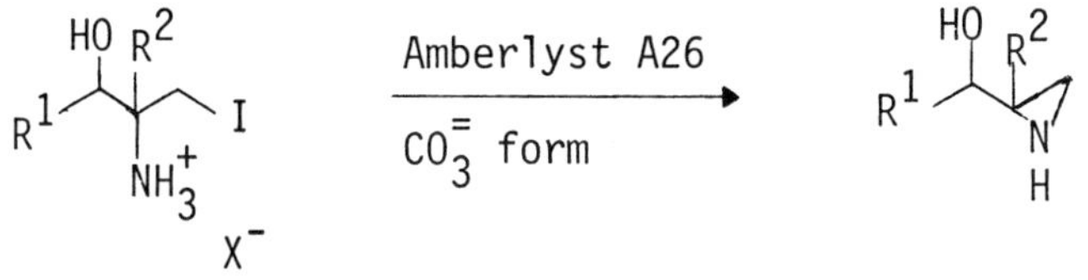

$$R^1\text{-CH(OH)-C}(R^2)(NH_3^+)\text{-CH}_2\text{I} \cdot X^- \xrightarrow[CO_3^= \text{ form}]{\text{Amberlyst A26}} R^1\text{-CH(OH)-C}(R^2)\text{(aziridine-NH)}$$

~95%

IV.A-4 C. M. Bladon and G. W. Kirby, J. Chem. Soc. Chem. Commun., 1402 (1982).

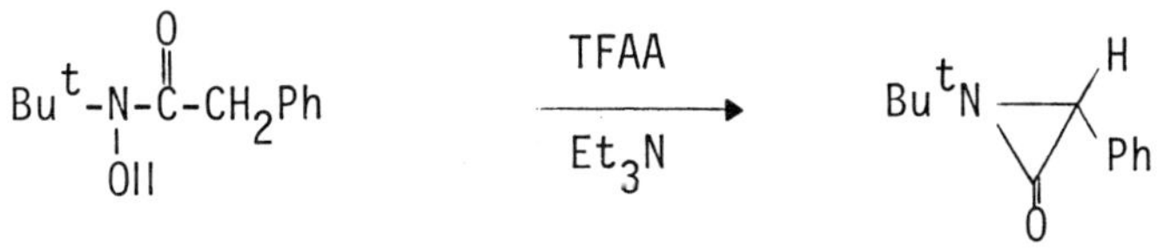

$$Bu^t\text{-N(OH)-C(=O)-CH}_2Ph \xrightarrow[Et_3N]{TFAA} Bu^t N\text{-aziridinone(Ph,H)}$$

97%

IV.A-5 S. Auricchio and O. V. De Pava, J. Chem. Res. Synop., 132 (1983).

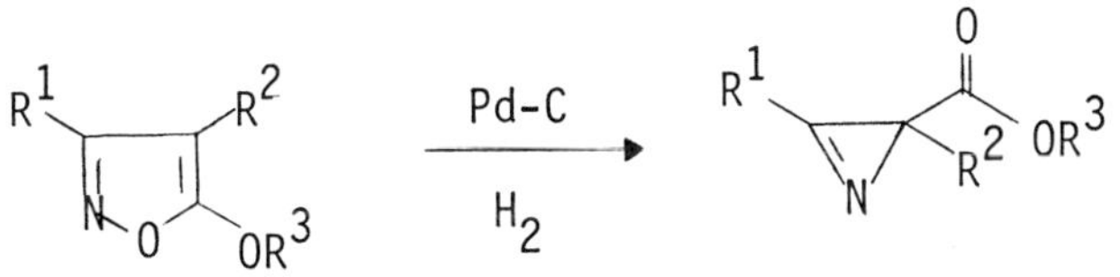

$$\underset{\text{isoxazole}(R^1,R^2,OR^3)}{} \xrightarrow[H_2]{\text{Pd-C}} \underset{\text{azirine}(R^1,R^2,C(=O)OR^3)}{}$$

91-100%

IV.B. Furans, etc.

IV.B-1 M. Kosugi et al., J. Chem. Soc. Chem. Commun., 989 (1983).

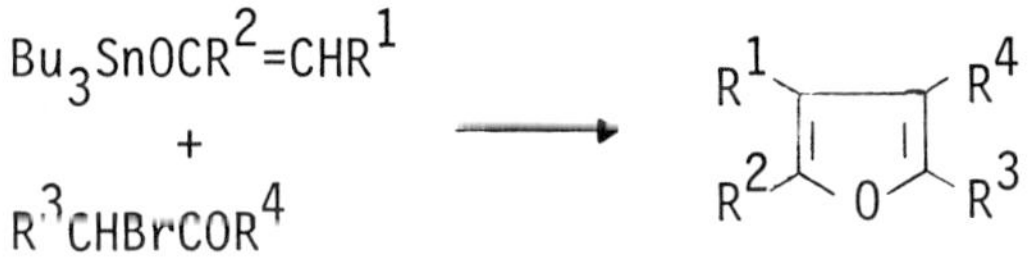

$$Bu_3SnOCR^2=CHR^1 + R^3CHBrCOR^4 \longrightarrow \underset{\text{furan}(R^1,R^2,R^3,R^4)}{}$$

30-92%

IV.B-2 H. Ishibashi et al., Tetrahedron Lett., 24, 3877 (1983).

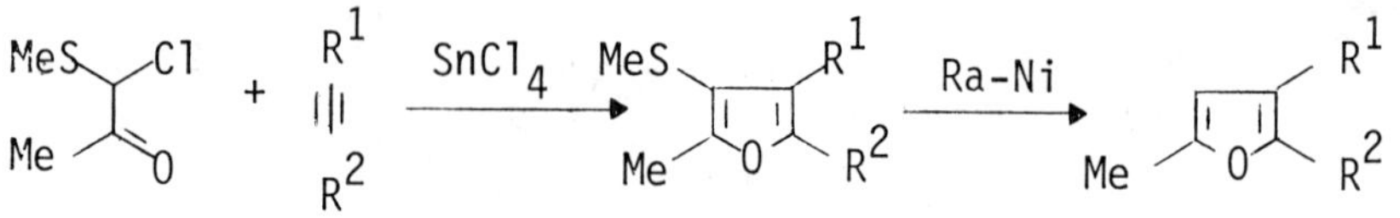

$MeS\text{-}CHCl\text{-}C(Me)=O$ + $R^1C{\equiv}CR^2$ $\xrightarrow{SnCl_4}$ [MeS, Me, R^1, R^2 furan] $\xrightarrow{Ra\text{-}Ni}$ [Me, R^1, R^2 furan]

IV.B-3 O. Tsuge, K. Matsuda and S. Kanemasa, Heterocycles, 20, 593 (1983).

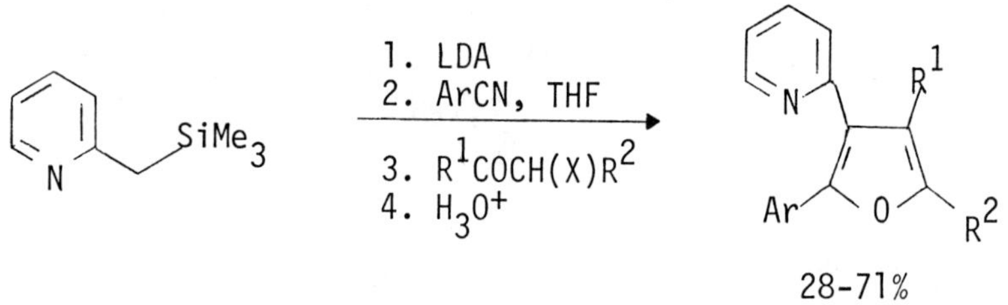

2-Py-CH$_2$-SiMe$_3$ $\xrightarrow{\begin{array}{l}1.\ LDA\\ 2.\ ArCN,\ THF\\ 3.\ R^1COCH(X)R^2\\ 4.\ H_3O^+\end{array}}$ [2-pyridyl, R^1, Ar, R^2 furan]

28-71%

IV.B-4 D. Liotta, M. Saindane and W. Ott., Tetrahedron Lett., 24, 2473 (1983).

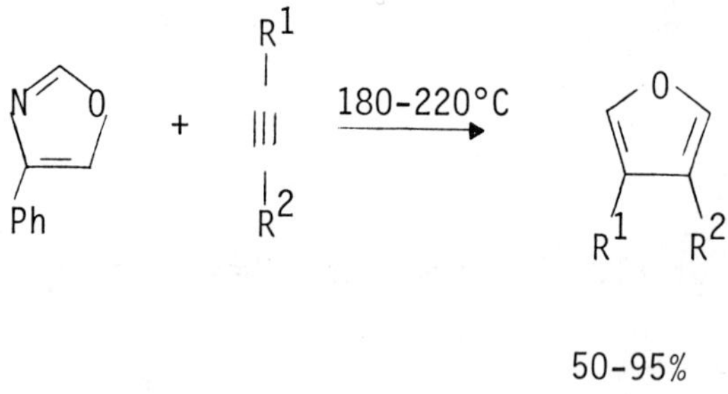

[4-Ph-oxazole] + $R^1C{\equiv}CR^2$ $\xrightarrow{180\text{-}220°C}$ [3,4-disubstituted furan with R^1, R^2]

50-95%

IV.B-5 M. E. Garst, T. A. Spencer, J. Org. Chem., 48, 2442 (1983).

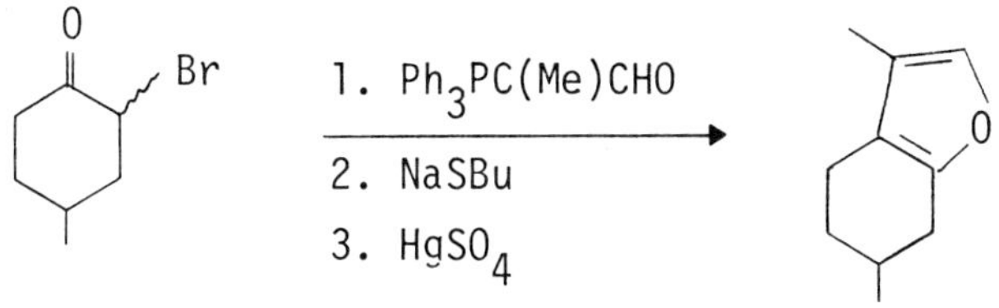

[cyclohexanone with α-Br and 4-Me] → 1. Ph$_3$PC(Me)CHO; 2. NaSBu; 3. HgSO$_4$ → [4,5,6,7-tetrahydrobenzofuran with 3-Me and 6-Me]

49%

IV.B-6 K. Eichinger et al., J. Chem. Res. Synop., 167 (1983).

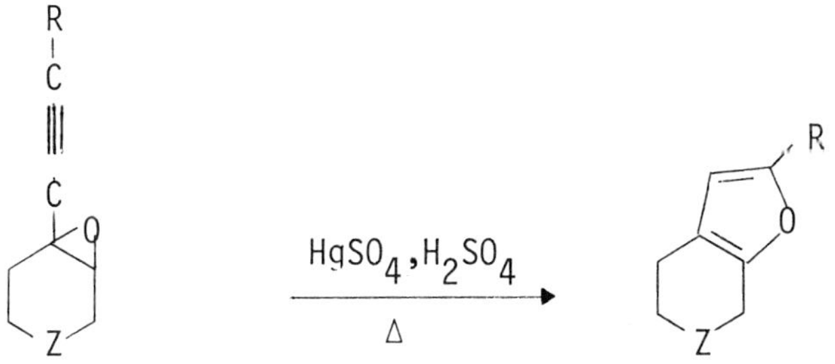

[epoxide with C≡C-R substituent and Z ring] → HgSO$_4$, H$_2$SO$_4$ / Δ → [2-R-substituted tetrahydrobenzofuran with Z]

17-95%

R = Me, aryl
Z = CR'R", NR', SO$_2$

IV.B-7 D. E. Ames and A. Opalko, Synthesis, 234 (1983)

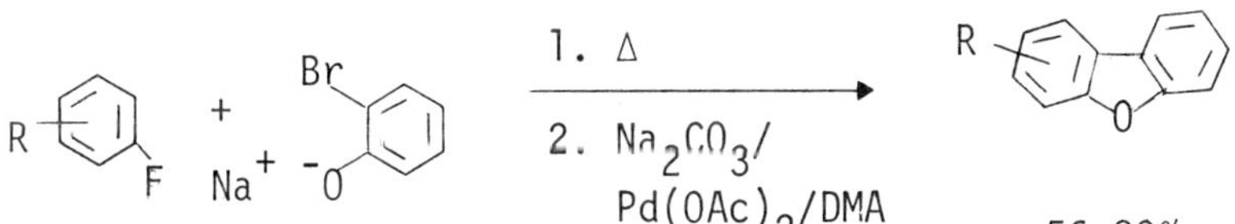

R-[C$_6$H$_3$]-F + Na$^+$ $^-$O-[C$_6$H$_4$]-Br → 1. Δ; 2. Na$_2$CO$_3$/ Pd(OAc)$_2$/DMA → R-[dibenzofuran]

56-80%

IV.B-8 G. Casiraghi *et al.*, *J. Chem. Soc. Perkin Trans. 1*, 1649 (1983).

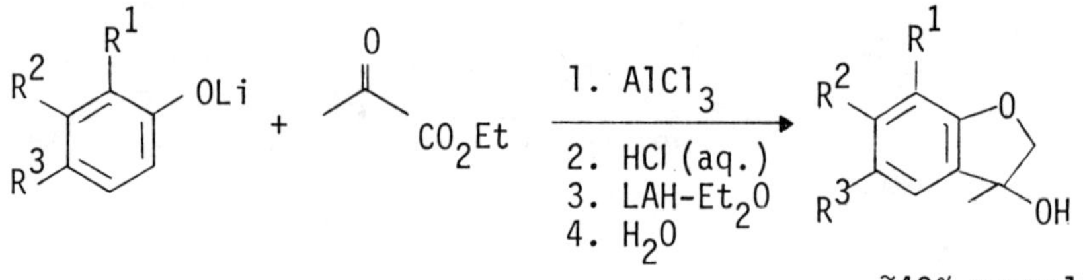

R1
R2
R3
OLi
+
O
CO2Et
1. AlCl3
2. HCl (aq.)
3. LAH-Et2O
4. H2O
O
OH

~40% overall

IV.B-9 K. Akiba *et al.*, *Tetrahedron Lett.*, *24*, 1715 (1983).

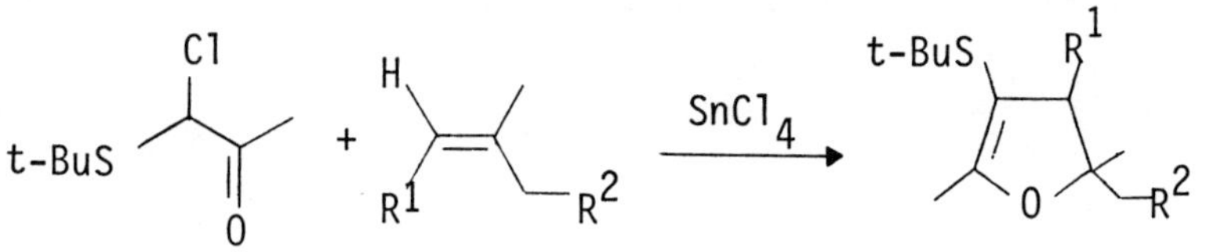

Cl
t-BuS
O
+
H
R1
R2
SnCl4
t-BuS
R1
O
R2

41-55%

IV.B-10 I. Kuwajima, S. Sugahara and J. Enda, *Tetrahedron Lett.*, *24*, 1061 (1983)

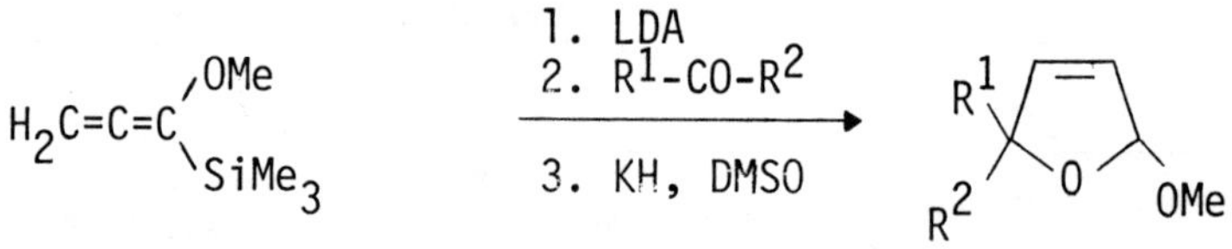

OMe
H2C=C=C
SiMe3
1. LDA
2. R1-CO-R2
3. KH, DMSO
R1
R2
O
OMe

56-87%

IV.B-11 S. H. Andersen, K. K. Sharma and K. B. G. Torssell, Tetrahedron, 39, 2241 (1983).

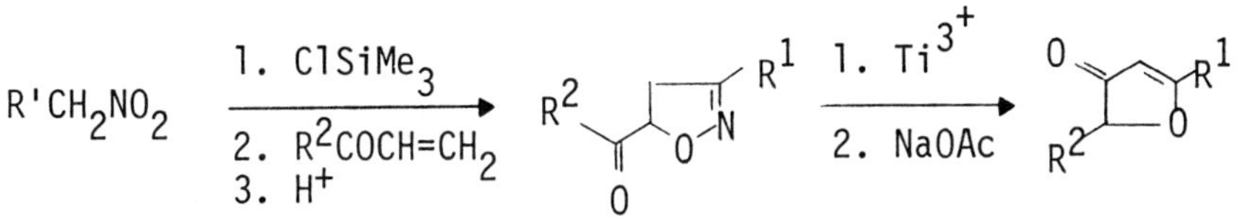

$$R'CH_2NO_2 \xrightarrow[\substack{2.\ R^2COCH=CH_2 \\ 3.\ H^+}]{1.\ ClSiMe_3} \underset{O}{\overset{R^2}{\bigg\backslash}}\!\!\!\text{[isoxazoline]}\!\!\!R^1 \xrightarrow[2.\ NaOAc]{1.\ Ti^{3+}} \text{[dihydrofuranone]}$$

48-87%

IV.B-12 D. P. Curran and D. H. Singleton, Tetrahedron Lett., 24, 2079 (1983).

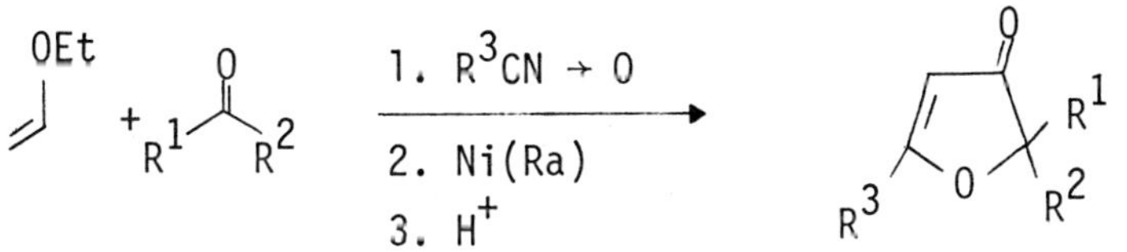

$$\underset{\substack{OEt}}{\diagup\!\!\!=}\ +\ \underset{R^1}{\overset{O}{\diagdown\!\!\diagup}}\!R^2 \xrightarrow[\substack{2.\ Ni(Ra) \\ 3.\ H^+}]{1.\ R^3CN \to O} \text{[dihydrofuranone product]}$$

50-77%

IV.B-13 C. N. Barry and S. A. Evans Jr., J. Org. Chem., 48, 2825, (1983).

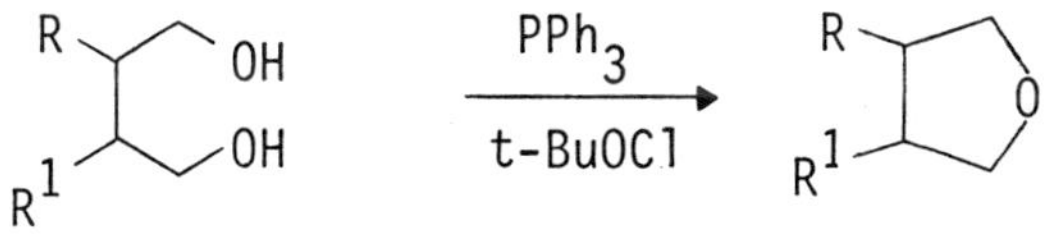

$$\underset{R^1}{\overset{R}{\text{diol}}}\!\!\!\!\!\xrightarrow[t\text{-BuOCl}]{PPh_3}\ \text{[tetrahydrofuran]}$$

61-100%

IV.B-14 E. Piers and V. Karunaratne, J. Org. Chem., 48, 1774 (1983).

$R^1R C=O$ + $Li-CH_2CH_2-C(=CH_2)-Cl$ $\xrightarrow{\text{1. -78°C, THF} \atop \text{2. HMPA}}$ tetrahydrofuran product with R^1, R, and exocyclic methylene

51-63%

IV.B-15 R. C. Ronald and T. S. Lillie, J. Am. Chem. Soc. 105, 5709 (1983).

cyclohexenyl-CH$_2$CH$_2$-OBs $\xrightarrow{\text{1. K}_2\text{HPO}_4 \cdot 3\text{H}_2\text{O, THF} \atop \text{2. H}_2\text{O}_2\text{, THF}}$ bicyclic ether with OOH

OBs = brosylate

47-84%

IV.C. Indoles

IV.C-1 A. K. Sinhababu and R. T. Borchardt, J. Org. Chem., 48, 3347 (1983).

substituted β-nitrostyrene (R_1-R_5, NO$_2$, NO$_2$) $\xrightarrow{\text{Fe/HOAc/silica gel} \atop \text{benzene + cyclohexane} \atop \text{or} \atop \text{toluene reflux 1 h.}}$ indole (R_1-R_5)

75-96%

IV.C-2 S. Raucher and G. A. Koolpe, J. Org. Chem., 48, 2066 (1983).

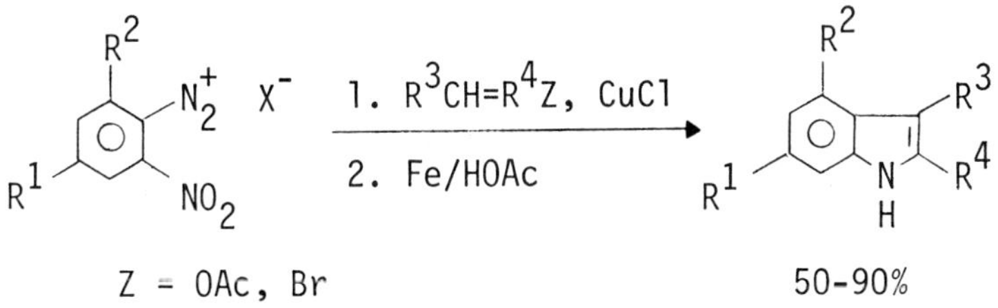

[Reaction scheme: 2,6-disubstituted aryl diazonium salt (R^2 ortho to N$_2^+$X$^-$, R^1 and NO$_2$ on ring) treated with 1. R^{3}CH=R^{4}Z, CuCl; 2. Fe/HOAc → indole with R^2, R^3, R^4, R^1 substituents]

Z = OAc, Br

50-90%

IV.C-3 D. H. Lloyd and D. E. Nichols, Tetrahedron Lett., 24, 4561 (1983).

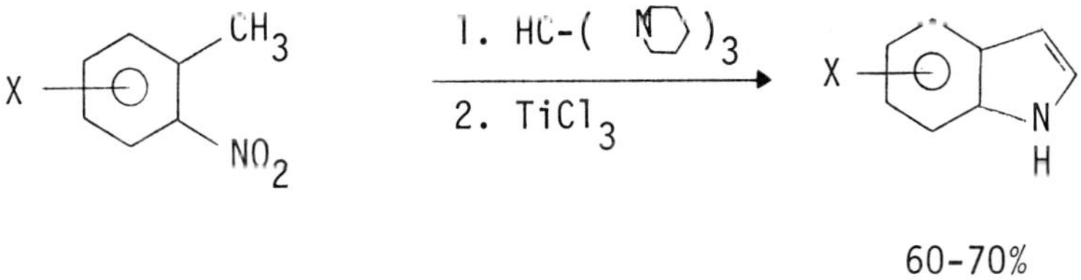

[Reaction scheme: X-substituted 2-methyl-nitrobenzene treated with 1. HC-(morpholine)$_3$; 2. TiCl$_3$ → indole]

60-70%

IV.C-4 K. Menri et al., Chem. Pharm. Bull., 30, 3097 (1982).

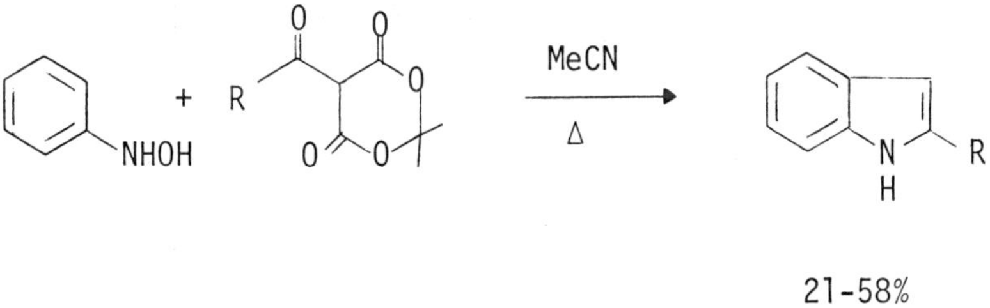

[Reaction scheme: PhNHOH + acyl Meldrum's acid derivative, MeCN, Δ → 2-R-indole]

21-58%

IV.C-5 J. Dijkink et al., Heterocycles, 20, 1255 (1983).

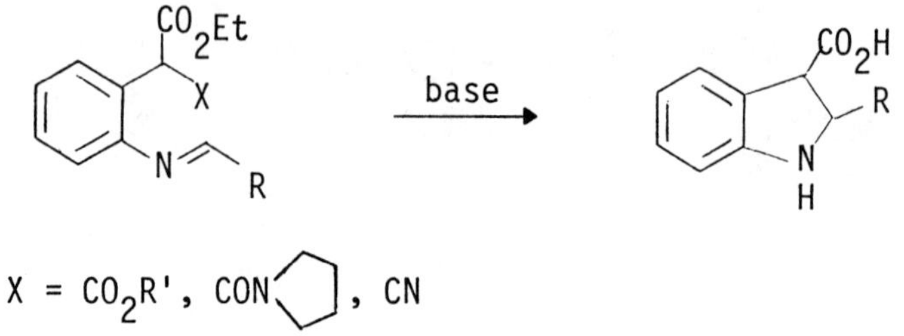

[Scheme: o-(CH(CO$_2$Et)X)-C$_6$H$_4$-N=CHR → indoline with CO$_2$H and R substituents, via base]

X = CO$_2$R', CON⟨ ⟩, CN

IV.C-6 B. Witkop, Heterocycles, 20, 2059 (1983).

Review: "Forty Years of Trypto-fun."

IV.D. Lactams

IV.D-1 H. Alper, F. Urso and D. J. Smith, J. Am. Chem. Soc., 105, 6737 (1983).

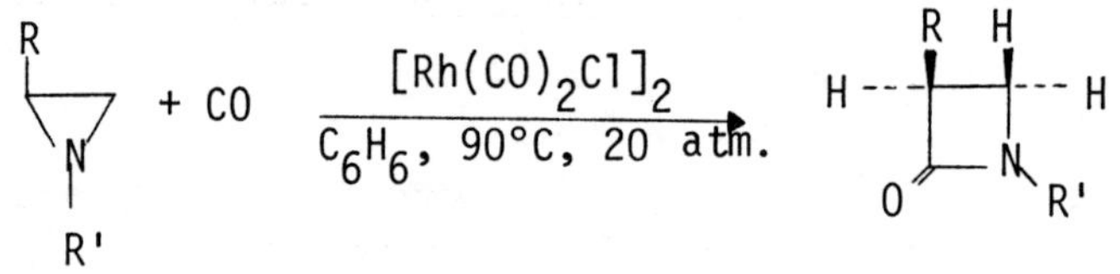

[Scheme: aziridine (R, N-R') + CO, [Rh(CO)$_2$Cl]$_2$, C$_6$H$_6$, 90°C, 20 atm → β-lactam]

quant.

IV. D-2 H. W. Moore and M. J. Arnold, *J. Org. Chem.*, **48**, 3365 (1983).

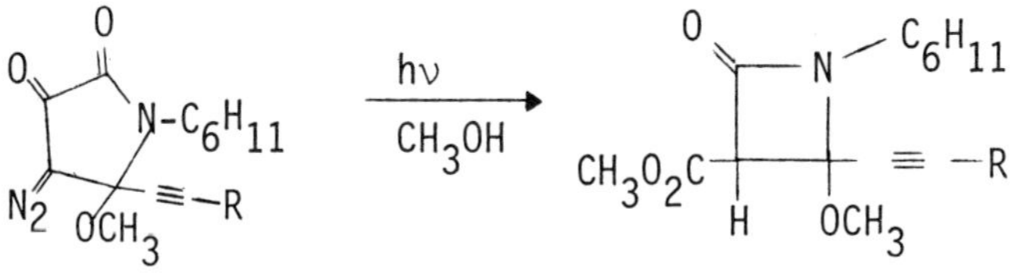

57-74%

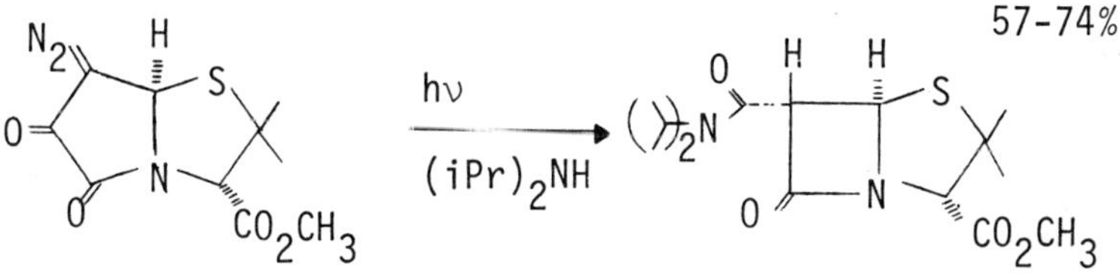

72%

IV.D-3 S. N. Ege *et al.*, *J. Chem. Soc. Perkin Trans. 1*, 1111 (1983).

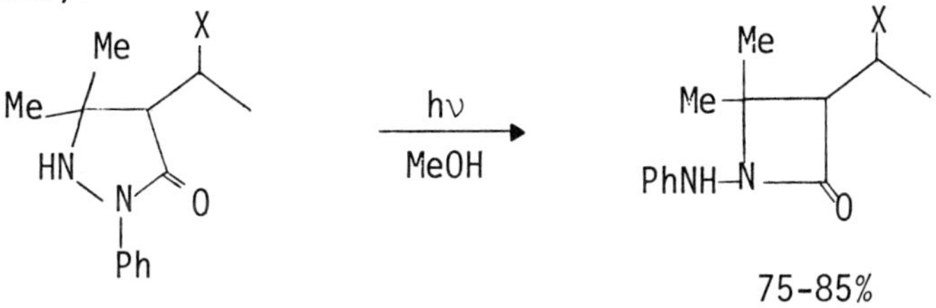

75-85%

X = OH, NH_2, NHAc

IV.D-4 N. Tokutake, M. Miyake and M. Kirisawa, Synthesis, 66 (1983).

$$\text{Tos-CH}_2\text{-}\overset{\overset{O}{\|}}{C}\text{-O-}\!\!\bigcirc\!\!\text{-NO}_2 + Ar^1\text{-CH=N-}Ar^2 \xrightarrow{\text{imidazole}} \begin{array}{c} \text{Tos} \quad Ar^1 \\ \square_N \\ O \quad Ar^2 \end{array}$$

45%

IV.D-5 A. Arrieta, J. M. Aizpurua and C. Palomo, Synth. Commun., 12, 967 (1982).

$$\underset{R^2}{\overset{R^1}{\diagdown}}\!\!\!\text{CH-CO}_2\text{H} \xrightarrow[\text{2. } Ar^1\text{-CH=N-}Ar^2, Et_3N]{1. \text{PhOP(O)Cl}_2 - \text{DMF, } 0°C} \begin{array}{c} R^1 \quad R^2 \; H \\ \square_N \; Ar^1 \\ O \quad Ar^2 \end{array}$$

R^1 = phthalimido, Br

R^2 = H, Me

51-75%

IV.D-6 M. Miyake, N. Tokutake and M. Kirisawa, Synthesis, 833 (1983).

$$R^1\text{-CH}_2\text{-COOH} \xrightarrow{\text{TosCl/(C}_2\text{H}_5)_3\text{N}} \xrightarrow[\text{(C}_2\text{H}_5)_3\text{N}]{R^2\text{-CH=N-}R^3} \begin{array}{c} R^1 \quad R^2 \\ \square_N \\ O \quad R^3 \end{array}$$

R^1 = -O-C$_6$H$_4$-Cl, phthalimido

39-64%

IV.D-7 M. S. Manhas et al., Tetrahedron Lett., 24, 2323 (1983).

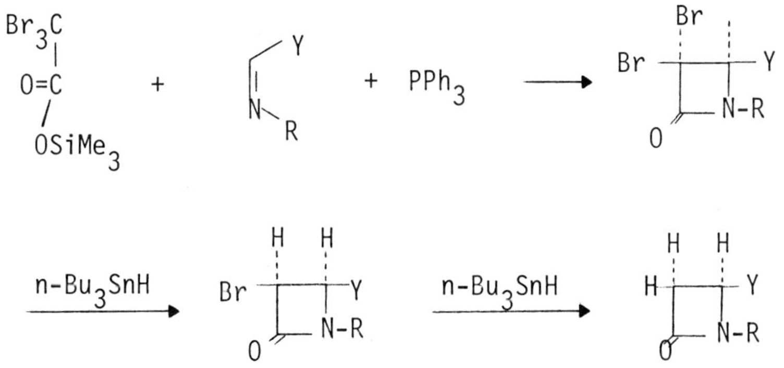

$$Br_3C-C(=O)-OSiMe_3 \;+\; \text{(imine CH=N-R with Y)} \;+\; PPh_3 \;\longrightarrow\; \text{β-lactam (Br, Br, Y, N-R)}$$

$$\xrightarrow{n-Bu_3SnH} \text{β-lactam (Br, H, H, Y, N-R)} \xrightarrow{n-Bu_3SnH} \text{β-lactam (H, H, H, Y, N-R)}$$

IV.D-8 K. Ikeda, K. Achiwa and M. Sekiya, Tetrahedron Lett., 24, 4707 (1983).

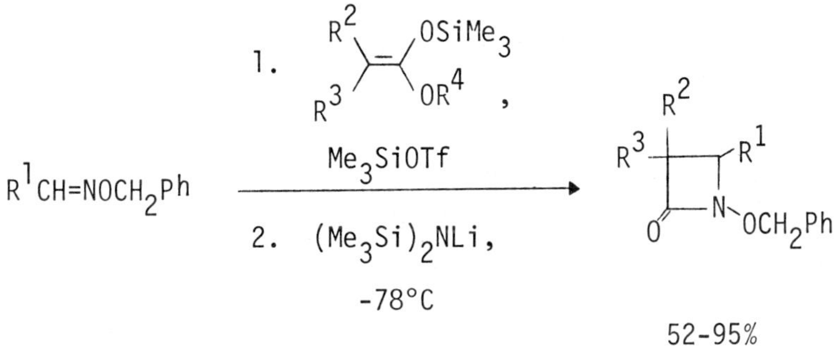

$$R^1CH=NOCH_2Ph \xrightarrow[\text{2. }(Me_3Si)_2NLi, \; -78°C]{\text{1. } R^2R^3C=C(OSiMe_3)(OR^4), \; Me_3SiOTf} \text{β-lactam with } R^2, R^3, R^1, N\text{-}OCH_2Ph$$

52-95%

IV.D-9 J. Marchand-Brynaert et al., J. Chem. Soc. Chem. Commun., 818 (1983).

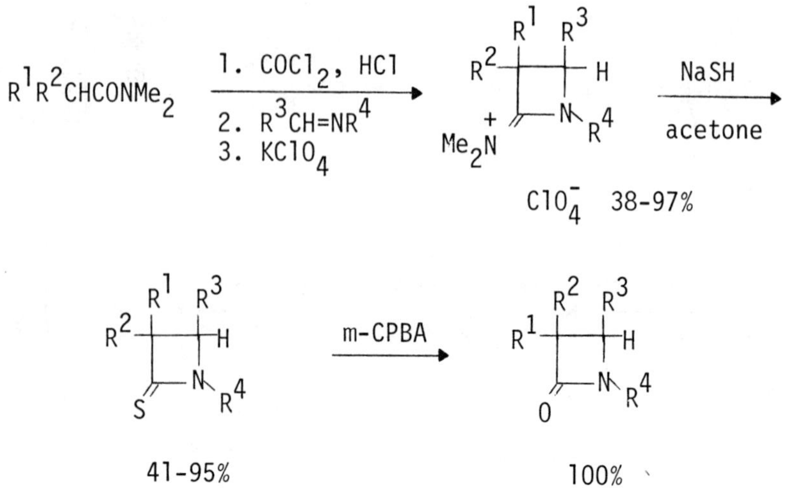

$R^1R^2CHCONMe_2$ $\xrightarrow{\begin{array}{l}1.\ COCl_2,\ HCl\\ 2.\ R^3CH=NR^4\\ 3.\ KClO_4\end{array}}$ [β-lactam iminium salt with $R^1, R^2, R^3, H, N-R^4, Me_2N^+$] ClO_4^- 38-97% $\xrightarrow{\begin{array}{c}NaSH\\ \text{acetone}\end{array}}$

[β-thiolactam with $R^1, R^2, R^3, H, S, N-R^4$] 41-95% $\xrightarrow{m\text{-CPBA}}$ [β-lactam with $R^2, R^3, R^1, H, O, N-R^4$] 100%

IV.D-10 R. M. Adlington et al., J. Chem. Soc. Perkin Trans. 1, 605 (1983).

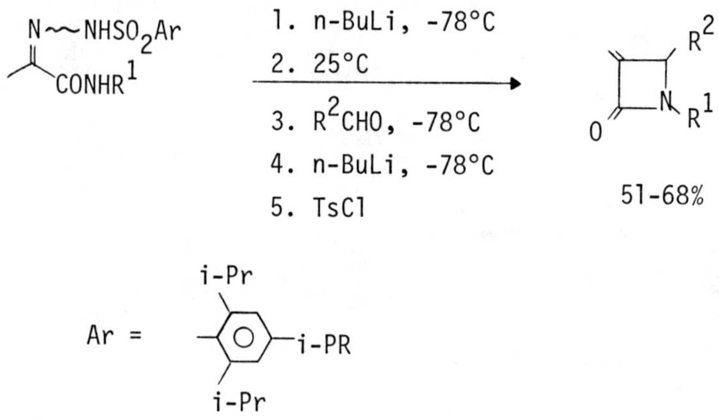

[Structure: $CH_3C(=N{\sim}NHSO_2Ar)CONHR^1$] $\xrightarrow{\begin{array}{l}1.\ n\text{-BuLi},\ -78°C\\ 2.\ 25°C\\ 3.\ R^2CHO,\ -78°C\\ 4.\ n\text{-BuLi},\ -78°C\\ 5.\ TsCl\end{array}}$ [β-lactam with R^2, $N-R^1$, O] 51-68%

Ar = 2,4,6-tri-i-Pr-phenyl (i-Pr, i-PR, i-Pr)

IV.D-11 A. Commercon and G. Ponsinet, Tetrahedron Lett., **24**, 3725 (1983).

$$\text{Br-}\underset{\text{Br-}}{\diagup}\text{-CONHR} \xrightarrow[\text{n-Bu}_4\text{NBr}]{\substack{\text{pulv. KOH} \\ \text{CH}_2\text{Cl}_2}} \text{[azetidinone with R on N, =CH}_2\text{ exocyclic]}$$

65-84%

IV.D-12 S. Sebti and A. Foucaud, Synthesis, 546 (1983).

$$R^1\text{-}\underset{\underset{O}{\parallel}}{C}\text{-}\underset{\underset{Cl}{|}}{CH}\text{-}R^2 + R^3\text{-CO}_2\text{H} + R^4\text{-NC} \xrightarrow{\text{CsF}} R^3\text{-CO}_2\text{-}[\beta\text{-lactam with }R^1, R^2, R^4]$$

78-97%

IV.D-13 D. J. Hart et al., J. Org. Chem. **48**, 289 (1983).

$$\text{RCHO} \xrightarrow[\substack{2.\ R^1R^2C(Li)CO_2Et \\ 3.\ HCl\ (aq.)}]{1.\ LiN(SiMe_3)_2} [\beta\text{-lactam with }R^1, R^2, R, NH]$$

53-79%

R^1= H, CH$_3$

R^2= CH$_3$, SPh

IV.D-14 D. M. Floyd et al., J. Org. Chem., 47, 5160 (1982).

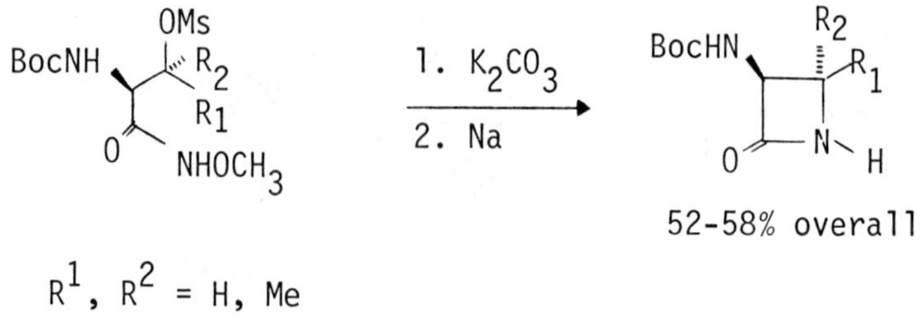

BocNH–CH(OMs)(R$_2$)–C(R$_1$)(NHOCH$_3$)–C(=O)–

1. K$_2$CO$_3$
2. Na

→ BocHN–[β-lactam with R$_2$, R$_1$, N–H]

52-58% overall

R^1, R^2 = H, Me

IV.D-15 M. Ihara et al., Heterocycles, 20, 137 (1983).

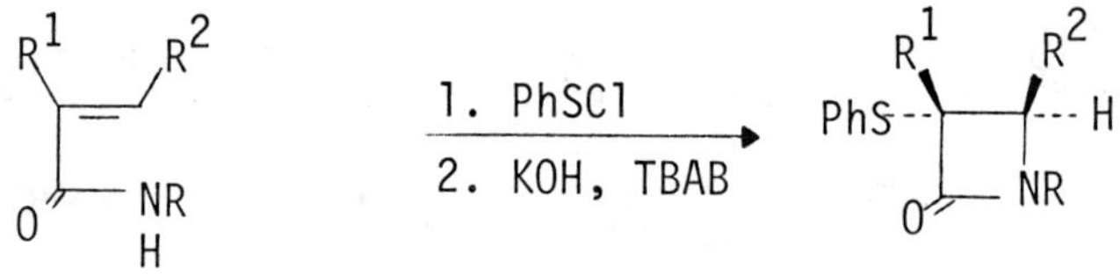

[α,β-unsaturated amide with R^1, R^2, C(=O)NRH]

1. PhSCl
2. KOH, TBAB

→ [β-lactam: PhS, R^1, R^2, H, NR]

IV.D-16 S. C. Shim and D.-W. Kim, Heterocycles, 20, 575 (1983).

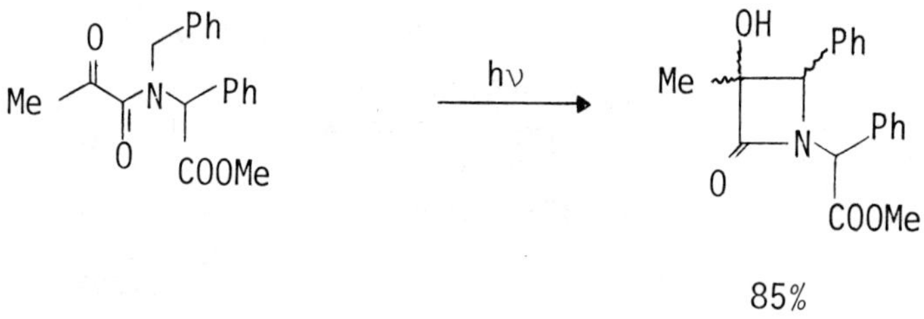

Me–C(=O)–C(=O)–N(CH$_2$Ph)(CH(Ph)COOMe)

hν →

[β-lactam: Me, OH, Ph, N–CH(Ph)COOMe]

85%

IV.D-17 H. Aoyama et al., J. Chem. Soc. Chem. Commun., 333 (1983).

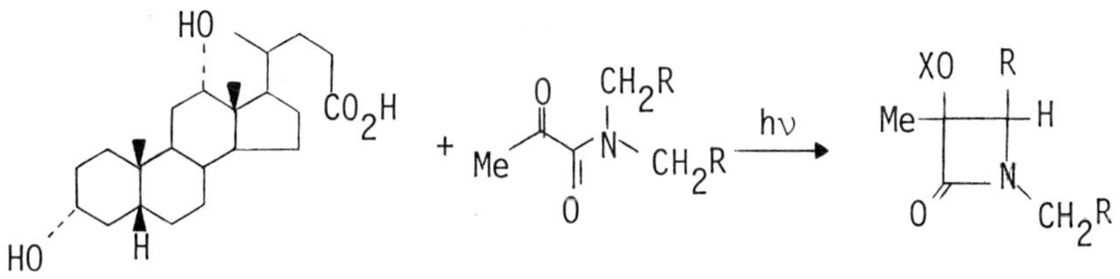

3-15% e.e.

IV.D-18 H. Matsunaga et al., Tetrahedron Lett., 24, 3009 (1983).

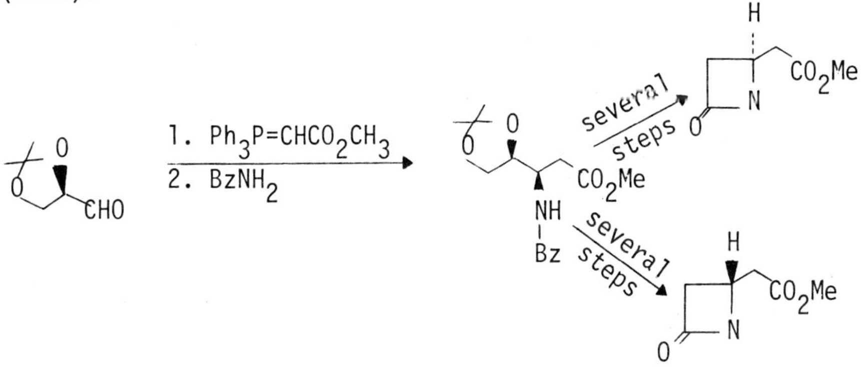

1. $Ph_3P=CHCO_2CH_3$
2. $BzNH_2$

IV.D-19 H.-G. Capraro, G. Rihs and P. Martin, Helv. Chim. Acta, 66, (1983).

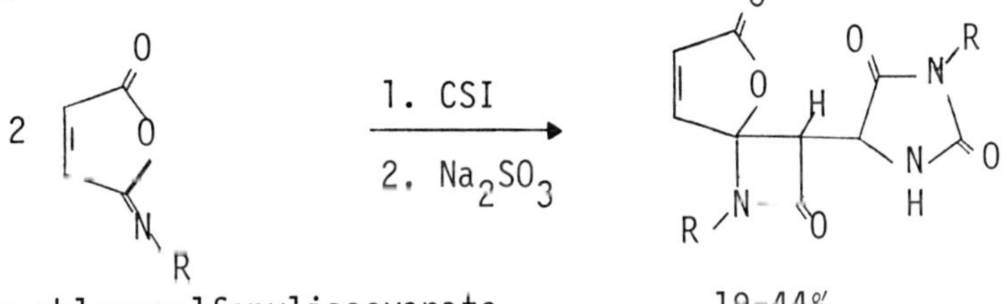

1. CSI
2. Na_2SO_3

CSI = chlorosulfonylisocyanate

19-44%

IV.D-20 E. M. Gordon and J. Pluscec, Tetrahedron Lett., 24, 3419 (1983).

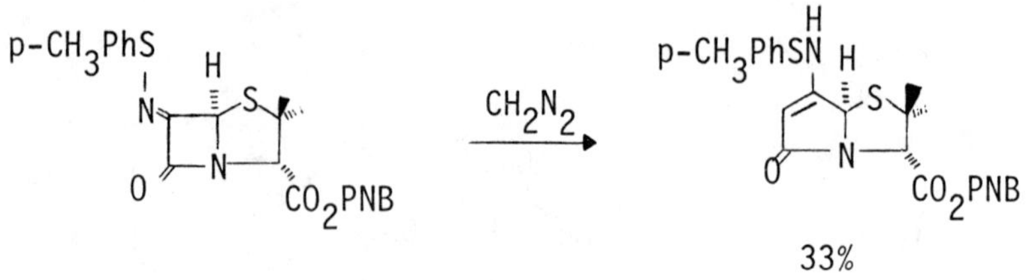

[Scheme: β-lactam with p-CH$_3$PhS-N= group and CO$_2$PNB, treated with CH$_2$N$_2$, gives product with p-CH$_3$PhSNH group, 33%]

IV.D-21 V. V. Kaminski et al., J. Org. Chem., 48, 2337 (1983).

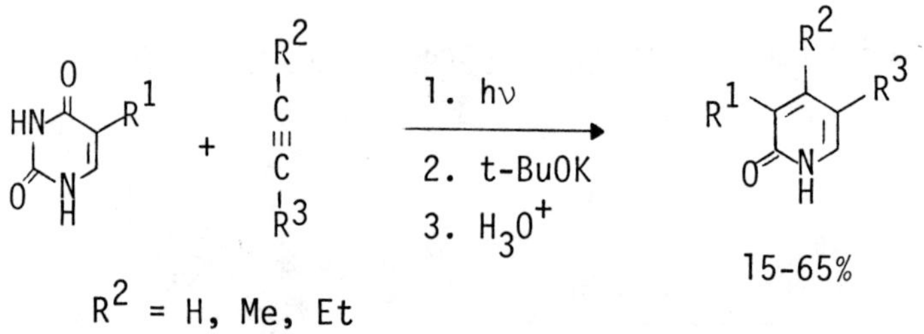

[Scheme: uracil derivative with R^1 + R^{2}C≡CR^3 alkyne, 1. hν, 2. t-BuOK, 3. H$_3$O$^+$, gives pyridinone, 15–65%]

R^2 = H, Me, Et

IV.D-22 O. Meth-Cohn and H. C. Taljaard, Tetrahedron Lett., 24, 4607 (1983).

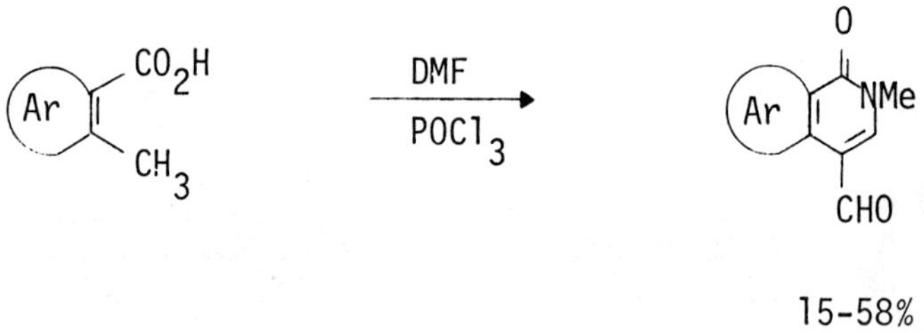

[Scheme: Ar ring with CO$_2$H and CH$_3$ substituents, DMF/POCl$_3$, gives Ar-fused N-methyl isoquinolinone with CHO, 15–58%]

IV.D-23 T. Ohnuma et al., Tetrahedron Lett., 24, 4249 (1983).

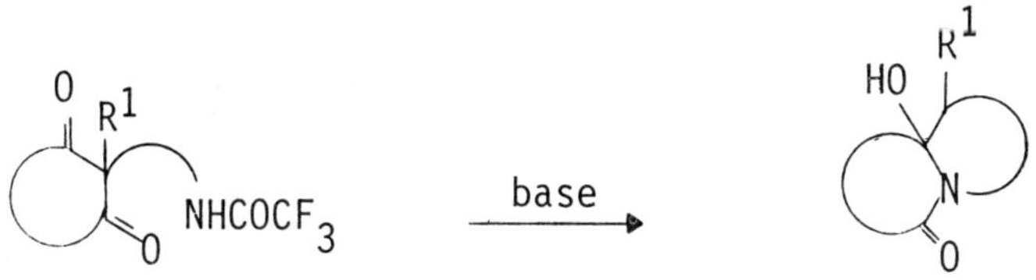

[Structure: cyclic ketone with R^1 and NHCOCF$_3$ substituents] →(base)→ [bicyclic structure with HO, R^1, N, O]

IV.D-24 L. Crombie et al., J. Chem. Soc. Chem. Commun., 959 (1983).

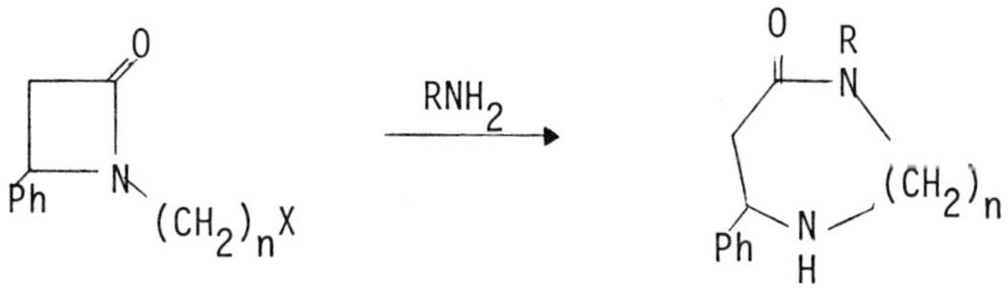

[β-lactam with Ph and N-(CH$_2$)$_n$X] →(RNH$_2$)→ [seven-membered ring product with Ph, NH, N-(CH$_2$)$_n$, N-R, C=O]

46-90%

IV.D-25 C. M. Cimarusti and R. B. Sykes, Chemistry, 19, 302 (1983).

Review: "Monobactams-Novel Antibiotics."

IV.E. Lactones

IV.E-1 D. Hoppe and A. Broenneke, Tetrahedron Lett., 24, 1687 (1983).

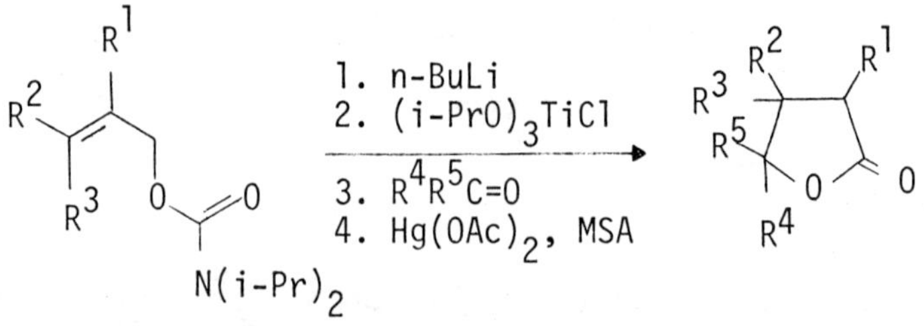

67-93%

IV.E-2 J. V. Comasseto and N. Petragnani, Synth. Commun., 13, 889 (1983).

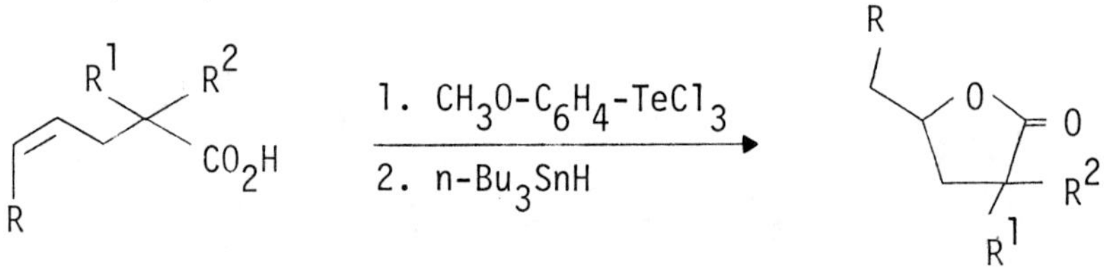

IV.E-3 G. Stork et al., J. Am. Chem. Soc., 105, 3741 (1983).

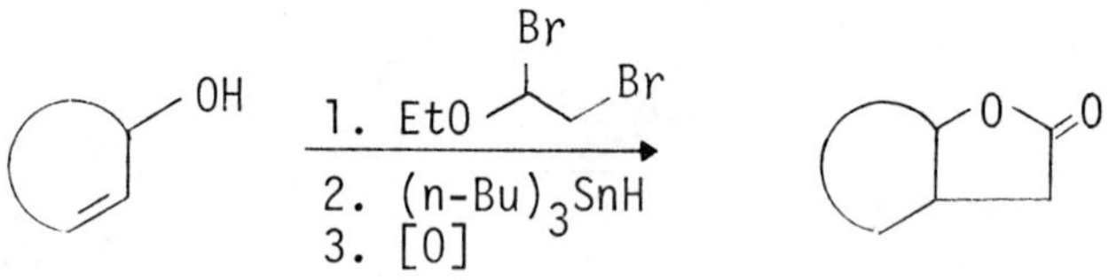

IV.E-4 L. Strekowski, M. Visnick and M. A. Battiste, Synthesis, 493 (1983).

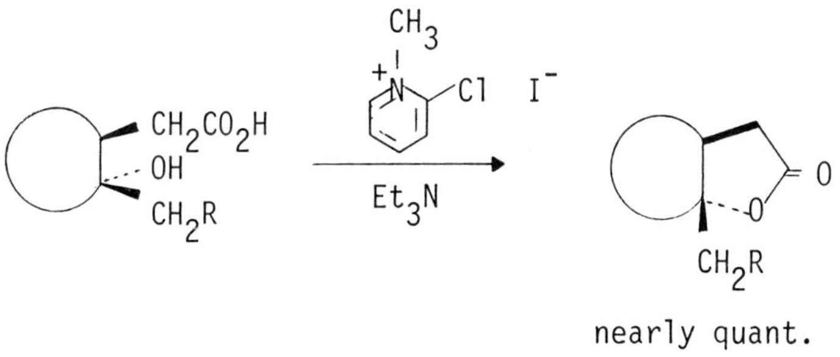

nearly quant.

IV.E-5 D. Ben-Ishai and S. Hirsch, Tetrahedron Lett., 24, 955 (1983).

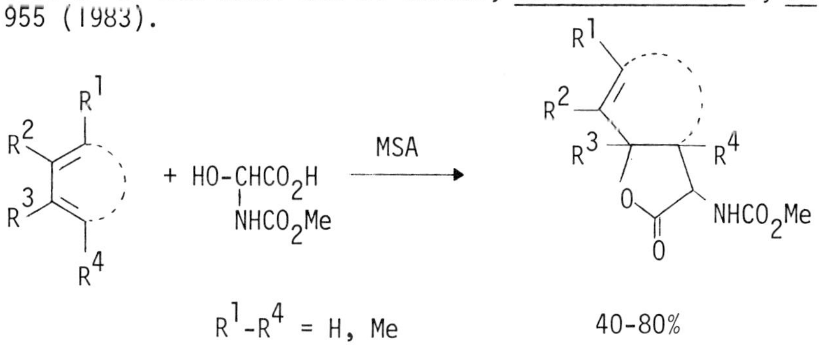

R^1-R^4 = H, Me   40-80%

IV.E-6 K. Akiba et al., Tetrahedron Lett., 24, 1715 (1983).

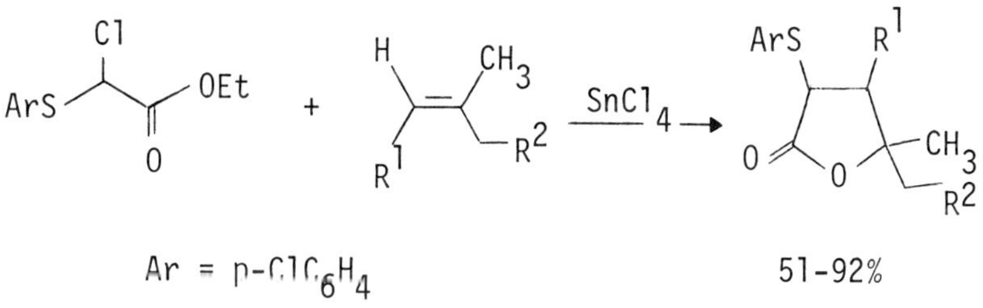

Ar = p-ClC$_6$H$_4$ 51-92%

IV.E-7 H. Nagashima et al., Tetrahedron Lett., 24, 2395 (1983).

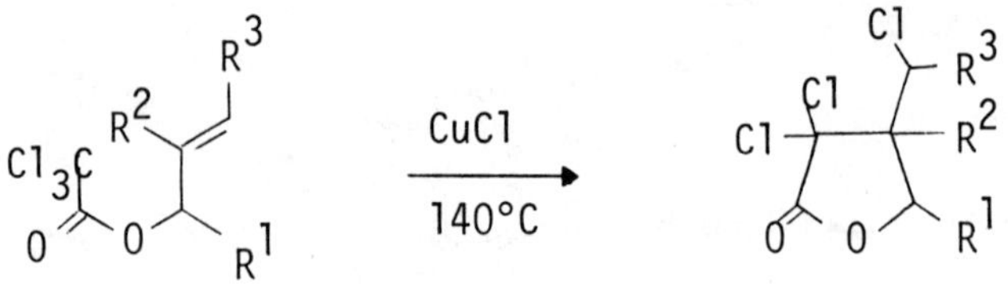

[Reaction scheme: Cl$_3$C-C(=O)-O-CHR^1-C(R^2)=CHR^3 → CuCl, 140°C → product with Cl, Cl, Cl, R^3, R^2, R^1 on lactone ring]

38-78%

IV. E-8 G. A. Garcia, H. Munoz and J. Tamariz, Synth. Commun., 13, 569 (1983).

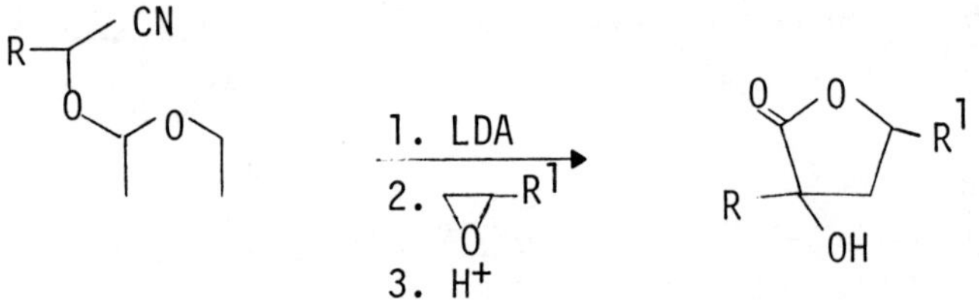

[Reaction scheme: R-CH(CN)-O-CH(CH$_3$)-O-Et → 1. LDA 2. epoxide-R^1 3. H$^+$ → γ-butyrolactone with R, OH, R^1]

68-95%

IV.E-9 F. Henin and J.-P. Pete, Tetrahedron Lett., 24, 4687 (1983).

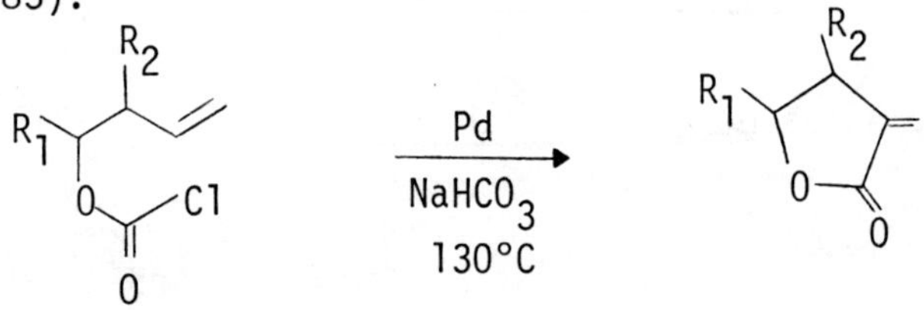

[Reaction scheme: R$_1$-CH(OC(=O)Cl)-CH(R$_2$)-CH=CH$_2$ → Pd, NaHCO$_3$, 130°C → α-methylene-γ-butyrolactone with R$_1$, R$_2$]

yield varies

IV.E-10 P. Barbier and C. Benezra, J. Org. Chem., 48, 2705 (1983).

C_6H_5S-C(CH$_3$)(COOEt) + RR'C(OAc)-CHO $\xrightarrow{\begin{array}{l}1.\ LDA\\2.\ Ba(OH)_2,\\3.\ [O]\\4.\ \Delta\end{array}}$ [lactone with R^1, R, HO, =CH$_2$]   50-65%

IV.E-11 A. P. Kozikowski and A. K. Ghosh, Tetrahedron Lett., 24, 2623 (1983).

[alkene with R^1, R^2, R] $\xrightarrow{\begin{array}{l}1.\ THPOCH_2\text{-}C\!\equiv\!\overset{+}{N}\text{-}O^-\\2.\ H_2,\ Ni(Ra)\\3.\ Zn\text{-}TiCl_4\text{-}CH_2Br_2\\4.\ MnO_2,\ CH_2Cl_2\end{array}}$ [fused lactone with R^1, R^2, R, =CH$_2$]

27-73%

IV.E-12 M. Kitaoka et al., Chem. Lett., 1065 (1983).

PhS-CH=C(CH$_3$)-CONH⧋ or PhS-CH$_2$-C(=CH$_2$)-CONH⧋ $\xrightarrow{\begin{array}{l}1.\ LDA,\ -80°C\\2.\ R^1R^2C=O\\3.\ \Delta\end{array}}$ [lactone with R^1, R^2, =C(H)(SPh)]

IV.E-13 R. W. Saalfrank, P. Schierling and P. Schaetzlein, Chem. Ber., 116, 1463 (1983); R. W. Saalfrank, P. Schierling and W. Hafner, Chem. Ber., 116, 3482 (1983).

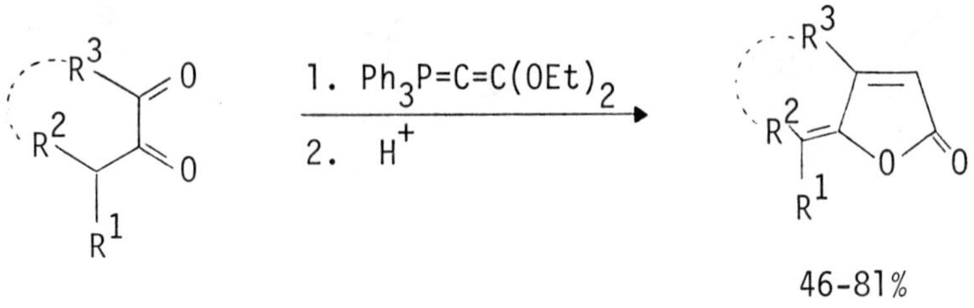

R3
O
R2
O
R1
1. $Ph_3P{=}C{=}C(OEt)_2$
2. H^+
R3
R2
R1
O
O
46-81%

IV.E-14 M. Ochiai, T. Ukita and E. Fujita, Tetrahedron Lett., 24, 4025 (1983).

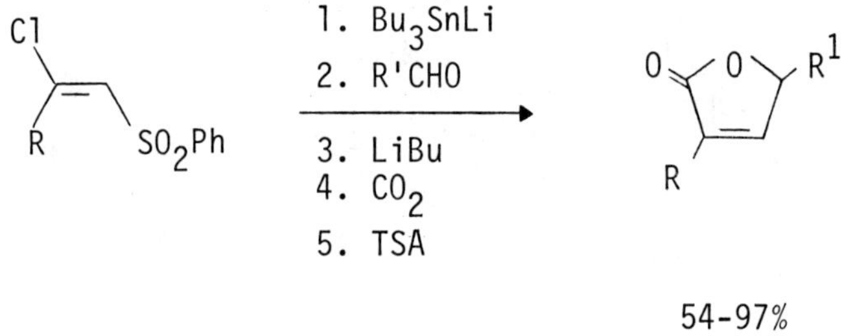

Cl
R
SO_2Ph
1. Bu_3SnLi
2. R'CHO
3. LiBu
4. CO_2
5. TSA
O
O
R^1
R
54-97%

IV.E-15 R. Tanikaga et al., Chem. Lett., 1703 (1982).

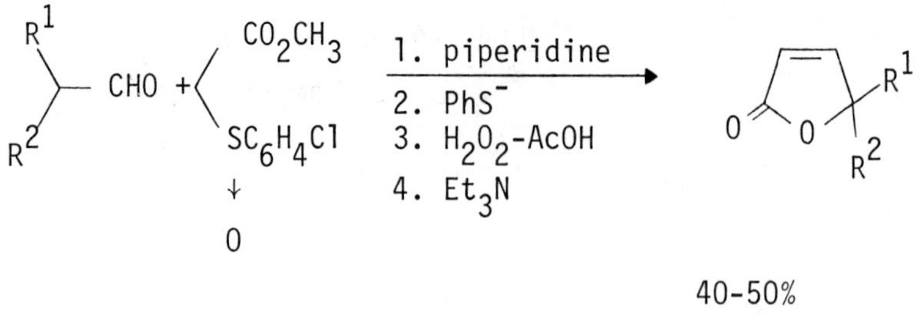

R1
R2
CHO +
CO_2CH_3
SC_6H_4Cl
O
1. piperidine
2. PhS^-
3. H_2O_2-AcOH
4. Et_3N
O
O
R1
R2
40-50%

IV.E-16 K. Tanaka et al., Chem. Lett., 633 (1983).

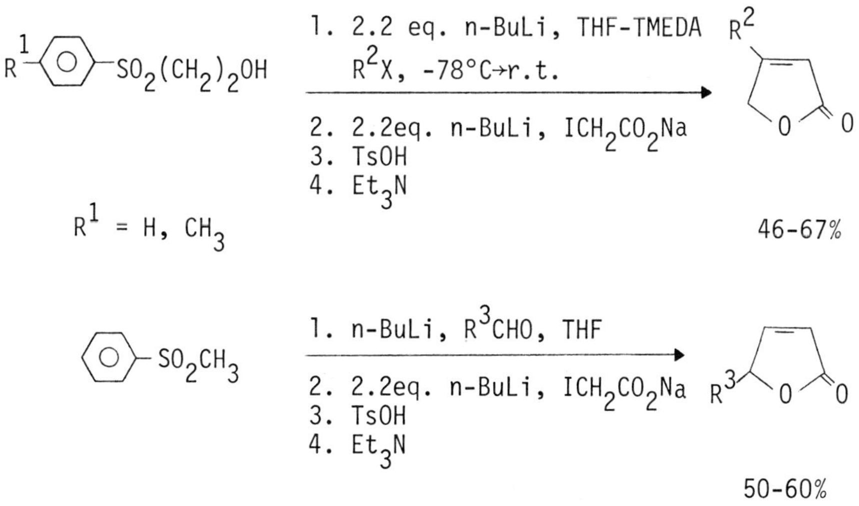

R^1-C$_6$H$_4$-SO$_2$(CH$_2$)$_2$OH

1. 2.2 eq. n-BuLi, THF-TMEDA
   R^2X, -78°C→r.t.
2. 2.2eq. n-BuLi, ICH$_2$CO$_2$Na
3. TsOH
4. Et$_3$N

R^1 = H, CH$_3$

46-67%

C$_6$H$_5$-SO$_2$CH$_3$

1. n-BuLi, R^3CHO, THF
2. 2.2eq. n-BuLi, ICH$_2$CO$_2$Na
3. TsOH
4. Et$_3$N

50-60%

IV.E-17 B. L. Feringa and W. Dannenberg, Synth. Commun., 13, 509 (1983).

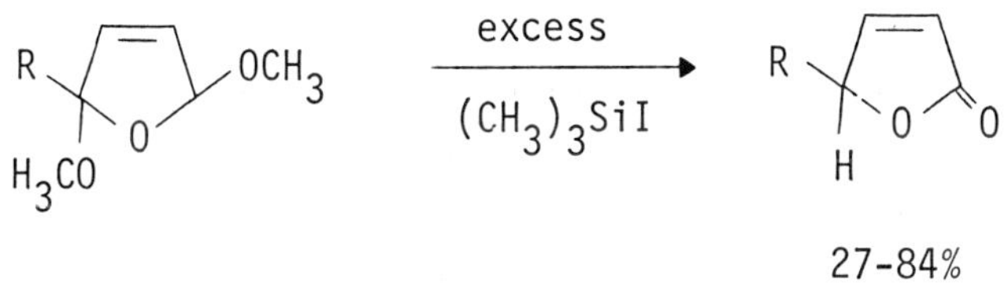

excess (CH$_3$)$_3$SiI

27-84%

IV.E-18 A. K. Sinhababu and R. T. Borchardt, J. Org. Chem., 48, 2356 (1983).

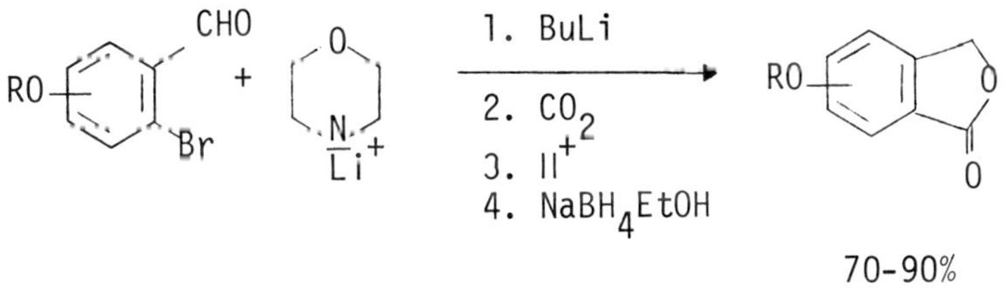

1. BuLi
2. CO$_2$
3. H$^+$
4. NaBH$_4$EtOH

70-90%

IV.E-19 A. I. Meyers et al., Tetrahedron, 39, 1991 (1983).

[Scheme: 2-bromobenzamide + 1. Et₃OBF₄; 2. HO-CH(Ph)-CH(NH₂)-CH₂OMe (aminoalcohol) → 2-bromoaryl-[OX]*]

[Scheme: 1. BuLi; 2. RCOY; 3. R¹MgX; 4. H⁺ → 3,3-disubstituted phthalide with R, R¹   40–80% e.e.]

[OX]* = 4,5-dihydro-oxazoline with 5-Ph and 4-CH₂OMe substituents

---

IV.E-20 J. Graumann and W. Kliegel, Chem. Ztg., 107, 136 (1983).

$$\text{Ar-CH=}\overset{+}{\underset{\underset{O^-}{|}}{N}}\text{-}\overset{R^1}{\underset{|}{C}}\text{H-CO}_2\text{H} \xrightarrow{Ac_2O} \text{Ar-oxazolinone with } R^1$$

IV.E-21   Y. Ishii et al., Tetrahedron Lett., 24, 2677 (1983).

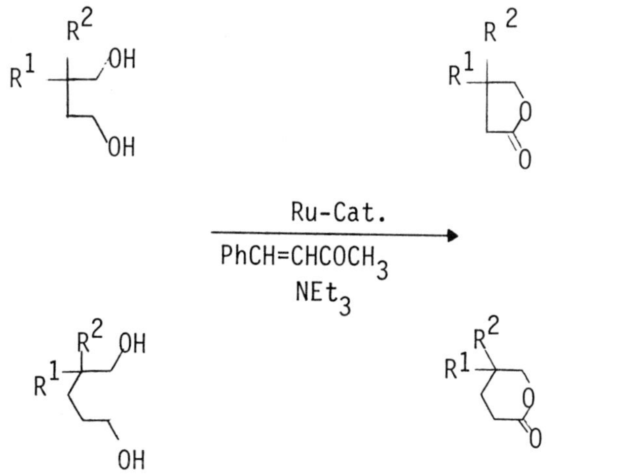

Ru-Cat.
―――――――→
PhCH=CHCOCH$_3$
NEt$_3$

major products

IV.E-22 R. M. Carlson and L. L. White, Synth. Commun., 13, 237 (1983).

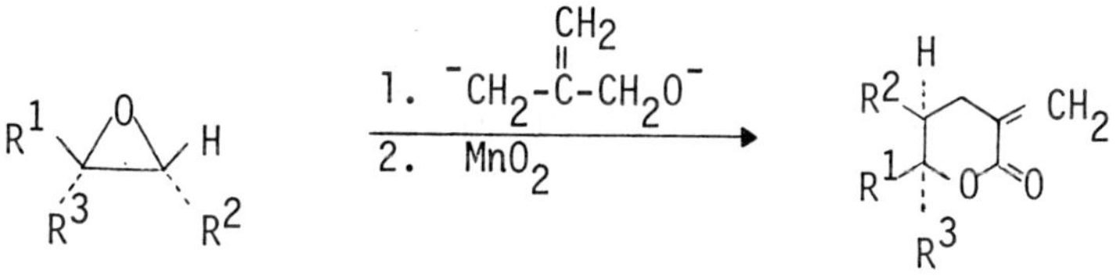

1. $^-CH_2-C(=CH_2)-CH_2O^-$
2. MnO_2

12-61%

IV.E-23   T. Yoshida and S. Saito, Chem. Lett., 1587 (1982).

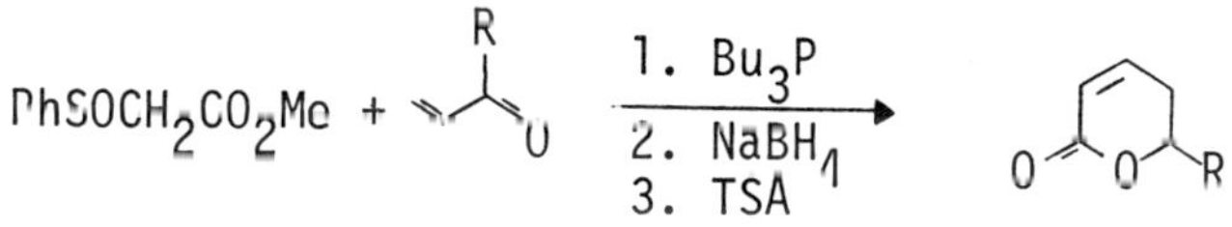

PhSOCH$_2$CO$_2$Me +

1. Bu$_3$P
2. NaBH$_4$
3. TSA

54-61%

IV.E-24 D. D. Chaudhari, Chem. Ind. 568 (1983).

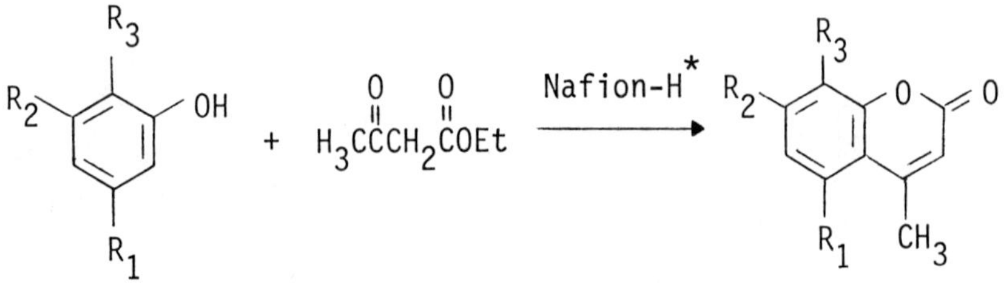
R3
R2
OH
+
O O
H_3CCCH_2COEt
Nafion-H*
R3
R2
O
O
R1
CH3
R1

25-90%

IV.E-25 J. T. A. Reuvers and A. De Groot, Synthesis, 1105 (1982).

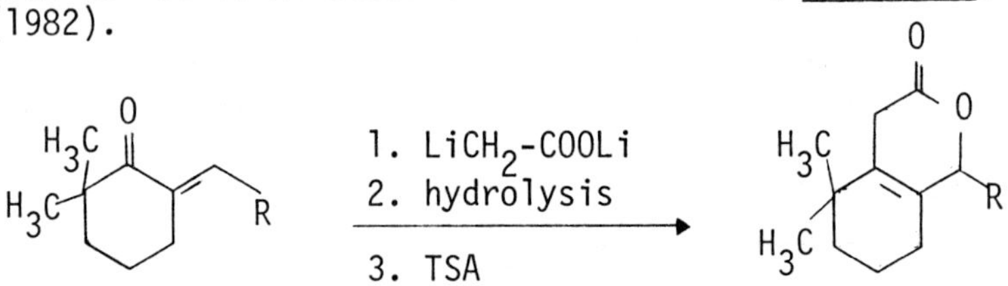
O
H3C
H3C
R
1. LiCH2-COOLi
2. hydrolysis
3. TSA
O
O
H3C
R
H3C

65-85%

IV.E-26 H.-W. Schmidt, R. Schipfer and H. Junek, Liebigs Ann. Chem., 695 (1983).

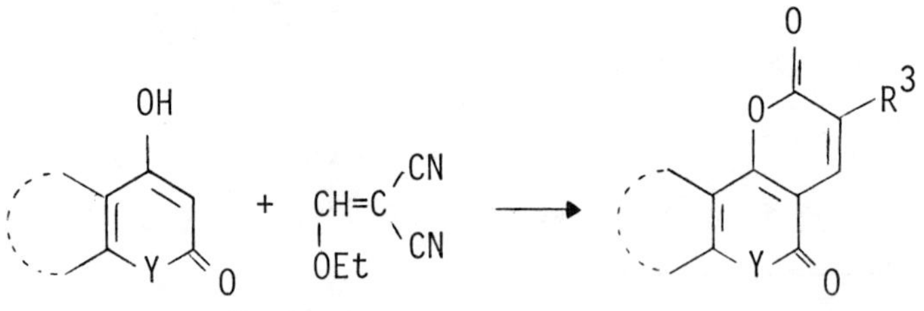
OH
Y
O
+
CH=C
CN
CN
OEt
O
O
R3
Y
O

Y = NH, O

R^3= CN, $CONH_2$

IV.E-27 Y. Kimura and S. L. Regen, J. Org. Chem. 48, 1533 (1983).

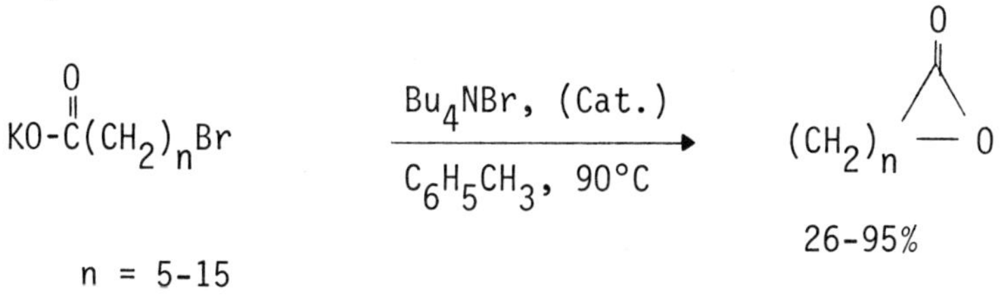

$$KO\text{-}\overset{O}{\underset{\|}{C}}(CH_2)_n Br \xrightarrow[C_6H_5CH_3,\ 90°C]{Bu_4NBr,\ (Cat.)} (CH_2)_n\text{—lactone}$$

n = 5-15

26-95%

IV.E-28 T. Kageyama et al., Chem. Lett., 1097 (1983).

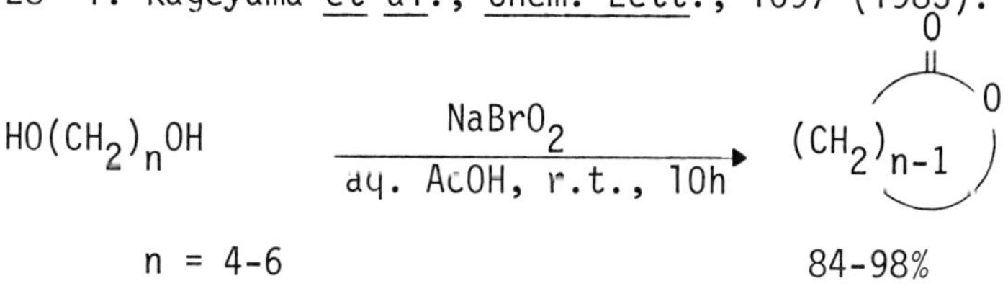

$$HO(CH_2)_nOH \xrightarrow[\text{aq. AcOH, r.t., 10h}]{NaBrO_2} (CH_2)_{n-1}\text{—lactone}$$

n = 4-6

84-98%

IV.E-29 S. L. Schrieber and W.-F. Liew, Tetrahedron Lett., 24, 2363 (1983).

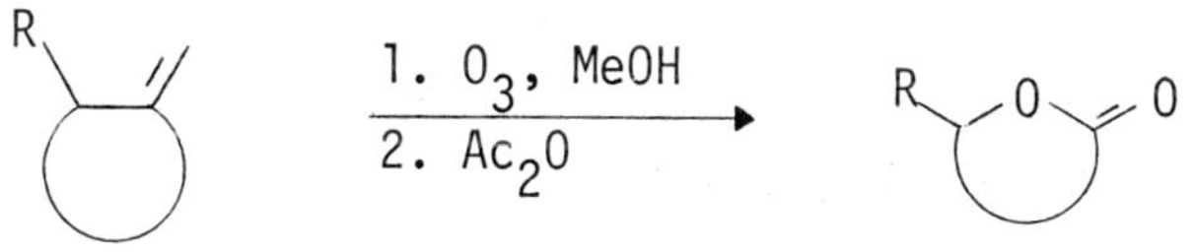

$$\text{R-methylenecycloalkane} \xrightarrow[\text{2. Ac}_2O]{1.\ O_3,\ MeOH} \text{R-lactone}$$

IV.E-30 K. Kostove and M. Hesse, Helv. Chim. Acta, 66, 741 (1983).

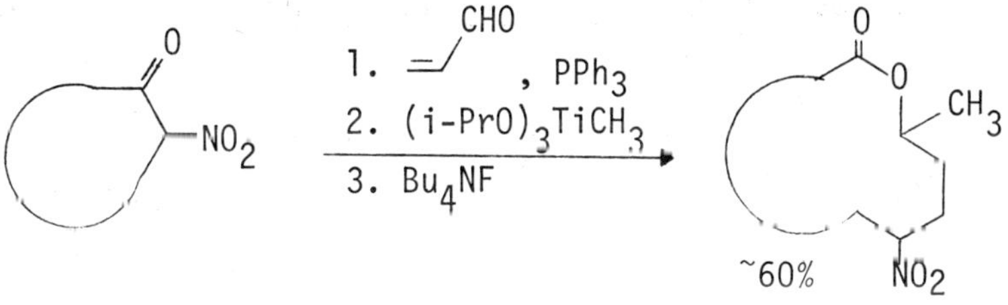

α-nitro cycloalkanone
1. CH$_2$=CHCHO, PPh$_3$
2. (i-PrO)$_3$TiCH$_3$
3. Bu$_4$NF
→ macrolactone product

~60%

IV.F. Pyridines, Quinolines, etc.

IV.F-1 Y. Watanabe et al., Synthesis, 761 (1983).

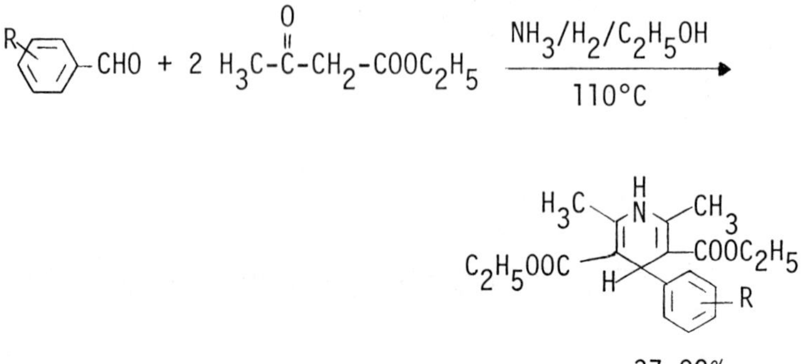

$$R\text{-C}_6H_4\text{-CHO} + 2\ H_3C\text{-CO-CH}_2\text{-COOC}_2H_5 \xrightarrow[110°C]{NH_3/H_2/C_2H_5OH}$$

3,5-bis(ethoxycarbonyl)-2,6-dimethyl-4-aryl-1,4-dihydropyridine

37-92%

IV.F-2 L. F. Tietze, A. Bergmann and K. Brueggemann, Tetrahedron Lett., 24, 3579 (1983).

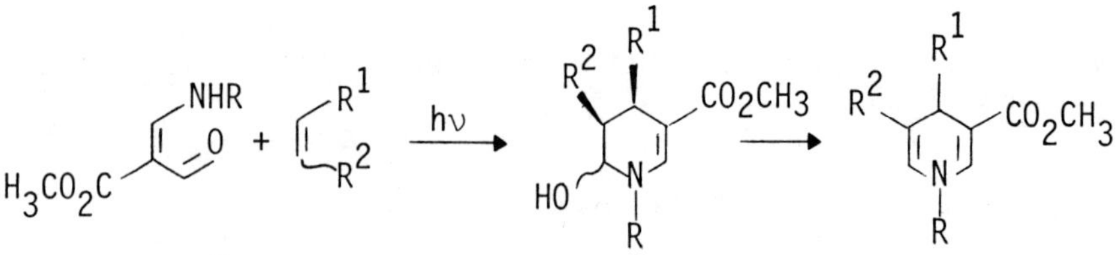

$$H_3CO_2C\text{-CH=CH-NHR} + \text{alkene}(R^1, R^2) \xrightarrow{h\nu} \text{tetrahydropyridine} \longrightarrow \text{dihydropyridine}$$

$R^2 \neq$ OTMS, O-Alkyl
R^1 = electron withdrawing group

IV.F-3 J. Barluenga et al., Synth. Commun., 13, 411 (1983).

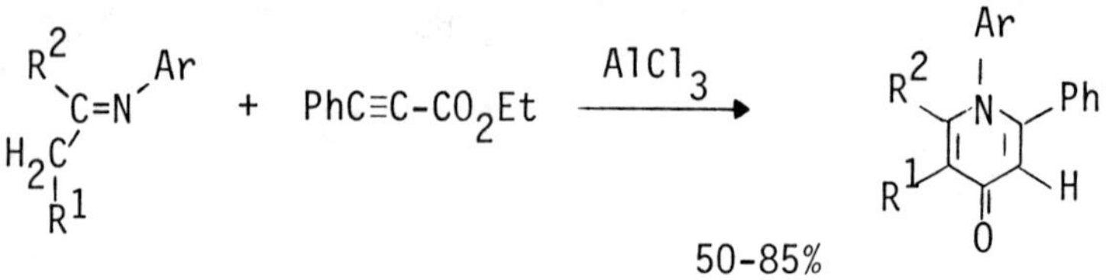

$$\underset{R^1}{\overset{R^2}{H_2C}}\text{-C(=N-Ar)} + PhC\equiv C\text{-CO}_2Et \xrightarrow{AlCl_3} \text{dihydropyridinone}$$

50-85%

IV.F-4 D. R. Adams, J. N. Dominguez and J. A. Perez, Tetrahedron Lett., 24, 517 (1983).

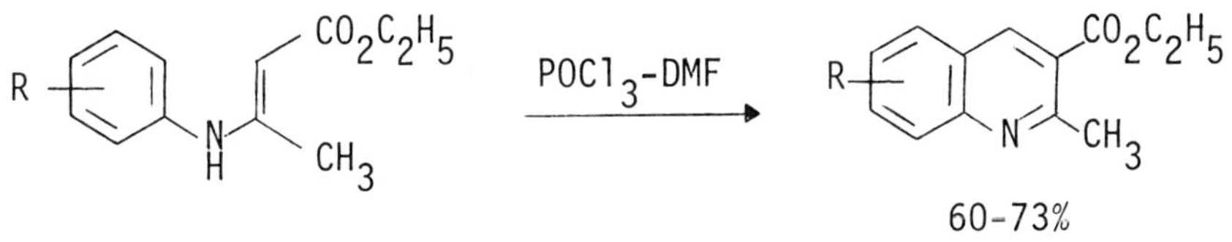

$$\text{R-C}_6\text{H}_4\text{-NH-C(CH}_3\text{)=CH-CO}_2\text{C}_2\text{H}_5 \xrightarrow{POCl_3\text{-DMF}} \text{quinoline-3-CO}_2\text{C}_2\text{H}_5\text{-2-CH}_3$$

60-73%

IV.F-5 B. M. Gutsulyak et al., J. Org. Chem. USSR, 18, 1123 (1982).

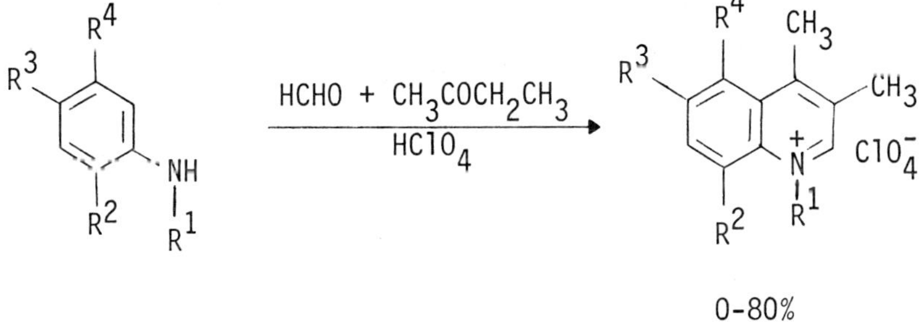

Aniline (R^1 on N, R^2, R^3, R^4 on ring) + $HCHO + CH_3COCH_2CH_3 \xrightarrow{HClO_4}$ quinolinium perchlorate (R^1–R^4 substituted, 3,4-dimethyl) ClO_4^-

0-80%

IV.F-6 T. Nishio and Y. Omote, J. Chem. Soc. Perkin Trans. 1, 1773 (1983).

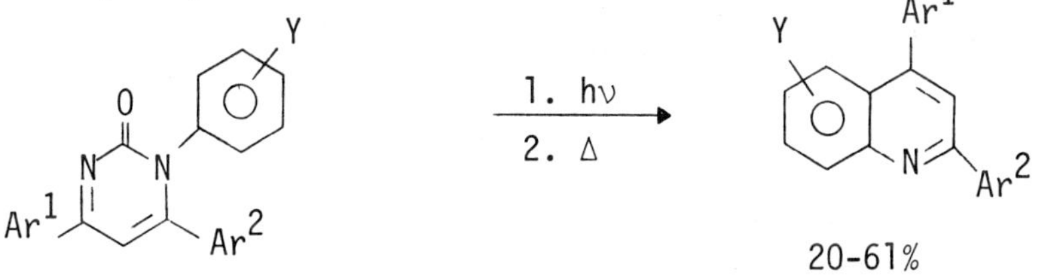

$$\text{pyrimidinone (Ar}^1\text{, Ar}^2\text{, N-cyclohexenyl-Y)} \xrightarrow[2.\ \Delta]{1.\ h\nu} \text{quinoline (Y, Ar}^1\text{, Ar}^2\text{)}$$

20-61%

IV.F-7 G. Adam, J. Andrieux and M. M. Plat, Tetrahedron Lett., 24, 3609 (1983).

$$\text{indene-R} \xrightarrow[\text{2. Ph}_3\text{CClO}_4,\ \text{AcOH}]{\text{1. NH}_3\text{-H}_2\text{SO}_4} \text{quinoline-R}$$

IV.F-8 T. Sato, K. Tamura and K. Nagayoshi, Chem. Lett., 791 (1983).

$$R^1\text{-C}_6\text{H}_4\text{-CH}_2\text{-CH=CH-}R^2 \xrightarrow[\substack{\text{2. }R^3\text{CN}\\ \text{3. KOH-CH}_3\text{OH}}]{\text{1. AgOTf-I}_2/\text{CH}_2\text{Cl}_2} \text{dihydroisoquinoline}$$

36-66%

IV.F-9 J. B. Hendrickson and C. Rodriguez, J. Org. Chem., 48, 3344 (1983).

$$\underset{Z}{\overset{X}{\underset{Y}{}}}\text{-CHO} \xrightarrow[\substack{\text{3. P(OCH}_3)_3\\ \text{4. TiCl}_4}]{\substack{\text{1. H}_2\text{NCH}_2\text{CH(OCH}_3)_2,\ \Delta\\ \text{2. ClCO}_2\text{Et, }-10°\text{C}}} \text{isoquinoline}$$

25-75%

X = H, CH_3, OCH_3
Y = H, OH, OCH_3
Z = H, Br, OCH_3

IV.G. Pyrroles, etc.

IV.G-1 M. M. Ito et al., Bull. Chem. Soc. Jpn., 56, 533 (1983).

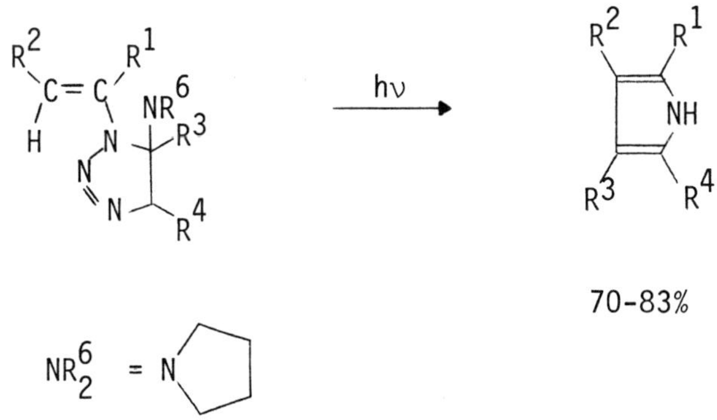

70-83%

IV.G-2 T. H. Chan and S. D. Lee, J. Org. Chem., 48, 3059 (1983).

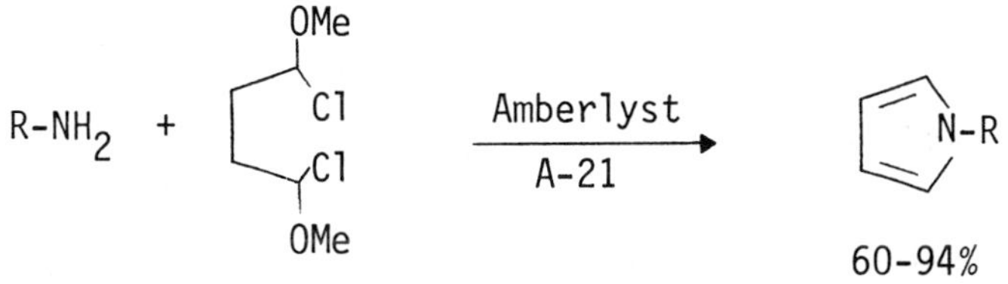

60-94%

IV.G-3 S. K. Mukerji, K. K. Sharma and K. B. G. Torssell, Tetrahedron, 39, 2231 (1983).

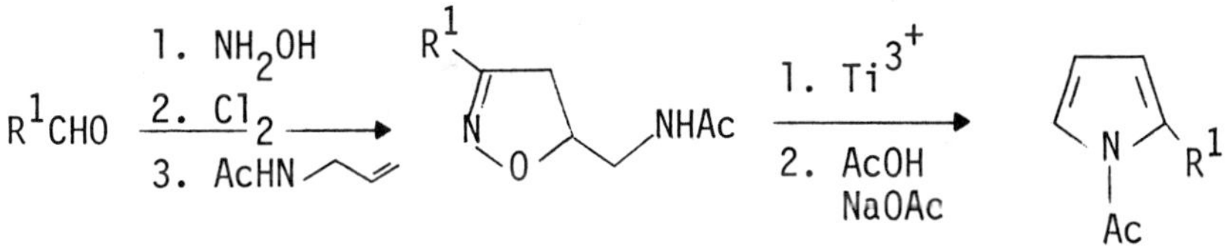

IV.G-4 W. Flitsch, K. Pandl and P. Russkamp, Liebigs Ann. Chem., 529 (1983).

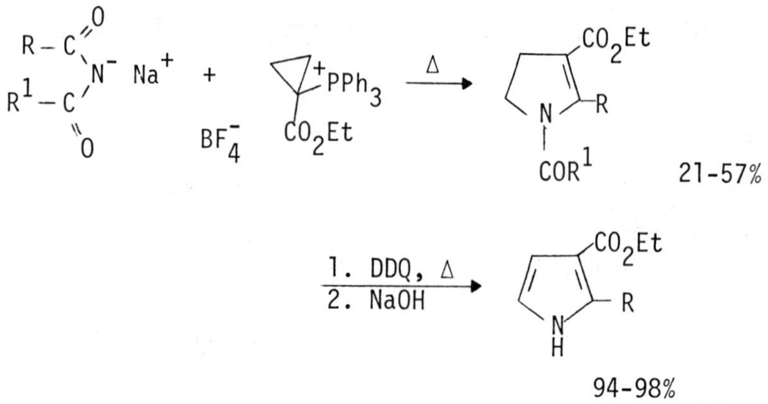

$$R-C(=O)-N^-\,Na^+ \atop R^1-C(=O)- \quad + \quad \triangle\text{-}PPh_3,\ CO_2Et,\ BF_4^- \xrightarrow{\Delta}$$ pyrroline with CO_2Et, R, N-COR^1 21-57%

$$\xrightarrow[\text{2. NaOH}]{\text{1. DDQ, }\Delta}$$ pyrrole with CO_2Et, R 94-98%

IV.G-5 O. Attanasi and S. Santeusanio, Synthesis, 742 (1983).

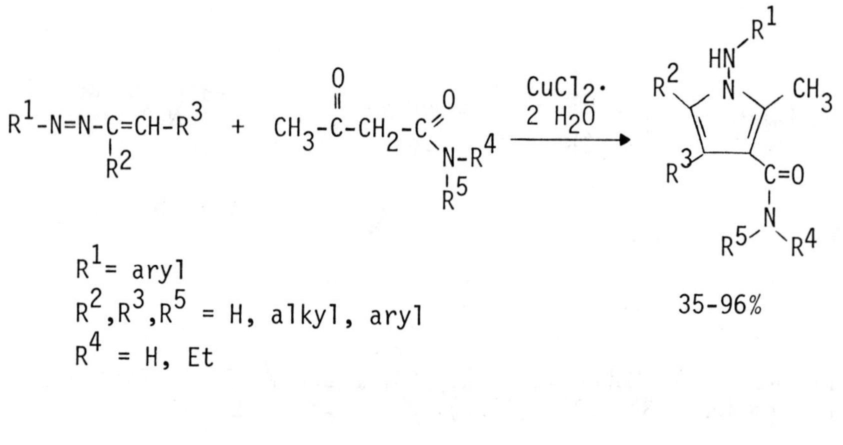

$$R^1-N=N-C(R^2)=CH-R^3 \ +\ CH_3-C(=O)-CH_2-C(=O)-N(R^5)R^4 \xrightarrow{CuCl_2\cdot 2H_2O}$$

pyrazole product   35-96%

R^1 = aryl
R^2, R^3, R^5 = H, alkyl, aryl
R^4 = H, Et

IV.G-6 R. Beugelmans, G. Negron and G. Roussi, J. Chem. Soc. Chem. Commun., 31 (1983).

$Me_3N \to O$ $\xrightarrow{\text{1. LDA, -78°C} \atop \text{2. } R^1CH=CHR^2}$ pyrrolidine with R^1, R^2 at 3,4-positions, N-Me

R^1, R^2 = alkyl, Ph   42-90%

IV.G-7 A. Padwa et al., Tetrahedron Lett., 24, 3447 (1983).

$NCCH_2\underset{\underset{CH_2Ph}{|}}{N}CH_2SiMe_3$ + $R_1CH=CHR_2$ $\xrightarrow[CH_3CN]{AgF}$ pyrrolidine with R_1, R_2 at 3,4-positions, N-CH_2Ph

R_1, R_2 = electron withdrawing groups   33-84%

IV.G-8 L. E. Overman et al., J. Am. Chem. Soc., 105, 6622 (1983).

$RH_2\overset{+}{N}$–CH_2–$C(CH_3)(OR^1)$–$CH=CH_2$, X^- + R^3–C(=O)–R^2 $\longrightarrow$ pyrrolidine with R^2, R^3 at 5-position, acetyl at 2-position, N-R

R^1 = H, Me   9-97%

IV.G-9 L. Stella, Angew. Chem. Int. Ed. Engl., 22, 337 (1983).

Review: "Homolytic Cyclizations of N-Chloroalkenyl-amines".

IV.H. Other Heterocycles with One Heteroatom

(see also: II.F.1, VI.A.9)

IV.H-1 J. B. Ousset, C. Mioskowski and G. Solladie, Tetrahedron Lett., 24, 4419 (1983).

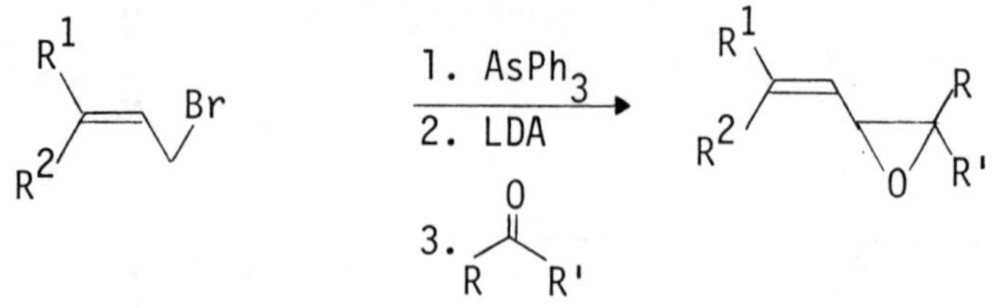

55-95%

IV.H-2 P. F. Vlad and N. D. Ungur, Synthesis, 216 (1983).

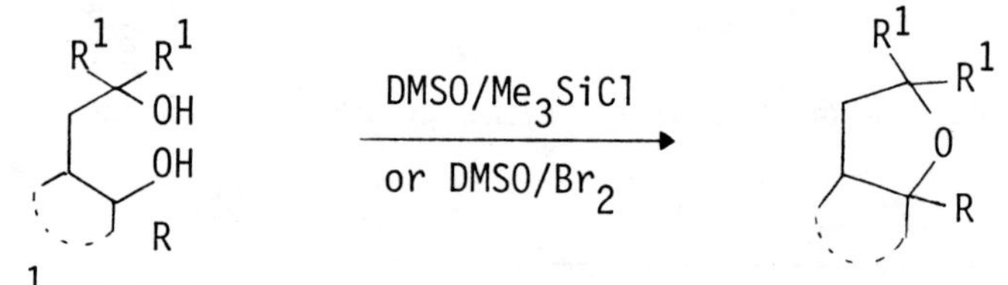

R^1 = H, Me

44-90%

IV.H-3 M. Yamauchi et al., J. Chem. Soc. Chem. Commun., 281 (1983).

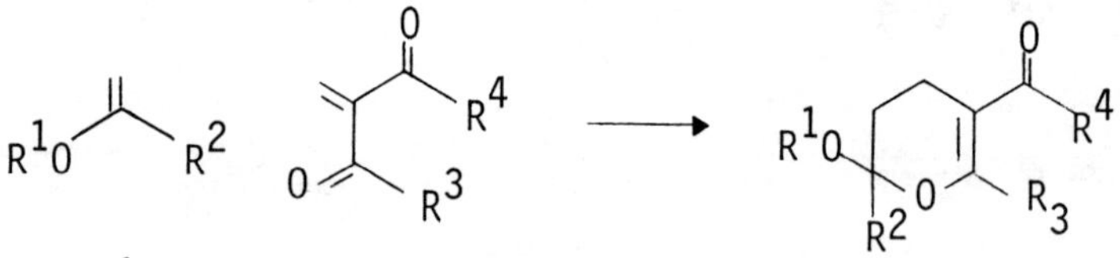

R^1 = Me, Et
R^2 = H, Me
R^3 = Me, Ph
R^4 = Ph, OEt

quant.

IV.H-4 U. K. Nadir and V. K. Koul, Synthesis, 554 (1983).

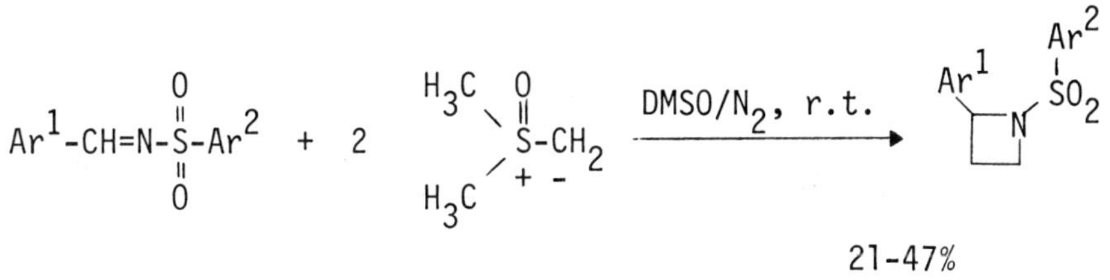

$$Ar^1-CH=N-\overset{O}{\underset{O}{\overset{\|}{S}}}-Ar^2 + 2 \; \overset{H_3C}{\underset{H_3C}{>}}\overset{O}{\underset{+ \; -}{\overset{\|}{S}}}-CH_2 \xrightarrow{DMSO/N_2, \; r.t.} \underset{\text{21-47\%}}{Ar^1\text{-azetidine-}SO_2Ar^2}$$

IV.H-5 S. Arseniyadis and J. Gore, Tetrahedron Lett., 24, 3997 (1983).

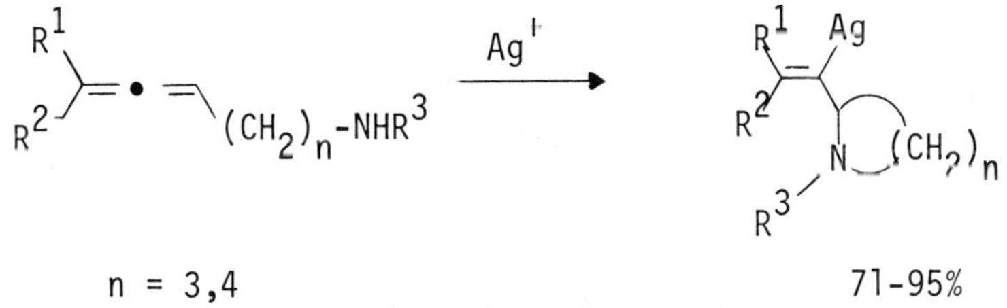

$$\underset{R^2}{\overset{R^1}{>}}=\bullet=\diagdown(CH_2)_n-NHR^3 \xrightarrow{Ag^+} \text{cyclic product}$$

n = 3,4   71-95%

IV.H-6 R. R. Webb, II. and S. Danishefsky, Tetrahedron Lett., 24, 1357 (1983).

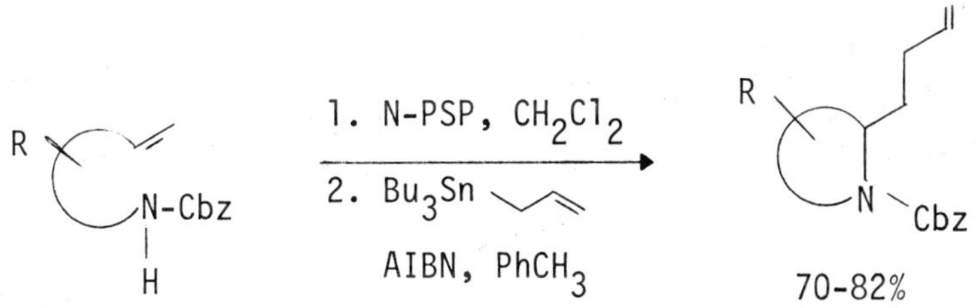

$$\text{R-alkene-N(H)-Cbz} \xrightarrow[\text{2. } Bu_3Sn\diagdown\diagup,\; AIBN, \; PhCH_3]{\text{1. N-PSP, } CH_2Cl_2} \text{product}$$

70-82%

N-PSP = N-(phenylseleno)phthalimide (phthalimide-N-SePh)

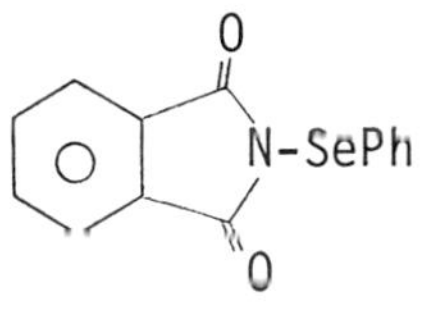

IV.H-7 C. N. Meltz and R. A. Volkmann, *Tetrahedron Lett.*, **24**, 4507 (1983).

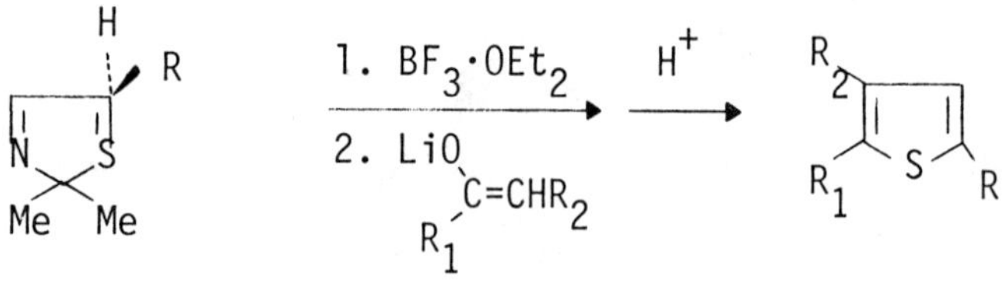

1. $BF_3 \cdot OEt_2$
2. $\underset{R_1}{LiO}C=CHR_2$; H^+

30-90%

IV.H-8 J. Liebscher, B. Abegaz and A. Areda, *J. Prakt. Chem.*, **325**, 168 (1983).

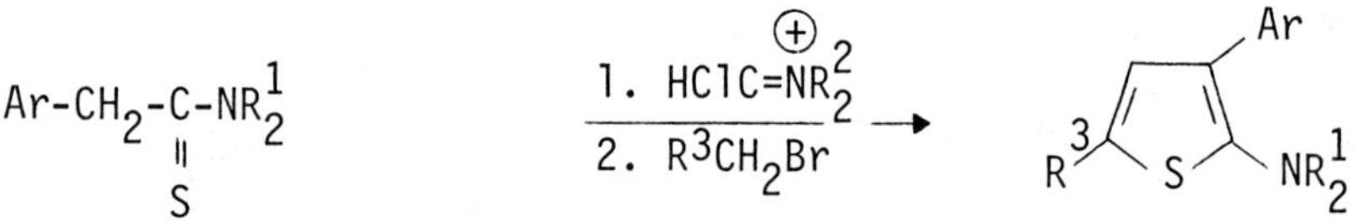

$Ar-CH_2-\underset{\underset{S}{\|}}{C}-NR^1_2$

1. $HClC=\overset{\oplus}{N}R^2_2$
2. R^3CH_2Br

R^3 = electron withdrawing group

IV.H-9 C. A. H. Rasmussen and A. De Groot, *Synthesis*, 575 (1983).

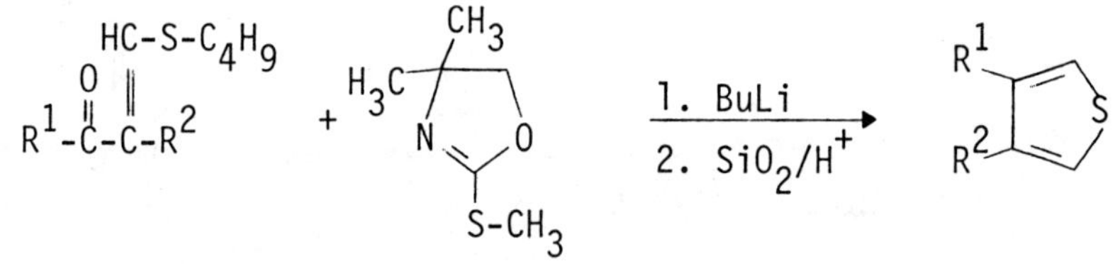

$HC-S-C_4H_9$
$R^1-\underset{\underset{}{\|}}{\overset{O}{C}}-\overset{\|}{C}-R^2$

+

(oxazoline with CH_3, H_3C, N, O, $S-CH_3$)

1. BuLi
2. SiO_2/H^+

R^1, R^2 = alkyl

32-63%

IV.H-10   H. Batzer, Chimia, 37, 329 (1983).

Review: "Epoxiverbindung -- ihre Synthese und Verwendung."

IV.H-11   J. K. Crandall and M. Apparu, Org. Reactions, 29, 345 (1983).

Review: "Base-Promoted Isomerizations of Epoxides."

IV.H-12   A. S. Rao, S. K. Paknikar and J. G. Kirtane, Tetrahedron, 39, 2323 (1983).

Review: "Recent Advances in the Preparation and Synthetic Applications of Oxiranes."

IV.H-13   R. W. Alder, Acc. Chem. Res., 16, 321 (1983).

Review: "Medium-Ring Bicyclic Compounds and Intrabridgehead Chemistry."

IV.H-14   V. Baliah, R. Jeyaraman and L. Chadrasekaran, Chem. Rev., 83, 379 (1983).

Review: "Synthesis of 2,6-Disubstituted Piperidines, Oxanes, and Thianes."

IV.H-15 B. Iddon, Heterocycles, 20, 1127 (1983).

Review: "Cycloaddition, Ring-Opening, and Other Novel Reactions of Thiophenes."

IV.H-16 G. A. Iacobucci and J. G. Sweeny, Tetrahedron, 39, 3005 (1983).

Review: "The Chemistry of Anthocyanins, Anthocyanidins and Related Flavylium Salts."

IV.H-17 A. G. Schultz, Acc. Chem. Res., 16, 210 (1983).

Review: "Photochemical Six-Electron Heterocyclization Reactions."

IV.I. Heterocycles with Two or More Heteroatoms

1.a. 5-Membered Heterocycles with 2 N's

IV.I.1.a-1 M. Casey, C. J. Moody and C. W. Rees, J. Chem. Soc. Chem. Commun., 1082 (1983).

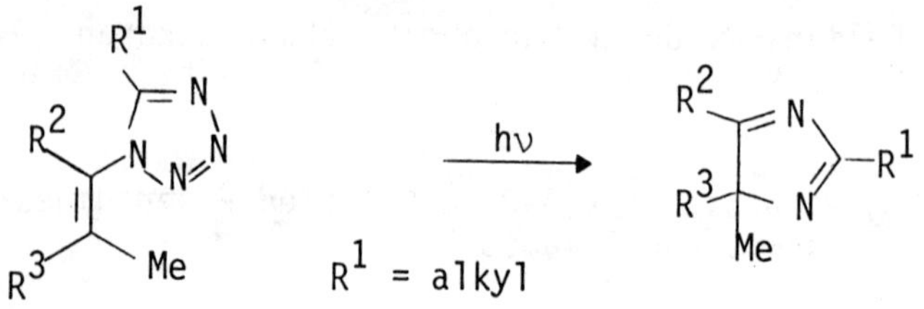

R^1 = alkyl

IV.I.1.a-2 S. Nakanishi, J. Nantaku and Y. Otsuji, Chem. Lett., 341 (1983).

$$R^1-\underset{NOH}{\underset{\|}{C}}-CH_2X + R^2-\underset{CH_3}{\underset{|}{C}}(=NH)-NC_6H_5 \xrightarrow{\text{Iron carbonyls}} \underset{R^1}{\overset{}{\text{imidazole}}} R^2$$

R^2 = Ph, m-Tol   31-79%

IV.I.1.a-3 G. Blotny, Synthesis, 391 (1983).

$$R^1-\underset{NH_2}{\underset{|}{CH}}-COOH + R^2-NH-\underset{\|}{\overset{S}{C}}-SCH_3 \xrightarrow{Et_3N, \nabla} \text{imidazolidinone}$$

76-97%

IV.I.1.a-4 N. Matsumura, A. Kunugihara and S. Yoneda, Tetrahedron Lett., 24, 3239 (1983).

$$\underset{Ph \quad CH_3}{EtO_2C-\underset{H}{\overset{H}{C}}=N-N} \xrightarrow[\begin{array}{l}\text{1. 2 n-BuLi, -78°C}\\ \text{2. } R_2COY, -78°C \\ \text{3. } H^+\end{array}]{} \text{pyrazole}$$

Y = Cl, NMe$_2$, R$_1$O, R$_1$CO$_2$ 25-77%

IV.I.1.b. 6-Membered Heterocycles with 2 N's

IV.I.1.b-1 A.-A. Pourzal, Synthesis, 717 (1983).

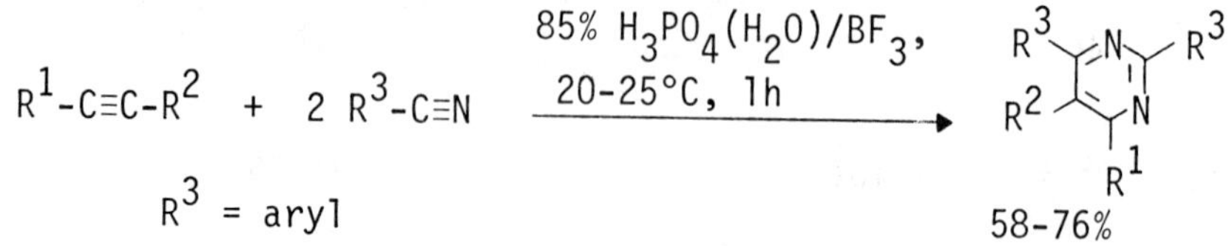

$$R^1-C\equiv C-R^2 + 2\ R^3-C\equiv N \xrightarrow[20-25°C,\ 1h]{85\%\ H_3PO_4(H_2O)/BF_3}$$

R^3 = aryl

58-76%

IV.I.1.b-2 C. J. Shishoo et al., Tetrahedron Lett., 24, 4611 (1983).

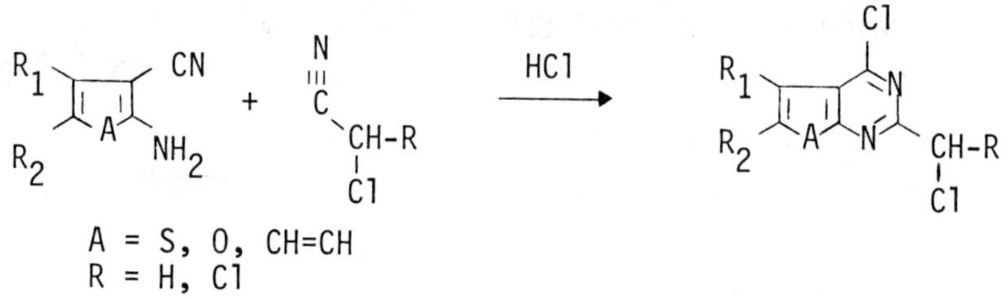

$$\text{(pyrrole/thiophene-CN + NH}_2\text{)} + \underset{\underset{Cl}{|}}{\overset{N}{\underset{|||}{C}}}\text{-CH-R} \xrightarrow{HCl} \text{product}$$

A = S, O, CH=CH
R = H, Cl

IV.I.1.b-3 S. K. Robev, Tetrahedron Lett., 24, 4351 (1983).

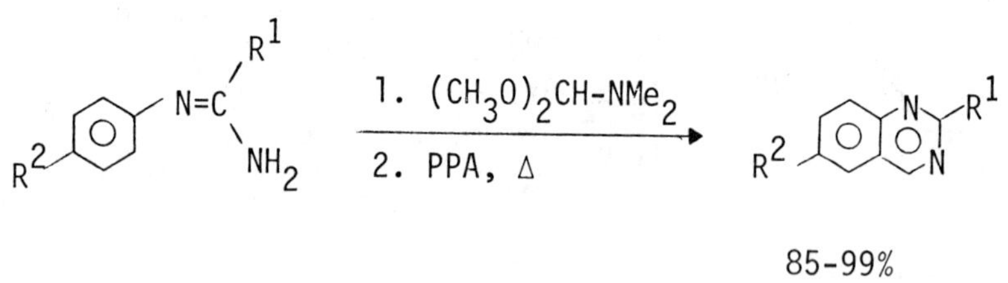

$$\xrightarrow[\text{2. PPA, }\Delta]{\text{1. }(CH_3O)_2CH-NMe_2}$$

85-99%

IV.I.1.b-4 C. Shin, Heterocycles, 20, 1407 (1983).

Review: "Stereoselective Synthesis of the Geometric and Optical Isomers of Unsaturated 3-Mono- and 3,6-Disubstituted 2,5-Piperazinediones."

IV.I.1.c. Other Heterocycles with 2 N's

IV.I.1.c-1 I. R. Robertson and J. T. Sharp, J. Chem. Soc. Chem. Commun., 1003 (1983).

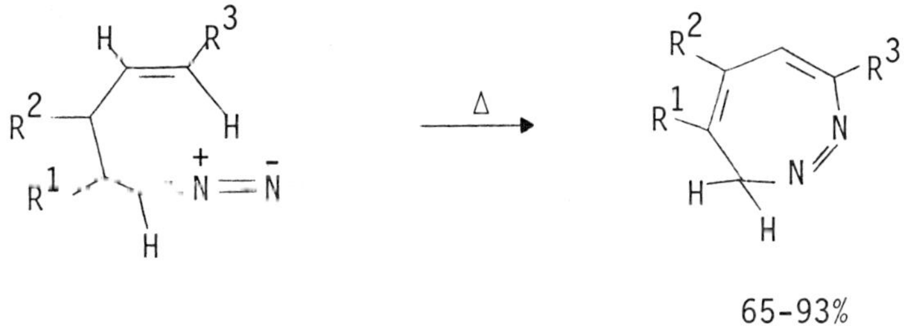

65-93%

IV.I.1.c-2 T. Tsuchiya, Yakugaku Zasshi (J. Pharm. Soc. Jpn.) 103, 373 (1983).

Review: "Ring Transformations of Cyclic Amine N-Imides."

IV.I.2. Heterocycles with 1 N and 1 O

IV.I.2-1 K. Issleib et al., Z. Chem., 23, 98 (1983).

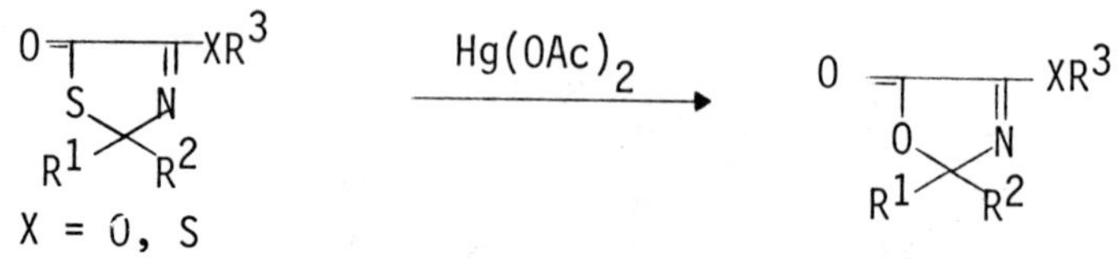

[Reaction scheme: 1,3-thiazetidinone with R^1, R^2, XR^3 substituents (X = O, S) converted by $Hg(OAc)_2$ to 1,3-oxazetidinone.]

IV.I.2-2 J. Barluenga et al., J. Org. Chem., 48, 1379 (1983).

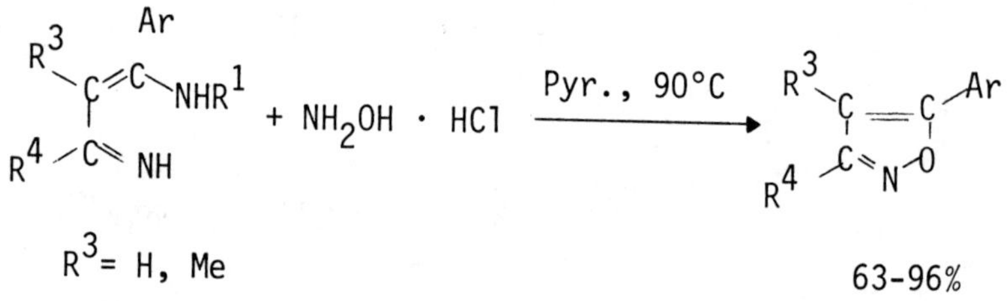

[Reaction scheme: $R^3(Ar)C=C(NHR^1)$ and $R^4-C(=NH)$ (with R^3 = H, Me) + $NH_2OH \cdot HCl$, Pyr., 90°C → isoxazole product, 63-96%.]

IV.I.2-3 J. F. W. Keana and G. M. Little, Heterocycles, 20, 1291 (1983).

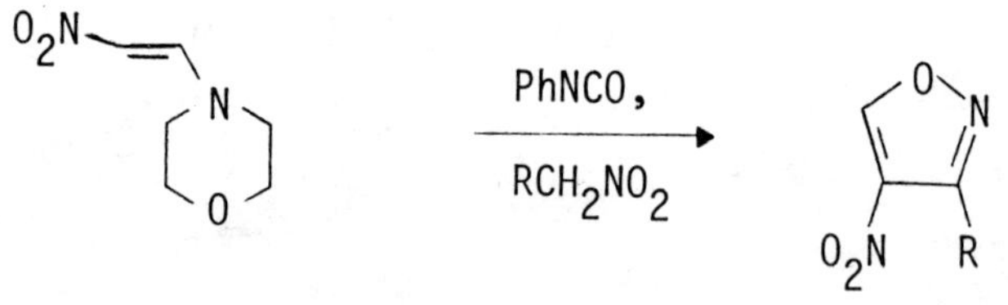

[Reaction scheme: O_2N-CH=CH-morpholine + PhNCO, RCH_2NO_2 → 4-nitro-3-R-isoxazole.]

IV.I.2-4 D. H. Hoskin and R. A. Olofson, *J. Org. Chem.*, **47**, 5222 (1982).

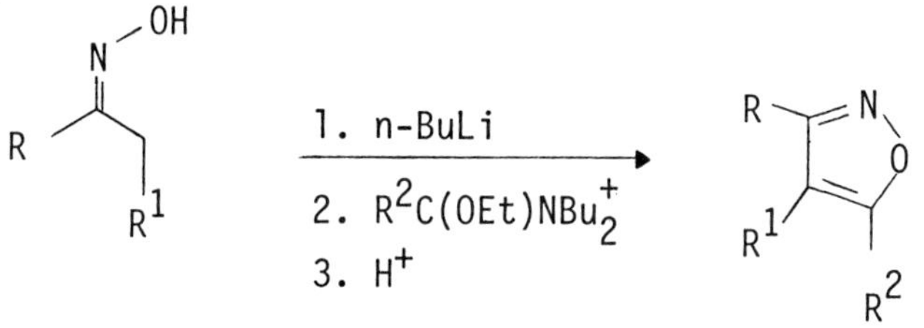

$$\underset{R^1}{\underset{R}{\text{C}}}=N-OH \xrightarrow[\substack{\text{2. } R^2C(OEt)NBu_2^+ \\ \text{3. } H^+}]{\text{1. n-BuLi}} \underset{R^2}{\underset{R^1}{\text{isoxazole}}}$$

IV.I.2-5 L. P. Vasileva *et al.*, *J. Org. Chem. USSR*, **18**, 1621 (1982).

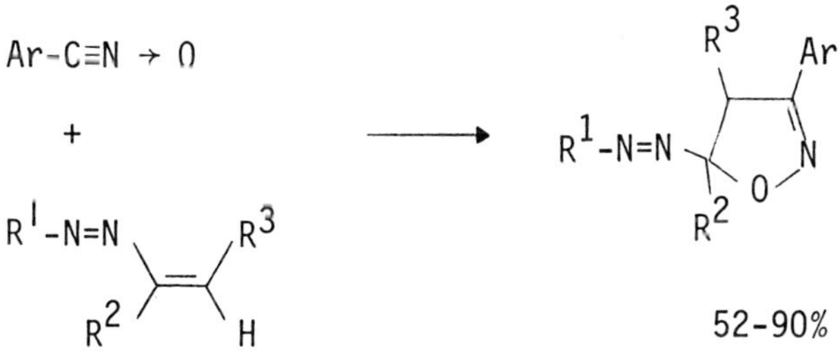

Ar-C≡N → O

+

R^1-N=N-C(R^2)=CH-R^3

⟶ product with R^1-N=N, R^2, R^3, Ar substituents on isoxazoline ring

52-90%

IV.I.2-6 P. A. Wade *et al.*, *J. Org. Chem.*, **48**, 1796 (1983).

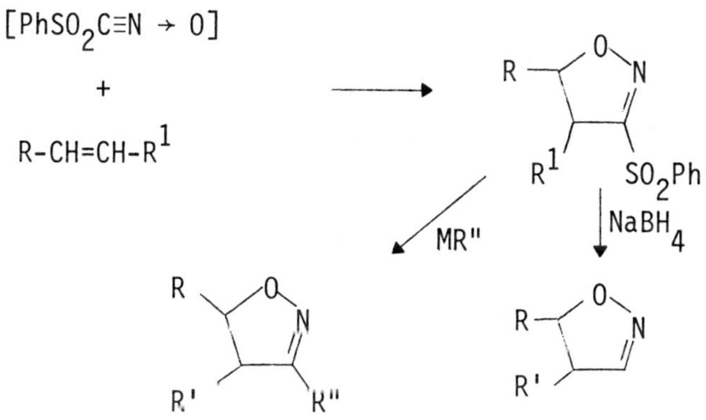

[PhSO₂C≡N → O]

+

R-CH=CH-R^1

⟶ isoxazoline with R, R^1, SO₂Ph substituents

MR″ → isoxazoline with R, R′, R″

NaBH₄ → isoxazoline with R, R′

62-94%

IV.I.2-7 S.-I. Murhashi et al., Tetrahedron Lett., 24, 1049 (1983).

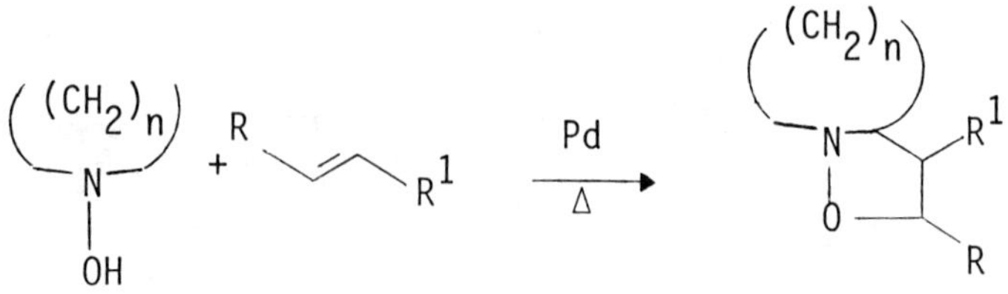

R = electron donating group; aryl; H
R^1 = electron withdrawing group; H

55-85%

IV.I.2-8 R. L. Funk et al., J. Org. Chem., 48, 2632 (1983).

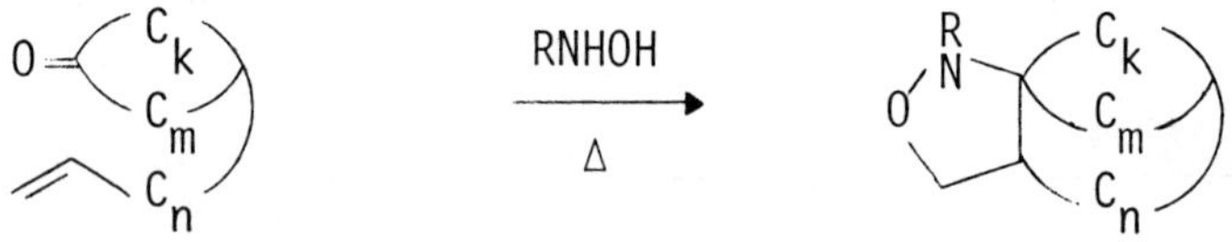

46-100%

IV.I.2-9 J. P. Freeman, Chem. Rev., 83, 241 (1983).

Review: "Δ^4-Isoxazolines (2,3-Dihydroisoxazoles)."

IV.I.2-10/IV.I.3-1 K. T. Potts, A. J. Ruffini and G. R. Titus, J. Org. Chem., 48, 623 (1983).

$$ArC(=O)-N=C\begin{cases}Y\\SEt\end{cases} \xrightarrow[\substack{2.\ HClO_4,\ Ac_2O\\ \text{or}\\ Na_2S, HClO_4, Ac_2O}]{1.\ R^2CO^-CH_2K^+}$$

[pyrimidinium salt with substituents Y (position 4), Ar (position 2), R^2 (position 6), ring atom X, charge (+)]

X = S, O
Y = SEt, NEt_2

IV.I.3. Heterocycles with 1 N and 1 S

IV.I.3-2 W. Koller et al., Tetrahedron Lett., 24, 2131 (1983).

[aziridine with R^2, R^3 on carbon and R^1 on N]
$\xrightarrow{\substack{1.\ H_2S\\ 2.\ Cl_2,\ EtOH\\ 3.\ Et_3N}}$

[β-aminoethyl thiosulfate R^1-NH-C(R^2)(R^3)-CH$_2$-S$_2$O$_3$H]
$\xrightarrow{\substack{1.\ I_2\\ 2.\ Cl_2,\ EtOH\\ 3.\ Et_3N}}$

[thiazetidine 1,1-dioxide product: $O_2S-N(R^1)$ ring with R^2, R^3]

26-96%

IV.I.3-3 E. Meyle and H.-H. Otto, Arch. Pharm., 316, 281 (1983).

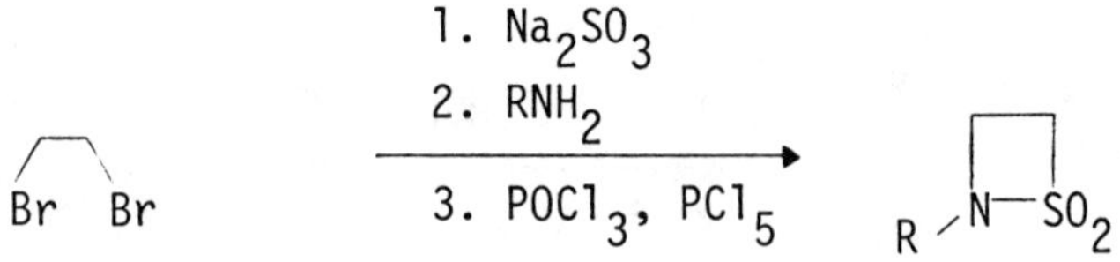

BrCH₂CH₂Br $\xrightarrow{\text{1. Na}_2\text{SO}_3 \quad \text{2. RNH}_2 \quad \text{3. POCl}_3, \text{PCl}_5}$ R–N–SO₂ (azetidine ring)

74-87%

IV.I.3-4 I. Thomsen et al., Chem. Lett., 809 (1983).

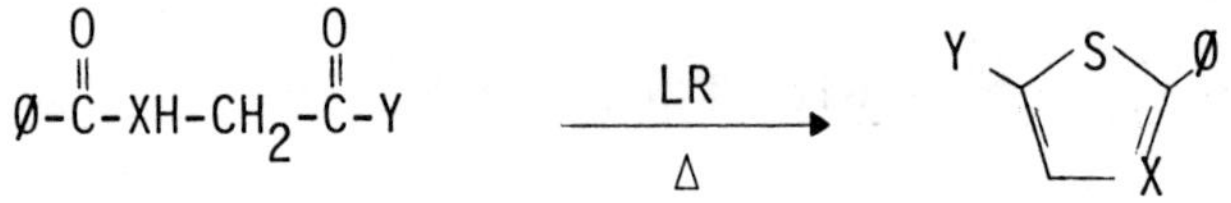

$$\phi\text{-C(=O)-XH-CH}_2\text{-C(=O)-Y} \xrightarrow[\Delta]{\text{LR}} \text{thiophene: } Y, S, \phi, X$$

X = N, CH
Y = OEt, N(piperidine)

LR =

CH₃O–C₆H₄–P(=S)(S)–S–P(=S)(S)–C₆H₄–OCH₃

~80%

IV.I.3-5 W. Friedrichsen and M. Koenig, Synthesis, 582 (1983).

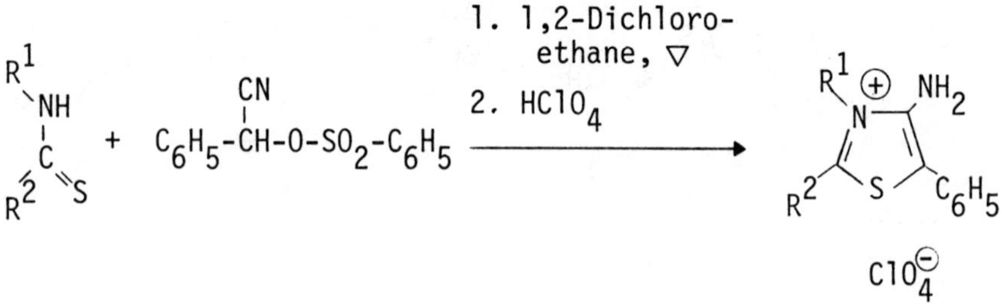

R^1–NH–C(=S)–R^2 + C₆H₅–CH(CN)–O–SO₂–C₆H₅ $\xrightarrow{\text{1. 1,2-Dichloroethane, }\triangledown \quad \text{2. HClO}_4}$ thiazolium: R^1–N⁺, NH₂, R^2, S, C₆H₅

ClO₄⁻

80%

IV.I.3-6 Y. Tanabe et al., Bull. Chem. Soc. Jpn., 56, 1255 (1983).

$Xn\text{-}C_6H_4\text{-}NHR \xrightarrow{ClCOSCl} \xrightarrow{Acid}$ benzoxazolinone-type product (Xn, N-R, C=O, S)

R = alkyl   58-78%

IV.I.3-7 W. Schroth, R. Spitzner and J. Freitag, Synthesis, 827 (1983).

1,3-diketone (R^1COCH$_2$COR^2)
$\xrightarrow{\text{1. NaOMe; 2. (PhO)}_2\text{POCl; 3. H}_2\text{NC(=S)R}^3\text{, AcOH, POCl}_3\text{, HClO}_4}$
thiazinium perchlorate (R^1, R^2, R^3, S, N$^+$) ClO_4^-

R^3 = aryl, alkyl, SR^4, $N(R^4)_2$ 39-97%

IV.I.3-8 D. K. Thakur and Y. D. Vankar, Synthesis, 223 (1983).

R^1, R^2-C$_6$H$_3$-S-CH$_2$-Cl + R^3-C≡N $\xrightarrow{SbCl_5/CH_2Cl_2}$ benzothiazine (R^1, R^2, R^3, N, S)

R^1, R^2 = H, CH$_3$ 19-57%

IV.I.3-9 W. D. Crow, I. Gosney and R. A. Ormiston,
J. Chem. Soc. Chem. Commun., 643 (1983).

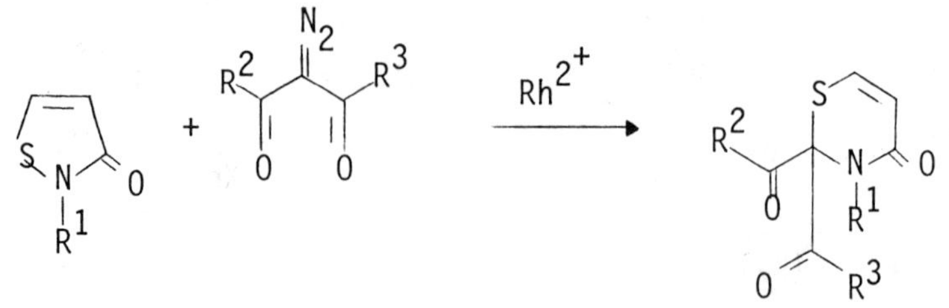

58-95%

IV.I.3-10 K. Schulze et al., Tetrahedron Lett., 23, 5529 (1982).

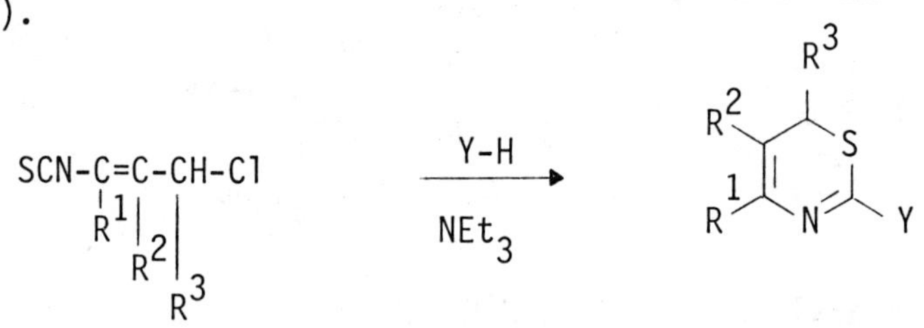

$$SCN-C=C-CH-Cl$$
$$R^1\ R^2\ R^3$$

$\xrightarrow[NEt_3]{Y-H}$

$Y = OR, NR^2$

$R^1, R^2, R^3 = H, alkyl$

60-85%

IV.I.3-11 D. Hellwinkel, R. Lenz and F. Laemmerzahl,
Tetrahedron, 39, 2073 (1983).

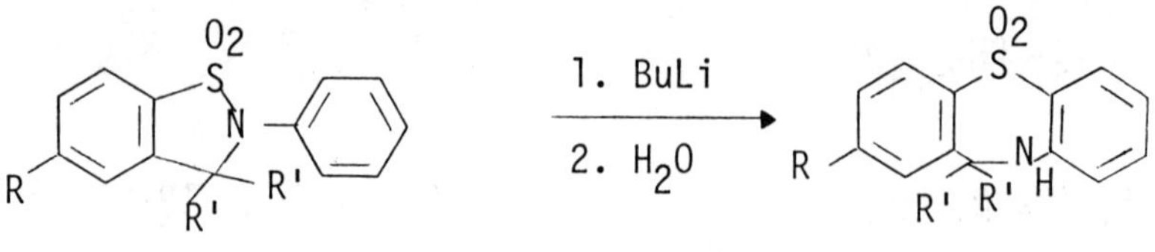

1. BuLi
2. H_2O

IV.I.5. Heterocycles with 3 N's

IV.I.5-1 Y. Hirai et al., Heterocycles, 20, 1243 (1983).

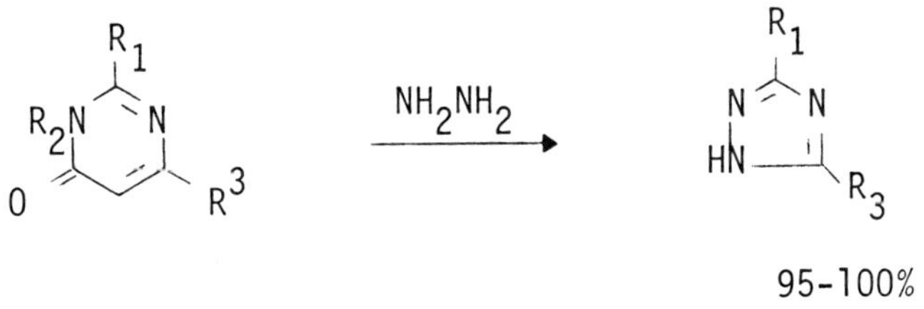

[Reaction: pyrimidinone with substituents R_1, R_2, R_3 + NH_2NH_2 → triazole product, 95-100%]

IV.I.5-2 T. Aoyama, K. Sudo and T. Shioiri, Chem. Pharm. Bull., 30, 3849 (1982).

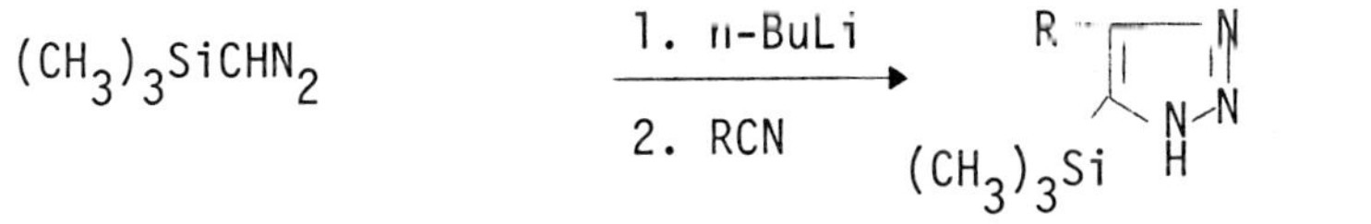

$(CH_3)_3SiCHN_2$ $\xrightarrow{\text{1. n-BuLi} \quad \text{2. RCN}}$ [1,2,3-triazole with R and $(CH_3)_3Si$ substituents]

R = alkyl, aryl, ethylthio, phenoxy

44-96%

IV.I.5-3 F. Risitano et al., J. Chem. Res. Synop., 52 (1983).

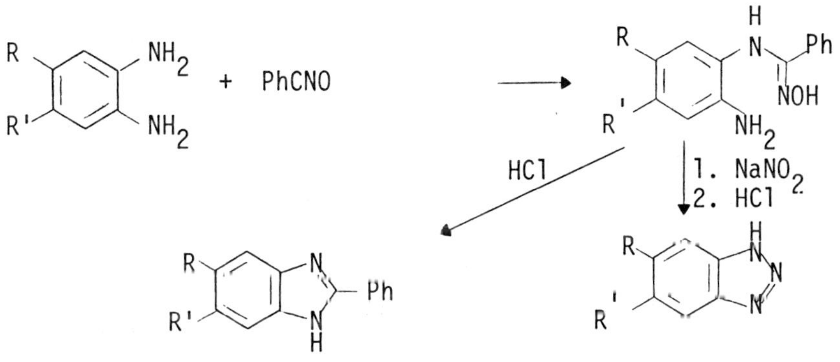

[Reaction: o-phenylenediamine (with R, R' substituents) + PhCNO → amidoxime intermediate; with HCl → benzimidazole (2-Ph); with 1. NaNO$_2$, 2. HCl → benzotriazole]

IV.I.5-4/IV.I.6-1 M. A. Perez, C. A. Dorado and J. L. Soto, Synthesis, 483 (1983).

$$R^1-C\underset{OR^2}{\overset{NH}{\diagup}} \xrightarrow[Et_3N]{R^3-\overset{O}{\overset{\|}{C}}-Cl} R^1-C\underset{OR^2}{\overset{N-C\diagdown R^3}{\diagup}}\diagdown_0$$

$R^4NH-NH_2 \longrightarrow$ triazole with R^1, R^3, R^4 55-87%

$HO-NH_2 \longrightarrow$ oxadiazole with R^1, R^3 38-74%

IV.I.6. Other Heterocycles with Two or More Heteroatoms.

IV.I.6-2 R. F. Smith, K. J. Coffman and S. M. Geer, J. Heterocycl. Chem., 20, 69 (1983).

$$R-C\underset{NH}{\overset{\overset{+}{N}N(CH_3)_3}{\diagup}} (-) + R'COCl \xrightarrow{\Delta}$$

oxadiazine product 40-85%

IV.I.6-3 R. Milcent and G. Barbier, J. Heterocycl. Chem. 20, 77 (1983).

$$Ar^1\underset{\overset{\|}{O}}{C}-NHN=CHAr^2 \xrightarrow[AcOH]{PbO_2}$$

oxadiazole with Ar^1, Ar^2 29-95%

IV.I.6-4 T. Aoyama and T. Shioiri, Chem. Pharm. Bull., 30, 3450 (1982).

$(CH_3)_3SiCHN_2$ →
1. LDA
2. RCO_2CH_3

gives tetrazole with $RCOCH_2$ and $Si(CH_3)_3$ substituents

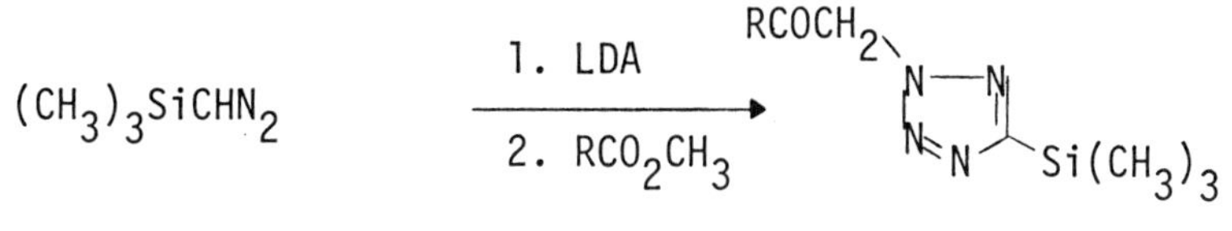

49-90%

IV.I.6-5 M. Sato et al., Chem. Pharm. Bull., 31, 1896 (1983).

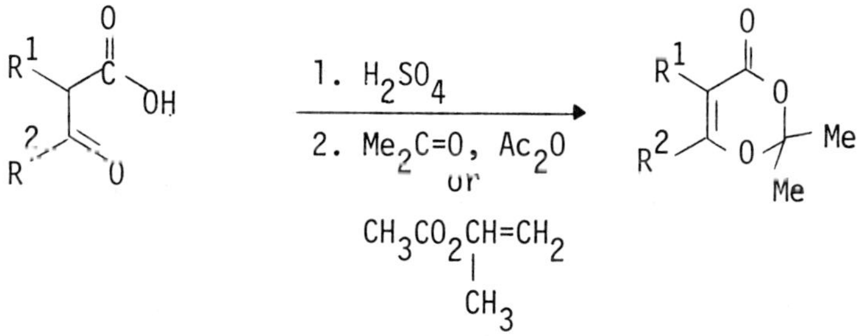

1. H_2SO_4
2. $Me_2C=O$, Ac_2O
or
$CH_3CO_2CH=CH_2$ with CH_3

47-89%

IV.I.6-6 C. W. Jefford et al., J. Am. Chem. Soc., 105, 6497 (1983).

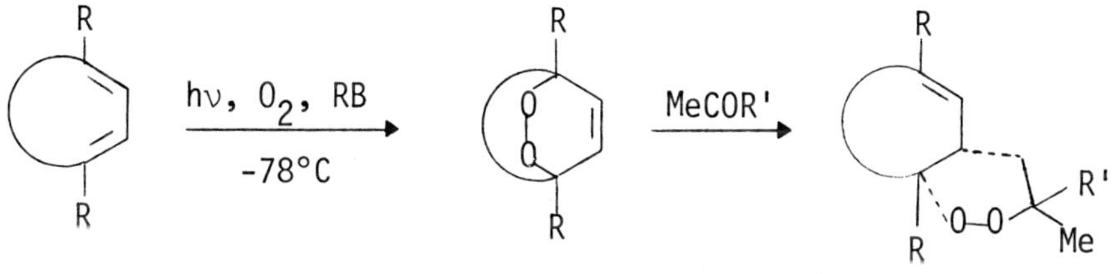

hν, O_2, RB, -78°C

MeCOR'

IV.I.6-7 W. Sliwa and T. Thomas, Heterocycles, 20, 71 (1983).

Review: "Fused 1,2,5-Thiadiazoles and Selenadiazoles."

IV.J. General Heterocyclic Reviews

IV.J-1 S.-J. Lee and J. M. Cook, Heterocycles, 20, 87 (1983).

Review: "Synthesis of Azaphenalenes."

IV.J-2 G. L'abbe, Tetrahedron, 38, 3537 (1982).

Review: "Some Ring-Transformation Reactions of Sulfur-Containing Heterocycles."

IV.J-3 A. Shanzer, Bull. Soc. Chim. Belg., 92, 411 (1983).

Review: "Large Ring Heterocycles: Template Synthesis and Structure."

IV.J-4 M. Tisler, Heterocycles, 20, 1591 (1983).

Review: "Heterocyclic Amidines and Hydroxyamidines as Synthons for Bi- and Polycyclic Heterocycles."

IV.J-5   D. L. Boger, Tetrahedron, 39, 2869 (1983).

   Review: "Diels-Alder Reactions of Azadienes."

IV.J-6   M. H. Elnagdi, H. A. Elfahham and G. E. H. Elgemeie, Heterocycles, 20, 519 (1983).

   Review: "Utility of α,β-Unsaturated Nitriles in Heterocyclic Synthesis."

V
PROTECTING GROUPS

V.A. Hydroxyl Protecting Groups   (see also: VI.A.10, 11)

V.A-1 M. Koreeda and L. Brown, J. Chem. Soc. Chem. Commun., 1113 (1983).

$$ROH \xrightarrow[-78°C]{PhC(=O)-OTf} ROC(=O)Ph$$

63-92%

V.A-2 M. Onaka, M. Kawai and Y. Izumi, Chem. Lett., 1101 (1983).

$$ROH + PhCH_2Cl \xrightarrow{Zeolite} ROCH_2Ph$$

R = alkyl   40-84%

V.A-3 M. Tordeux, R. Dorme and C. Wakselman, Synth. Commun., 13, 629 (1983).

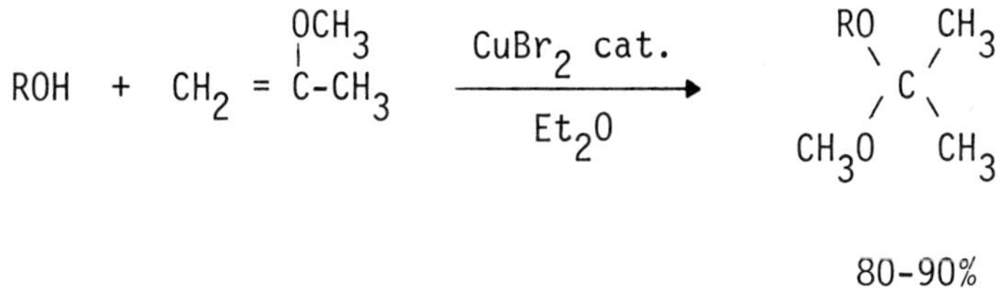

$$ROH + CH_2=C(OCH_3)-CH_3 \xrightarrow[Et_2O]{CuBr_2 \text{ cat.}} (RO)(CH_3O)C(CH_3)_2$$

80-90%

V.A-4 R. O. Hutchins and K. Learn, J. Org. Chem., 47, 4380 (1982).

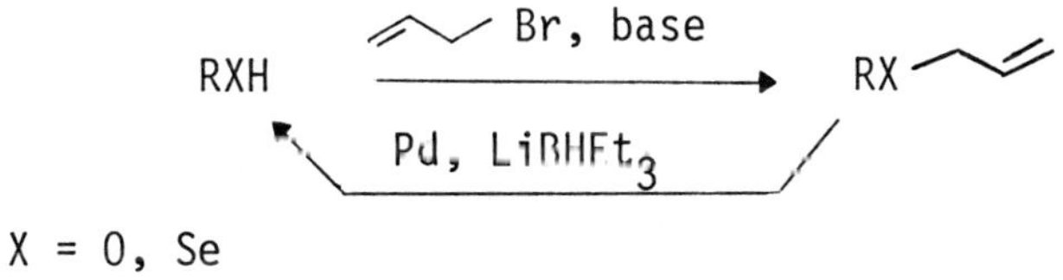

$$RXH \xrightarrow[Pd, LiBHEt_3]{\text{allyl-Br, base}} RX\text{-allyl}$$

X = O, Se

V.A-5 M. Sekine and T. Hata, J. Am. Chem. Soc., 105, 2044 (1983).

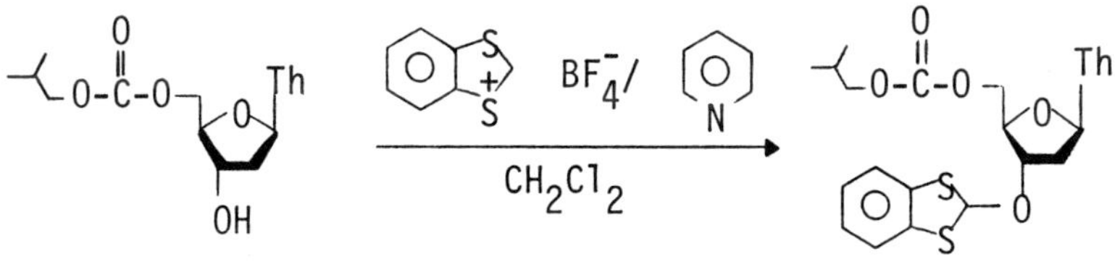

[Reaction of a 5'-O-carbonate-3'-OH thymidine nucleoside with benzodithiolylium BF$_4^-$ / pyridine in CH$_2$Cl$_2$ to give the 3'-O-(benzodithiol-2-yl) protected nucleoside]

V.A-6 G. Cardillo et al., Chem. Ind., 643 (1983).

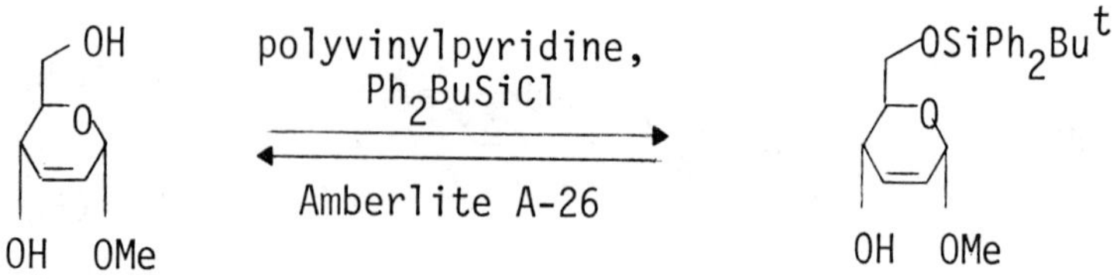

[Reaction: pyranose diol with OH, OMe groups + polyvinylpyridine, $Ph_2BuSiCl$ / Amberlite A-26 → silyl ether $OSiPh_2Bu^t$, ~90%]

V.A-7 M. Sekine and T. Hata, J. Org. Chem., 48, 3011 (1983).

$$R\text{-}CH_2OH \underset{OH^-}{\overset{TBTrBr}{\rightleftarrows}} TBTrOCH_2R$$

$TBTr = (PhCO_2\text{-}C_6H_4\text{-})_3C\text{-}$

42-91%

V.A-8 E. Mohacsi, Synth. Commun., 13, 827 (1983).

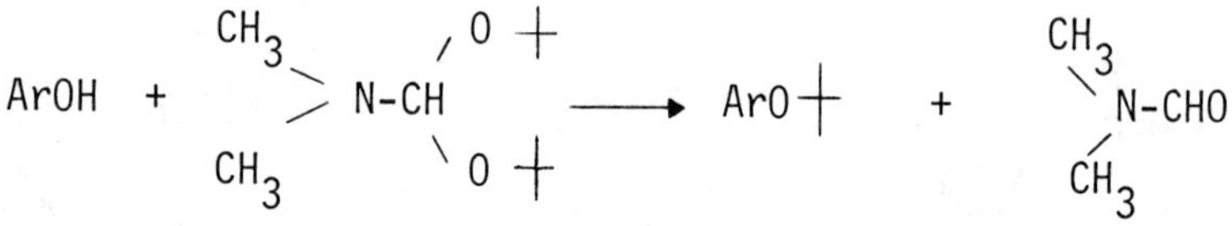

$$ArOH + \underset{CH_3}{\overset{CH_3}{>}}N\text{-}CH\underset{\diagdown O{+}}{\diagup^{O{+}}} \longrightarrow ArO{+} + \underset{CH_3}{\overset{CH_3}{>}}N\text{-}CHO$$

16-59%

V.A-9 Y. Oikawa, T. Nishi and O. Yonemitsu, Tetrahedron Lett., 24, 4037 (1983).

[Scheme: 1,3-diol with R^1, R^2, R^3 substituents + p-MeO-C$_6$H$_4$-CH$_2$OMe / DDQ → p-methoxybenzylidene 1,3-dioxane]

[Scheme: 1,2-diol with R^1, R^2 substituents + p-MeO-C$_6$H$_4$-CH$_2$OMe / DDQ → p-methoxybenzylidene 1,3-dioxolane]

V.A-10 Y. Kita et al., Tetrahedron Lett., 24, 1273 (1983)

[Scheme: $R^1R^2Si(OC(=CH_2)OMe)_2$ + HX—/HX— → cyclic $R^1R^2Si(X)(Y)$]

X,Y = O,S,NR   57-98%

V.A-11 Y. Quindon, H. E. Morton and C. Yoakim, Tetrahedron Lett., 24, 3969 (1983).

$$R-O-R' \xrightarrow[-78^\circ C]{Me_2BBr} R-OH$$

R' = MEM, MOM, MTM   90-97%

V.A-12 H. Monti et al., Synth. Commun., 13, 1021 (1983).

$$\text{R-OR'} \xrightarrow{\text{PhNH}_3^{\oplus} \ {}^{\ominus}\text{OTs}} \text{R-OH}$$

R' = MEM, MOM

V.A-13 N. C. Barua, R. P. Sharma and J. N. Baruah, Tetrahedron Lett., 24, 1189 (1983)

$$\text{R-OR'} \xrightarrow[\text{Ac}_2\text{O}]{\text{Me}_3\text{SiCl}} \text{R-OAc}$$

R' = Me, MTM   60-95%

V.A-14 P. K. Chowdhury, R. P. Sharma and J. N. Baruah, Tetrahedron Lett., 24, 4485 (1983).

$$\text{R-OMTM} \xrightarrow{\text{Ph}_3\text{CBF}_4} \text{ROH}$$

R = alkyl   95%

V.A-15 D. R. Williams and S. Sakdarat, <u>Tetrahedron Lett.</u>, <u>24</u>, 3965 (1983).

R-OMEM 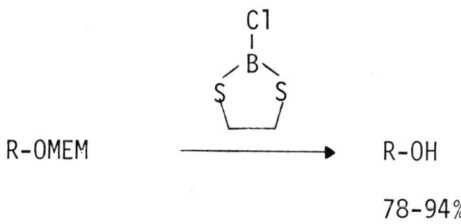 R-OH

78-94%

V.A-16 O. Attanasi, P. Filippone and Serra-Zanetti, F., <u>Synth. Commun.</u>, <u>12</u>, 1155 (1982).

$$R\text{-OAc} \xrightarrow[\text{MeOH/H}_2\text{O}]{\text{CuCl}_2 \cdot 2\text{H}_2\text{O}, \Delta} R\text{-OH}$$

70-95%

V.A-17 A. Klemer, M. Bieber and H. Wilbers, <u>Liebigs Ann. Chem.</u> 1416 (1983).

$$\text{ROZ} \xrightarrow{\text{ISiMe}_3} \text{ROSiMe}_3 \longrightarrow \text{ROH}$$

Z = CH_2Ph, CPh_3

V.A-18 C. A. A. van Boeckel and T. Beetz, Tetrahedron Lett., 24, 3775 (1983).

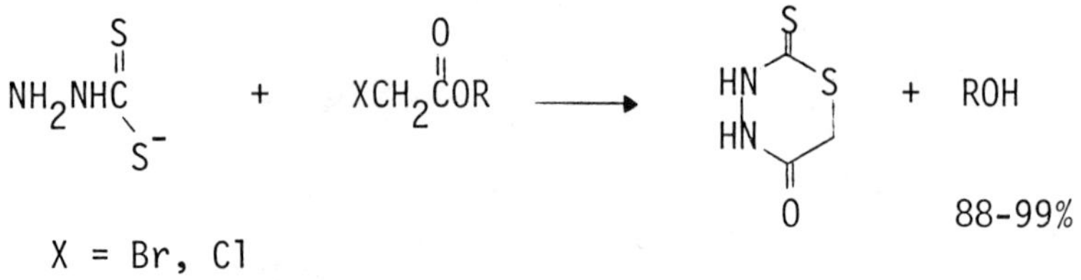

$$NH_2NHC(=S)S^- + XCH_2COR \longrightarrow \underset{O}{\underset{HN}{HN}}\overset{S}{\rceil}S + ROH$$

X = Br, Cl

88-99%

V.A-19/V.B-1 R. Balasuriya et al., Tetrahedron Lett., 24, 1385 (1983).

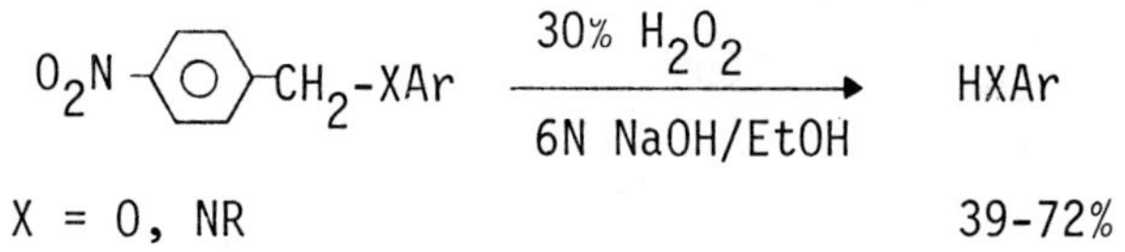

$$O_2N-C_6H_4-CH_2-XAr \xrightarrow[\text{6N NaOH/EtOH}]{30\% \ H_2O_2} HXAr$$

X = O, NR

39-72%

V.B. Amine Protecting Groups (see also: VI.A.4)

V.B-2 J. A. Andres and C. Palomo, Bull. Soc. Chim. Fr., II-369 (1982).

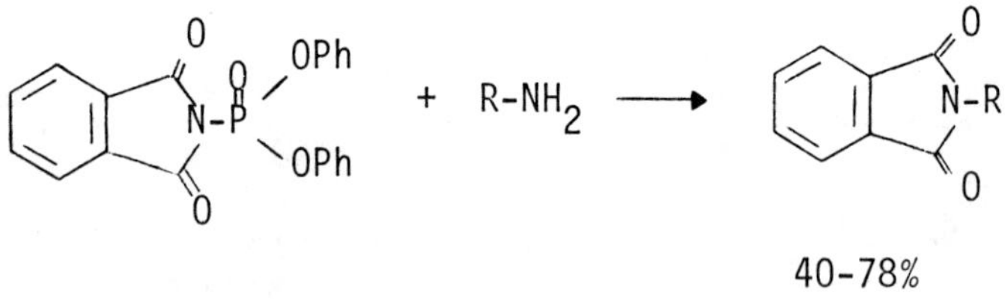

Phthalimide-N-P(=O)(OPh)$_2$ + R-NH$_2$ $\longrightarrow$ Phthalimide-N-R

40-78%

V.B-3 C. R. McArthur, P. M. Worster and A. U. Okon, Synth. Comm., 13, 393 (1983).

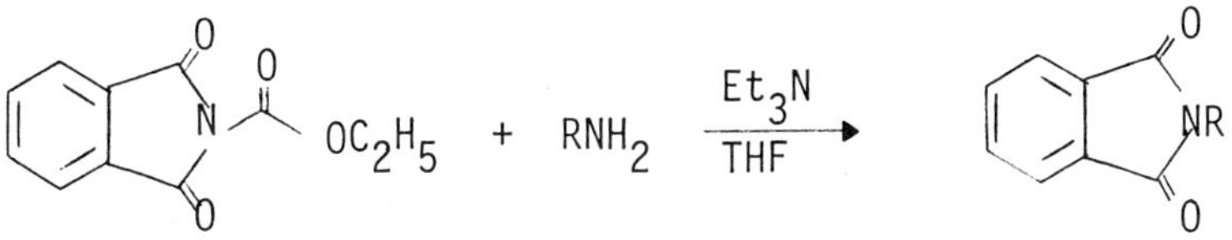

[Phthalimide-N-CO-OC$_2$H$_5$] + RNH$_2$ $\xrightarrow{\text{Et}_3\text{N}}_{\text{THF}}$ [Phthalimide-NR]

V.B-4 T. Mukaiyama et al., Chem. Lett., 879 (1983).

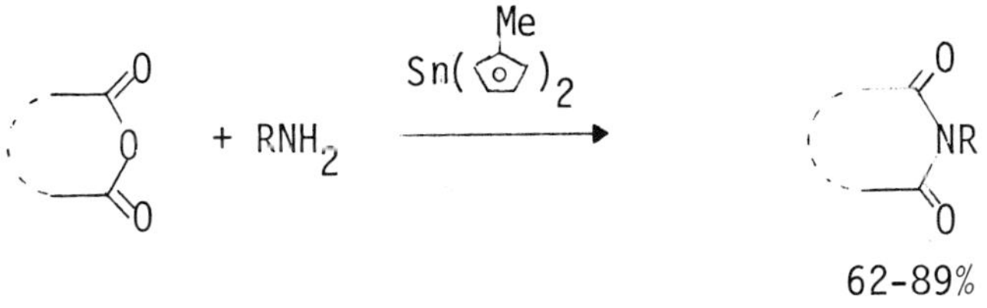

(anhydride) + RNH$_2$ $\xrightarrow{\text{Sn(Cp-Me)}_2}$ (imide-NR)

62-89%

V.B-5 A. Arrieta and C. Palomo, Synthesis, 1050 (1982).

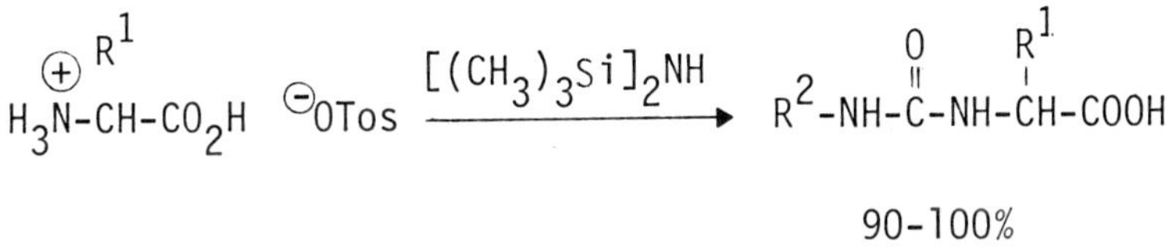

$\overset{\oplus}{H_3N}$-CH(R^1)-CO$_2$H $\overset{\ominus}{\text{OTos}}$ $\xrightarrow{[(CH_3)_3Si]_2NH}$ R^2-NH-C(=O)-NH-CH(R^1)-COOH

90-100%

V.B-6 W. Voelter and J. Mueller, Liebigs Ann. Chem., 248 (1983).

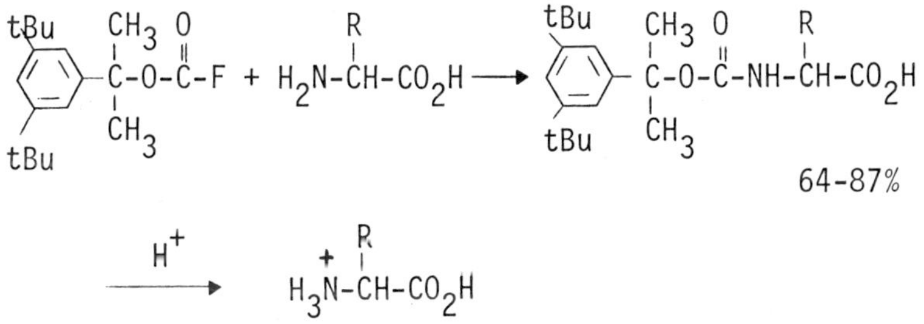

(2,6-di-tBu-4-)C(CH$_3$)$_2$-O-C(=O)-F + H$_2$N-CH(R)-CO$_2$H ⟶ (2,6-di-tBu-4-)C(CH$_3$)$_2$-O-C(=O)-NH-CH(R)-CO$_2$H

64-87%

$\xrightarrow{H^+}$ $\overset{+}{H_3N}$-CH(R)-CO$_2$H

V.B-7   L. Lapatsanis et al., Synthesis, 671 (1983).

$$Cl_3C-CH_2-O-\underset{O}{\overset{O}{C}}-O-N\begin{pmatrix}O\\O\end{pmatrix} + H_2N-CH-COO^{\ominus}\ Na^{\oplus}$$

$$\xrightarrow{\text{dioxan/H}_2\text{O/NaHCO}_3} Cl_3C-CH_2-O-\underset{}{\overset{O}{C}}-NH-\underset{}{\overset{R}{C}H}-COOH$$

77-94%

$$\text{Fmoc-Cl} + H_2N-\overset{R}{\underset{}{C}H}-COO^{-}\ Na^{+} \xrightarrow[Na_2CO_3,\ 7\text{-}20h]{\text{dioxan/H}_2O/} \text{Fmoc-NH-}\overset{R}{\underset{}{C}H}\text{-COOH}$$

79-96%

Fmoc = fluorenyl-$CH_2-O-\overset{O}{C}-$

V.B-8   J. Martinez and J. Laur, Synthesis, 979 (1982).

$$\text{H-Met-Leu-Phe}-O-CH_2C_6H_5$$
$$\downarrow 1\ \text{or}\ 2$$
$$O=CH-\text{Met-Leu-Phe}-O-CH_2C_6H_5$$

92%

1: $\underset{H}{\overset{O}{\underset{\|}{C}}}-O-C_6H_4-NO_2$

2: $\underset{H}{\overset{O}{\underset{\|}{C}}}-O-C_6H_2Cl_3$

V.B-9 E. J. Vlietstra *et al.*, Recl. J. R. Neth. Chem. Soc., 101, 460 (1982).

$$R\text{-}NH_2 \xrightarrow{Me_3CCO\text{-}O\text{-}CHO} R\text{-}NHCHO$$

V.B-10 J. M. Aizpurua and C. Palomo, Synth. Commun., 13, 745 (1983).

$$R\text{-}NH_2 \xrightarrow{DMF/HCOOH} R\text{-}NH\text{-}\overset{O}{\overset{\|}{C}}\text{-}H$$

61-97%

V.B-11 Y. S. Goldberg and M. V. Shimanskaya, J. Org. Chem. USSR., 18, 1789 (1982).

$$\begin{matrix} R \\ R^1 \end{matrix}\!\!>NH \xrightarrow[TEBA]{CHCl_3/NaOH} \begin{matrix} R \\ R^1 \end{matrix}\!\!>N\text{-}CHO$$

7-78%

V.B-12 N. S. Nudelman and D. Perez, J. Org. Chem., 48, 133 (1983)

$$R_2NLi \quad + \quad CO \xrightarrow[LiBr]{THF} R_2NCHO$$

V.B-13   G. Giesemann and I. Ugi, Synthesis, 788 (1983).

$$RNH_2 + Cl_3CHO \longrightarrow R-NH-\overset{OH}{\underset{|}{CH}}-CCl_3 \underset{Mg(ClO_4)_2}{\overset{SOCl_2}{\longrightarrow}} \begin{array}{l} R-N=CH-CCl_3 \\ 89-97\% \\ \\ R-NH-CHO \\ 65-77\% \end{array}$$

(upper arrow: SOCl$_2$; lower arrow: DMF, Mg(ClO$_4$)$_2$)

V.B-14   B. P. Branchaud, J. Org. Chem., 48, 3538 (1983).

$$\underset{R_3}{\overset{NHR_4}{R_1 \! \! \nearrow \! \! \nwarrow \! R_2}} + TrSCl \longrightarrow \underset{R_3}{\overset{NR_4Str}{R_1 \! \! \nearrow \! \! \nwarrow \! R_2}} \underset{\text{or Me}_2SiI}{\overset{CuCl_2}{\underset{\text{or HI}}{\longrightarrow}}} \underset{R_3}{\overset{NHR_4}{R_1 \! \! \nearrow \! \! \nwarrow \! R_2}}$$

85-97%   77-96%

V.B-15   D. Picq et al., Tetrahedron Lett., 24, 1399 (1983).

85%

PROTECTING GROUPS

V.B-16 J. S. Sawyer and B. A. Narayanan, <u>Synth. Commun.</u>, <u>13</u>, 135 (1983).

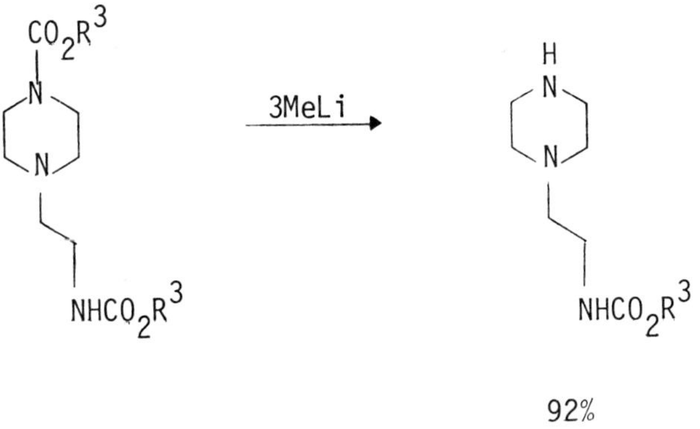

[Structure: piperazine with N-CO_2R^3 and N-propyl-$NHCO_2R^3$ substituents] → 3MeLi → [piperazine with N-H and N-propyl-$NHCO_2R^3$]

92%

Selective deprotection of secondary amines

V.B-17 R. G. K. Schneiderwind and I. Ugi, <u>Tetrahedron</u>, <u>39</u>, 2207 (1983).

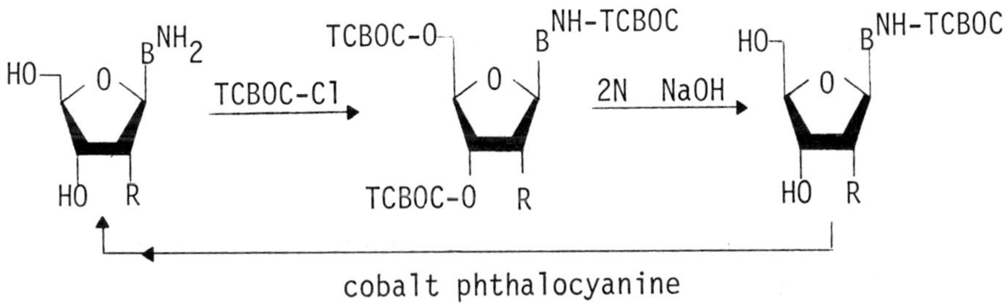

[Sugar with B^{NH_2}, HO, HO, R] → TCBOC-Cl → [Sugar with $B^{NH\text{-}TCBOC}$, TCBOC-O, TCBOC-O, R] → 2N NaOH → [Sugar with $B^{NH\text{-}TCBOC}$, HO, HO, R]

cobalt phthalocyanine

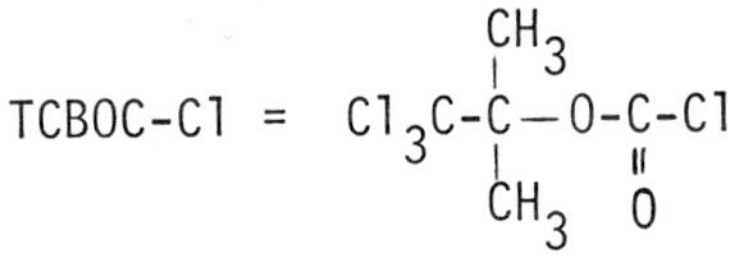

$$\text{TCBOC-Cl} = Cl_3C-\underset{\underset{CH_3}{|}}{\overset{\overset{CH_3}{|}}{C}}-O-\underset{\overset{\|}{O}}{C}-Cl$$

V.B-18 F. Himmelsbach and W. Pfleiderer, Tetrahedron Lett., 24, 3583 (1983).

V.B-19 L. J. McBride and M. H. Caruthers, Tetrahedron Lett., 24, 2953 (1983).

$(MeO)_2Tr$ = di-p-anisylphenylmethyl

80%

V.C. Sulfhydryl Protecting Groups   (see also: VI.A.19)

V.C-1 R. O. Hutchins and K. Learn, J. Org. Chem., 47, 4380 (1982).

$$RSH \xrightarrow[\text{Pd, LiBHEt}_3]{\text{$\sim$Br, base}} RS\sim$$

V.C-2 M. Ueki and K. Shinozaki, Bull. Chem. Soc. Jpn., 56, 1187 (1983).

$$\text{Boc-L-Ala-L-Cys-OH} \xrightleftharpoons[\text{AgNO}_3 \text{ or Hg(OAc)}_2]{\begin{array}{c}1.\ \text{ClSP(CH}_3)_2/\text{OH}^- \\ 2.\ \text{H}^+\end{array}} \text{Boc-L-Ala-L-Cys-OH} \overset{\text{P(S)(CH}_3)_2}{|}$$

V.C-3 Y. Keta et al., Tetrahedron Lett., 24, 1273 (1983).

$$\begin{matrix}R^1\\ \diagdown \\ Si \\ \diagup \\ R^2\end{matrix}\ (O\overset{\displaystyle\diagup\!\!\!\diagdown}{}OMe)_2\ +\ \begin{matrix}HX\\ HY\end{matrix}\Big) \longrightarrow \begin{matrix}R^1\\ \diagdown\ \ \diagup X\\ Si\\ \diagup\ \ \diagdown\\ R^2\ \ \ \ \ Y\end{matrix}\Big)$$

57-98%

X, Y = O, S, NR

V.D. Carboxyl Protecting Groups (see also: VI.A.4, VI.A.10)

V.D-1 U. Widmer, *Synthesis*, 135 (1983).

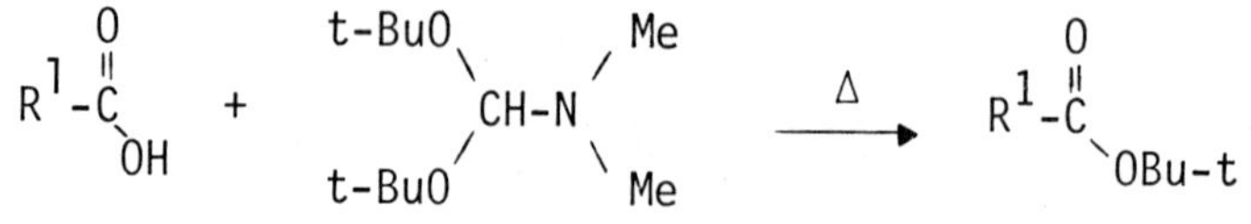

$$R^1-\underset{OH}{\underset{\|}{C}}=O \;+\; \underset{t-BuO}{\overset{t-BuO}{\diagdown}}CH-N\overset{Me}{\underset{Me}{\diagup}} \xrightarrow{\Delta} R^1-\underset{OBu-t}{\underset{\|}{C}}=O$$

55-83%

V.D-2 M. Tordeux, R. Dorme and C. Wakselman, *Synth. Commun.*, **13**, 629 (1983).

$$RCO_2H \;+\; CH_2=CH-OEt \xrightarrow[Et_2O]{CuBr_2} RCO_2-CH(OEt)CH_3$$

65-80%

V.D-3 H. Kessler and R. Siegmeier, *Tetrahedron Lett.*, **24**, 281 (1983).

$$FmOH \;+\; RCO_2H \xrightarrow[DMAP]{DCC,} RCO_2Fm \xrightarrow{Et_2NH} RCO_2H$$

50-90%

FmOH = 9-fluorenyl-CH$_2$OH

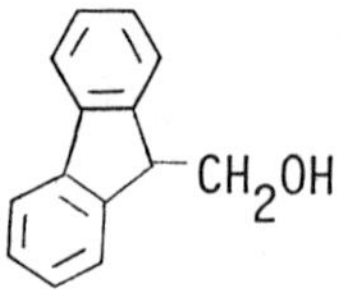

V.D-4 H. Kunz and H. Waldmann, Angew. Chem. Int. Ed., 22, 62 (1983).

$$H_2N\text{-}CH(R)\text{-}CO_2CH_3 + HO\text{-}CH_2\text{-}\overset{S}{\underset{S}{\diagdown}}\!\!\diagup \xrightarrow{(i\text{-}PrO)_3Al}$$

$$H_2N\text{-}CH(R)\text{-}CO_2\text{-}CH_2\text{-}\overset{S}{\underset{S}{\diagdown}}\!\!\diagup \xrightarrow[\text{2. pH=8/H}_2O,\ 1hr.]{1.\ H_2O_2,\ (NH_4)_2MoO_4} H_2N\text{-}CH(R)\text{-}CO_2H$$

V.D-5 T. Kametani, H. Sekine and T. Honda, Chem. Pharm. Bull., 30, 4545 (1982).

$$R\text{-}CO_2CHPh_2 + HCO_2H \longrightarrow R\text{-}CO_2H$$

R = Cephalosporin nucleus   70-97%

V.D-6 G. H. L. Nefkens and B. Zwanenburg, Tetrahedron, 39, 2995 (1983).

$$R\text{-}CH(NH_2)\text{-}CO_2H \xrightarrow{Et_3B} \begin{array}{c} R\quad O \\ H_2N\diagdown\!\!\!\diagup\!\!\!\diagdown\!\!\!O \\ \quad B \\ Et\quad Et \end{array}$$

(H⁺ reverses)

50-100%

V.E. Protecting Groups for Ketones and Aldehydes

(see also: VI.A.18)

V.E-1 V. Kumar and S. Dev, *Tetrahedron Lett.* **24**, 1289 (1983).

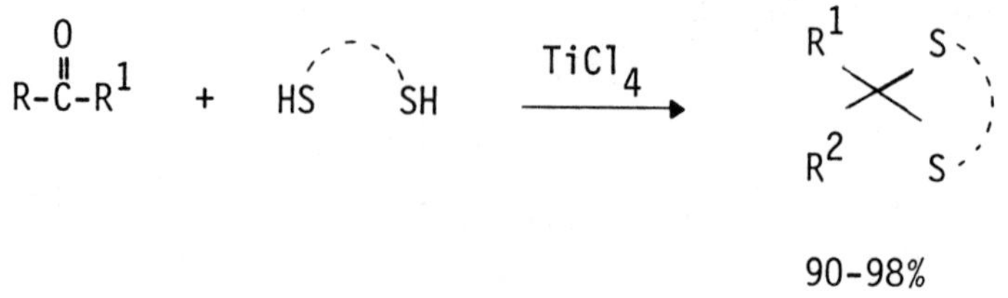

$$R-\overset{O}{\underset{\|}{C}}-R^1 + HS\frown SH \xrightarrow{TiCl_4} \underset{R^2}{\overset{R^1}{>}}\!\!\!\!\!\!\underset{S}{\overset{S}{<}}$$

90-98%

V.E-2 E. J. Corey and K. Shimoji, *Tetrahedron Lett.*, **24**, 169 (1983).

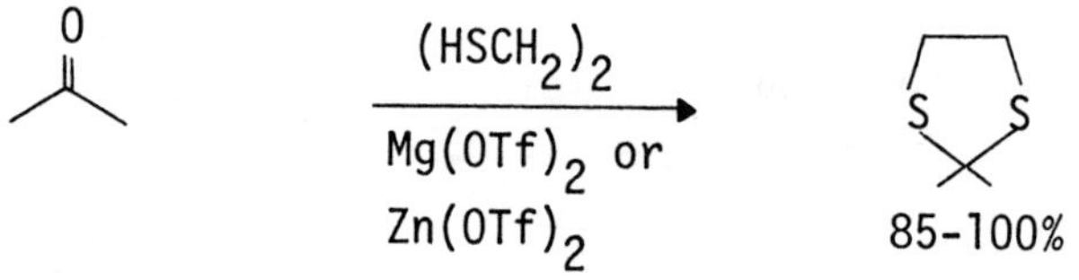

$$\underset{}{\overset{O}{\underset{\|}{\wedge}}} \xrightarrow[\text{Mg(OTf)}_2 \text{ or } \text{Zn(OTf)}_2]{(HSCH_2)_2} \underset{}{\overset{S\frown S}{\times}}$$

85-100%

V.E-3 R. S. Musavirov *et al.*, *J. Gen. Chem. USSR*, **52**, 1229 (1982).

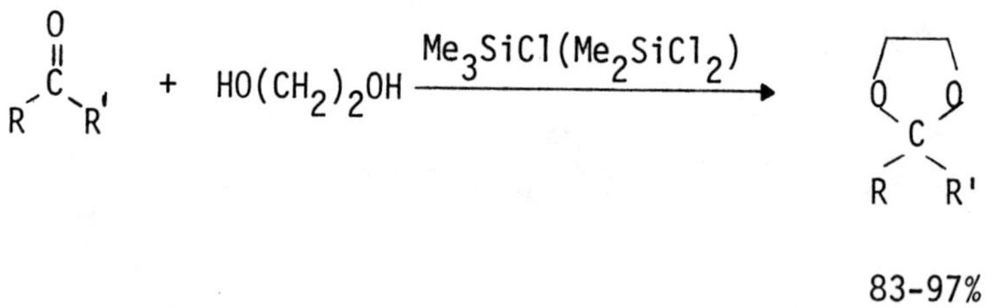

$$R-\overset{O}{\underset{\|}{C}}-R' + HO(CH_2)_2OH \xrightarrow{Me_3SiCl(Me_2SiCl_2)} \underset{R\ \ R'}{\overset{O\frown O}{\underset{C}{}}}$$

83-97%

V.E-4 T. H. Chan, M. A. Brook and T. Chaly, Synthesis, 203 (1983).

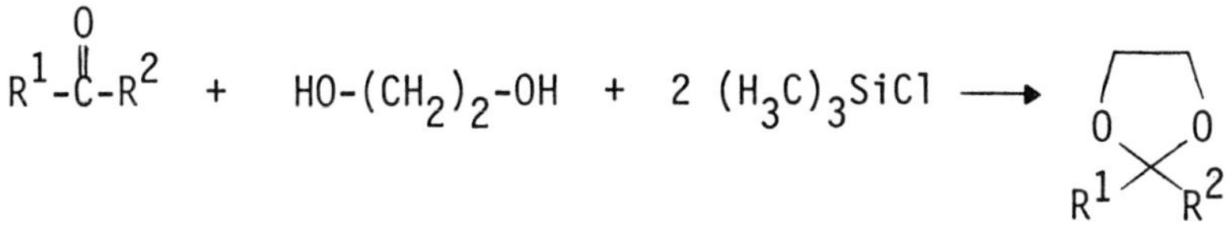

$$R^1-\underset{\underset{O}{\|}}{C}-R^2 + HO-(CH_2)_2-OH + 2\ (H_3C)_3SiCl \longrightarrow \underset{R^1\ \ R^2}{\underset{O\diagup\diagdown O}{\bigcap}}$$

71-98%

V.E-5 K. S. Kochhar et al., J. Org. Chem., 48, 1765 (1983).

$$R-CHO \xrightarrow[FeCl_3]{Ac_2O} R-CH(OAc)_2$$

57-93%

V.E-6 J. N. Barua, Tetrahedron Lett., 24, 3383 (1983).

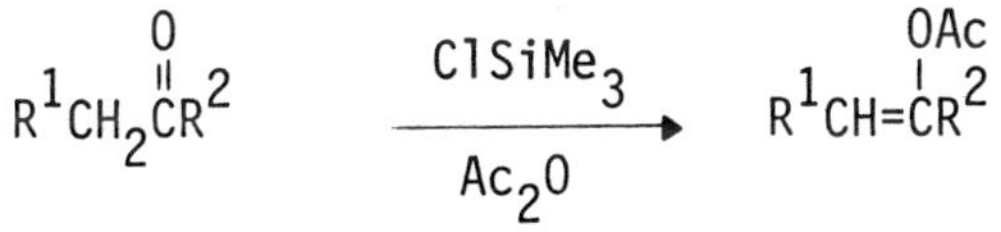

$$R^1CH_2\underset{\underset{O}{\|}}{C}R^2 \xrightarrow[Ac_2O]{ClSiMe_3} R^1CH=\underset{\underset{OAc}{|}}{C}R^2$$

70-80%

V.E-7 M. T. Reetz, B. Wenderoth and R. Peter, J. Chem. Soc. Chem. Commun., 406 (1983).

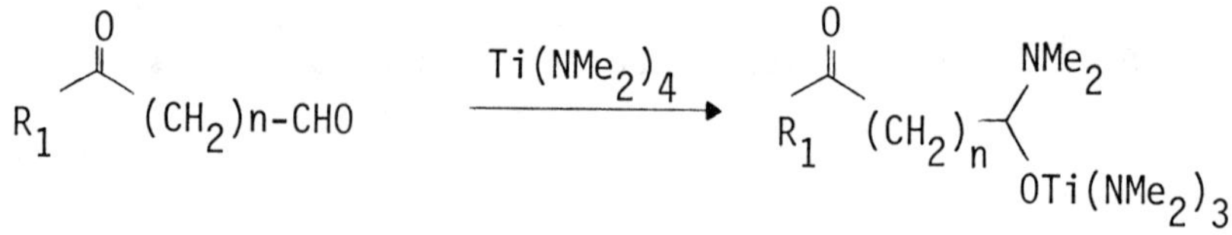

$$R_1-CO-(CH_2)_n-CHO \xrightarrow{Ti(NMe_2)_4} R_1-CO-(CH_2)_n-CH(NMe_2)(OTi(NMe_2)_3)$$

V.E-8 D. J. Ager, J. Chem. Soc. Perkin Trans. 1, 1131 (1983).

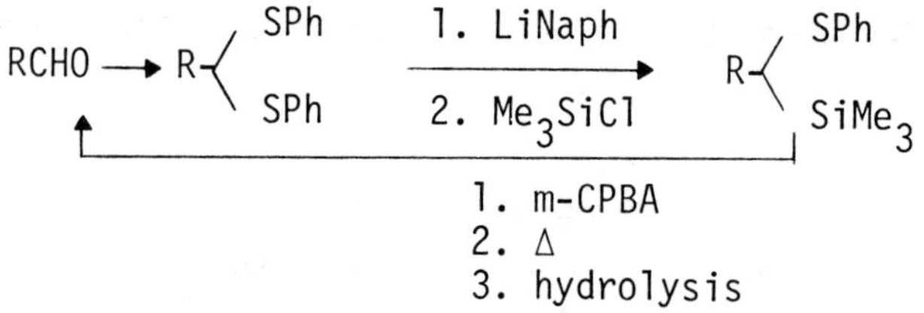

$$RCHO \rightarrow R-CH(SPh)_2 \xrightarrow[2.\ Me_3SiCl]{1.\ LiNaph} R-C(SPh)(SiMe_3)$$

1. m-CPBA
2. Δ
3. hydrolysis

V.E-9 M. T. El-Wassimy, K. A. Jorgensen and S.-O. Lawesson, J. Chem. Soc. Perkins Trans. 1, 2201 (1983).

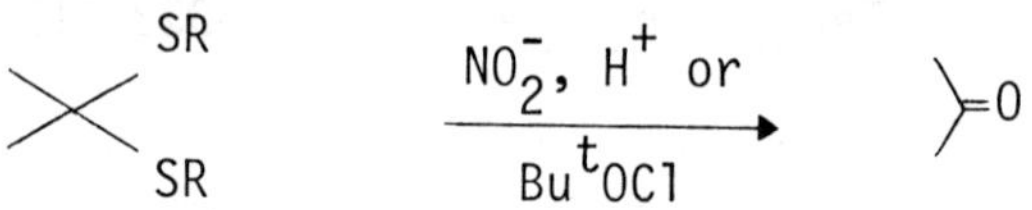

$$C(SR)_2 \xrightarrow[Bu^tOCl]{NO_2^-,\ H^+\ or} C=O$$

V. E-10 G. Balme and J. Gore, J. Org. Chem., 48, 3336 (1983).

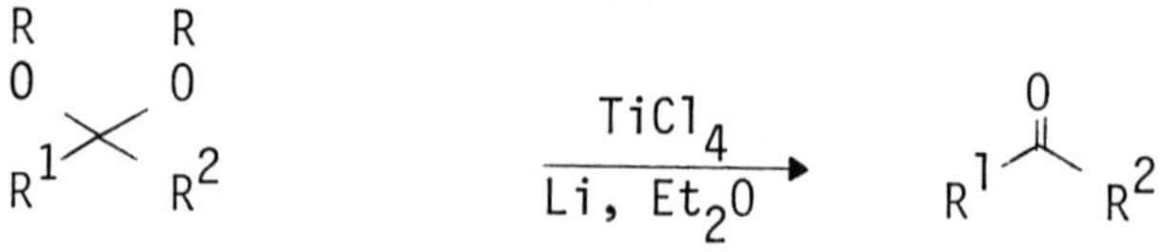

$$R^1C(OR)(OR)R^2 \xrightarrow[Li,\ Et_2O]{TiCl_4} R^1-CO-R^2$$

65-90%

V.E-11 Y. Quindon, H. E. Morton and C. Yoakim, Tetrahedron Lett., 24, 3969 (1983).

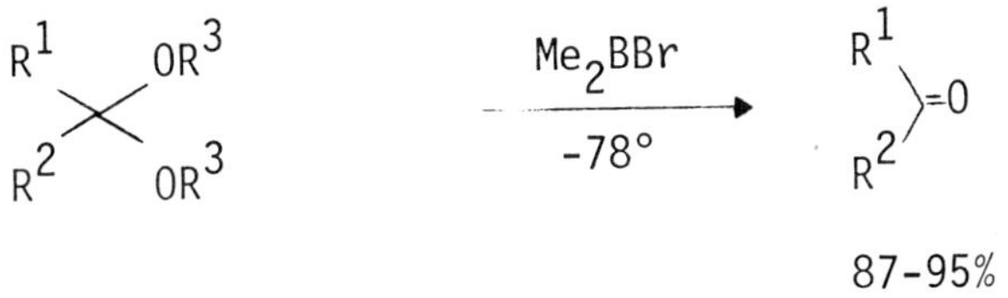

$$\underset{R^2}{\overset{R^1}{>}}\underset{OR^3}{\overset{OR^3}{<}} \xrightarrow[-78°]{Me_2BBr} \underset{R^2}{\overset{R^1}{>}}=O$$

87-95%

V. E-12 H. Urabe, Y. Takano and I. Kuwajima, J. Am. Chem. Soc., 105, 5703 (1983)

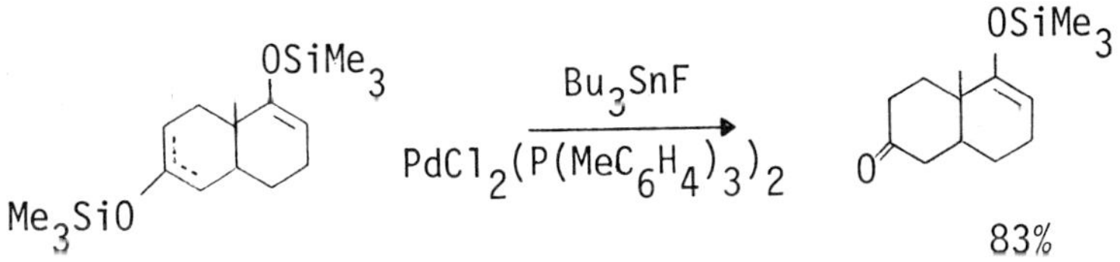

$$\xrightarrow[PdCl_2(P(MeC_6H_4)_3)_2]{Bu_3SnF}$$

83%

V.E-13 J. M. Aizpurua and C. Palomo, Tetrahedron Lett., 24, 4367 (1983).

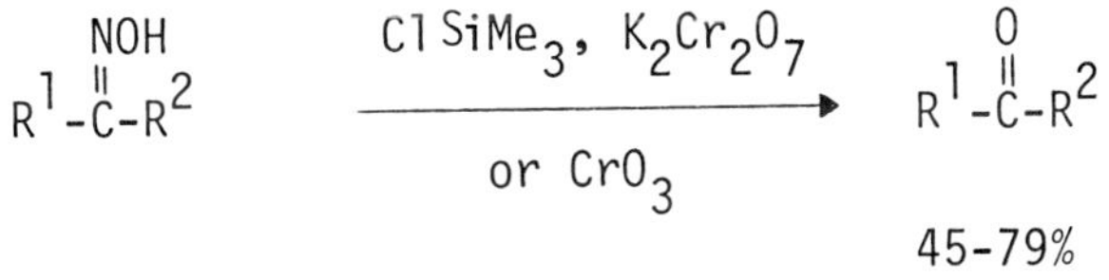

$$R^1-\underset{\overset{\|}{NOH}}{C}-R^2 \xrightarrow[\text{or } CrO_3]{ClSiMe_3,\ K_2Cr_2O_7} R^1-\underset{\overset{\|}{O}}{C}-R^2$$

45-79%

V. E-14 C. G. Rao et al., Synthesis, 808 (1983).

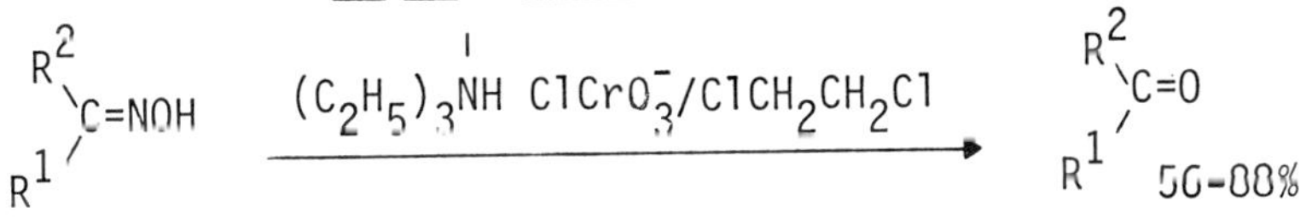

$$\underset{R^1}{\overset{R^2}{>}}C=NOH \xrightarrow{(C_2H_5)_3NH\ ClCrO_3^-/ClCH_2CH_2Cl} \underset{R^1}{\overset{R^2}{>}}C=O$$

56-88%

V.E-15 E.-C. Lin and M. R. Van De Mark, J. Chem. Soc. Chem. Commun., 1176 (1982).

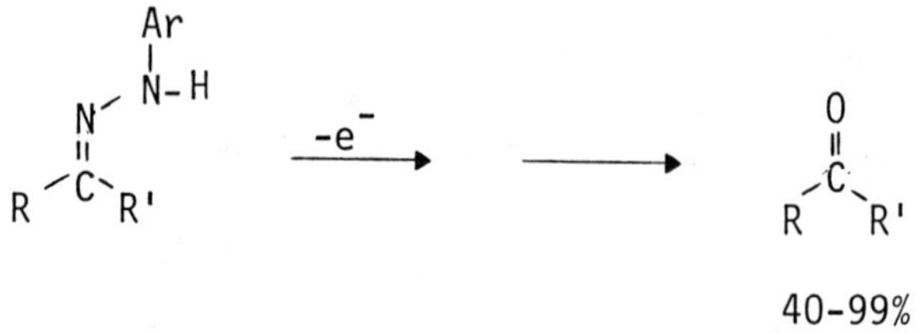

$$\underset{RR'}{\overset{\underset{|}{N-H}}{\underset{||}{N}}\diagdown C} \xrightarrow{-e^-} \longrightarrow \underset{RR'}{\overset{O}{\underset{||}{C}}}$$

40-99%

V.F. Phosphate Protecting Groups

V.F-1 J. J. Vasseur, B. Rayner and J. L. Imbach, Tetrahedron Lett., 24, 2753 (1983).

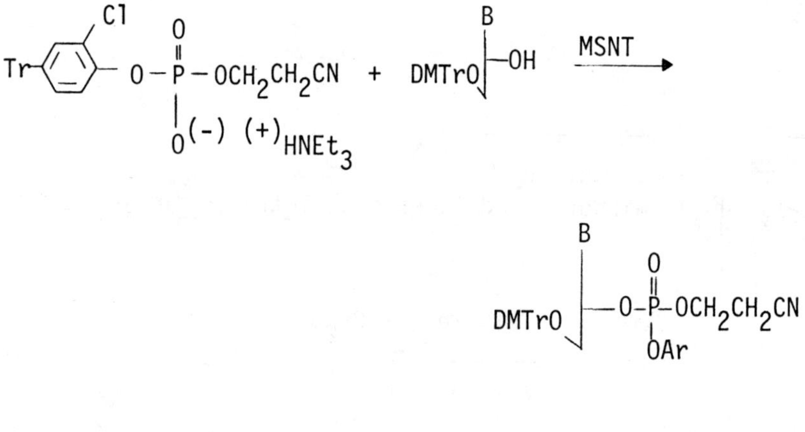

Tr—⟨Cl⟩—O—P(=O)(OCH$_2$CH$_2$CN)—O$^{(-)}$ $^{(+)}$HNEt$_3$ + DMTrO—[B]—OH $\xrightarrow{MSNT}$

DMTrO—[B]—O—P(=O)(OCH$_2$CH$_2$CN)—OAr

MSNT = 1-mesitylsulfonyl-3-nitro-1,2,4-triazole

DMTr = 4,4'-dimethoxytrityl

V.F-2 A. Wolter and H. Koester, Tetrahedron Lett., 24, 873 (1983).

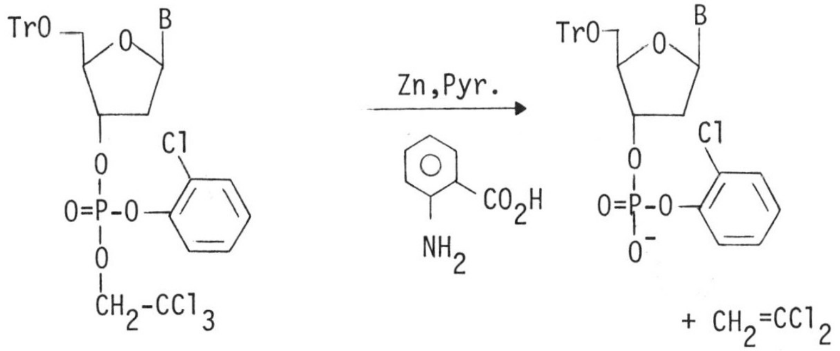

[Reaction: 5'-TrO-protected nucleoside with 3'-phosphate bearing o-chlorophenyl and 2,2,2-trichloroethyl (CH$_2$-CCl$_3$) groups, treated with Zn, Pyridine, and anthranilic acid (2-aminobenzoic acid, with CO$_2$H and NH$_2$) to give the phosphate with o-chlorophenyl group and O$^-$, plus CH$_2$=CCl$_2$, >90%]

V.F-3 L. Jacob et al., Synthesis, 451 (1983).

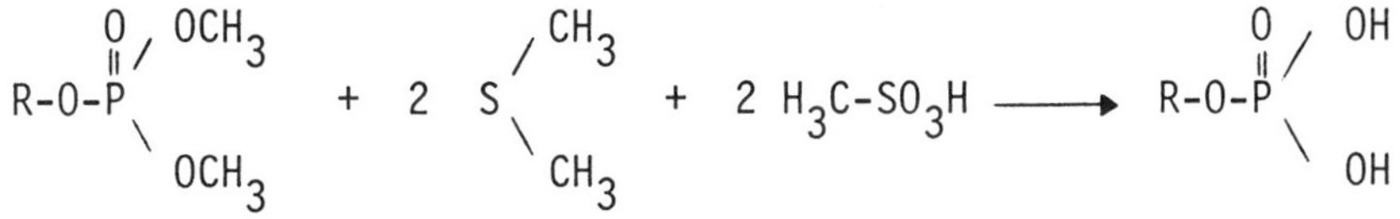

$$R-O-\overset{O}{\underset{\underset{OCH_3}{|}}{P}}-OCH_3 \;+\; 2\; S(CH_3)_2 \;+\; 2\; H_3C-SO_3H \longrightarrow R-O-\overset{O}{\underset{\underset{OH}{|}}{P}}-OH$$

66-83%

V.G. Pi-Bond Protecting Groups

V.G-1 V. Janout and P. Cefelin, Tetrahedron Lett., 24, 3913 (1983).

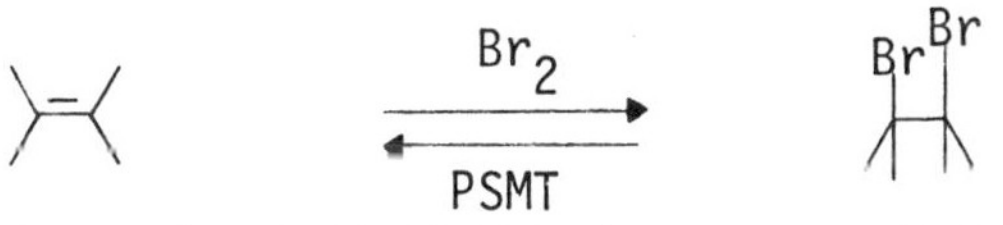

[2,3-dimethyl-2-butene + Br$_2$ ⇌ (PSMT) dibromide, 62-100%]

PSMT = poly(styrylmethylthiolate)

V.H. Miscellaneous Protecting Groups

V.H-1 T. Kamimura et al., Chem. Lett., 1051 (1983).

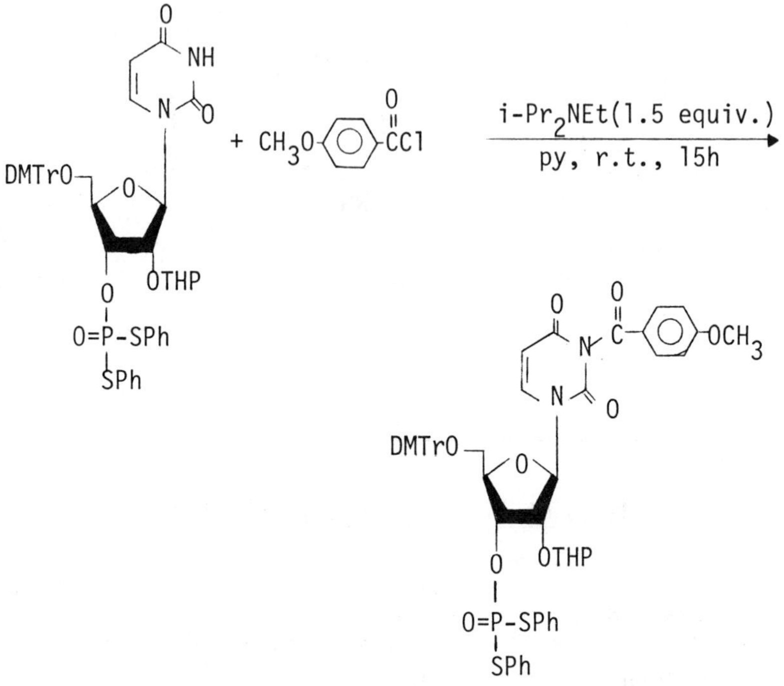

[Reaction scheme: DMTrO-protected nucleoside with O=P(SPh)₂ phosphate and OTHP group, plus CH₃O-C₆H₄-COCl, with i-Pr₂NEt (1.5 equiv.), py, r.t., 15h, giving the N-anisoyl protected product in 92%]

V.H-2 B. S. Schulz and W. Pfleiderer, Tetrahedron Lett., 24, 3587 (1983).

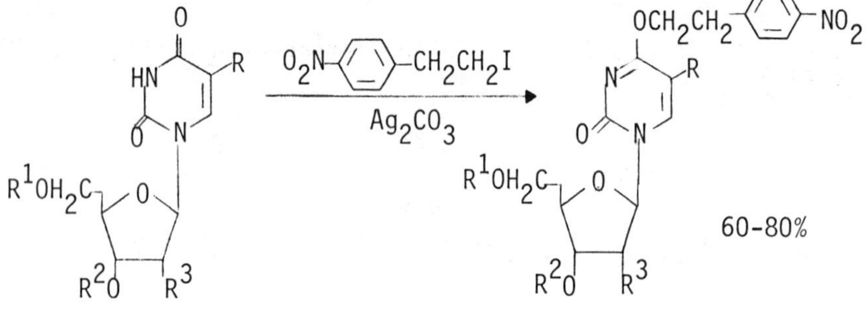

[Reaction scheme: nucleoside with R¹OH₂C, R²O, R³ substituents on sugar, reacting with O_2N-C₆H₄-CH₂CH₂I / Ag_2CO_3 to give O-(4-nitrophenethyl) protected product, 60–80%]

V.H -3 T. Kamimura *et al.*, *Tetrahedron Lett.*, *24*, 2775 (1983).

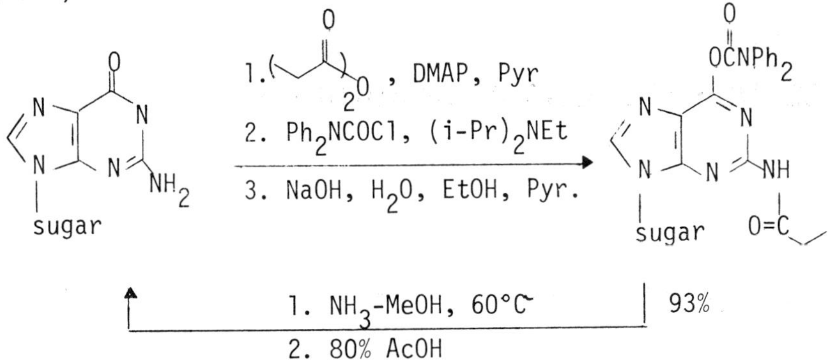
sugar
1. , DMAP, Pyr
2. Ph_2NCOCl, $(i\text{-}Pr)_2NEt$
3. NaOH, H_2O, EtOH, Pyr.
$OCNPh_2$
sugar
1. NH_3-MeOH, 60°C
2. 80% AcOH
93%

V.H-4 J. V. Comasseto and N. Petragnani, *Synth. Commun.*, *13*, 889 (1983).

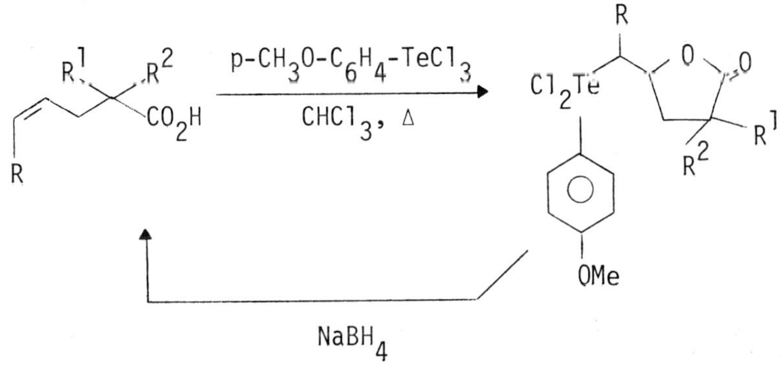
CO_2H
p-CH_3O-C_6H_4-$TeCl_3$
$CHCl_3$, Δ
Cl_2Te
OMe
$NaBH_4$

V.H-5 N. Fujii, *Yakugaku Zasshi (J. Pharm. Soc. Jpn.)*, *103*, 805 (1983).

Review: "Synthetic Studies on Peptides Using Organosulfonic Acids-Deprotecting Procedure."

VI
USEFUL SYNTHETIC PREPARATIONS

VI.A. Functional Group Preparations

1. Acids, Acid Halides, Anhydrides   (see also: II.A.2)

VI.A.1-1 M. Tiecco et al., Synth. Commun., 13, 617 (1983)

$$C_6H_5CO_2R + MeSeLi \xrightarrow{DMF} MeSeR + C_6H_5CO_2Li$$

$$\text{R = Me, Et, i-Pr} \qquad\qquad 95\%$$

VI.A.1-2 G. A. Olah, R. Karpeles and S. C. Narang, Synthesis, 963 (1982).

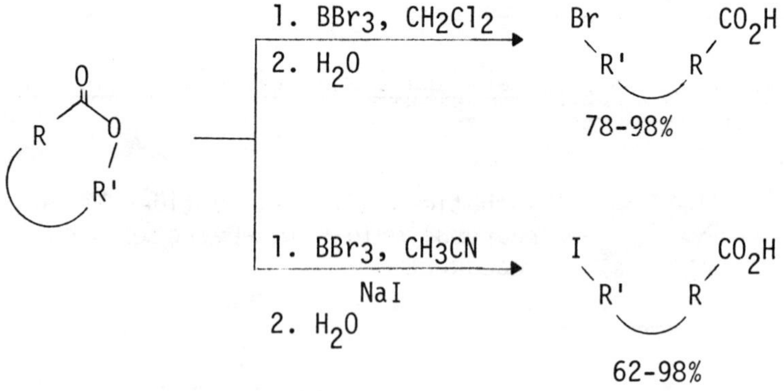

$$\underset{R\frown R'}{\overset{O}{\underset{\|}{C}}\!\!\!-\!O} \quad\xrightarrow[\text{2. H}_2\text{O}]{\text{1. BBr}_3,\ \text{CH}_2\text{Cl}_2}\quad \underset{R'\frown R}{Br\quad CO_2H}$$

$$78\text{-}98\%$$

$$\xrightarrow[\text{2. H}_2\text{O}]{\substack{\text{1. BBr}_3,\ \text{CH}_3\text{CN} \\ \text{NaI}}}\quad \underset{R'\frown R}{I\quad CO_2H}$$

$$62\text{-}98\%$$

VI.A.1-3 D. L. Flynn, R. E. Zelle and P. A. Grieco, J. Org. Chem., 48, 2424 (1983).

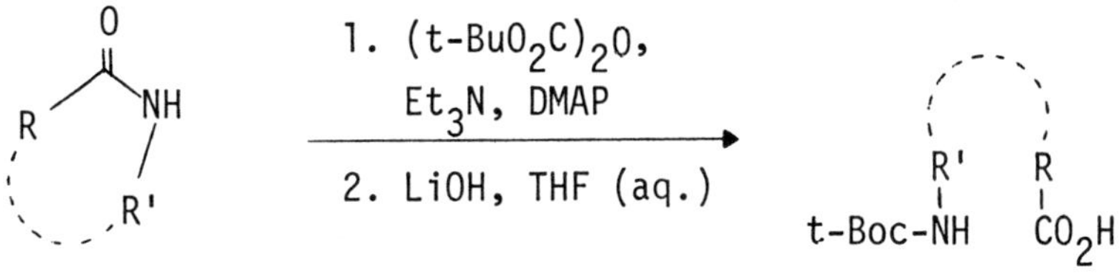

$$\text{cyclic } R\text{-C(=O)-NH-R'} \xrightarrow[\text{2. LiOH, THF (aq.)}]{\text{1. }(t\text{-BuO}_2C)_2O,\ Et_3N,\ DMAP} \underset{t\text{-Boc-NH}\quad CO_2H}{R'\quad R}$$

80-96%

VI.A.1-4 G. Cainelli et al., Synthesis, 306 (1983).

$$R\text{-CO}_2H \xrightarrow[(CH_2Cl)_2]{\text{Amberlite IRA 93 } \cdot PCl_5} R\text{-COCl}$$

51-91%

VI.A.1-5 F. Roulleau, D. Plusquellec and E. Brown, Tetrahedron Lett., 24, 4195 (1983).

$$RCOCl \xrightarrow[P.T.C.]{NaOH} (RCO)_2O$$

60-90%

VI.A.1-6 A. J. L. Cooper, J. Z. Ginos and A. Meister, Chem. Rev., 83, 321 (1983).

Review: "Synthesis and Properties of the ∇-Keto Acids."

VI.A.1-7 S. R. Ramadas *et al.*, *Synthesis*, 605 (1983).

Review: "Methods of Synthesis of Dithiocarboxylic Acids and Esters."

VI.A.2. Alcohols and Phenols (see also: II.B.1, III.A, III,F.1)

VI.A.2-1 P. Molina, M. Alajarin and M. J. Perez de Vega, *Synth. Commun.*, 13, 501 (1983).

$$RCH_2NH_2 \xrightarrow[\text{2. } Hg(OAc)_2 \quad \text{3. } \Delta]{\text{1. } \underline{A}} RCH_2OAc \quad 35\text{-}94\%$$

A =

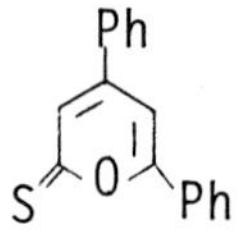

VI.A.2-2 J. Barluenga *et al.*, *Synthesis*, 53 (1983).

1. HgO/35% $HBF_4(H_2O)$
2. $NaHCO_3$
3. 3 normal $KOH(H_2O)$

$$R^1\text{-}Br \longrightarrow R^1OH$$

R^1 = alkyl 65-95%

VI.A.2-3 Y. Guidon, C. Yoakim and H. E. Morton, Tetrahedron Lett., 24, 2969 (1983).

$$R-O-R^1 \xrightarrow[\text{2. NaHCO}_3]{\text{1. (ClH}_2\text{C})_2, \text{Me}_2\text{BBr, Et}_3\text{N}} R-OH + R^1-Br$$

49-96%

R_1 = alkyl, aryl
R^1 = Me, benzyl

VI.A.2-4 S. Matsubara, K. Takai and H. Nozaki, Tetrahedron Lett., 24, 3741 (1983).

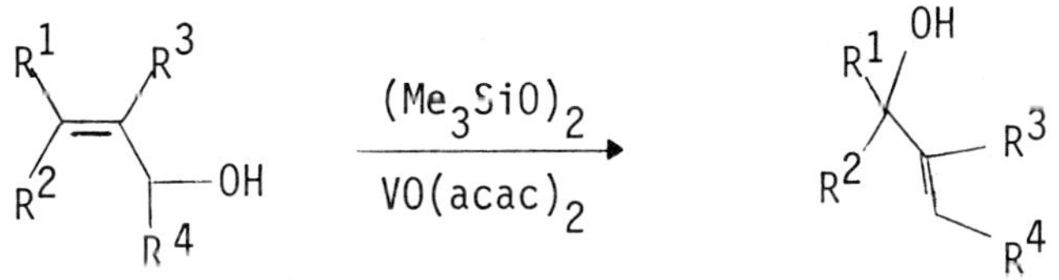

R^1, R^2 = alkyl
R^3, R^4 = H, alkyl

35-97%

VI.A.2-5 S. Yamamoto et al., J. Am. Chem. Soc., 105, 2908 (1983).

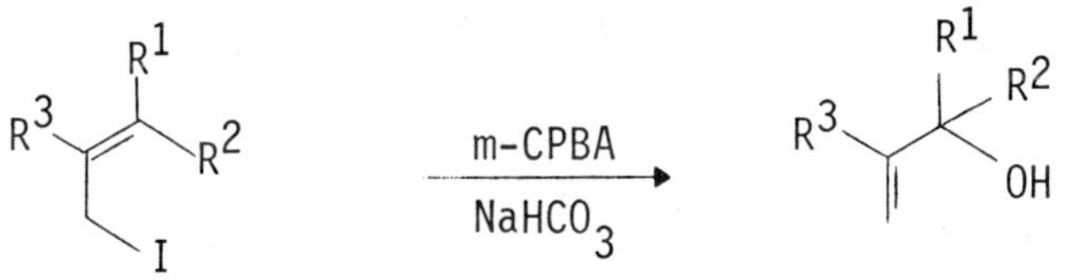

82-93%

VI.A.2-6 T. Di Giamberardino et al., Tetrahedron Lett., 24, 3413 (1983).

$$\underset{R-CH-SeR^1}{\overset{SeR^1}{|}} \xrightarrow[\text{3. }CH_2=PPh_3]{\text{1. BuLi} \quad \text{2. DMF}} \underset{R-CH-SeR^1}{\overset{CH=CH_2}{|}} \xrightarrow[\text{NaIO}_4/\text{MeOH}/H_2O]{H_2O_2/\text{Pyr} \text{ or}} R-CH=CH-CH_2OH$$

R = alkyl
R^1 = Me, PH

70-85%

VI.A.2-7 G. M. Pickles et al., J. Chem. Soc. Perkin Trans. 1, 2949 (1982).

$$ArTl(OCOCF_3)_2 \;+\; B_2H_6 \xrightarrow[\begin{array}{l}\text{3. NaOH, }H_2O_2 \\ \text{4. }H^+\end{array}]{\begin{array}{l}\text{1. }\Delta \\ \text{2. }H_2O\end{array}} ArOH$$

VI.A.2-8 R. W. Hoffmann and K. Ditrich, Synthesis, 107 (1983).

$$Ar\text{-}MgBr \xrightarrow[\begin{array}{l}\text{2. }\Delta \\ \text{3. }H^+\end{array}]{\text{1. }\overset{O}{\underset{O}{\bigcirc}}B\text{-}O\text{-}O\text{-}t\text{-}Bu} Ar\text{-}OH$$

14-94%

VI.A.2-9 A. K. Sinhababu and R. T. Borchardt, J. Org. Chem., 48, 1941 (1983).

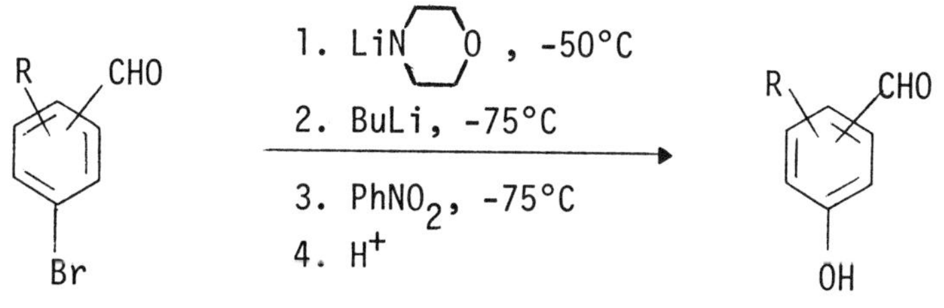
R
CHO
Br
1. LiN O , -50°C
2. BuLi, -75°C
3. PhNO2, -75°C
4. H+
R
CHO
OH

55-65%

VI.A.2-10 D. S. Matteson and K. M. Sadhu, J. Am. Chem. Soc., 105, 2077 (1983).

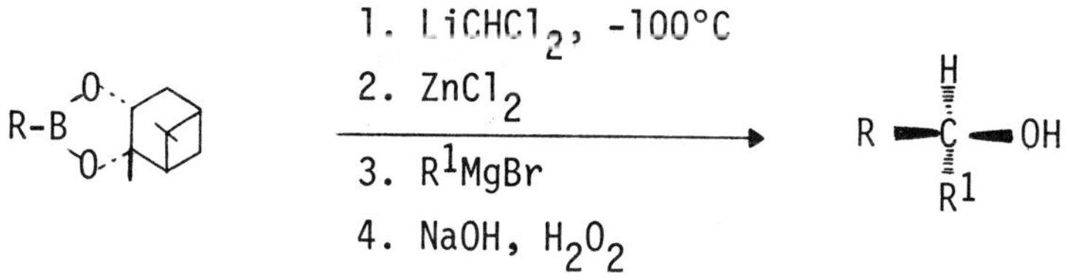
R-B
O
O
1. LiCHCl2, -100°C
2. ZnCl2
3. R1MgBr
4. NaOH, H2O2
H
R
C
OH
R1

82-99%
95.7-99.5% e.e.

VI.A.2-11 H. C. Brown, P. K. Jadhav and A. K. Mandal, J. Org. Chem., 47, 5074 (1982).

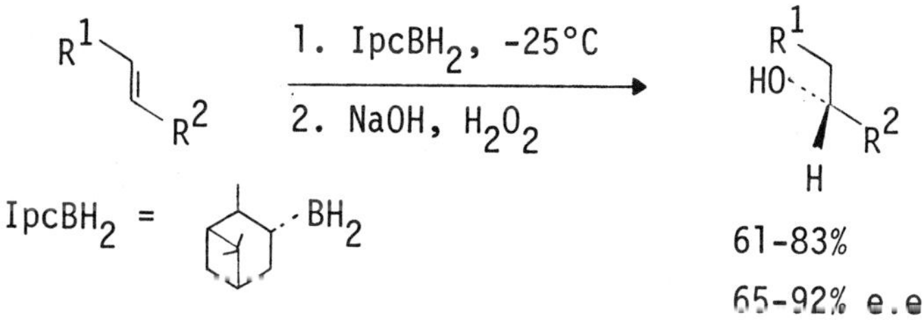
R1
R2
1. IpcBH2, -25°C
2. NaOH, H2O2
R1
HO
R2
H
IpcBH2 =
BH2

61-83%
65-92% e.e.

VI.A.2-12 H. C. Brown, M. C. Desai and P. K. Jadhav. J. Org. Chem., 47, 5065 (1982).

$$\underset{R^2}{\overset{R^1}{\diagup}}\diagdown \quad \xrightarrow[\text{2. MeOH, NaOH, H}_2\text{O}_2]{\text{1. Ipc}_2\text{BH, }-25°\text{C}} \quad R^1\text{-CH}_2\text{-C}(R^2)(H)(OH)$$

Ipc$_2$BH = (structure))$_2$BH

63-81%
60-98% e.e.

VI.A.2-13 W. C. Still and J. C. Barrish, J. Am. Chem. Soc., 105, 2487 (1983).

$$\underset{RO}{\overset{R^1}{\diagup}}\overset{CH_3}{\underset{R^3}{\diagdown}}R^2 \quad \xrightarrow[\text{2. H}_2\text{O}_2\text{NaOH}]{\text{1. BH}} \quad \text{product}$$

$R^2, R^3 = H;$ BH = 9-BBN
$R^2, R^3 \neq H;$ BH = ThBH$_2$

VI.A.2-14 H. Ziffer et al., J. Org. Chem., 48, 3017 (1983).

$$\underset{R}{\overset{Ar}{\diagup}}\overset{H}{\underset{OAc}{\diagdown}} \quad \xrightarrow{\text{Rhizopus nigricans}} \quad \underset{R}{\overset{Ar}{\diagup}}\overset{H}{\underset{OAc}{\diagdown}} + \underset{R}{\overset{Ar}{\diagup}}\overset{OH}{\underset{H}{\diagdown}}$$

enantioselective hydrolysis

VI.A.2-15 J. W. Huffman and R. C. Desai, Synth. Commun., 13, 553 (1983).

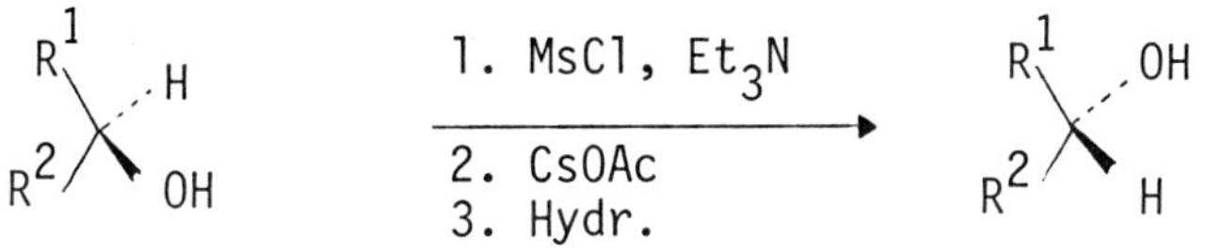

$R^1\cdots H$ / R^2—OH   →(1. MsCl, Et$_3$N; 2. CsOAc; 3. Hydr.)→ $R^1\cdots OH$ / R^2—H

VI.A.2-16 H. L. Holland, Chem. Soc. Rev., 11, 371 (1982).

Review: "The Mechanism of the Microbial Hydroxylation of Steroids."

VI.A.3. Alkyl and Aryl Halides   (see also: II.B.2)

VI.A.3-1 T. Imamoto et al., Synthesis, 460 (1983).

$$R\text{-}OH \xrightarrow{NaI/PPSE} R\text{-}I$$

62-99%

PPSE = trimethylsilyl polyphosphate
R = alkyl

VI.A.3-2 R. Richter and B. Tucker, J. Org. Chem., 48, 2625 (1983).

$$R\text{-}CH_2OH \xrightarrow[2.\ DMF,\ \Delta]{1.\ COCl_2} R\text{-}CH_2Cl$$

76-100%

VI.A.3-3 A. Takadate, T. Tahara and S. Goya, Synthesis, 806 (1983).

$$Ar\text{-}CHN_2 \xrightarrow{HBF_4/CH_2Cl_2} ArCH_2F + ArCH_2OH$$
$$\phantom{Ar\text{-}CHN_2 \xrightarrow{HBF_4/CH_2Cl_2}} 31\text{-}50\% \quad 24\text{-}40\%$$

Ar must have electronegative groups

VI.A.3-4 W. Buijs, P. van Elburg and A. van der Gen, Synth. Commun., 13, 387 (1983).

$$HOCHR(CH_2)_nCH_2OH \xrightarrow{P_2I_4} HOCHR(CH_2)_nCH_2I$$
$$\phantom{HOCHR(CH_2)_nCH_2OH \xrightarrow{P_2I_4}} 42\text{-}87\%$$

VI.A.3-5 M. Lissel and K. Drechsler, Synthesis, 314 (1983).

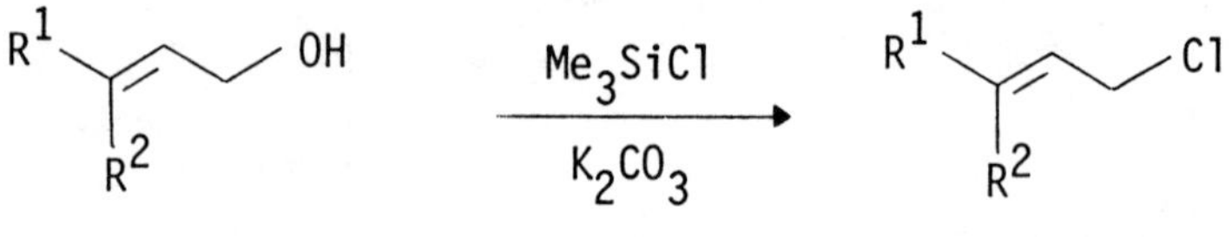

83-97%

VI.A.3-6 G. A. Olah et al., J. Org. Chem., 48, 2766 (1983).

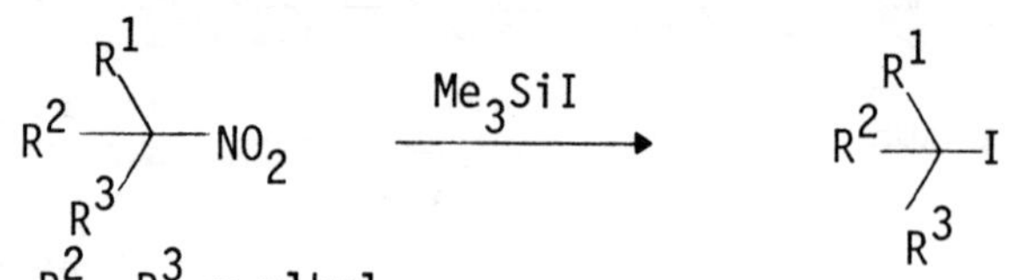

R^1, R^2, R^3 = alkyl

VI.A.3-7 T. Azuhata and Y. Okamoto, Synthesis, 461 (1983).

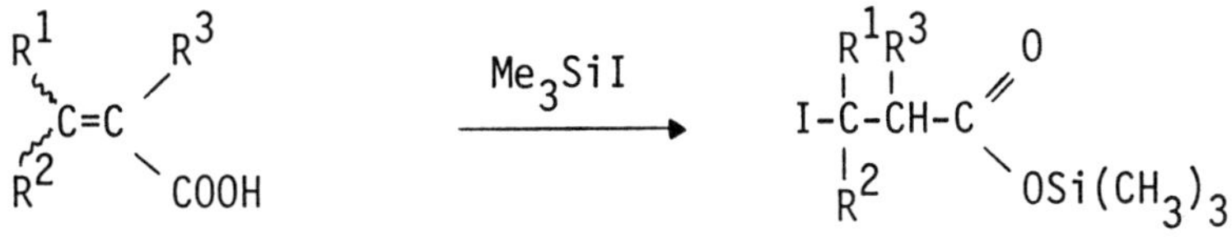

$$\underset{R^2}{\overset{R^1}{>}}C=C\underset{COOH}{\overset{R^3}{<}} \quad \xrightarrow{Me_3SiI} \quad I-\underset{R^2}{\overset{R^1}{\underset{|}{C}}}-\underset{}{\overset{R^3}{\underset{|}{CH}}}-C\underset{OSi(CH_3)_3}{\overset{O}{<}}$$

$R^2, R^3 = H, CH_3$ 74-92%

VI.A.3-8 J. N. Marx, Tetrahedron, 39, 1529 (1983).

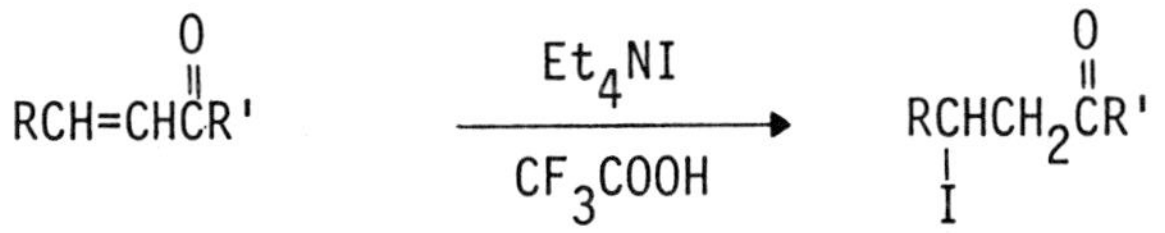

$$RCH=CH\overset{O}{\underset{\|}{C}}R' \quad \xrightarrow[CF_3COOH]{Et_4NI} \quad RCH\underset{I}{\underset{|}{C}}H_2\overset{O}{\underset{\|}{C}}R'$$

50-97%

VI.A.3-9 N. Yoneda et al., Chem. Lett., 1135 (1983).

$$R^1CH=CHR^2 \quad \xrightarrow[THF, 0°C]{melamine/HF} \quad R^1CH_2-CHFR^2$$

$R^1, R^2 = H, alkyl$ 83-98%

VI.A.3-10 L. S. Boguslavskaya, I. Y. Panteleeva and
N. N. Chuvatkin, J. Org. Chem. USSR, 18, 198 (1982).

$$R\text{-}\underset{\underset{O}{\|}}{C}\text{-}OCH_3 \xrightarrow[-60°C]{ClF,\ HF} R\text{-}CF_2\text{-}OCH_3$$

VI.A.3-11 F. Bellsia et al., J. Chem. Res. Synop., 16, (1983).

$$R^1\text{-}\underset{\underset{R^2}{|}}{C}=N\text{-}NH\text{-}Ts \xrightarrow[CH_2Cl_2,\ -15°C]{SO_2Cl_2} R^1\text{-}\underset{\underset{R^2}{|}}{C}Cl_2$$

R^1, R^2 = alkyl   76-82%

$$R^3\text{-}CH_2\text{-}\underset{\underset{H}{|}}{C}=N\text{-}NH\text{-}Ts \xrightarrow[CH_2Cl_2,\ -15°C]{SO_2Cl_2} R^3\text{-}CCl_2\text{-}CHO$$

59-75%

VI.A.4. Amides   (see also: IV.D, VI.A.17)

VI.A.4-1 J. M. Lago and C. Palomo, Synth. Commun., 13, 653 (1983).

Phthalimide-N$^-$K$^+$ + RCOO$^-$K$^+$ $\xrightarrow{\text{solvent/PTC/ PhOP(O)Cl}_2}$ Phthalimide-N-C(O)-R

45-78%

VI.A.4-2 A. N. Mandal, S. R. Raychaudhuri and A. Chatterjee, Synthesis, 727 (1983).

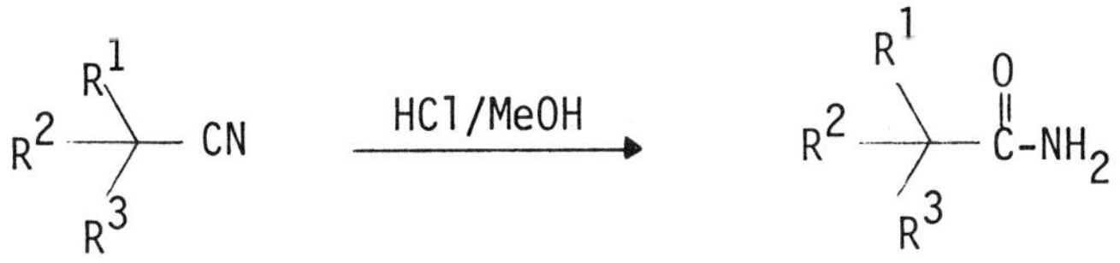

$$R^2-\underset{R^3}{\overset{R^1}{C}}-CN \xrightarrow{HCl/MeOH} R^2-\underset{R^3}{\overset{R^1}{C}}-\overset{O}{\underset{}{C}}-NH_2$$

R^1, R^2, R^3 = alkyl   66-75%

VI.A.4-3 R. Appel and E. Hiester, Chem. Ber., 116, 2037 (1983).

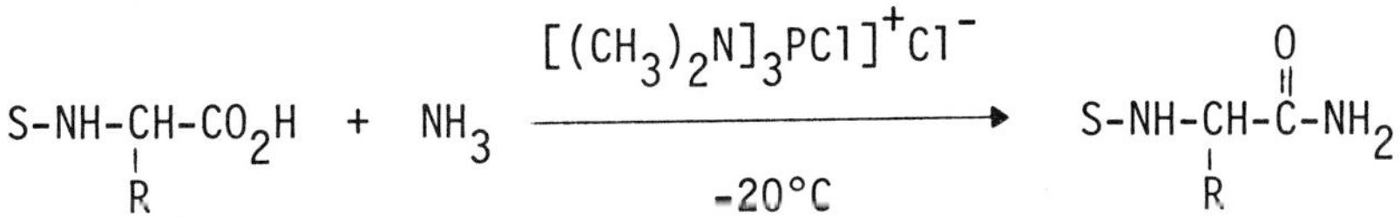

$$S-NH-\underset{R}{CH}-CO_2H + NH_3 \xrightarrow[-20°C]{[(CH_3)_2N]_3PCl]^+Cl^-} S-NH-\underset{R}{CH}-\overset{O}{\underset{}{C}}-NH_2$$

S = protecting group   73-98%

VI.A.4-4 M. E. Jung and Z. Long-Mei, Tetrahedron Lett., 24, 4533 (1983).

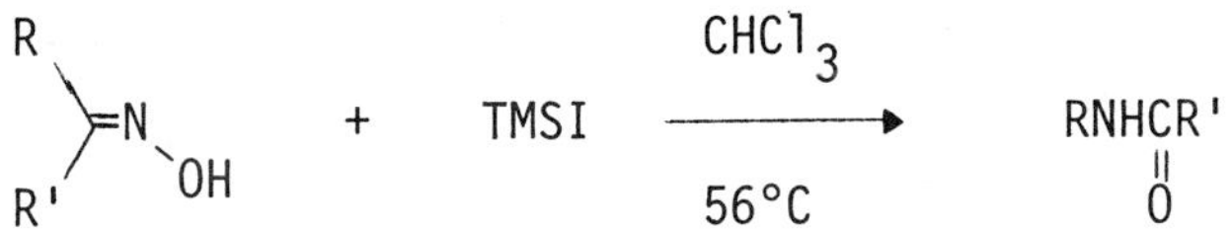

$$\underset{R'}{\overset{R}{>}}C=N_{OH} + TMSI \xrightarrow[56°C]{CHCl_3} RNH\overset{O}{\underset{}{C}}R'$$

R = aryl   55-80%

VI.A.4-5 T. Mukaiyama, J. Ichikawa and M. Asami, Chem. Lett., 683 (1983).

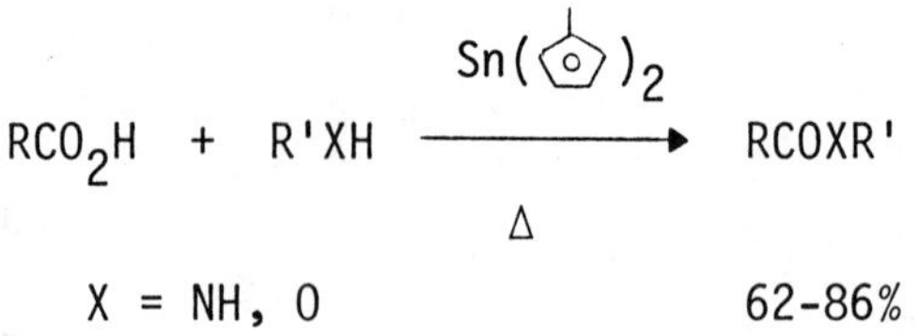

$$RCO_2H + R'XH \xrightarrow[\Delta]{Sn(Cp)_2} RCOXR'$$

X = NH, O   62-86%

VI.A.4-6 H. Suzuki et al., Chem. Lett., 449 (1983).

$$R-COOH + R'-NH_2 \xrightarrow[CCl_4-CH_2Cl_2, \Delta]{P_2I_4} R-CO-NH-R'$$

R^1 = aryl, benzyl   20-100%

VI.A.4-7 M. Ueda and N. Kawaharasaki, Synthesis, 933 (1982).

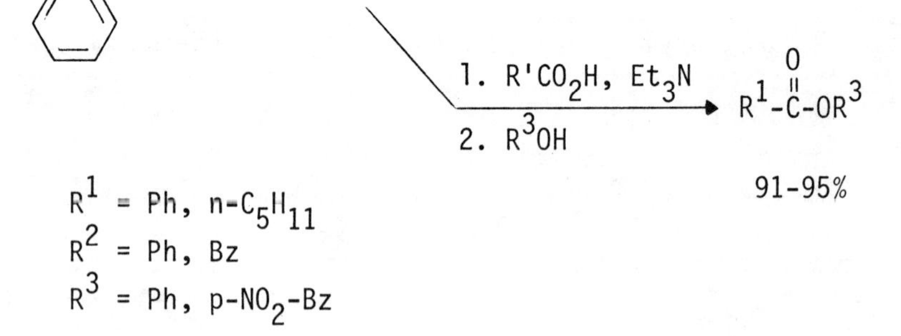

$$\xrightarrow[2.\ R^2NH_2]{1.\ R'CO_2H,\ Et_3N} R^1-\overset{O}{\underset{\|}{C}}-NH-R^2 \quad 81\text{-}93\%$$

$$\xrightarrow[2.\ R^3OH]{1.\ R'CO_2H,\ Et_3N} R^1-\overset{O}{\underset{\|}{C}}-OR^3 \quad 91\text{-}95\%$$

R^1 = Ph, n-C_5H_{11}
R^2 = Ph, Bz
R^3 = Ph, p-NO_2-Bz

VI.A.4-8 R. J. Lahoti and D. R. Wagle, Indian J. Chem. Sect. B, 20B, 1007 (1981).

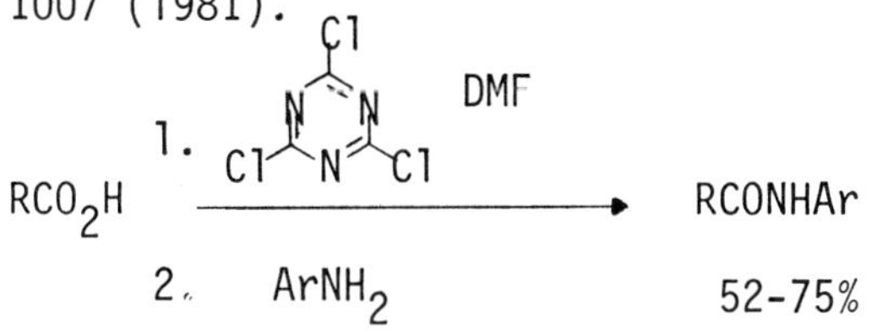

$$RCO_2H \xrightarrow[\text{2. } ArNH_2]{\text{1. } \underset{\text{DMF}}{\text{cyanuric chloride}}} RCONHAr$$

52-75%

VI.A.4-9 T. Kametani, H. Sekine and T. Honda, Heterocycles, 20, 1577 (1983).

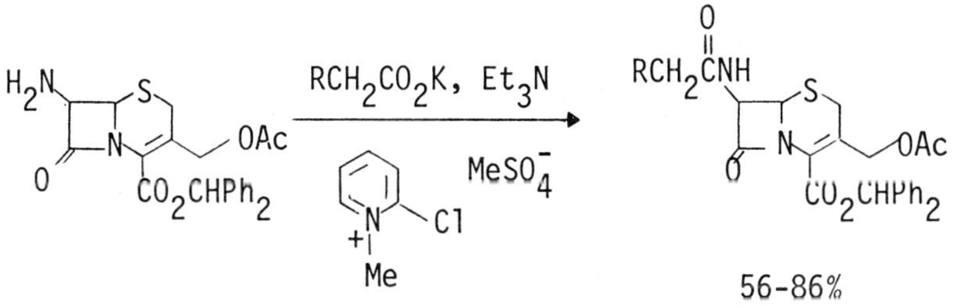

[cephalosporin H_2N-substrate with -OAc and CO_2CHPh_2] $\xrightarrow[\underset{Me}{\overset{}{N^+{-}Cl}} \ MeSO_4^-]{RCH_2CO_2K, \ Et_3N}$ [product: $RCH_2C(O)NH$-cephalosporin with -OAc and CO_2CHPh_2]

56-86%

VI.A.4-10 R. N. Ram, R. Ashare and A. K. Mukerjee, Chem. Ind. (London), 569 (1983).

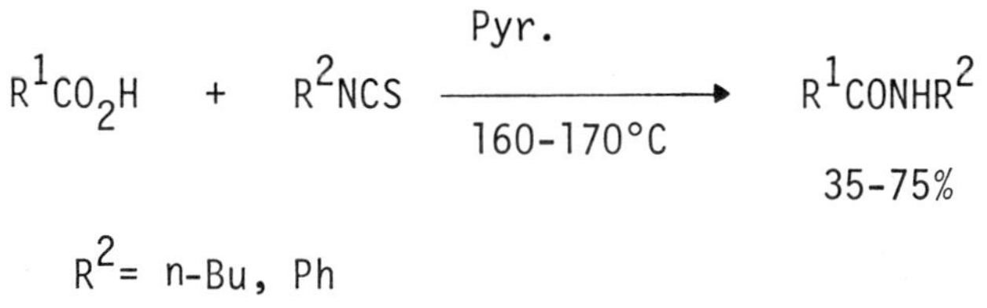

$$R^1CO_2H \ + \ R^2NCS \xrightarrow[160-170°C]{Pyr.} R^1CONHR^2$$

35-75%

R^2 = n-Bu, Ph

VI.A.4-11 J. Barluenga et al., J. Chem. Soc. Perkin Trans. 1, 591 (1983).

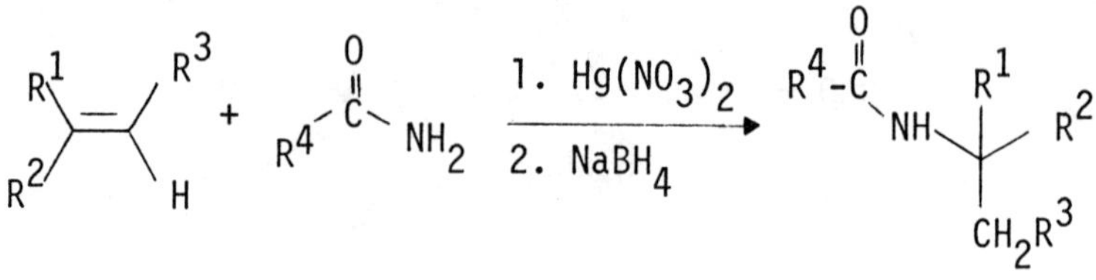

$$R^1R^2C=C(R^3)(H) + R^4C(O)NH_2 \xrightarrow[\text{2. NaBH}_4]{\text{1. Hg(NO}_3)_2} R^4C(O)NH-C(R^1)(R^2)(CH_2R^3)$$

R^4 = alkyl, aryl, alkoxy, NH_2 17-99%

VI.A.4-12 B. M. Trost and W. H. Pearson, J. Am. Chem. Soc., 105, 1054 (1983).

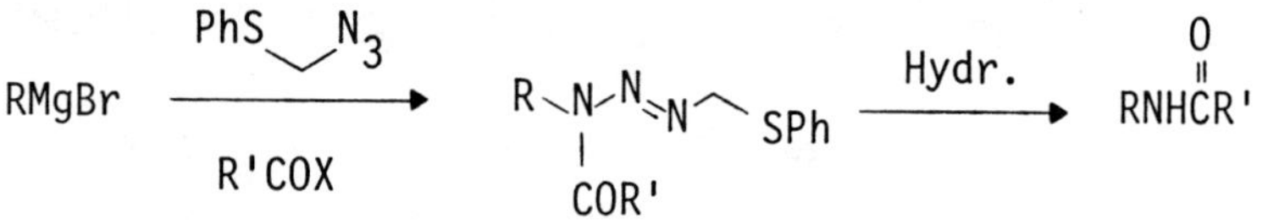

$$RMgBr \xrightarrow[\text{R'COX}]{\text{PhS-CH}_2\text{-N}_3} R-N(COR')-N=N-CH_2-SPh \xrightarrow{\text{Hydr.}} RNHC(O)R'$$

R = alkyl   64-98%

VI.A.4-13 I. Ganboa and C. Palomo, Synth. Commun., 13, 941 (1983).

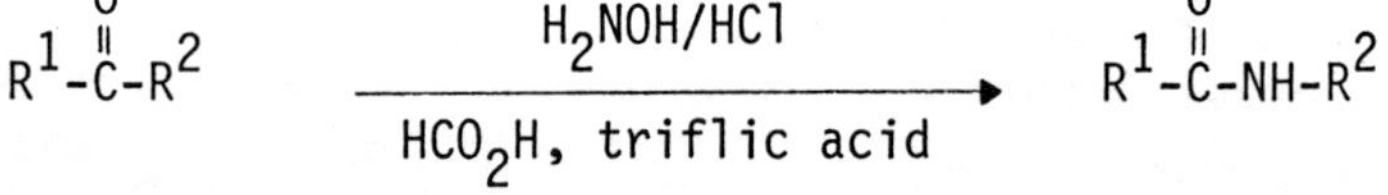

$$R^1-C(O)-R^2 \xrightarrow[\text{HCO}_2\text{H, triflic acid}]{\text{H}_2\text{NOH/HCl}} R^1-C(O)-NH-R^2$$

60-90%

VI.A.4-14   A. Costa, R. Mestres and J. M. Riego, Synth. Commun., 12, 1003 (1982).

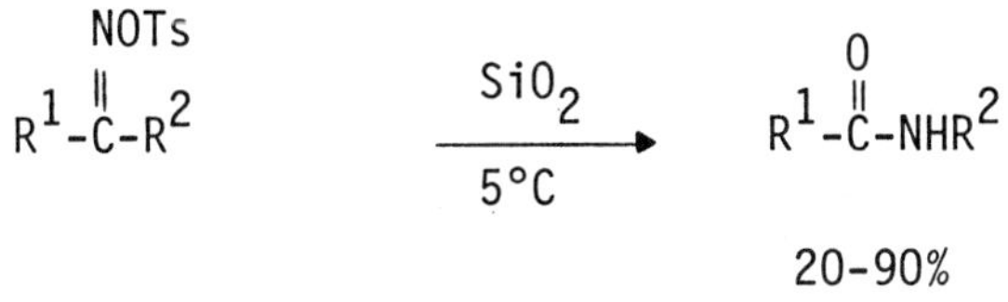

$$R^1-\underset{\underset{NOTs}{\|}}{C}-R^2 \xrightarrow[5°C]{SiO_2} R^1-\underset{\underset{O}{\|}}{C}-NHR^2$$

20-90%

VI.A.4-15   K. S. Kochhar, D. A. Cottrell and H. W. Pinnick, Tetrahedron Lett., 24, 1323 (1983).

$$R-\underset{\underset{S}{\|}}{C}-N\underset{R^2}{\overset{R^1}{\diagdown}} \xrightarrow{m-CPBA} R-\underset{\underset{O}{\|}}{C}-N\underset{R^2}{\overset{R^1}{\diagdown}}$$

72-98%

VI.A.4-16   A. Ohta et al., Heterocycles, 20, 797 (1983).

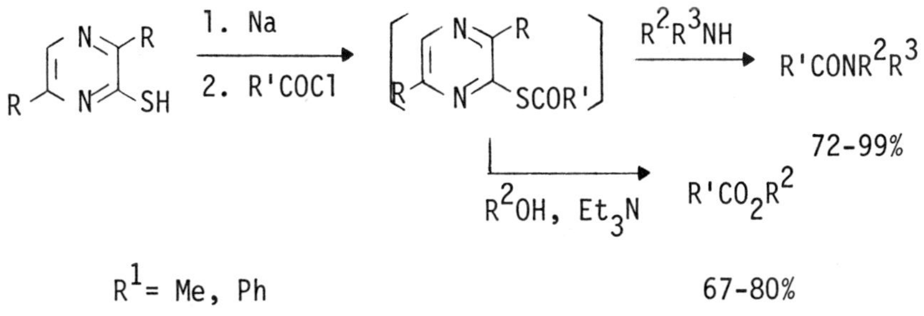

R^1 = Me, Ph

72-99%

67-80%

VI.A.4-17   K. Iizuka, Chem. Pharm. Bull., 30, 4242 (1982).

$$\text{RC(=O)-N} \underset{\underline{}}{\overset{\frown}{}} \text{N} \quad \xrightarrow[\text{2. R'R''NH}]{\text{1. PhCH}_2\text{X}} \quad \text{RC(=O)-NR'R''}$$

80-95%

R = Me, Ph
R' = aryl, alkyl
R'' = alkyl, H

VI.A.4-18   J. I. Levin, E. Turos and S. M. Weinreb, Synth. Commun., 12, 989 (1982).

$$\text{RR'NH} \cdot \text{HCl} \quad \xrightarrow[\text{2. R}^2\text{CO}_2\text{R}^3]{\text{1. (CH}_3)_3\text{Al}} \quad \text{R}^2\text{C(=O)NRR'}$$

R, R' = H, Me   31-100%
R^3 = Me, Et

VI.A.4-19   A. Ricci et al., Synthesis, 319 (1983).

$$(\text{H}_2\text{C})_n\overset{}{\underset{}{\square}}\text{O} \; + \; (\text{H}_3\text{C})_3\text{Sn-N}\genfrac{}{}{0pt}{}{R^1}{R^2} \quad \xrightarrow{\text{H}_2\text{C}(\text{COOH})_2} \quad \text{HO-CH}_2\text{-(CH}_2)_n\text{-C(=O)-N}\genfrac{}{}{0pt}{}{R^1}{R^2}$$

n = 1-4

63-94%

VI.A.5. Amines (see also: III.B.2, III.D)

VI.A.5-1 J. P. Genet et al., Tetrahedron Lett., 24, 2745 (1983).

$$\underset{R^2\ \ R^4}{\overset{R^1\ \ R^3}{RO\diagup\!\!\!\diagdown OAc}} \quad \xrightarrow[\text{Pd}^\circ - \text{Phosphine}]{\underset{R^6}{\overset{R^5}{\diagdown}}\!\!\text{CH-NH}} \quad \underset{R^3\ \ R^4}{\overset{R^1\ \ R^2}{RO\diagup\!\!=\!\!\diagdown NR^5R^6}}$$

R = H, Ac, THP

45-89%

VI.A.5-2 M. J. Calverley, Synth. Commun., 13, 601 (1983).

$$R\text{-}NH_2 \xrightarrow[Lt_3N]{Me_2ClSi\!+\!\!\!\!+} R\text{-}NHSiMe_2\!\!+\!\!\!\!+ \xrightarrow[\substack{2.\ R'I \\ 3.\ H^+}]{1.\ n\text{-}BuLi} R\text{-}NHR'$$

65-91%

VI.A.5-3 M. Kosugi, M. Kameyama and T. Migita, Chem. Lett., 927 (1983).

$$n\text{-}Bu_3SnNEt_2 + ArBr \xrightarrow[\text{in PhCH}_3]{PdCl_2(o\text{-}tolyl_3P)_2} ArNEt_2$$

16-81%

VI.A.5-4 G. Boche, M. Bernheim and W. Schrott, Tetrahedron Lett., 23, 5399 (1982).

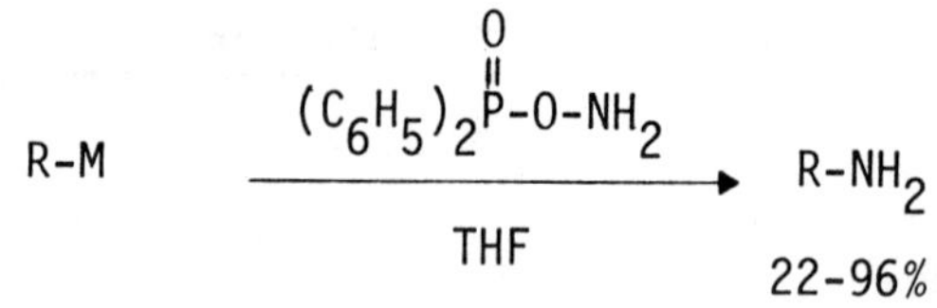

$$R-M \xrightarrow[\text{THF}]{(C_6H_5)_2\overset{\text{O}}{\underset{\|}{P}}-O-NH_2} R-NH_2 \quad 22\text{-}96\%$$

M = Li, MgX
R = aryl, benzyl

VI.A.5-5 G. Boche and W. Schrott, Tetrahedron Lett., 23, 5403 (1982).

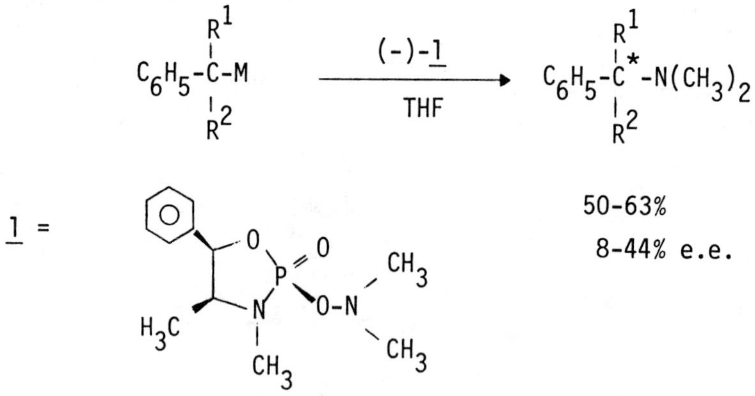

$$C_6H_5-\underset{R^2}{\overset{R^1}{\underset{|}{\overset{|}{C}}}}-M \xrightarrow[\text{THF}]{(-)-\underline{1}} C_6H_5-\underset{R^2}{\overset{R^1}{\underset{|}{\overset{|}{C^*}}}}-N(CH_3)_2$$

50-63%
8-44% e.e.

$\underline{1}$ = [oxazaphospholidine with phenyl, N-CH₃, C-CH₃, P=O, O-N(CH₃)₂ substituents]

VI.A.5-6 B. J. Kokko and P. Beak, Tetrahedron Lett., 24, 561 (1983).

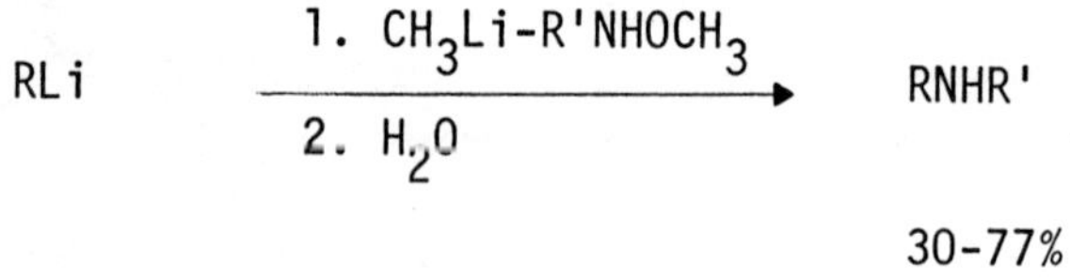

$$RLi \xrightarrow[\text{2. } H_2O]{\text{1. } CH_3Li\text{-}R'NHOCH_3} RNHR'$$

30-77%

VI.A.5-7 K. A. Parker and R. P. O'Fee, J. Org. Chem., 48, 1547 (1983).

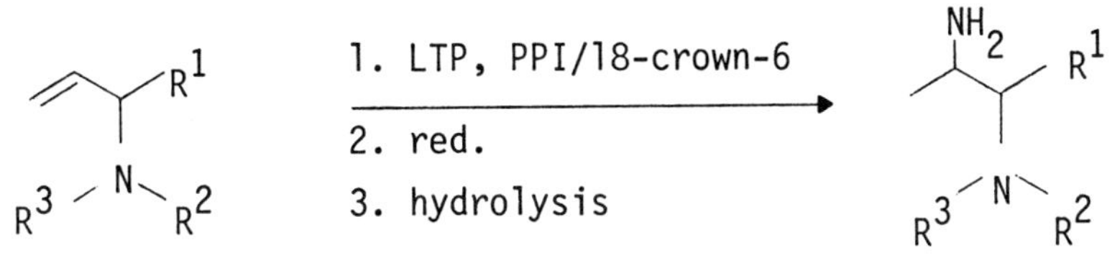

$$\text{CH}_2=\text{CH-CHR}^1\text{-N(R}^2\text{)R}^3 \xrightarrow[\substack{\text{2. red.} \\ \text{3. hydrolysis}}]{\text{1. LTP, PPI/18-crown-6}} \text{H}_2\text{N-CH(CH}_3\text{)-CHR}^1\text{-N(R}^2\text{)R}^3$$

LTP = Lithium tetrachloropalladate
PPI = Potassium phthalimide   60-99%

VI.A.5-8 T. S. Manoharan et al., Synthesis, 809 (1983).

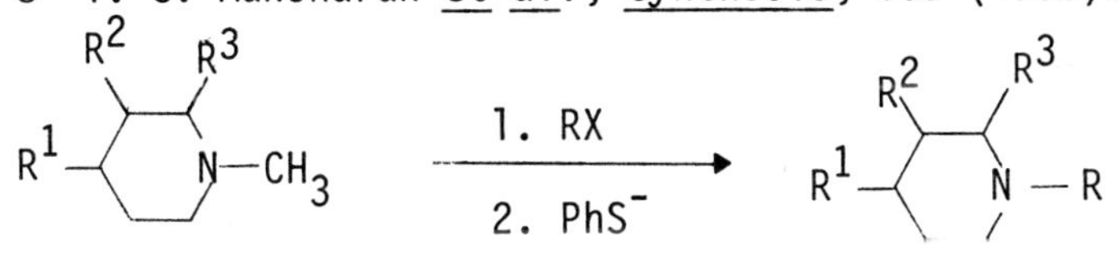

$$\underset{\text{piperidine-}R^1,R^2,R^3\text{-N-CH}_3}{} \xrightarrow[\text{2. PhS}^-]{\text{1. RX}} \underset{\text{ring-opened } R^1,R^2,R^3\text{-N-R}}{}$$

40-75%

VI.A.5-9 S. Sasatani et al., Tetrahedron Lett., 24, 4711 (1983).

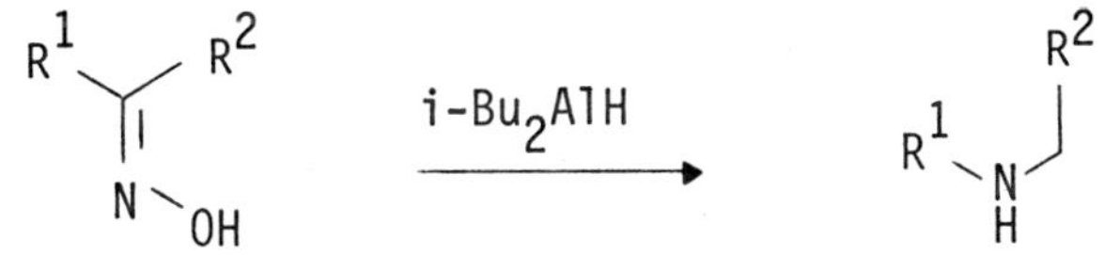

$$R^1\text{-C(=N-OH)-}R^2 \xrightarrow{i\text{-Bu}_2\text{AlH}} R^1\text{-NH-CH}_2\text{-}R^2$$

70-92%

VI.A.5-10 K. Suzuki et al., Synthesis, 723 (1983).

$$R^1-\underset{R^2}{\underset{|}{C}}-\overset{Cl}{\underset{}{}}\overset{O}{\underset{}{\overset{\|}{C}}}-N\overset{R^3}{\underset{R^4}{\diagdown}} \xrightarrow{LiAlH_4} R^1-\underset{R^2}{\underset{|}{C}}-\overset{CH_3}{\underset{}{}}N\overset{R^3}{\underset{R^4}{\diagdown}}$$

R^1 = alkyl, benzyl
R^2 = H, Me

15-76%

VI.A.5-11 A. S. Radhakrishna et al., Synthesis, 538 (1983).

$$\text{C}_6\text{H}_5\text{-I=O} \xrightarrow[\text{2. R-C(=O)-NH}_2]{\text{1. HCO}_2\text{H, H}_2\text{O}} \text{R-NH}_2$$

70-90%

R = alkyl, benzyl

VI.A.5-12 J. R. Pfister and W. E. Wymann, Synthesis, 38 (1983).

$$R-\overset{O}{\underset{}{\overset{\|}{C}}}-N_3 \xrightarrow[\text{MeOH}]{\substack{\text{1. F}_3\text{CCO}_2\text{H, CH}_2\text{Cl}_2 \\ \text{2. K}_2\text{CO}_3\text{, H}_2\text{O}}} R-NH_2$$

80-97%

VI.A.5-13   Z. Eckstein, E. Lipczynska-Kochany and
J. Krzeminski, Heterocycles, 20, 1899 (1983).

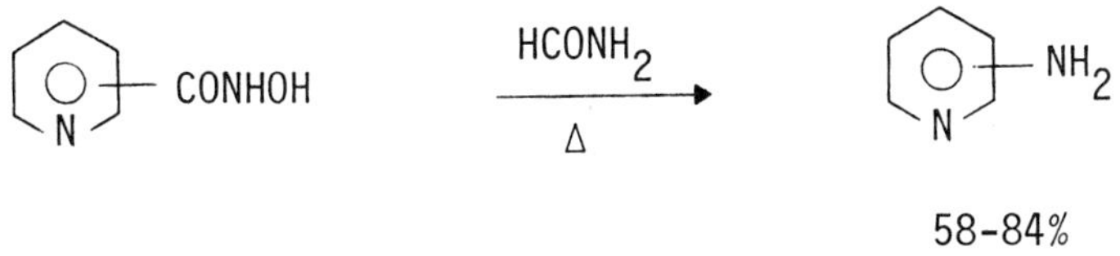

$$\text{Py-CONHOH} \xrightarrow[\Delta]{HCONH_2} \text{Py-NH}_2$$

58-84%

VI.A.5-14   J. T. Gupton et al., J. Org. Chem., 48, 2933 (1983).

$$ArX \xrightarrow{HMPA} ArN(CH_3)_2$$

Ar = activated aryl
X = halide, NO_2

VI.A.5-15   R. B. Cheikh et al., Synthesis, 685 (1983).

Review: "Synthesis of Primary Allylic Amines."

VI.A.5-16   M. B. Gasc, A. Lattes and J. J. Perie, Tetrahedron, 39, 703 (1983).

Review: "Amination of Alkenes."

VI.A.6. Amino Acids and Derivatives   (see also: III.E.1, VI.A.4, VI.A.10)

VI.A.6-1 R. M. Freidinger et al., J. Org. Chem., **48**, 77 (1983).

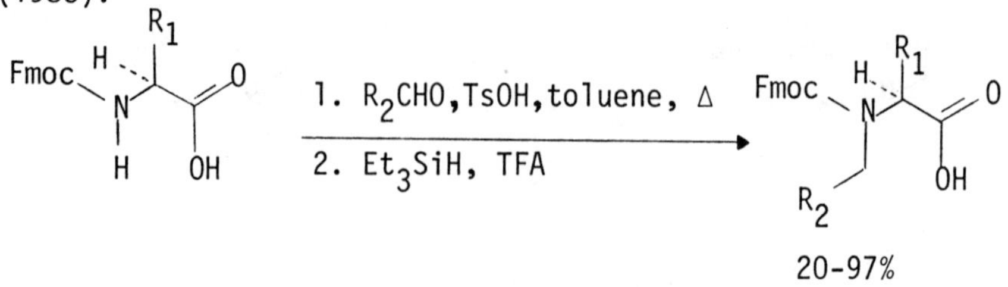

[Reaction scheme: Fmoc-NH-CH(R_1)-COOH reacts with 1. R_2CHO, TsOH, toluene, Δ; 2. Et_3SiH, TFA to give Fmoc-N(CH_2R_2)-CH(R_1)-COOH, 20-97%]

R_1 = alkyl
R_2 = alkyl, H

VI.A.6-2 F. Effenberger, U. Burkard and J. Willfahrt, Angew. Chem. Int. Ed. Engl., **22**, 65 (1983).

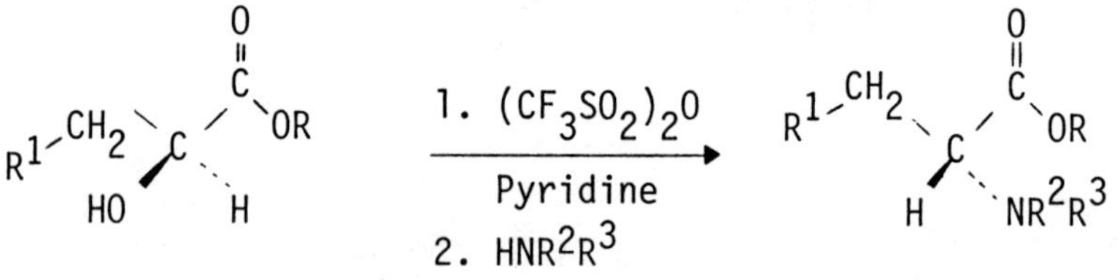

[Reaction scheme: R^1-CH_2-C(OH)(H)-COOR reacts with 1. $(CF_3SO_2)_2O$, Pyridine; 2. HNR^2R^3 to give R^1-CH_2-C(H)(NR^2R^3)-COOR, 75-94%]

R = Me, Et
R^1 = H, Ph, CO_2Et
R^2 = H, Me

VI.A.6-3 K. Nakajima et al., Bull. Chem. Soc. Jpn., 55, 3049 (1982).

$$\begin{array}{c} R_1 \\ | \\ CH \\ Z-N \diagdown \\ | * \\ CH-CO_2-R_2 \end{array} \xrightarrow[BF_3 \cdot OEt_2]{R_3-OH} \begin{array}{c} R_1-CH-O-R_3 \\ | \\ Z-NH-CH-CO_2-R_2 \\ * \end{array}$$

27-100%

R_1 = H, Me

R_2 = H, Me, Bz

VI.A.6-4 P. Scrimin, F. D'Angeli and G. Cavicchioni, Synthesis, 1092 (1982).

$$R^1-\underset{Br}{\underset{|}{C}}-\overset{CH_3}{\underset{}{\overset{|}{C}}}\overset{O}{\diagup}_{NH-R^2} + HN\diagdown_{R^4}^{R^3} \xrightarrow[NaOH/H_2O/CH_2Cl_2/(n-C_4H_9)_4 \overset{+}{N} Br^-]{NaH/THF\ or} R^1-\underset{|}{\overset{CH_3}{\overset{|}{C}}}-\overset{O}{\underset{N}{\overset{\diagup}{C}}}\diagdown_{NH-R^2}$$

R^1 = Me, Et

R^2 = Bz, Me, Ph, t-Bu

66-88%

VI.A.6-5 R. Andruszkiewicz and A. Czerwinski, Synthesis, 968 (1983).

$$\begin{array}{c} R^2-CH-OH \\ R^1-NH-CH-COOCH_3 \end{array} \xrightarrow[(C_2H_5)_3N]{\underset{O}{\overset{N\diagdown N\diagup C\diagdown N\diagup N}{||}}} \begin{array}{c} R^2-CH \\ | \\ R^1-NH-\underset{||}{C}-COOCH_3 \\ O \end{array}$$

R^2 = H, Me

65-85%

VI.A.6-6 L. Somekh and A. Shanzer, J. Org. Chem., 48, 907 (1983).

$$\underset{R^1}{\overset{HO}{\underset{R^2\cdots}{\text{C}}}}-\underset{NHR^4}{\overset{H}{\text{C}}}-CO_2R^3 \xrightarrow[CH_2Cl_2,\ 0°C]{Et_2NSF_3,\ Pyr.} \underset{R^1}{\overset{R^2}{\text{C}}}=\underset{NHR^4}{\overset{CO_2R^3}{\text{C}}}$$

$R^1, R^2 = H, Me, CHMe_2$ 65-90%
$R^3 = CH_2Ph, CH_2Me$
$R^4 = CO_2Bz, CO_2\text{-}t\text{-}Bu$

VI.A.6-7 T. Kolasa, Synthesis, 539 (1983).

$$\underset{OH}{\underset{|}{\text{Ac-N-CH-COOR}^2}}\overset{R^1\text{-}CH_2}{\underset{|}{}} \xrightarrow{TosCl/(C_2H_5)_3N} \underset{}{Ac\text{-}NH\text{-}\overset{R^1\text{-}CH}{\underset{\|}{C}}\text{-}COOR^2}$$

82-97%

$R^1 = H, CO_2Me, CO_2Et$
$R^2 = Me, Et, t\text{-}Bu$

VI.A.6-8 V. Rachina and I. Blagoeva, Synthesis, 967 (1982).

$$\underset{R^1\ \ O}{\overset{R^3}{\text{ring}}} \xrightarrow{NaOH/H_2O} \xrightarrow[\text{cation exchanger}]{H^+\ \text{(strongly acidic)}} \underset{R^1}{\overset{R^3}{\underset{|}{R^2\text{-}C\text{-}NH_3^+}}}\\ HC\text{-}COO^-$$

$R^1, R^2 = H, Me$
$R^3 = H, Me, Ph$ 85-90%

USEFUL SYNTHETIC PREPARATIONS

VI.A.8. Enamines

VI.A.8-1 N. De Kimpe and N. Schamp, <u>Org. Prep. Proced. Int.</u>, **15**, 71 (1983).

Review: "Reactivity of β-Haloenamines."

VI.A.9. Epoxides   (see also: II.F.1, IV.H)

VI.A.9-1 S. Sebti and A. Foucaud, <u>Synthesis</u>, 546 (1983).

$$CH_3-\underset{\underset{O}{\|}}{C}-\underset{\underset{Cl}{|}}{CH}-R^2 \;+\; R^3-CO_2H \;+\; R^4-NC \longrightarrow$$

$$R^3-CO_2-\underset{\underset{R^2}{|}}{\overset{\overset{CH_3}{|}}{C}}-\underset{\underset{}{|}}{\overset{\overset{O}{\|}}{C}}-NH-R^4 \quad \xrightarrow[\text{THF}]{\text{KOH powder}} \quad \underset{H}{\overset{R^2}{\diagup}}\!\!\!\underset{O}{\triangle}\!\!\!\overset{CH_3}{\diagdown}\underset{\underset{O}{\|}}{C}-NHR^4$$

         70-90%   91-98%

R^2 = H, Me

R^3 = Me, Ph

R^4 = t-Bu, i-Pr

VI.A.9-2 C. N. Barry and S. A. Evans Jr., Tetrahedron Lett., 24, 661 (1983).

$$\begin{array}{c} R \\ \diagdown \\ R^1\diagup CHCH_2OH \\ | \\ OH \end{array} \xrightarrow{TPP-CCl_4-K_2CO_3} \begin{array}{c} R \\ \diagdown \\ CH-CH_2 \\ R^1\diagup \diagdown\diagup \\ O \end{array}$$

R, R^1 = alkyl $\qquad\qquad\qquad\qquad$ 45-50%

VI.A.10. Esters

VI.A.10-1 R. A. Amos, R. W. Emblidge and N. Havens, J. Org. Chem., 48, 3598 (1983).

$$\text{(P)}-\underset{Ph}{\overset{Ph}{P}} + ROH + R'CO_2H + EtO_2CN=NCO_2Et \xrightarrow{THF}$$

$$R'-CO_2-R + EtO_2CNHNHCO_2Et + \text{(P)}-\underset{Ph}{\overset{Ph}{P}}=O$$

R = alkyl, benzyl $\qquad\qquad\qquad\qquad$ 64-99%

VI.A.10-2 V. Balasubramaniyan, V. G. Bhatia and S. B. Wagh, Tetrahedron, 39, 1475 (1983).

$$RCO_2H \xrightarrow[1:6]{P_4O_{10}:R'OH} RCO_2R^1$$

VI.A.10-3 M. A. Brook and T. H. Chang, Synthesis, 201 (1983).

$$R^1\text{-}CO_2H \xrightarrow[\text{2. ClSiMe}_3]{\text{1. } R^2OH} R^1CO_2R^2$$

R^2 = Me, Et, Bz

67-98%

VI.A.10-4 A. Arrieta et al., Synth. Commun., 13, 471 (1983).

$$R\text{-}CO_2H \xrightarrow[\substack{POCl_3 \text{ or} \\ PhOP(O)Cl_2}]{R^1XH,} R\text{-}COXR^1$$

X = O, S

60-95%

VI.A.10-5 S. Takimoto et al., Bull. Chem. Soc. Jpn., 56, 639 (1983).

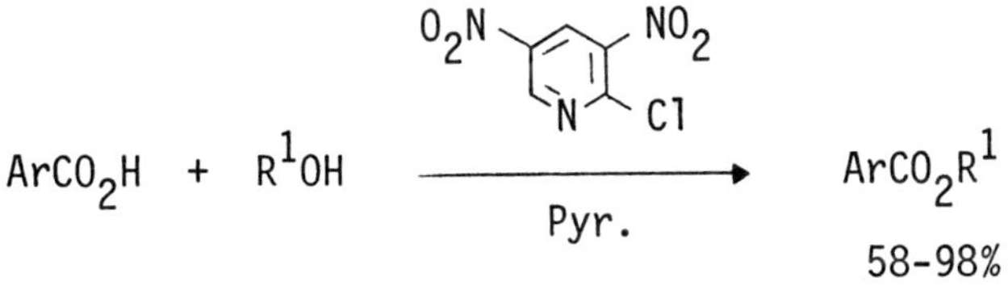

$$ArCO_2H + R^1OH \xrightarrow[\text{Pyr.}]{\underset{N}{\overset{O_2N\diagup\diagdown NO_2}{\diagdown\diagup}}Cl} ArCO_2R^1$$

58-98%

R^1 = alkyl, benzyl

VI.A.10-6 S. Kim, Y. C. Kim and J. I. Lee, Tetrahedron Lett., 24, 3365 (1983).

$$\text{RCOOH} + \text{ClCOOR'} + \text{Et}_3\text{N} \xrightarrow[\text{CH}_2\text{Cl}_2,\ 0°\text{C}]{\text{cat. DMAP}} \text{RCOOR'}$$

R = alkyl, aryl, benzyl
R' = alkyl, benzyl

47-98%

VI.A.10-7 J. Barluenga *et al.*, Synthesis, 649 (1983).

1. HgO/HX/solvent, 20 or 80°C, 1-3 h.
2. Na_2CO_3
3. 1 normal aq. KOH

$$R^1\text{-COOH} + R^2\text{-Br} \longrightarrow R^1\text{-COOR}^2$$

32-79%

X = F, BF_4
R^2 = alkyl, benzyl

VI.A.10-8 M. S. Manhas *et al.*, Synthesis, 549 (1983).

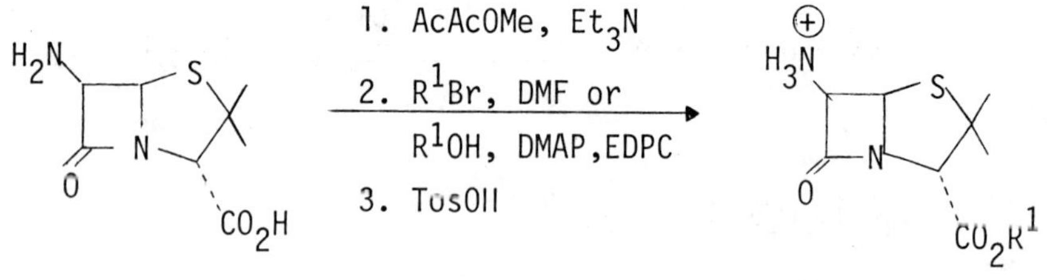
H2N
S
N
O
CO2H
1. AcAcOMe, Et3N
2. R1Br, DMF or
R1OH, DMAP,EDPC
3. TosOH
H3N
S
N
O
CO2R1

EDPC = Et-N=C=N ~~~ NMe_2

45-65%

VI.A.10-9   T. Kageyama et al., Chem. Lett., 1097 (1983).

$$2\ RCO_2H \xrightarrow[\text{AcOH(aq.)}]{NaBrO_2} RCO_2CH_2R$$

R = alkyl, benzyl

71-96%

VI.A.10-10   T. Mukaiyama, J. Ichikawa and M. Asami, Chem. Lett., 293 (1983).

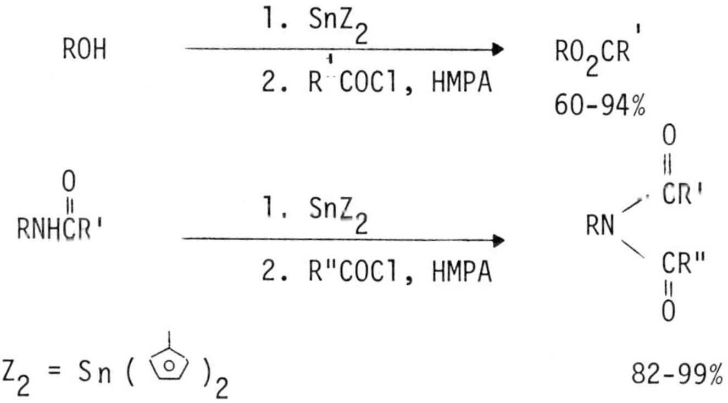

$$ROH \xrightarrow[\text{2. R'COCl, HMPA}]{\text{1. }SnZ_2} RO_2CR'$$

60-94%

$$RNHCR' \xrightarrow[\text{2. R''COCl, HMPA}]{\text{1. }SnZ_2} RN(COR')(COR'')$$

$$SnZ_2 = Sn(\text{pyrrolyl})_2$$

82-99%

VI.A.10-11   A. Arrieta, T. Garcia and C. Palomo, Synth. Commun., 12, 1139 (1982).

$$R^1CO_2H \xrightarrow{SOCl_2/DMAP} R^1\text{-C(=O)-Cl} \xrightarrow[-20°C]{DMAP/R^2OH/} R^1\text{-C(=O)-}OR^2$$

60-92%

VI.A.10-12 J. W. Fitch, W. G. Payne and D. Westmoreland, J. Org. Chem., 48, 751 (1983).

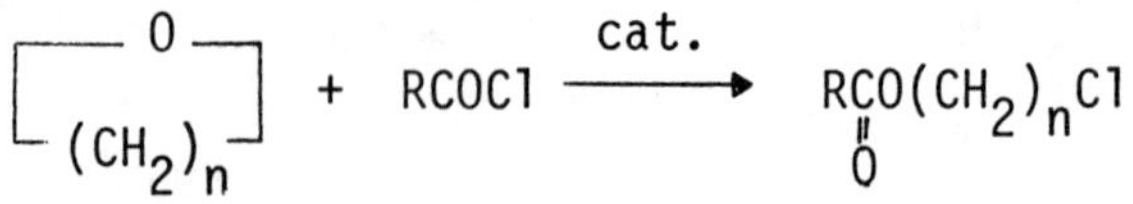

$$\left[\begin{array}{c} O \\ (CH_2)_n \end{array}\right] + RCOCl \xrightarrow{cat.} R\underset{\underset{O}{\|}}{C}O(CH_2)_nCl$$

47-83%

n = 3-5
R = Me, Ph

cat. = $K[Pt(C_2H_4)Cl_3]$, $[Pt(C_2H_4)Cl_2]_2$

VI.A.10-13 T. Kamijo et al., Chem. Pharm. Bull., 30, 4242 (1982).

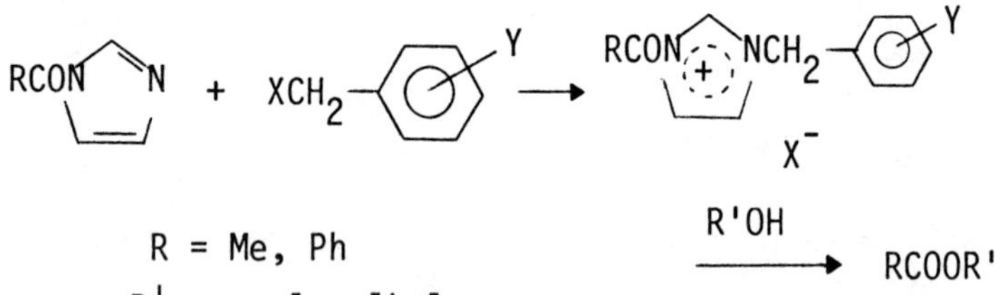

$$RCON\underset{\diagdown}{\diagup}N + XCH_2-\!\!\!\bigcirc\!\!\!-Y \longrightarrow RCON\underset{\diagdown}{\overset{+}{\diagup}}NCH_2-\!\!\!\bigcirc\!\!\!-Y \quad X^-$$

R = Me, Ph
R' = aryl, alkyl

$\xrightarrow{R'OH}$ RCOOR'

75-99%

VI.A.10-14 J. Gallos and A. Varvoglis, J. Chem. Soc. Perkin Trans. 1, 1999 (1983).

$$R^1I + PhI(O_2CR)_2 \xrightarrow{CHCl_3} RCO_2R^1$$

R^1 = alkyl
R = CF_3, CH_3, Ph

32-100%

VI.A.10-15 S. D. Higgins and C. B. Thomas, J. Chem. Soc. Perkin Trans. 1, 1483 (1983).

$$\text{Ar-}\underset{\underset{}{\|}}{\text{C}}\text{-CH}_2\text{R} \xrightarrow[\text{3. R}^1\text{CO}_3\text{H}]{\substack{\text{1. ICl}\\ \text{2. HC(OMe)}_3}} \text{Ar-}\underset{\underset{R}{|}}{\text{CH}}\text{-CO}_2\text{Me}$$

VI.A.10-16 A. Goosen and C. W. McCleland, J. Chem. Soc. Chem. Commun., 1311 (1982).

R^1 = H, Me
R^2 = H, Me, I

82-93%

VI.A.10-17 C. Giordano et al., J. Chem. Soc. Perkin Trans. 1, 2575 (1982).

$$\text{Ar-}\underset{\underset{R^1}{|}}{\underset{\|}{\overset{O}{\text{C}}}}\text{-CH-X} \xrightarrow[\text{2. AgY}]{\substack{\text{1. H}^+, \text{R}^2\text{OH}\\ \text{HC(OR}^2)_3}} \text{Ar-}\underset{\underset{R^1}{|}}{\text{CH}}\text{-CO}_2\text{R}^2$$

R^1 = H, Me
R^2 = Me, Et
Y = BF$_3$, NO$_3$

VI.A.11. Ethers (see also: V.A)

VI.A.11-1 J. Barluenga et al., Synthesis, 53 (1983).

$$R^1\text{-Br} + R^2\text{-OH} \xrightarrow[\text{2. 3N KOH}]{\text{1. HgO/TBFA}} R^1\text{-O-}R^2 \quad 55\text{-}80\%$$

R^1, R^2 = alkyl

VI.A.11-2 A. Afzali, H. Firouzabadi and A. Khalafi-nejad, Synth. Commun., 13, 335 (1983).

$$Ar^1OH + Ar^2Br \xrightarrow[\text{2. CuC≡CPh}]{\text{1. }\Delta\text{, Pyr.}} Ar^1OAr^2 \quad 41\text{-}77\%$$

VI.A.11-3 F. Ogura et al., Bull. Chem. Soc. Jpn., 56, 1257 (1983).

$$O=C(CH_2SO_2Cl)_2 \xrightarrow{ROH} O=C(CH_2SO_3R)_2 \xrightarrow{PhOH} PhOR$$

R = Me : 97%
R = Et : 82%

VI.A.11-4 M. D. Bachi and A. Gross, J. Chem. Soc. Perkin Trans. 1, 1157 (1983).

$$\underset{O}{\overset{R}{\underset{\diagdown}{\bigsqcup}}}\overset{Y}{\underset{NH}{\diagup}} \xrightarrow{R^1OH, \Delta} \underset{O}{\overset{R}{\underset{\diagdown}{\bigsqcup}}}\overset{OR^1}{\underset{NH}{\diagup}}$$

R = Phth, Cl
Y = S(O)Me, S(O)CH$_2$CH$_2$CO$_2$Me,

$S-\underset{N}{\overset{S}{\diagdown}}\bigcirc$

VI.A.12. Ketones and Aldehydes (see also: I.A.2, II.A.1, III.F.1)

VI.A.12-1 K. Uneyama et al., Tetrahedron Lett., 24, 2857

$$R_3 \overset{R_2}{\underset{O}{\diagdown}} R_1 \xrightarrow{\text{Electrogenerated Acid}} R_3 \overset{R_2}{\underset{O}{\diagdown}} R_1$$

67-91%

VI.A.12-2 D. Momose and Y. Yamada, Tetrahedron Lett., 24, 2669 (1983).

$$R \overset{OH}{\underset{Br}{\diagdown}} R^1 \xrightarrow[Et_3N]{CoCl(PPh_3)_3} R \overset{O}{\diagdown} R^1$$

85-90%

VI.A.12-3   K. A. Jorgensen, M. T. El-Wassimy and
S.-O. Lawesson, Tetrahedron, 39, 469 (1983).

$$R^1\underset{S}{\overset{}{-}}\!\!\!\!\!\!C\!\!-\!\!R^2 \xrightarrow[\text{or piperidine-N-NO}]{t\text{-BuOCl}} R^1\underset{O}{\overset{}{-}}\!\!\!\!\!\!C\!\!-\!\!R^2$$

varying yields

when $R^2 = NH_2 \rightarrow$ cyclized products.

VI.A.12-4   H. Firouzabadi and A. Sardarian, Synth. Commun., 13, 863 (1983).

$$R^1\underset{}{\overset{NOH}{-}}\!\!\!\!\!\!C\!\!-\!\!R^2 \xrightarrow{BPSP} R^1\underset{}{\overset{O}{-}}\!\!\!\!\!\!C\!\!-\!\!R^2$$

BPSP = bispyridinesilver permanganate

50-90%

VI.A.12-5   T. Shono and S. Kashimura, J. Org. Chem., 48, 1939 (1983).

$$R\!\!-\!\!\underset{OAc}{C}\!\!=\!\!\underset{R^1}{\overset{R^2}{C}} \xrightarrow[\begin{array}{l}2.\ NaBH_4,\ MeOH\\3.\ TsCl\\4.\ acetone\ (aq.),\ \Delta\end{array}]{1.\ -e,\ AcOH\text{-}MeOH,\ Et_4NOTs} R\!\!-\!\!CH_2\!\!-\!\!\underset{}{\overset{R^2}{C}}\!\!=\!\!\underset{O}{\overset{}{C}}\!\!-\!\!R^1$$

30-75% overall

VI.A.12-6 V. V. Kane et al., Tetrahedron, 39, 345 (1983).

Review: "The Chemistry of 1,2-Carbonyl Transposition."

VI.A.12-7 D. G. Morris, Chem. Soc. Rev., 11, 397 (1982).

Review: "Carbonyl Group Transpositions."

VI.A.12-8 D. J. Ager, Chem. Soc. Rev., 11, 493 (1982).

Review: "Silicon-containing Carbonyl Equivalents."

VI.A.12-9 P. Brownbridge, Synthesis, 1 (1983).

Review: "Silyl Enol Ethers in Synthesis - Part I."

VI.A.12-10 P. Brownbridge, Synthesis, 85 (1983).

Review: "Silyl Enol Ethers in Synthesis - Part II."

VI.A.12-11 J. K. Whitesell and M. A. Whitesell, Synthesis, 517 (1983).

> Review: "Alkylation of Ketones and Aldehydes via Their Nitrogen Derivatives."

VI.A.12-12 F. Serratosa, Acc. Chem. Res., 16, 170 (1983).

> Review: "Acetylene Diethers: A Logical Entry to Oxocarbons."

VI.A.13. Nitriles

VI.A.13-1 G. A. Olah et al., J. Org. Chem., 48, 2766 (1983).

$$ArCH_2NO_2 \xrightarrow{Me_3SiI} Ar\text{-}CN$$
$$90\text{-}96\%$$

VI.A.13-2 H. Shinozaki, M. Imaizumi and M. Tajima, Chem. Lett., 929 (1983).

$$R\text{-}CH=NOH \xrightarrow[n\text{-}Bu_4N^+HSO_4^-,\ aq.\ NaOH]{CS_2,\ C_6H_6} R\text{-}CN$$
$$64\text{-}90\%$$

VI.A.13-3 O. Attanasi, P. Palma and F. Serra-Zanetti, Synthesis, 741 (1983).

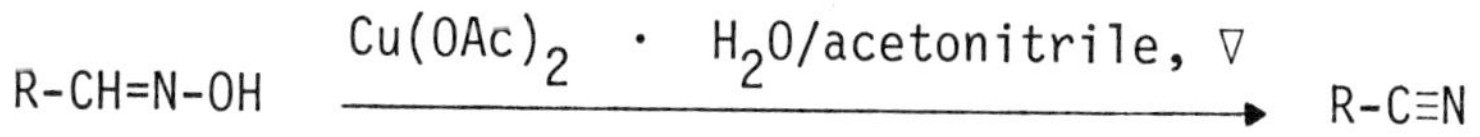

$$R-CH=N-OH \xrightarrow{Cu(OAc)_2 \cdot H_2O/acetonitrile, \nabla} R-C\equiv N$$

80-98%

VI.A.13-4 A. Saednya, Synthesis, 748 (1983).

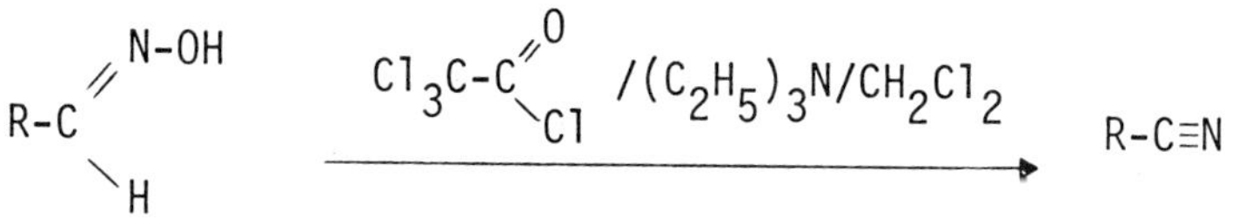

$$R-C\underset{H}{\overset{N-OH}{\diagup\!\!\!\diagdown}} \xrightarrow{Cl_3C-C(\!=\!O)Cl \ /(C_2H_5)_3N/CH_2Cl_2} R-C\equiv N$$

75-95%

VI.A.13-5 A. Arrieta and C. Palomo, Synthesis, 472 (1983).

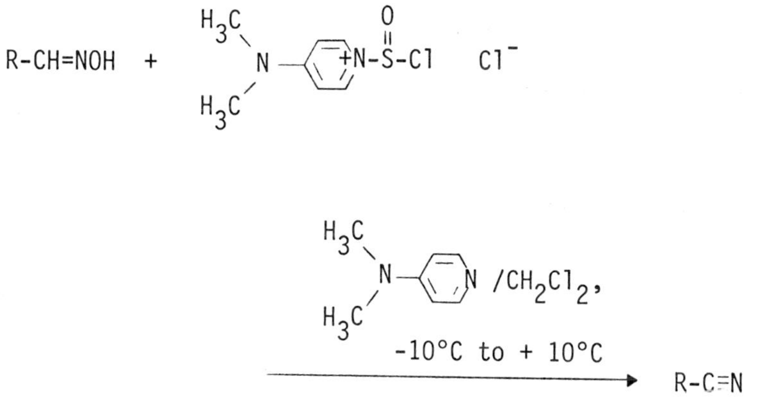

$$R-CH=NOH + (H_3C)_2N-C_6H_4-\overset{+}{N}-\underset{O}{\overset{\parallel}{S}}-Cl \quad Cl^-$$

$$\xrightarrow[-10°C \text{ to } +10°C]{(H_3C)_2N-C_6H_4-N \ /CH_2Cl_2,} R-C\equiv N$$

70-100%

VI.A.13-6 M. E. Jung and Z. Long-Mei, Tetrahedron Lett., 24, 4533 (1983).

$$RCH=NOH + Me_3SiI \xrightarrow[CHCl_3]{(TMS)_2NH \quad 56°C} RCN$$

R = alkyl, benzyl

84-88%

VI.A.13-7 I. Ganboa and C. Palomo, Synth. Commun., 13, 219 (1983).

$$Ar-CHO \xrightarrow{H_2NOH \cdot HCl/MgSO_4/TosOH} Ar-C\equiv N$$

65-97%

VI.A.13-8 I. Ganboa and C. Palomo, Synth. Commun., 13, 999 (1983).

Method A: PPA/AcOH/NH$_2$OH · HCl
ArCHO ⎯⎯⎯⎯⎯⎯⎯⎯⎯⎯⎯⎯⎯⎯⎯⎯⎯⎯⎯⎯⎯⎯→ ArCN
Method B: PPA/AcOH/CH$_3$NO$_2$

PPA = polyphosphoric acid   70-90%

VI.A.13-9 T. Imamoto, T. Takaoka and M. Yokoyama, Synthesis, 142 (1983).

$$R\text{-}COOH + NH_3 \xrightarrow{PPE/CHCl_3} R\text{-}C\equiv N$$

61-90%

PPE = ethyl polyphosphate

VI.A.14. Nitro-compounds

VI.A.14-1 N. Kornblum, H. K. Singh and W. J. Kelly, J. Org. Chem., 48, 332 (1983).

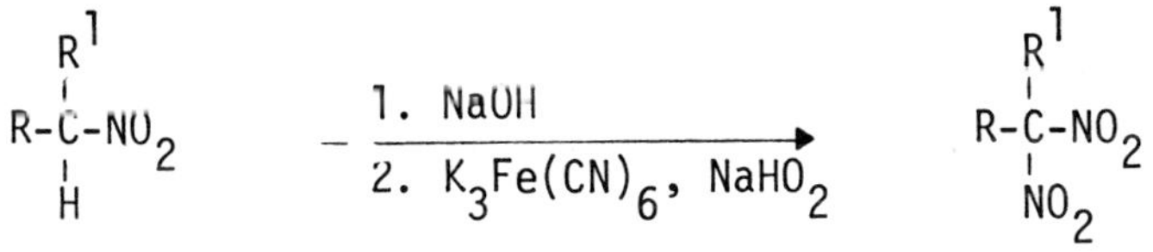
R1
R-C-NO2
H
1. NaOH
2. K3Fe(CN)6, NaHO2
R1
R-C-NO2
NO2

64-90%

VI.A.14-2 P. Dampawan and W. W. Zajac, Jr., Synthesis, 545 (1983).

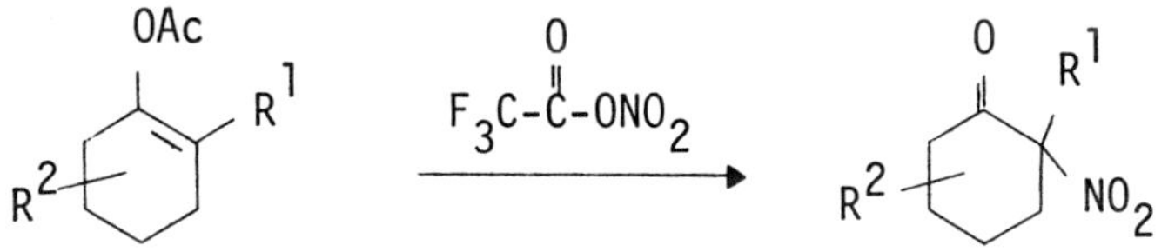
OAc
R1
R2
O
F3C-C-ONO2
O
R1
R2
NO2

R^1 = prim. alkyl

41-100%

VI.A.15 Nucleotides, etc.   (see also: IV.I.1a, b; V.F)

VI.A.15-1 B. C. Froehler and M. D. Matteucci, <u>Tetrahedron Lett.</u>, 24, 3171 (1983).

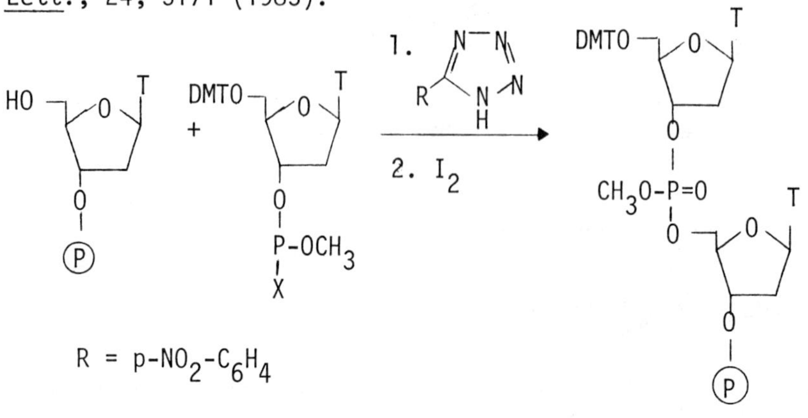

R = p-NO$_2$-C$_6$H$_4$

VI.A.15-2 Y. Hayakawa et al., <u>Tetrahedron Lett.</u>, 24, 1165 (1983).

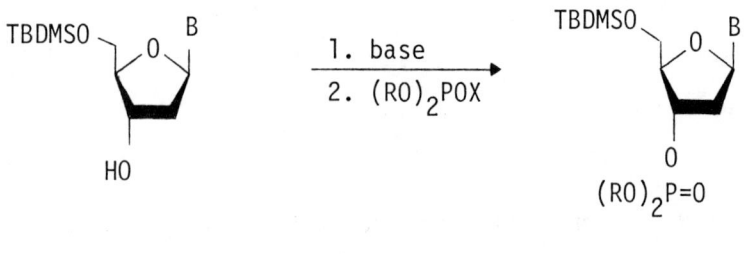

79-97%

VI.A.16. Olefins, Acetylenes   (see also: I.B, I.C, II.J, III.G)

VI.A.16-1   D. R. Anton and R. H. Crabtree, Tetrahedron Lett., 24, 2449 (1983).

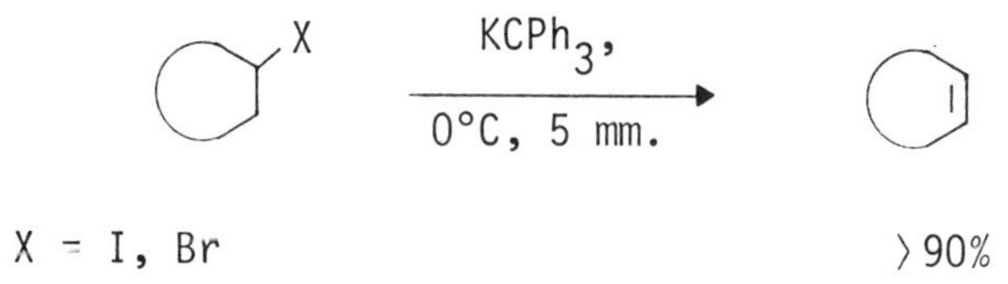

$$\text{cyclohexyl-X} \xrightarrow[0°C, 5 \text{ mm.}]{KCPh_3} \text{cyclohexene}$$

X = I, Br   >90%

VI.A.16-2   B. M. Trost, M. Lautens and B. Peterson, Tetrahedron Lett., 24, 4525 (1983).

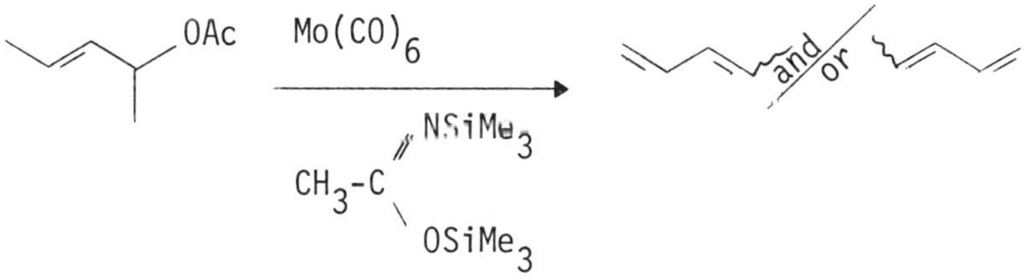

$$\text{CH}_3\text{CH=CHCH(CH}_3\text{)OAc} \xrightarrow[\substack{CH_3-C(=NSiMe_3)OSiMe_3}]{Mo(CO)_6} \text{dienes (and/or)}$$

50-95% combined

VI.A.16-3   N. C. Barua et al., Chem. Ind. (London), 956 (1982).

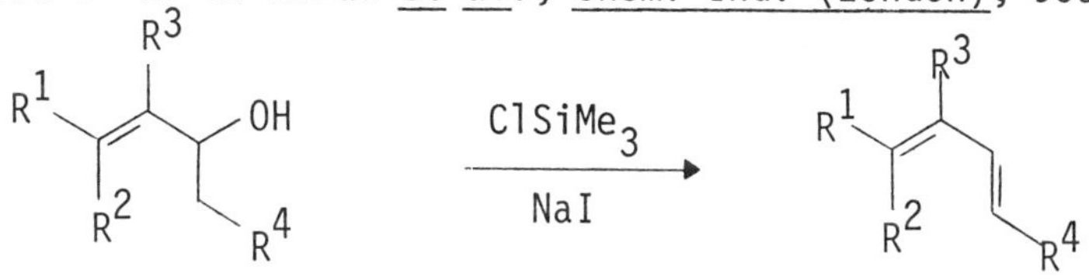

$$\underset{R^2}{\overset{R^1}{>}}C=C\overset{R^3}{\underset{}{-}}CH(OH)CH_2R^4 \xrightarrow[NaI]{ClSiMe_3} \underset{R^2}{\overset{R^1}{>}}C=C\overset{R^3}{\underset{}{-}}CH=CHR^4$$

75-90%

VI.A.16-4 H. J. Reich and S. Wollowitz, *J. Am. Chem. Soc.*, 104, 7051 (1982).

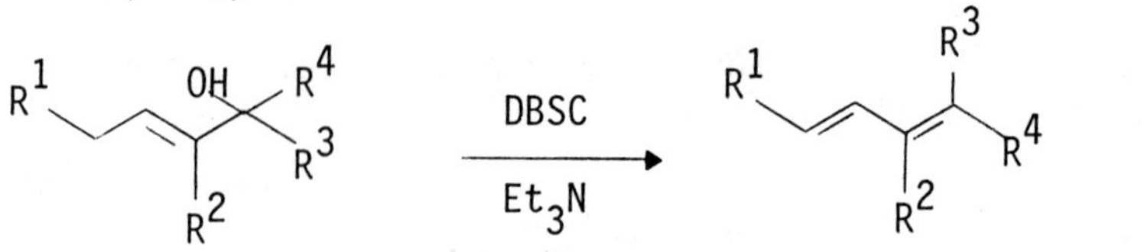

$R^1\text{-CH=C}(R^2)\text{-C}(R^3)(R^4)\text{-OH}$ $\xrightarrow[\text{Et}_3\text{N}]{\text{DBSC}}$ $R^1\text{-CH=CH-C}(R^2)\text{=C}(R^3)(R^4)$

DBSC = 2,4-dinitrobenzenesulfenyl chloride (SCl with NO$_2$ groups)

varying yields

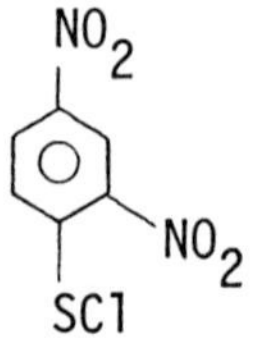

VI.A.16-5 K. B. Becker, *Synthesis*, 341 (1983).

Review: "Synthesis of Stilbenes."

VI.A.16-6 R. M. Adlington and A. G. M. Barrett, *Acc. Chem. Res.* 16, 55 (1983).

Review: "Recent Applications of the Shapiro Reaction."

VI.A.17. Peptides (see also: V.B, V.C, V.D, VI.A.4)

VI.A.17-1 K. Takeda et al., Tetrahedron Lett., 24, 4451 (1983).

$$\text{ZHNCHCOOH} + R^1O-\underset{\underset{}{}}{\overset{O}{\overset{\|}{C}}}-\underset{\underset{}{}}{\overset{O}{\overset{\|}{C}}}-OR^1 \xrightarrow{\text{Pyridine}} \text{ZHNCH}\overset{O}{\overset{\|}{C}}-OR^1$$
with R^2 substituent on the α-carbon.

$$\xrightarrow{H_2NCHCOOR^4 \; (R^3)} \text{ZHNCHCNHCHCOOR}^4$$
with R^2, R^3 substituents.

$R^1 = $ succinimidyl, benzotriazolyl, 6-chlorobenzotriazolyl, norbornene-dicarboximidyl, phthalimidyl

---

VI.A.17-2 M. Furukawa, N. Hokama and T. Okawara, Synthesis, 42 (1983).

$$Z-NH-\underset{R^1}{CH}-CO_2H + \text{(6-NO}_2\text{-benzotriazolyl-O-SO}_2\text{-C}_6\text{H}_4\text{-NO}_2\text{)}$$

$$\xrightarrow[\text{2. } H_2N-CH(R^2)-CO_2R^3]{\text{1. Et}_3N, \text{ CH}_3\text{CN}} Z-NH-\underset{R^1}{CH}-\overset{O}{\overset{\|}{C}}-NH-\underset{R^2}{CH}-CO_2R^3$$

$R^2 = H, Ph$ 69-91%

$R^3 = Et, Bz$

VI.A.18. Vinyl Halides, Vinyl Ethers, Vinyl Esters

VI.A.18-1 M. Marsi and J. A. Gladysz, Organometallics, $\underline{1}$, 1467 (1982).

$$\underset{R^2}{\underset{|}{H_3CO}}\!\!\!\underset{R^1}{\overset{OCH_3}{\diagdown\!\!\!\diagup}} \xrightarrow[CH_3CN,\ 50°C]{(CO)_5MnSi(CH_3)_3} \underset{R^2}{\overset{OCH_3}{\diagdown}}\!\!C\!=\!\!\underset{}{\overset{}{C}}\!\!\diagdown\!R^1$$

56->95%

VI.A.18-2 S. Hara et al., Tetrahedron Lett., $\underline{24}$, 731 (1983).

$$RC\equiv CH \xrightarrow[\text{2. AcOH}]{\text{1. B-X-9-BBN}} \underset{X}{\overset{R}{\diagdown}}C\!=\!C\underset{H}{\overset{H}{\diagup}}$$

X = Br, I   65-100%

VI.A.18-3 G. W. Kabalka et al., Synth. Commun., $\underline{13}$, 1027 (1983).

$$RC\equiv CH \xrightarrow[\text{2. H}_2O]{1.\ \text{catecholborane}} \underset{H}{\overset{R}{\diagdown}}C\!=\!C\underset{B(OH)_2}{\overset{H}{\diagup}} \xrightarrow[NCS]{NaBr} \underset{H}{\overset{R}{\diagdown}}C\!=\!C\underset{H}{\overset{Br}{\diagup}}$$

R = alkyl   40-75%

VI.A.18-4 M. C. Pirrung and J. R. Hwu, Tetrahedron Lett., 24, 565 (1983).

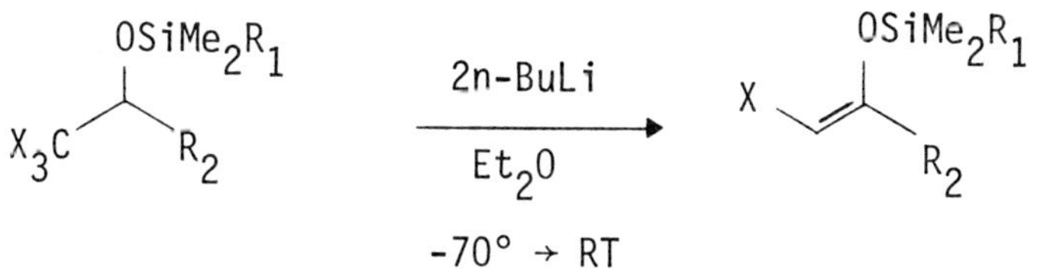

$X_3C-C(OSiMe_2R_1)(R_2)H$ $\xrightarrow[-70° \to RT]{\text{2n-BuLi, Et}_2O}$ $X-CH=C(OSiMe_2R_1)(R_2)$

R_2 = H, CH$_3$
X = Br, Cl

68-96%

VI.A.18-5 T. Proll and W. Walter, Chem. Ber., 116, 1564 (1983).

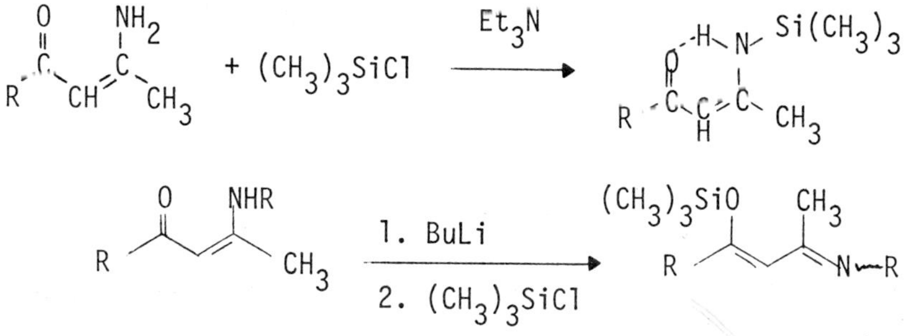

$R-C(=O)-CH=C(NH_2)-CH_3$ + $(CH_3)_3SiCl$ $\xrightarrow{Et_3N}$ [enol-imine with O-H···N-Si(CH$_3$)$_3$ hydrogen bond]

$R-C(=O)-CH=C(NHR)-CH_3$ $\xrightarrow[\text{2. (CH}_3)_3\text{SiCl}]{\text{1. BuLi}}$ $R-C(OSi(CH_3)_3)=CH-C(CH_3)=N-R$

VI.A.18-6 J. E. McMurry and W. J. Scott, Tetrahedron Lett., 24, 979 (1983)

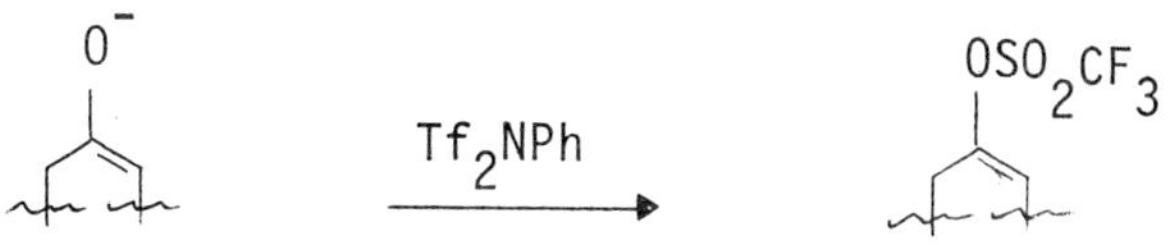

[cyclohexenolate O$^-$] $\xrightarrow{Tf_2NPh}$ [cyclohexenyl OSO$_2$CF$_3$]

65-97%

VI.A.19. Sulfur Compounds (see also: II.E, III.C)

VI.A.19-1 T. Oida et al., Bull. Chem. Soc. Jpn. **56**, 959 (1983).

$$(RS)_2CHCHR'R'' \xrightarrow[(i-Pr_2)_2NCH_2CH_2NH_2]{CuCl_2} RSCH=CR'R''$$

30-90%

VI.A.19-2 R. D. Miller and D. R. McKean, Tetrahedron Lett., **24**, 2619 (1983).

$$RCH_2CH_2\overset{O}{\underset{\|}{S}}Ph \xrightarrow[i-Pr_2NEt]{Me_3SiI} RCH=CHSPh$$

75-91%

VI.A.19-3 O. Huel, E. Guittet and S. Julia, Tetrahedron Lett., **24**, 61 (1983).

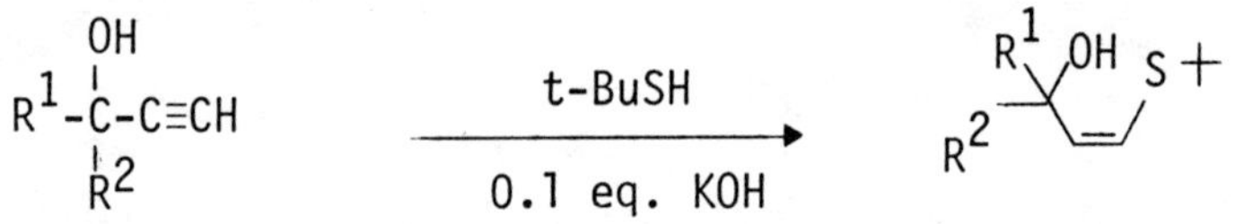

$$R^1-\underset{R^2}{\overset{OH}{C}}-C\equiv CH \xrightarrow[0.1\ eq.\ KOH]{t-BuSH} \underset{R^2}{\overset{R^1}{\diagdown}}\!\!\!\overset{OH}{\underset{}{C}}\!\!\diagup\!\!\diagdown S+$$

75-90%

VI.A.19-4 K. Suzuki, A. Ikegawa and T. Mukaiyama, Bull. Chem. Soc. Jpn. 55, 3277 (1982).

ArSH + [cyclic enone] $\xrightarrow{\text{catalyst}}$ [cyclic β-thioether ketone]
PhCH$_3$, -5°C

64-85%
11-88% e.e.

catalyst = [pyrrolidine structure with HO, N-C$_2$H$_5$, CH$_2$NHPh substituents]

VI.A.19-5 C. B. Reese and H. P. Sanders, J. Chem. Soc. Perkin Trans 1, 2719 (1982).

$R^1\text{-CHX-C(O)-}R^2$ $\xrightarrow{\begin{array}{l}1.\ \text{TsNHNH}_2\\ 2.\ \text{PhYH, Et}_3\text{N, -78°C}\\ 3.\ \text{BF}_3\cdot\text{Et}_2\text{O}\end{array}}$ $R^1\text{-CH(YPh)-C(O)-}R^2$

X = Br, Cl

Y=S : 72-83%
Y=Se: 64-72%

VI.A.19-6 N. De Kimpe et al., Synthesis, 632 (1983).

$\underset{R^3}{\overset{R^2}{\diagdown}}\text{C(Cl)-CH=N-}R^1$ $\xrightarrow{\text{NaS-}R^4/\text{CH}_3\text{OH}}$ $\underset{R^3}{\overset{R^2}{\diagdown}}\text{C(S}R^4\text{)-CH=N-}R^1$

R^1 = t-Bu, i-Pr, c-Hex
R^2, R^3 = Me, Et

65-92%

VI.A.19-7 S. Kato et al., Synthesis, 552 (1983).

$$Ar-\overset{O}{\underset{\|}{C}}-SAg \xrightarrow{\begin{array}{c}Br_2 \text{ or}\\ NBS\end{array}} Ar-\overset{O}{\underset{\|}{C}}-SBr \quad 4-71\%$$

$$\xrightarrow{I_2} Ar-\overset{O}{\underset{\|}{C}}-SI \quad 12-60\%$$

VI.A.19-8 S. M. M. Elsafie, Org. Prep. Proced. Int., 15, 225 (1983).

$$RNH_2 \xrightarrow[\text{2. } K\overset{S}{\underset{\|}{S}}COEt]{\text{1. } \underline{A}} R\overset{S}{\underset{\|}{S}}COEt \xrightarrow[\text{2 } H^+]{\text{1. } OH^-} RSH$$

43-85% 58-85%

$\underline{A}$ = [benzo-fused pyrylium with Ph substituents] BF_4^-

---

VI.A.19-9 S. Takano, K. Hiroya and K. Ogasawara, Chem. Lett., 255 (1983).

$$p-Ts^- Na^+ \xrightarrow[\begin{array}{c}\text{2. Amberlyst A-26}\\ \text{3. RX}\end{array}]{\text{1. } S_8, \text{ Pyr.}} p-Ts-SR$$

R = alkyl, allyl, benzyl   61-100%

VI.A.19-10 R. R. Schmidt and M. Stumpp, Liebigs Ann. Chem., 1249 (1983).

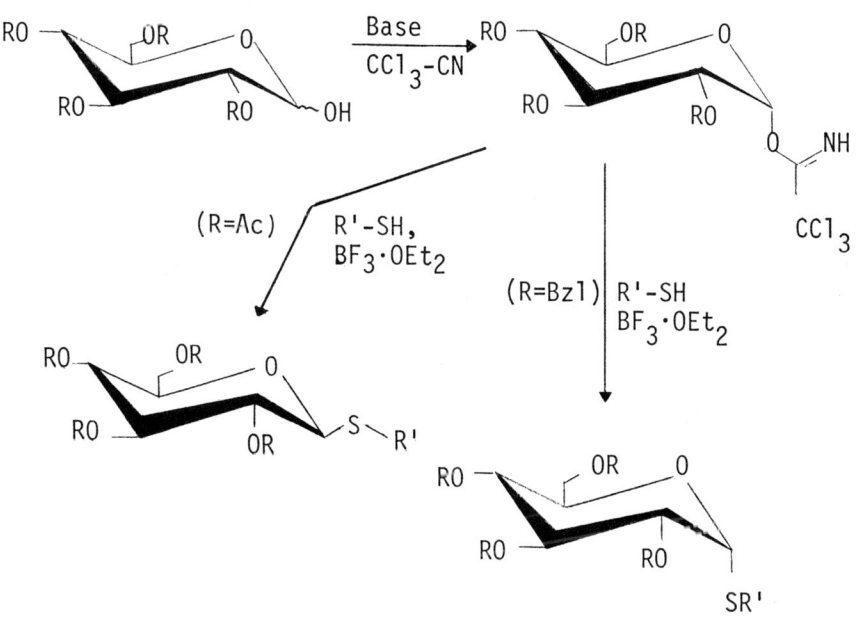

VI.A.19-11 I. Nakagawa, K. Aki and T. Hata, J. Chem. Soc. Perkin Trans. 1, 1315 (1983).

25-96%

VI.A.19-12   I. Degani, R. Fochi and V. Regondi, Synthesis, 630 (1983).

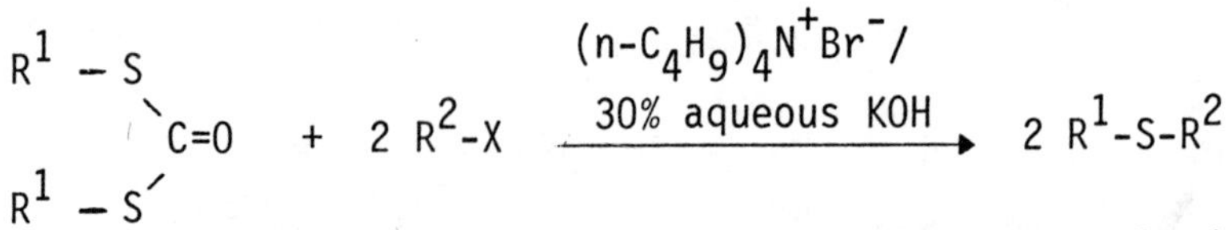

$$R^1-S\diagdown C=O + 2\ R^2-X \xrightarrow[30\%\ \text{aqueous KOH}]{(n-C_4H_9)_4N^+Br^-/} 2\ R^1-S-R^2$$
$$R^1-S\diagup$$

93-100%

VI.A.19-13   H. Singh, N. Malhotra and M. S. Batra, Indian J. Chem. Sect. B $\underline{21}$, 272 (1982).

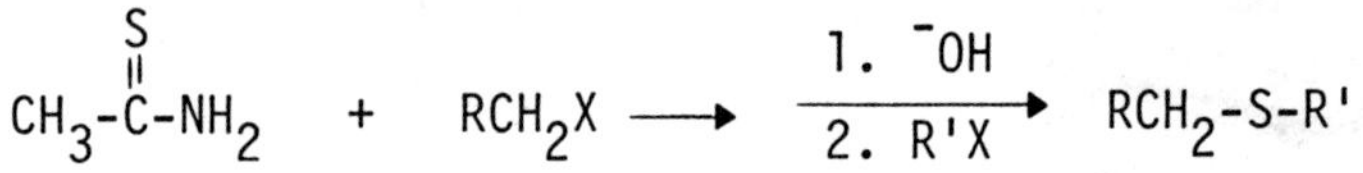

$$CH_3-\overset{S}{\underset{\|}{C}}-NH_2 + RCH_2X \longrightarrow \xrightarrow[2.\ R'X]{1.\ ^-OH} RCH_2-S-R'$$

R = H, Ph

VI.A.19-14   J. Nakayama, T. Fujita and M. Hoshino, Chem. Lett., 249 (1983).

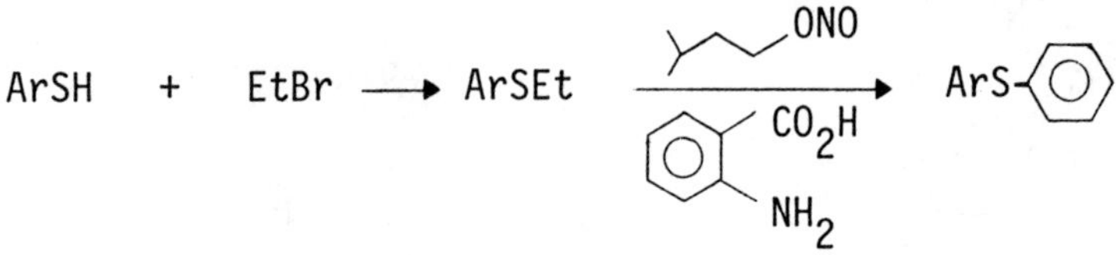

$$ArSH + EtBr \longrightarrow ArSEt \xrightarrow{\text{iso-amyl ONO, anthranilic acid}} ArS-C_6H_5$$

87-97%

VI.A.19-14 J. Nakayama, T. Fujita and M. Hoshino, *Chem. Lett.*, 249 (1983).

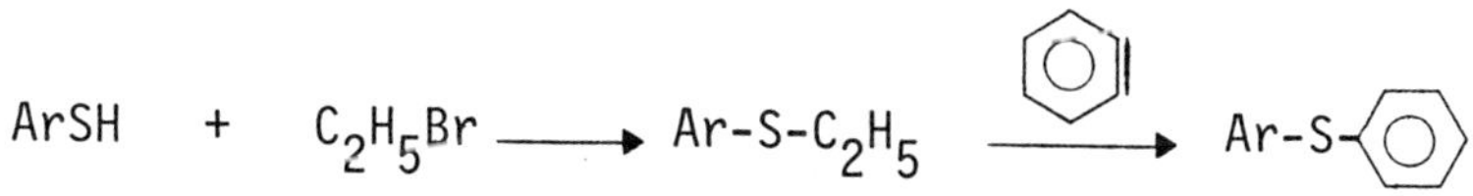

$$ArSH + C_2H_5Br \longrightarrow Ar\text{-}S\text{-}C_2H_5 \xrightarrow{C_6H_6} Ar\text{-}S\text{-}C_6H_5$$

87-97%

VI.A.19-15 Y. Guindon *et al.*, *J. Org. Chem.*, **48**, 1357 (1983).

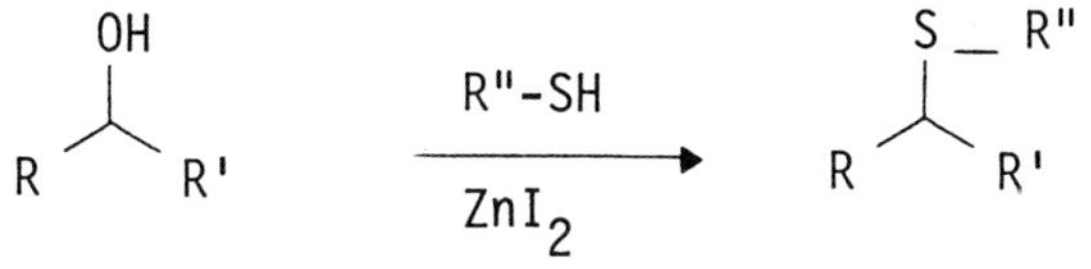

$$\underset{R\quad R'}{\overset{OH}{>\!\!<}} \xrightarrow[ZnI_2]{R''\text{-}SH} \underset{R\quad R'}{\overset{S-R''}{>\!\!<}}$$

R = aryl, vinyl   60-99%

VI.A.19-16 H. Yamada *et al.*, *Bull. Chem. Soc. Jpn.*, **56**, 949 (1983).

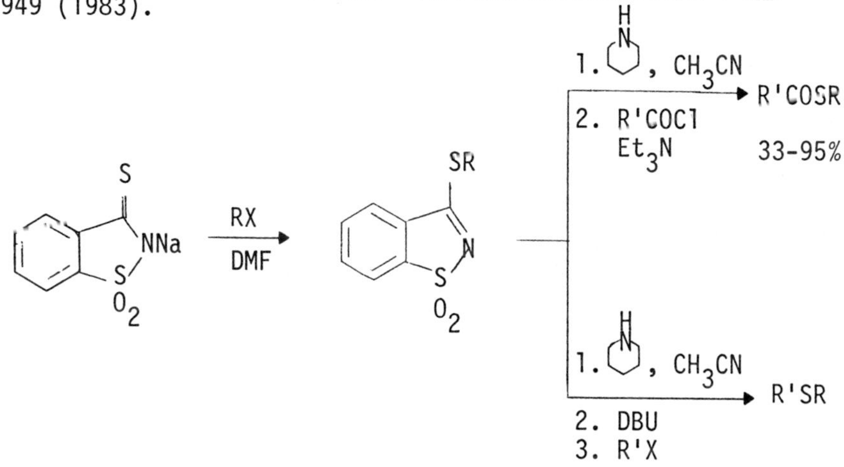

Benzisothiazole-S,S-dioxide-NNa → (RX/DMF) → 3-SR-benzisothiazole-S,S-dioxide

1. piperidine, CH_3CN
2. R'COCl, Et_3N → R'COSR   33-95%

1. piperidine, CH_3CN
2. DBU
3. R'X → R'SR   81-100%

VI.A.19-17 K. Soai, H. Hayashi and A. Ookawa, J. Chem. Res. Synop., 20 (1983).

$$\underset{O}{\overset{R^1}{\underset{|}{C}}}=C-N\underset{S}{\overset{S}{\rangle}} + HSR^2 \xrightarrow{Et_3N} R^1\text{-CO-SR}^2$$

R^1 = Ph, Bz   62-100%

VI.A.19-18 V. Reutrakul and P. Poochaivatananon, Tetrahedron Lett., 24, 535 (1983).

$$\underset{R^2}{\overset{R^1}{}}C=C\underset{H}{\overset{SPh}{}} \xrightarrow[PhSH]{70\% HClO_4} \underset{R^2}{\overset{R^1}{}}CH-CH\underset{SPh}{\overset{SPh}{}}$$

67-87%

$$\underset{R^2}{\overset{R^1}{}}C=C\underset{Cl}{\overset{SPh}{}} \xrightarrow[PhSH]{70\% HClO_4} \underset{R^2}{\overset{R^1}{}}CH-C\underset{SPh}{\overset{O}{}}$$

62-76%

VI.A.19-19 V. Jigajinni, R. H. Wightman and M. M. Campbell, J. Chem. Res. Synop., 187 (1983).

$$ArSCHR \atop \underset{Cl}{|} \xrightarrow[\Delta]{Al_2O_3} (ArS)_2CHR$$

VI.A.19-20 T. Satoh, S. Uwaya and K. Yamakawa, Chem. Lett., 667 (1983).

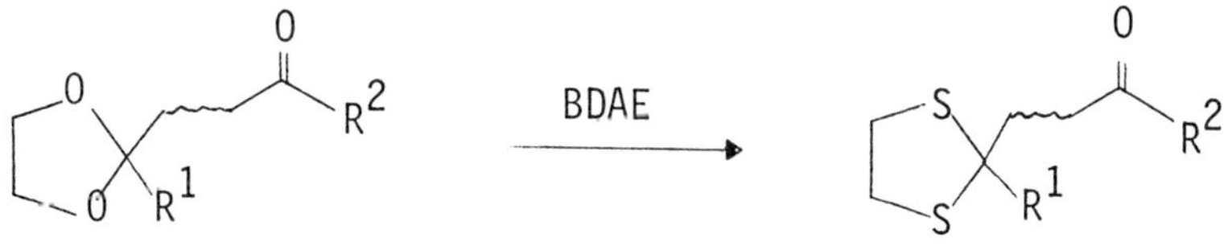

[reaction scheme: 1,3-dioxolane ketone → 1,3-dithiolane ketone via BDAE]

BDAE = $(iso\text{-}C_4H_9)_2AlS(CH_2)_2SAl(iso\text{-}C_4H_9)_2$

18-89%

VI.A.19-21 E. Schaumann, U. Wriede and G. Ruehter, Angew. Chem. Int. Ed. Engl., 22, 55 (1983).

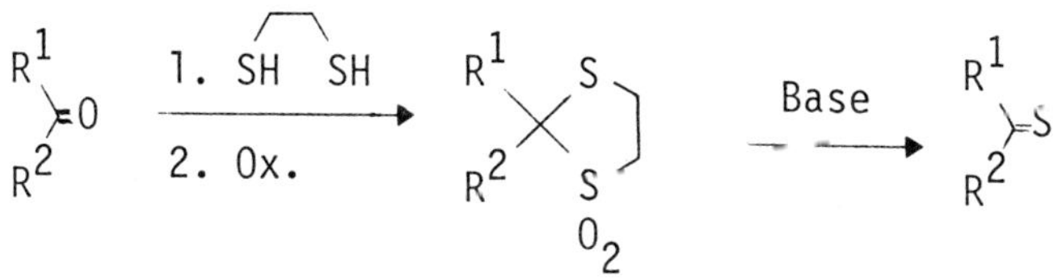

[reaction scheme: $R^1R^2C=O$ → dithiolane S,S-dioxide → $R^1R^2C=S$ via (1. SH SH, 2. Ox.) then Base]

VI.A.19-22 I. Cutting and P. J. Parsons, Tetrahedron Lett., 24, 4463 (1983).

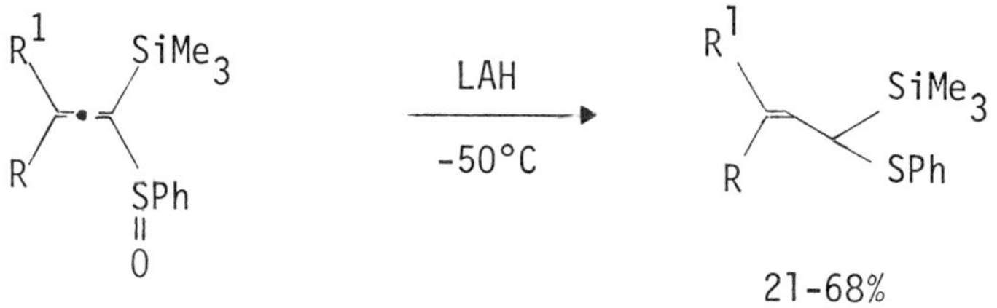

[reaction scheme: allene with SiMe₃ and S(=O)Ph groups → vinyl silane with SPh via LAH, -50°C]

R,R' = H, alkyl

21-68%

VI.A.19-23 R. Sato et al., Chem. Lett., 535 (1983).

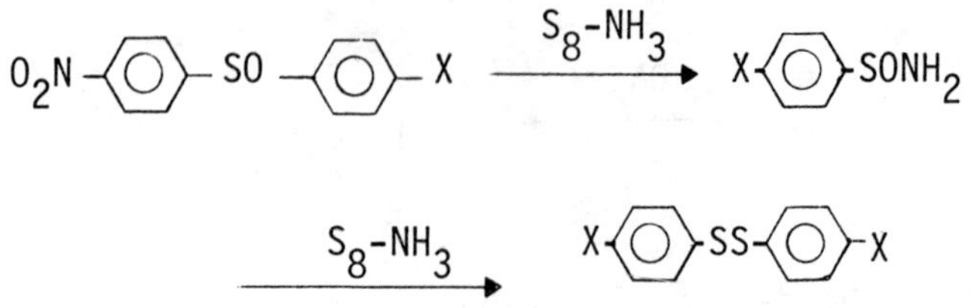

$O_2N\text{-}C_6H_4\text{-}SO\text{-}C_6H_4\text{-}X \xrightarrow{S_8\text{-}NH_3} X\text{-}C_6H_4\text{-}SONH_2$

$\xrightarrow{S_8\text{-}NH_3} X\text{-}C_6H_4\text{-}SS\text{-}C_6H_4\text{-}X$

X = H, CH_3, Cl, NO_2

VI.A.19-24 T. G. Back, S. Collins and R. G. Kerr, J. Org. Chem., 48, 3077 (1983).

$RC\equiv CH \xrightarrow[\text{2. MCPBA}]{\text{1. PhSeSO}_2\text{Ar, AIBN}} RC\equiv CSO_2Ar$
3. KOH

~90%

VI.B.1. Ring Enlargement

VI.B.1-1 L. A. Paquette et al., Tetrahedron, 39, 913 (1983).

$$\underset{R^2}{\overset{R^1}{\triangleright}}\!\!\!\!\!\!\!\!\diagup\!\!\!\!\diagdown SiMe_3 \quad \xrightarrow[2.\ E^+]{1.\ \text{Pyrolysis}} \quad \underset{R^2}{\overset{R^1}{\bigcirc}}\!\!\!E$$

VI.B.1-2 J. Salaun and Y. Almirantis, Tetrahedron, 39, 2421 (1983)

$$\overset{OSiMe_3}{\underset{R^1}{\triangleright}}\!\!\!\!=\!\!\!R^2 \quad \xrightarrow[2.\ Pd(OAc)_2]{1.\ 600°C} \quad \underset{R^2}{\overset{O}{\bigcirc}}\!\!R^1$$

VI.B.1-3 E. Nakamura, J.-i. Shimada and I. Kuwajima, J. Chem. Soc. Chem. Commun., 498 (1983).

$$\square\!\!\!\!\overset{OSiMe_3}{\underset{OSiMe_3}{|}} + RCH(OMe)_2 \quad \xrightarrow[\substack{1.\ BF_3 \cdot Et_2O,\ -70°C \\ 2.\ Ph_3P=CH_2,\ KH \\ 3.\ TFA}]{} \quad \overset{O}{\bigcirc}\!\!R$$

55-75%

R = vinyl, aryl

VI.B.1-4 R. C. Gadwood, *J. Org. Chem.*, **48**, 2098 (1983).

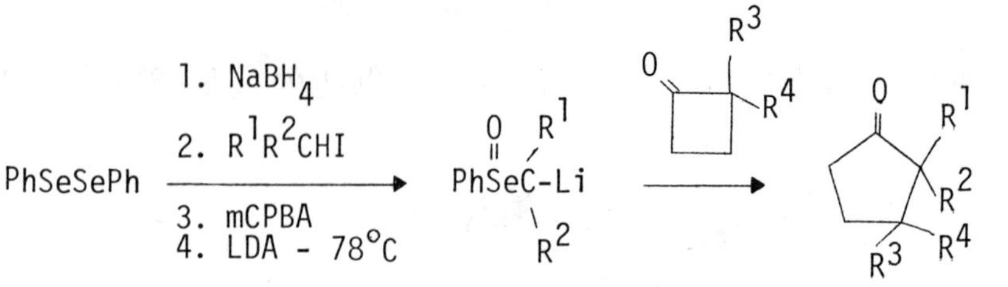

PhSeSePh $\xrightarrow{\begin{array}{l}1.\ NaBH_4\\ 2.\ R^1R^2CHI\\ 3.\ mCPBA\\ 4.\ LDA\ -78°C\end{array}}$ PhSeC(O)-CHR^{1}R^2 → [cyclobutanone with R^3, R^4] → [cyclopentanone with R^1, R^2, R^3, R^4]

$R^1, R^2 = H, Me$

VI.B.1-5 Y. Nakashita and M. Hesse, *Helv. Chim. Acta*, **66**, 845 (1983).

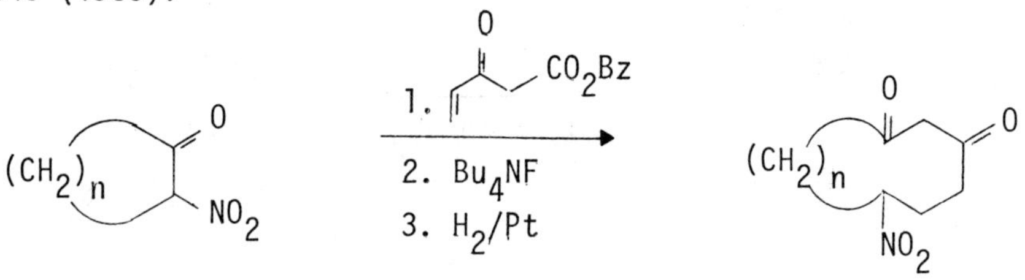

[cycloalkanone-NO$_2$] $\xrightarrow{\begin{array}{l}1.\ CH_2=C(O)CH_2CO_2Bz\\ 2.\ Bu_4NF\\ 3.\ H_2/Pt\end{array}}$ [ring-expanded diketone with NO$_2$]

45-70%

VI.B.2. Ring Contraction

VI.B.2-1 M. Albeck, T. Tamari and M. Sprecher, *J. Org. Chem.*, **48**, 2276 (1983).

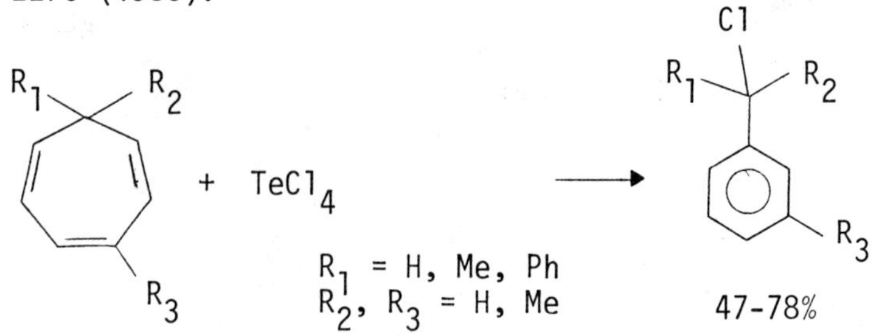

[cycloheptatriene with R$_1$, R$_2$, R$_3$] + TeCl$_4$ → [benzene with CCl(R$_1$)(R$_2$) and R$_3$]

$R_1 = H, Me, Ph$
$R_2, R_3 = H, Me$

47-78%

VII
MISCELLANEOUS REVIEWS

VII-1   V. G. S. Box, Heterocycles, 20, 677 (1983).

   Review: "The Role of Lone Pair Interactions
   in the Selective Functionalization
   of Some 4,6-O-Benzylidene-hexopyrano-
   sides by Both the Phase Transfer
   Esterification Reactions and the Tin-
   Mediated Esterification and Alkylation
   Reactions."

VII-2   V. G. S. Box, Heterocycles, 20, 1641 (1983).

   Review: "The Role of Lone Pair Interactions in
   the Chemistry of Monosaccharides. The
   Selective Esterification of 4,6-O-Benzyl-
   idene-hexopyranosides."

VII-3 D. Ginsburg, Tetrahedron, 39, 2095 (1983).

Review: "The Role of Secondary Orbital Interactions in Control of Organic Reactions."

VII-4 R. Gleiter and L. A. Paquette, Acc. Chem. Res., 16, 328 (1983).

Review: "σ/π Interaction as a Controlling Factor in the Stereoselectivity of Addition Reactions."

VII-5 S. Oae and K. Shinhama, Org. Prep. Proced. Int., 15, 165 (1983).

Review: "Organic Thionitrites and Related Substances."

VII-6 M. V. Bhatt and S. U. Kulkarni, Synthesis, 249 (1983).

Review: "Cleavage of Ethers."

VII-7 N. H. Werstiuk, Tetrahedron, 39, 205 (1983).

Review: "Homoenolate Anions and Homoenolate Anion Equivalents. Mechanistic Aspects and Synthetic Applications."

VII-8   F. M. Menger, Tetrahedron, 39, 1013 (1983).

Review: "Directionality of Organic Reactions in Solution."

VII-9   H. A. Daboun and S. E. Abdou, Heterocycles, 20, 1615 (1983).

Review: "Recent Developments in the Chemistry of Ylidene Azolones."

VII-10   Z. J. Witczak, Heterocycles, 20, 1435 (1983).

Review: "Synthesis and Preparative Applications of Monosaccharide Thiocyanates."

VII-11   P. Cocagne, J. Elguero and R. Gallo, Heterocycles, 20, 1379 (1983).

Review: "The Present Use and the Possibilities of Phase Transfer Catalysis in Drug Synthesis."

VII-12   R. C. Sheppard, Chemistry (Britain) 19, 402 (1983).

Review: "Continuous Flow Methods in Organic Synthesis."

VII-13   J. Klein, Tetrahedron, 39, 2733 (1983).

Review: "Directive Effects in Allylic and Benzylic Polymetalations: The Question of U-Stabilization, Y-Aromaticity and Cross-Conjugation."

VII-14   T. Shono, Yakugaku Zasshi (J. Pharm. Soc. Jpn.), 102, 995 (1982).

Review: "Electroorganic Chemistry as Tools of Bioactive Compounds."

VII-15   F. Minisci and A. Citterio, Acc. Chem. Res., 16, 27 (1983).

Review: "Electron-Transfer Processes: Peroxydisulfate, a Useful and Versatile Reagent in Organic Chemistry."

VII-16   B. R. Castro, Org. Reactions, 29, 1 (1983).

Review: "Replacement of Alcoholic Hydroxyl Groups by Halogens and Other Nucleophiles via Oxyphosphonium Intermediates."

VII-17   R. A. McClleland and L. J. Santry, Acc. Chem. Res.,
16, 394 (1983).

Review: "Reactivity of Tetrahedral Intermediates."

VII-18   W. Kaim, Angew. Chem. Int. Ed. Engl., 22, 171 (1983).

Review: "The Versatile Chemistry of 1,4-Diazines:
Organic, Inorganic and Biochemical Aspects."

VII-19   H.-D. Martin and B. Mayer, Angew. Chem. Int. Ed.
Engl., 22, 283 (1983).

Review: "Proximity Effects in Organic Chemistry --
The Photoelectron Spectroscopic Investigation of Non-Bonding and Transannular
Interactions."

VII-20   G.K. S. Prakash, T. N. Rawdah and G. A. Olah,
Angew. Chem. Int. Ed. Engl., 22, 390 (1983).

Review: "Stable Carbodications."

VII-21   H. K. Hall, Jr., Angew. Chem. Int. Ed. Engl., 22,
440 (1983).

Review: "Bond-Forming Initiation in Spontaneous Addition
and Polymerization Reactions of Alkenes."

VII-22   J. I. Seeman, Chem. Rev., 83, 83 (1983).

Review: "Effect of Confromational Change on Reactivity in Organic Chemistry. Evaluations, Applications, and Extensions of Curtin-Hammett/Winstein-Holness Kinetics."

VII-23   W. Smadja, Chem. Rev., 83, 263 (1983).

Review: "Electrophilic Addition to Allenic Derviatives: Chemo-, Regio-, and Stereochemistry and Mechanisms."

VII-24   T. L. Gilchrist, Chem. Soc. Rev., 12, 53 (1983).

Review: "Nitroso-alkenes and Nitroso-alkynes."

VII-25   E. F. V. Scriven, Chem. Soc. Rev., 12, 129 (1983).

Review: "4-Dialkylaminopyridines: Super Acylation and Alkylation Catalysts."

VII-26   S. A. Narang, Tetrahedron, 39, 3 (1983).

Review: "DNA Synthesis."

VII-27 A. P. Croft and R. A. Bartwch, Tetrahedron, 39, 1417 (1983).

Review: "Synthesis of Chemically Modified Cyclodextrins."

VII-28 W. S. Murphy and S. Wattanasin, Chem. Soc. Rev., 12, 213 (1983).

Review: "Anionic Cyclization of Phenols."

VII-29 H. N. C. Wong, T.-K. Ng and T.-Y. Wong, Heterocycles, 20, 1815 (1983).

Review: "Arene Syntheses by Extrusion of Heteroatoms from 7-Heteroatombicyclo(2.2.1)heptene Systems."

VII-30 C. Santelli-Rouvier and M. Santelli, Synthesis, 429 (1983).

Review: "The Nazarov Cyclisation."

VII-31 G. Guchbauer and E. M. Stifter, Chem. Ztg., 106, 383 (1982).

Review: "Mesityloxide as a Dienophile in Diels-Alder Reactions."

VII-32 W. Wolfsberger, Chem. Ztg., 107, 77 (1983).

Review: "Synthesis, Properties and Reactions of Chlorodiethylphosphine."

VII-33 H. W. Pinnick, Org. Prep. Proced. Int., 15, 199 (1983).

Review: "Potassium Hydride in Organic Synthesis."

VII-34 G. G. Yakobson and N. E. Akhmetova, Synthesis, 169 (1983).

Review: "Alkali Metal Fluorides in Organic Synthesis."

VII-35 F. B. Mallory and C. W. Mallory, Org. Reactions, 30, 1 (1983).

Review: "Photocyclization of Stilbenes and Related Molecules."

VII-36 R. G. Salomon, Tetrahedron, 39, 485 (1983).

Review: "Homogeneous Metal-Catalysis in Organic Photochemistry."

VII-37 M. A. Fox, Acc. Chem. Res., 16, 314 (1983).

Review: "Organic Heterogeneous Photocatalysis: Chemical Conversions Sensitized by Irradiated Semiconductors."

VII-38 G. G. Wubbels, Acc. Chem. Res., 16, 285 (1983).

Review: "Catalysis of Photochemical Reactions."

VII-39 W. H. Laarhoven, Recl. J. R. Neth. Chem. Soc., 102, 185 (1983).

Review: "Photochemical Cyclizations and Intramolecular Cycloadditions of Conjugated Arylolefins. Part I.: Photocyclization with Dehydrogenation."

VII-40 W. H. Laarhoven, Recl. J. R. Neth. Chem. Soc., 102, 241 (1983).

Review: "Photochemical Cyclizations and Intramolecular Cycloadditions of Conjugated Arylolefins. Part 2. Photocyclizations Without Dehydrogenation and Photocycloadditions."

VII-41 W. H. Okanura, Acc. Chem. Res. 16, 81 (1983).

Review: "Pericyclic Reactions of Vinylallenes: From Calciferols to Retinoids and Drimanes."

VII-42 L. N. Mander, Acc. Chem. Res., 16, 48 (1983).

Review: "New Strategies for the Construction of Highly Functionalized Organic Molecules: Applications to C_{19} Gibberellin Synthesis."

VII-43 J. Redpath and F. J. Zeelen, Chem. Soc. Rev., 12, 75 (1983).

Review: "Stereoselective Synthesis of Steroid Side-chains."

VII-44 G. Quinkert and H. Stark, Angew. Chem. Int. Ed. Engl., 22, 637 (1983).

Review: "Stereoselective Synthesis of Enantiomerically Pure Natural Products-Estron as Example."

VII-45 S.-I. Hashimoto, Yakugaku Zasshi (J. Pharm. Soc. Jpn.) 102, 1103 (1982).

Review: "Highly Selective Asymmetric Synthesis Based on the Strategy of Fixing the Conformation. Asymmetric Carbon-Carbon Bond Forming Reactions via Chelate Intermediates."

VII-46 L. Weber, Angew. Chem. Int. Ed. Engl., 22, 516 (1983).

Review: "Metal Complexes of Sulfur Ylides: Coordination Chemistry, Preparative Organic Chemistry, and Biochemistry."

VII-47 B. Weidmann and D. Seebach, Angew. Chem. Int. Ed. Engl. 22, 31 (1983).

Review: "Organometrallic Compounds of Titanium and Zirconium as Selective Nucleophilic Reagents in Organic Synthesis."

VII-48 I. Omae, Angew. Chem. Int. Ed. Engl. 21, 889 (1982).

Review: "Organometallic Intramolecular π-Olefin-Metal Coordination Compounds."

VII-49   M. D. Johnson, Acc. Chem. Res. 16, 343 (1983).

   Review: "Bimolecular Homolytic Displacement
             of Transition-Metal Complexes from
             Carbon."

VII-50   D. Walther and E. Dinjus, Z. Chem., 23, 237 (1983).

   Review: "Activation of Carbon Dioxide on Transition
             Metal Centers: New Routes for Organic
             and Metallorganic Synthesis."

VII-51   J. Org. Chem., 48, 1394 (1983).

   "Recent Reviews. 11"

VII-52   J. Org. Chem., 48, 4160 (1983).

   "Recent Reviews. 12"

AUTHOR INDEX

AUTHOR INDEX

Abrams, S. R. - 140
Adam, J. M. - 218
Adams, D. R. - 341
Adlington, R. M. - 324, 432
Agami, C. - 89
Ager, D. J. - 5, 20, 76, 209, 384, 425
Agosta, W. C. - 94
Ahlbrecht, H. - 9, 70, 163
Ahmad, M. - 141
Akhrem, I. S. - 229
Akiba, K. - 33, 331
Akiyama, S. - 139
Albeck, M. - 446
Alberola, A. - 287
Albright, J. D. - 7
Alder, R. W. - 349
Alexakis, A. - 132
Alper, H. - 126, 129, 140, 222, 223, 290, 300, 320
Alston, P. V. - 162
Amos, R. A. - 416
Amvaitre, G. - 312

Amupitan, J. D. - 235
Ando, K. - 101
Ando, T. - 37
Ando, W. - 62, 99, 257
Adrieux, J. - 342
Andruszkiewicz, R. - 413
Aneja, R. - 307
Annen, K. - 44
Anselme, J. P. - 97
Anton, D. R. - 118, 431
Antonioletti, R. - 260
Aoyama, H. - 327
Aoyama, T. - 97
Apeloig, Y. - 143
Apparu, M. - 145
Appel, R. - 401
Arai, K. - 193
Arase, A. - 131
Arias, L. A. - 261
Aristoff, P. A. - 9, 115
Armand, J. - 59
Armesto, D. - 217
Arrieta, A. - 417

Arseniyadis, S. - 347
Arumugam, N. - 235
Astruc, D. - 12
Attanasi, O. - 43, 344, 371, 427
Atwal, K. S. - 85
Aubry, J. M. - 193
Auge, J. - 116
Auricchio, S. - 313
Ayorinde, F. O. - 245
Babler, J. H. - 101, 144
Bachi, M. D. - 423
Back, T. G. - 268, 444
Backvall, J. E. - 15, 101, 275
Badanyan, S. O. - 128
Baird, M. S. - 150
Baker, R. - 35, 42, 230
Balasubramaniyan, V. - 416
Balasuriya, R. - 372
Baldwin, J. E. - 24
Baliah, V. - 349
Banerjee, U. K. - 241
Banwell, M. G. - 141

Baraldi, P. G. - 74
Barluenga, J. - 30, 64, 99, 144, 283, 340, 354, 392, 404, 418, 422
Barrett, A. G. M. - 121
Barriere, J. C. - 71
Bartlett, P. A. - 34
Barton, D. H. R. - 53, 273, 306
Barton, T. J. - 135
Bartsch, R. A. - 118, 267
Barua, J. N. - 383
Barua, N. C. - 304, 431
Baruah, R. N. - 304
Bates, G. S. - 1
Battiste, M. S. - 9, 331
Batzer, H. - 349
Bauld, N. L. - 153, 154, 170
Beak, P. - 26, 157, 408
Becker, D. - 173
Becker, H. D. - 169
Becker, K. B. - 130, 432
Beckwith, A. L. J. - 92, 297
Beebe, T. R. - 240, 272

AUTHOR INDEX

Beletskaya, I. P. - 227
Bellassoued, M. - 241
Belleau, B. - 4
Bellesia, F. - 244, 400
Belletire, J. L. - 117, 119, 308
Bellus, D. - 183
Benezra, C. - 333
Ben-Ishai, D. - 331
Benkeser, R. A. - 290
Berchtold, G. A. - 184
Bergbreiter, D. E. - 4, 15
Bergman, R. G. - 266
Bergstrom, D. E. - 205
Berlan, J. - 79
Bertrand, M. - 85, 92, 182
Bertz, S. H. - 71, 151
Beslin, P. - 183
Bestmann, H. J. - 58, 107, 112, 139, 220
Bettolo, R. M. - 87
Beugelmans, R. - 18
Bhaduri, A. P. - 13, 52, 193

Bhatt, M. V. - 448
Bhattacharya, A. J. - 214
Bhide, G. V. - 47
Bickelhaupt, F. - 34, 63
Billington, D. C. - 224
Binger, P. - 175
Birkofer, L. - 157, 215
Black, T. H. - 96
Blackburn, G. M. - 22
Blagoeva, I. - 414
Blandy, C. - 266
Block, E. - 119, 311
Block, R. - 120, 165
Blotny, G. - 351
Blunt, J. W. - 60
Boche, G. - 408
Bock, M. G. - 48
Boeckman, R. K., Jr. - 166, 214
Boger, D. L. - 153, 215, 365
Boguslavskaya, L. S. - 400
Bohme, H. - 90
Bonet, J. J. - 95
Bongini, A. - 262

Borchardt, R. T. - 193, 206, 287, 318, 335, 395

Borschberg, H. J. - 181

Bowman, W. R. - 18

Box, V. G. S. - 447

Bozzini, S. - 83

Brady, W. T. - 145

Branchaud, B. P. - 283, 376

Brassard, P. - 213

Braun, M. - 31, 42

Breitmaier, E. - 192

Bridges, A. J. - 81, 124, 162

Brisdon, B. J. - 169

Broadbent, A. D. - 72

Brockmann, H. - 126

Brook, M. A. - 417

Brookhart, M. - 147

Brooks, D. W. - 189

Brown, E. - 197, 391

Brown, H. C. - 68, 203, 219, 220, 221, 280, 297, 309, 395, 396

Brownbridge, P. - 36, 425

Buchi, G. - 182

Burke, S. D. - 166, 168, 184

Burton, D. J. - 115, 146

Butsugan, Y. - 24

Cacchi, S. - 201, 218

Caine, D. - 51, 130, 165

Cainelli, G. - 51, 391

Calligaris, M. - 175

Calo, V. - 24

Calverley, M. J. - 407

Camps, F. - 114, 260

Canonne, P. - 61

Cantacuzene, D. - 115

Caporusso, A. M. - 194, 281, 288

Capraro, H. G. - 327

Capuano, L. - 58

Cardillo, G. - 267, 313, 368

Caristi, C. - 249

Carlson, R. M. - 27, 205, 337

Carpita, A. - 140

Carrie, R. - 142, 146

Carturan, G. - 175

AUTHOR INDEX

Caruthers, M. H. - 378
Casey, M. - 350
Casiraghi, G. - 189
Casnati, G. - 196
Cassani, G. - 131
Castedo, L. - 199, 255
Castro, B. - 46
Castro, B. R. - 450
Cate, L. A. - 206
Catellani, M. - 148, 218
Caubere, P. - 216, 222, 311
Cava, M. P. - 91
Cavinato, G. - 221
Cella, J. A. - 269
Chalchat, J. C. - 147
Chamberlin, A. R. - 121
Chambers, R. D. - 124
Chan, T. H. - 204, 209, 343, 383
Chandler, M. - 101
Chandrasekaran, S. - 284
Chapleur, Y. - 43
Chatterjee, A. - 145, 401

Chaudhari, D. D. - 338
Cheikh, R. B. - 413
Chen, E. Y. - 74
Chen, G. J. - 206
Chenevert, R. - 90
Cheng, C. H. - 198
Chikashita, H. - 287
Childs, R. F. - 176
Chiusoli, G. P. - 218
Chou, T. S. - 66
Chowdhury, P. K. - 370
Christl, M. - 163
Chu, D. T. W. - 274
Chukovskaya, E. C. - 313
Chung, S. K. - 92
Cimarusti, C. M. - 329
Cinquini, M. - 55
Citterio, A. - 92, 263
Claesson, A. - 130, 135
Clark, J. H. - 80
Clark, T. - 31
Clerici, A. - 67
Clive, D. L. J. - 92, 99

Coates, R. M. - 186, 236
Cocagne, P. - 104, 449
Cohen, N. - 55
Cohen, T. - 147
Comasseto, J. V. - 22, 330, 389
Comins, D. L. - 207
Conia, J. M. - 85, 149, 150, 152, 174
Consiglio, G. - 24, 218, 228
Contreras, R. - 276, 294, 309
Cook, J. M. - 151
Cooke, M. P., Jr. - 125
Cooney, J. V. - 89
Cooper, A. J. L. - 105, 391
Cooper, G. F. - 121
Corey, E. J. - 58, 127, 138, 382
Cornelis, A. - 258
Corriu, R. J. P. - 37, 72, 277, 293
Costa, A. - 405
Costa, P. R. R. - 52

Coutrot, P. - 58
Coward, J. K. - 58
Coxon, J. M. - 61
Crandall, J. K. - 349
Craven, B. M. - 168
Crawford, R. J. - 43, 241
Croft, A. P. - 453
Crombie, L. - 329
Crossland, I. - 70
Croteau, A. A. - 44
Crow, W. D. - 360
Curran, D. P. - 51, 317
Current, S. P. - 224
Cutting, I. - 136, 443
Cuvigny, T. - 301
Czernecki, S. - 190
Daboun, H. A. - 449
Dalla Cort, A. - 238
D'Angeli, F. - 413
d'Angelo, J. - 1, 45, 61
Dangyan, Y. M. - 210
Danheiser, R. L. - 173
Danishefsky, S. - 74, 161, 347

Das, K. G. - 159
Daub, G. H. - 195
Daub, G. W. - 181, 182
Daub, J. - 177
Dauben, W. G. - 73, 74
Davies, G. D., Jr. - 218
Davies, S. G. - 32
Davis, F. A. - 259
Dean, F. M. - 126
De Clercq, P. J. - 87, 169
Degani, I. - 440
Degrand, C. - 43
De Groot, A. - 54, 113, 348
De Kimpe, N. - 105, 129, 144, 145, 415, 437
Delmas, M. - 111
De Lucchi, O. - 164
de Meijere, A. - 163
Demuth, M. - 88, 152
Denmark, S. E. - 63, 127, 182
des Abbayes, H. - 222
Descotes, G. - 226
De Shong, P. - 155

Dev, S. - 85, 128, 382
Dieck, H. A. - 15
Dieter, R. K. - 78, 81
Di Giamberavdino, T. - 394
Dijink, J. - 320
Dinjus, E. - 225, 227
Djerassi, C. - 152
Dormand, A. - 51
Dotz, K. H. - 228
Doyle, M. P. - 97
Drago, R. S. - 269
Dreiding, A. S. - 169, 170, 171, 187
Drewes, S. E. - 12
Dryanska, V. - 83
Dubois, J. E. - 299
Duboudin, F. - 90
Duhamel, L. - 5, 121
Dunogues, J. - 65, 115
Dupas, G. - 277
Durst, T. - 167
Duthaler, R. O. - 210
Dyke, S. F. - 191

Dyusenova, Z. I. - 72
Dzhemilev, U. M. - 216
Eaton, P. E. - 49
Effenberger, F. - 190, 412
Ege, S. N. - 321
Eilbracht, P. - 224
Eisch, J. J. - 66, 224
Elliott, J. D. - 89
Elnagdi, M. H. - 365
Elphimoff-Felkin, I. - 146
Elsafie, S. M. M. - 438
Enders, D. - 4, 76
Endo, Y. - 249
Engel, Ch. R. - 145
Epsztajn, J. - 209
Erden, I. - 171
Erdik, E. - 35
Erker, G. - 67
Ermolov, A. F. - 57
Eugster, C. H. - 121
Evans, S. A., Jr. - 317, 416
Fabrissin, S. - 72
Falck, J. R. - 30

Faller, J. W. - 225
Farina, F. - 215
Favorskaya, I. A. - 2
Fehr, C. - 57
Fehrentz, J. A. - 303
Feiring, A. E. - 18
Feit, B. A. - 121
Feringa, B. L. - 335
Fernandez, S. - 293
Feuer, H. - 247
Ficini, J. - 52, 145, 174
Fiecchi, A. - 146
Fife, W. K. - 218
Fiorenza, M. - 264
Firouzabadi, H. - 258, 422, 424
Fischer, A. - 46
Fischer, F. - 55
Fitch, J. W. - 420
Fitjer, L. - 170
Fleet, G. W. J. - 115
Fleming, I. - 5, 36, 119, 146
Flitsch, W. - 344
Flood, T. C. - 5

Florio, S. - 42
Florjanczyk, Z. - 80
Floyd, D. M. - 326
Flynn, G. A. - 115
Foa, M. - 222
Ford, W. T. - 7, 111
Foucaud, A. - 102, 325, 415
Fox, M. A. - 149, 455
Franck, R. W. - 158
Franck-Neumann, M. - 228
Frechet, J. M. J. - 63
Freeman, J. P. - 356
Freidinger, R. M. - 412
Frejd, T. - 44, 204
Fried, J. - 53
Friedrichsen, W. - 358
Fringuelli, F. - 165
Fristad, W. E. - 263, 306
Froehler, B. C. - 430
Fuchs, P. L. - 52, 74, 76, 115, 125
Fuganti, C. - 69
Fuji, K. - 54
Fujii, N. - 389
Fujimori, K. - 178
Fujisawa, T. - 10, 25, 29, 53, 64, 123, 145, 183, 294, 295
Fujise, Y. - 186
Fujita, E. - 14, 21, 81, 123, 152, 188, 301, 334
Fujita, T. - 93
Fujiwara, Y. - 216
Funabiki, T. - 91, 129, 218
Fung, S. - 195
Funk, R. L. - 356
Furukawa, M. - 433
Furukawa, S. - 208
Gadwood, R. C. - 446
Gaede, B. - 164
Gandolfi, C. A. - 110
Ganem, B. - 31, 236, 255
Garcia, G. A. - 7, 332
Gard, G. L. - 232
Garst, M. E. - 11, 83, 176, 315
Gasc, M. B. - 411

Gasparrini, F. - 259
Gassman, P. G. - 94
Gaudemar, M. - 9, 41
Geivandov, R. K. - 158
Genet, J. P. - 407
Gesson, J. P. - 157, 213
Ghatak, U. R. - 87, 217, 271
Ghera, E. - 13
Ghosez, L. - 170
Giacomelli, G. - 139
Giese, B. - 91, 92
Gilbert, A. - 148
Gilbert, J. C. - 97, 112
Gilchrist, T. L. - 452
Ginsberg, D. - 153, 448
Giordano, C. - 84, 421
Gladiali, S. - 222
Gladysz, J. A. - 434
Glass, R. S. - 158
Gleiter, R. - 153, 163
Gnanadoss, L. M. - 205
Goel, E. P. - 307
Goering, H. L. - 25

Gogte, V. N. - 129
Gold, V. - 106, 205
Goldberg, Y. S. - 375
Gololobov, Y. G. - 10
Golse, R. - 32
Gonzalez, A. G. - 87
Goodwin, T. E. - 78
Gordon, E. M. - 328
Gore, J. - 68, 180, 226, 384
Gorelik, A. M. - 193
Gosselin, P. - 88
Goya, S. - 398
Graumann, J. - 336
Gravel, D. - 2
Graziano, M. L. - 272
Greene, A. E. - 28, 171
Greenlee, W. J. - 90
Gribble, G. W. - 214, 276
Grieco, P. A. - 75, 156, 180, 391
Grigg, R. - 222
Grigorieva, N. Y. - 44
Grubbs, R. H. - 66, 107, 136

AUTHOR INDEX

Guanti, G. - 48
Guchbauer, G. - 454
Guindon, Y. - 393, 441
Gupton, J. T. - 411
Gutsulyak, B. M. - 341
Guy, H. - 242
Guyot, J. - 43
Guziec, F. S. - 146
Hall, H. K., Jr. - 228, 451
Hallberg, A. - 140
Hamana, H. - 65
Hamer, N. K. - 94
Hara, S. - 434
Harayama, T. - 89
Harris, T. M. - 31, 48
Hart, D. J. - 130, 325
Hart, H. - 206
Hartman, G. D. - 199, 256
Hartmann, H. - 50
Harvey, R. G. - 89, 208, 239
Hashimoto, S. I. - 457
Hassner, A. - 173
Hata, N. - 218

Hata, T. - 367, 368, 439
Hauser, F. M. - 215
Hayakawa, Y. - 430
Hayashi, T. - 19, 34, 62, 119
Heasley, V. L. - 265, 267
Heathcock, C. M. - 41, 42, 62, 73, 82
Hebert, E. - 8
Heck, R. F. - 127, 128, 201
Hegedus, L. S. - 15, 201
Heilmann, S. M. - 65
Hellwinkel, D. - 360
Helmchen, G. - 10
Hendrickson, J. B. - 165, 342
Henin, F. - 127, 332
Henning, H. G. - 217
Henning, R. - 82
Hesse, M. - 145, 339
Hewson, A. T. - 76, 80
Himbert, G. - 138
Hirai, Y. - 361
Hirama, M. - 239, 282
Hirao, T. - 123, 298, 299

Hiyama, T. - 49, 60, 68, 124, 261
Ho, T. L. - 13, 27, 51, 91, 141, 182
Hoberg, H. - 139, 224
Hodge, P. - 76, 165
Hofer, E. - 303
Hoffmann, H. M. R. - 124, 177
Hoffman, R. V. - 255
Hoffmann, R. W. - 68, 394
Hojo, M. - 293, 294
Holland, H. L. - 27, 397
Holmes, A. B. - 101
Holton, R. A. - 1
Hoornaert, G. J. - 185
Hoppe, D. - 59, 136, 330
Hornback, J. M. - 166
Horner, L. - 139
Horton, D. - 163
Hosokawa, T. - 269
Hosomi, A. - 114, 120
Houk, K. N. - 105, 176, 177
House, H. O. - 52

Howell, H. G. - 283
Hoye, T. R. - 10
Huche, M. - 24
Hudlicky, T. - 97
Hudson, A. T. - 126
Huffman, J. W. - 311, 397
Hutchins, R. O. - 367, 379
Iacobucci, G. A. - 350
Ibuka, T. - 150
Iddon, B. - 350
Iguchi, M. - 159
Ihara, M. - 326
Iizuka, K. - 406
Ikeda, K. - 11, 323
Ikeda, M. - 172
Ikehira, H. - 22
Ikekawa, N. - 144
Ila, H. - 174
Imamoto, T. - 42, 231, 397, 429
Imbach, J. L. - 386
Inamoto, N. - 108, 117
Inch, T. D. - 205
Inesi, A. - 144

AUTHOR INDEX

Innocenti, S. - 51
Ireland, R. E. - 184, 215
Ishibashi, H. - 314
Ishihara, T. - 40, 115, 299
Ishii, Y. - 337
Isoe, S. - 94, 168, 213
Issleib, K. - 354
Ito, K. - 226
Ito, M. M. - 343
Ito, Y. - 167
Itoh, K. - 33, 101
Itokawa, H. - 86
Itsuno, S. - 279
Ivanov, I. C. - 70
Iwakuma, T. - 282
Iwamoto, R. T. - 31
Iwao, M. - 209
Iyoda, M. - 226
Jackman, L. M. - 2
Jackson, W. R. - 129
Jacob, L. - 387
Jacob, P., III - 192
Jager, V. - 104

Jahngen, E. G. E. - 148
Jalander, L. - 79
James, B. R. - 226
Janout, V. - 118, 387
Jaworski, T. - 122
Jefford, C. W. - 363
Jendralla, H. - 171
Johnson, M. D. - 458
Johnson, R. P. - 160
Johnson, W. S. - 34, 88, 89, 137
Jonczyk, A. - 21, 149
Jones, D. N. - 76
Jones, M., Jr. - 72
Jones, R. C. F. - 121
Jorgensen, K. A. - 384
Joullie, M. M. - 108
Julia, M. - 15, 23, 56, 119, 120, 302
Julia, S. A. - 121, 147, 436
Jung, M. E. - 28, 127, 181, 401, 428
Junjappa, H. - 174

Jurczak, J. - 155
Kabalke, G. W. - 434
Kagan, H. B. - 65, 100, 244
Kageyama, T. - 339, 419
Kaim, W. - 451
Kaji, A. - 38, 70, 112
Kajmoto, T. - 224
Kakisawa, H. - 87
Kallmerten, J. - 183
Kalvoda, J. - 61
Kametani, T. - 167, 174, 381, 403
Kamijo, T. - 420
Kamimura, T. - 388, 389
Kaminski, V. V. - 328
Kamitori, Y. - 277, 278
Kamogawa, H. - 129
Kandil, A. A. - 27
Kane, V. V. - 425
Kanematsu, K. - 141, 154
Kantlehner, W. - 3
Kapil, R. S. - 84
Karavan, V. S. - 300

Karimian, K. - 91
Kasahara, A. - 221
Kashima, C. - 78, 300
Kashin, A. N. - 237
Kato, S. - 438
Kato, T. - 29
Katritzky, A. R. - 18, 31, 251
Katzenellenbogen, J. A. - 62
Kawabata, N. - 257
Kawakami, Y. - 123
Kawana, M. - 61
Kawashima, T. - 117
Keana, J. F. W. - 155, 354
Keinan, E. - 15, 32, 304
Kellogg, R. M. - 37
Kelly, T. R. - 61, 213
Kende, A. S. - 103, 124
Kessler, H. - 380
Keta, Y. - 379
Khenkina, T. V. - 50
Khusid, A. K. - 141
Kice, J. L. - 265
Kikukawa, K. - 200, 201
Kim, S. - 64, 297, 418

Kim, S. W. - 97
Kirby, G. W. - 313
Kirmse, W. - 87
Kishi, Y. - 47, 61
Kita, Y. - 369
Kitaoka, M. - 333
Kjonaas, R. A. - 198
Klaveness, J. - 24
Klein, J. - 26, 450
Klemer, A. - 371
Klumpp, G. W. - 24, 30
Knight, D. W. - 109
Knorr, R. - 295
Knowles, W. S. - 310
Kobayashi, H. - 214
Kochetkov, N. K. - 64
Kochhar, K. S. - 383
Kocienski, P. - 55, 86
Kohl, F. X. - 51
Kohn, H. - 266, 282, 296
Kolasa, T. - 117, 414
Kolb, M. - 17
Koller, W. - 357

Komarov, N. V. - 134
Komiyama, M. - 192
Koreeda, M. - 30, 184, 366
Kornblum, N. - 246, 429
Kostikov, R. R. - 141
Kosugi, M. - 313
Kotake, H. - 21
Kozerski, L. - 19
Kozikowski, A. P. - 33, 82, 108, 175, 333
Kozima, S. - 41
Kraus, G. A. - 38, 157, 167, 215
Kremlev, M. M. - 124
Krepski, L. R. - 65
Kresze, G. - 248
Krief, A. - 81, 108, 119, 161, 304
Krohn, K. - 66, 196
Kulinkovich, O. G. - 152
Kulkarni, G. H. - 151
Kulkarni, S. N. - 47
Kumada, M. - 23, 62, 130, 135
Kumamoto, T. - 216

Kumar, B. - 217
Kumar, Y. - 286
Kundig, E. P. - 202, 223
Kunz, H. - 381
Kurokawa, S. - 193
Kurosawa, H. - 18
Kurozumi, S. - 125
Kurth, M. J. - 183
Kuwajima, I. - 6, 40, 127, 139,
    145, 152, 180, 385, 445
Kuzuya, M. - 173
Kvita, V. - 162
Laarhoven, W. H. - 148, 170
Laatsch, H. - 126
L'abbe, G. - 364
Ladner, W. - 45
Lahoti, R. J. - 403
Lallemand, J. Y. - 160
Lambert, J. B. - 142
La Mattina, J. L. - 48
Landor, S. R. - 147
Landry, D. W. - 165
Lapatsanis, L. - 374

Larcheveque, M. - 49
Larock, R. C. - 200, 218
Larson, G. L. - 45
Lattes, A. - 111
Lau, K. S. Y. - 200
Laurent, E. - 243
Lawesson, S. O. - 424
Lazzaroni, R. - 221
Le Bigot, Y. - 108
Lee, S. J. - 364
Lee, T. J. - 216
Lee, T. V. - 5
Lee-Ruff, E. - 174
Lehmkuhl, H. - 226
Lelandais, D. - 96
Levisalles, J. - 28
Levy, L. A. - 214
Ley, S. V. - 2, 113, 160
Leyendecker, F. - 79
Lhommet, G. - 50
Liebeskind, L. S. - 3
Liebscher, J. - 192, 348
Lieto, J. - 104

Linstrumelle, G. - 135, 137
Liotta, D. - 314
Lipczynska-Kochany, E. - 411
Lipshutz, B. H. - 28, 77
Lissel, M. - 398
Little, R. D. - 80, 98, 254
Liu, H. J. - 24
Liu, S. H. - 85
Loewenthal, H. J. E. - 6
Lombardo, L. - 74
Longone, D. T. - 178
Lounasmaa, M. - 73
Luche, J. L. - 205
Luh, T. Y. - 218
Lupi, A. - 87
Luttke, W. - 129
Mackor, A. - 170
Magnus, P. - 224
Majchrzak, M. W. - 142
Majetich, G. - 82
Makin, S. M. - 1, 161
Makosza, M. - 205
Malherbe, R. - 183

Mali, R. S. - 206
Mallory, R. S. - 454
Mander, L. N. - 49, 74, 106, 456
Manhas, M. S. - 323, 418
Manoharan, T. S. - 409
Marchand-Brynaert, J. - 324
Marchese, G. - 64
Marino, J. P. - 132
Marinovic, N. M. - 92
Markgraf, J. H. - 160
Marshall, J. A. - 29
Martens, H. J. - 185
Martin, H. D. - 451
Martin, J. C. - 209
Martin, P. - 145
Martin, S. F. - 166
Martinez, J. - 374
Maruyama, K. - 74, 82
Marvell, E. N. - 186
Marx, J. N. - 399
Maryanoff, B. E. - 109, 110
Masamune, S. - 163

Masuda, S. - 187
Matsuda, I. - 62
Matsumoto, M. - 270
Matsumoto, T. - 87
Matsumura, N. - 44, 50, 351
Matsunaga, H. - 327
Matteson, D. S. - 220, 395
Matthews, R. S. - 27
Matui, S. - 45
Maulding, D. R. - 192
Mayr, H. - 85, 128, 160, 170
McArthur, C. R. - 373
McCleland, C. W. - 421
McCleland, R. A. - 451
McEwen, W. E. - 109
McGarvey, G. J. - 10
McKervey, M. A. - 190
McKillop, A. - 192, 256
McMurry, J. E. - 99, 203, 435
McOmie, J. F. W. - 212
Melikyan, G. G. - 118
Menger, F. M. - 37, 449
Menri, K. - 319

Merchant, J. R. - 193
Meth-Cohn, O. - 328
Meyers, A. I. - 4, 30, 111, 206, 207, 336
Mezheritskii, V. V. - 193
Miginiac, P. - 23, 78, 138
Migita, T. - 130, 202, 407
Miller, M. J. - 9
Miller, R. B. - 122
Miller, R. D. - 436
Miller, S. I. - 161
Minami, T. - 83, 158
Minisci, F. - 450
Minowa, N. - 83
Mioskowski, C. - 30, 111, 346
Mirbach, M. J. - 221
Mironov, V. A. - 153
Mitani, M. - 78, 148
Mitchell, R. H. - 31
Mitscher, L. A. - 196
Miwa, T. - 15
Miyoshi, N. - 249
Mizuno, K. - 172, 173

AUTHOR INDEX

Mohacsi, E. - 368
Moisseenkov, A. M. - 109
Moiseev, I. K. - 84
Mol, J. C. - 226, 227
Molander, G. A. - 133, 200
Molina, P. - 392
Montanari, F. - 55
Monti, D. - 285
Monti, H. - 370
Moore, H. W. - 321
Moore, J. A. - 154
Moreno-Manas, M. - 116
Morgans, D. J., Jr. - 149
Mori, A. - 278
Morizur, J. P. - 95
Morris, D. G. - 425
Morrison, J. D. - 104
Mosher, H. S. - 104
Motohashi, S. - 242
Mukaiyama, T. - 14, 36, 41, 50, 61, 62, 77, 169, 263, 274, 373, 402, 419, 437
Mukherjee, D. - 34, 97

Muller, N. - 96
Mulzer, J. - 71, 119, 146, 308
Murai, S. - 100, 144, 221
Murhashi, S. I. - 356
Murphy, W. S. - 196, 453
Murray, D. F. - 110
Musavirov, R. S. - 382
Musso, H. - 100
Muzart, J. - 35, 272
Nadir, U. K. - 347
Magarajan, M. - 76
Magashima, H. - 332
Nair, P. M. - 246
Nakai, T. - 63, 185
Nakajima, K. - 413
Nakajima, R. - 197
Nakamura, A. - 27, 67
Nakamura, K. - 292
Nakanishi, S. - 351
Nakashita, Y. - 446
Nakata, T. - 280
Nakayama, J. - 118, 305, 440
Nakayama, M. - 151

Napolitano, E. - 207
Narang, S. A. - 452
Narshimhan, N. S. - 251
Naruta, Y. - 61
Nasipuri, D. - 280
Naso, F. - 199
Natale, N. R. - 227, 285
Negishi, E. I. - 1, 3, 69, 126, 131, 133, 201, 220
Nemoto, H. - 152
Neubert, M. E. - 191
Neuenschwander, M. - 140
Nevrekar, N. B. - 188
Newcomb, M. - 4
Newkome, G. R. - 148
Newman, M. S. - 195
Newton, R. F. - 30, 42, 84
Nicholaides, D. N. - 212
Nicholas, K. M. - 140
Nichols, D. E. - 319
Nishida, S. - 173, 186
Nishimura, J. - 86
Nishio, T. - 60, 341

Nishizawa, M. - 88
Noels, A. F. - 142
Nojima, M. - 31
Nokami, J. - 62, 112, 253
Noland, W. E. - 161
Nomura, Y. - 80
Normant, J. F. - 24, 124, 131, 132, 149
Norris, R. K. - 12
Novak, J. - 206
Noyori, R. - 5, 39, 311
Nudulman, N. S. - 375
Nugent, W. A. - 62, 75
Nunomoto, S. - 123
Oae, S. - 448
Obushak, N. D. - 200
Ochiai, M. - 274
Oda, H. - 310
Oda, M. - 158, 226
Oehlschlager, A. C. - 136
Ogilvie, K. K. - 291
Ogino, T. - 172
Oguni, N. - 69

AUTHOR INDEX

Ogura, F. - 20, 422
Ogura, K. - 22, 302
Ohashi, M. - 217
Ohnuma, T. - 329
Ohshiro, Y. - 252
Ohta, A. - 405
Ohta, S. - 48, 234
Ohtsuka, Y. - 54
Oida, T. - 11, 436
Oikawa, Y. - 369
Oishi, T. - 54, 208
Ojima, I. - 100, 222, 296
Okamoto, Y. - 111, 155, 399
Okamura, W. H. - 109, 134, 185, 456
Oku, A. - 98, 137, 143, 150, 158
Olah, G. A. - 89, 90, 192, 244, 259, 307, 390, 398, 426, 451
Ollis, W. D. - 185
Olofson, R. A. - 355
Omae, I. - 457
Onaka, M. - 366
Ongania, K. H. - 298
Ono, A. - 284
Ono, N. - 38, 70, 112, 303
Oppolzer, W. - 71, 79, 101, 163
Orchin, M. - 221
Orsini, F. - 9
Ortar, G. - 185
Oshima, K. - 127, 133, 231
Oguka, A. - 111, 202, 298
Otera, J. - 62, 161
Otsuji, Y. - 1/2, 226
Otto, H. H. - 358
Ourari, A. - 268
Ourisson, G. - 21
Overman, L. E. - 162, 180, 226, 345
Ozasa, S. - 197
Padwa, A. - 148, 164, 217, 345
Paloma, C. - 233, 257, 322, 372, 373, 375, 385, 400, 404, 419, 427, 428
Palumbo, G. - 264

Panetta, C. A. - 43
Panunzi, A. - 188
Paquette, L. A. - 2, 34, 142, 152, 153, 156, 163, 164, 181, 445, 448
Paraskewas, S. M. - 275
Parker, K. A. - 409
Parrick, J. - 212
Parsons, P. J. - 84, 136
Pasternak, M. - 94
Paterson, I. - 11, 33, 42, 45
Pattenden, G. - 75, 127, 172
Patterson, J. W. - 134
Pearson, A. J. - 8, 12, 28
Pelizzoni, F. - 157
Pellacani, L. - 247
Pelter, A. - 45, 81, 219
Penco, S. - 208
Perez-Ossorio, R. - 60
Perichon, J. - 199
Peruzzo, V. - 65
Peterson, J. L. - 98
Pfister, J. R. - 410

Pfleiderer, W. - 378, 388
Piacenti, F. - 221
Piancatelli, G. - 242
Pickles, G. M. - 394
Picq, D. - 376
Piers, E. - 125, 187, 318
Pietrusiewicz, K. M. - 115
Pillai, N. - 227
Pinhey, J. T. - 201
Pinnick, H. W. - 36, 405, 454
Pirrung, M. C. - 435
Pivnitskii, K. K. - 12
Pochini, A. - 196
Pommer, H. - 107
Ponaras, A. A. - 184
Poncini, L. - 23
Porta, O. - 67
Potts, K. T. - 357
Pouvzal, A. A. - 352
Pozdnyakovich, Y. V. - 188
Prasad, J. V. N. V. - 227
Prochazka, M. - 129
Proll, T. - 435

AUTHOR INDEX

Pucci, S. - 221
Purohit, P. C. - 150
Purrington, S. T. - 243
Queguiner, G. - 209
Quindon, Y. - 369, 385
Quinkert, G. - 106, 456
Quintard, J. P. - 62
Rabinovitz, M. - 111
Radhakrishna, A. S. - 410
RajanBabu, T. V. - 165
Rajaram, J. - 289
Ram, R. N. - 403
Ramadas, S. R. - 63, 392
Ranganathan, D. - 164
Rao, A. S. - 193, 349
Rao, A. V. R. - 17, 161, 195, 196
Rao, C. G. - 385
Rao, D. V. - 48
Raphael, R. A. - 163
Rapoport, H. - 64
Rasmussen, J. K. - 65, 281
Rathke, M. W. - 47

Raucher, S. - 159, 319
Razdan, R. K. - 159
Redpath, J. - 106, 456
Reese, C. B. - 437
Reetz, M. T. - 6, 39, 67, 384
Regen, S. L. - 7, 118, 339
Regitz, M. - 132, 162
Reich, H. J. - 432
Reissig, H. U. - 148
Renaud, R. N. - 50
Reusch, W. - 89, 146, 250
Reutrakul, V. - 20, 21, 305, 442
Reuvers, J. T. A. - 338
Rezende, M. C. - 118
Rhodes, Y. E. - 41
Ricca, A. - 211
Ricci, A. - 406
Rich, D. H. - 16
Richey, H. G., Jr. - 77, 103
Richter, R. - 397
Rickborn, B. - 159
Ridley, D. D. - 56

Ried, W. - 172

Rieke, R. D. - 63, 69, 198

Rigby, J. H. - 154, 194

Risitano, F. - 361

Rizzi, G. P. - 190

Roberts, B. W. - 38

Robev, S. K. - 352

Robins, M. J. - 291

Rocca, J. R. - 115

Rodrigo, R. - 159

Ronald, R. C. - 318

Rosenberger, M. - 55

Rosenblum, M. - 13, 35

Rosini, G. - 53, 171, 233, 302

Rossi, R. - 140

Roulet, R. - 163

Roush, W. R. - 27, 68

Rousseau, G. - 69, 142

Roussi, G. - 345

Rozen, S. - 243

Rubottom, G. M. - 64, 237

Ruchardt, C. - 100

Ruminski, J. K. - 190

Rumyamtseva, K. S. - 12

Russell, R. A. - 128

Russkikh, S. A. - 126

Russkikh, V. S. - 90

Saalfrank, R. W. - 47, 334

Saavedra, J. E. - 19

Sabourin, E. T. - 200

Saednya, A. - 427

Saegusa, T. - 167

Saigo, K. - 4

Sainsbury, M. - 199

Saito, I. - 270

Sakakibara, T. - 57

Sakamoto, M. - 65, 90

Sakan, K. - 168

Sakurai, H. - 6, 33, 62, 114, 120

Salaun, J. - 144, 445

Saljoughian, M. - 66

Salomon, R. G. - 228, 455

Sanchez, F. - 81

Sanchez, I. H. - 8

Sands, R. D. - 210

AUTHOR INDEX

Santaniello, E. - 233, 258
Santelli, M. - 74, 84, 137, 453
Santelli-Rouvier, C. - 84, 453
Sardina, F. J. - 24
Sarma, A. S. - 108
Sarti-Fantoni, P. - 140
Sartori, G. - 189
Sasaki, H. - 17
Sasatani, S. - 409
Sato, F. - 66, 84, 133
Sato, K. - 21
Sato, M. - 29, 363
Sato, R. - 444
Sato, S. - 46
Sato, T. - 281, 342
Sato, Y. - 45, 123
Sauer, J. - 165, 173
Sauvetre, R. - 124
Savignac, P. - 58
Sawicki, R. A. - 222
Sawyer, J. S. - 377
Schaden, G. - 120

Schank, K. - 75, 240
Schaumann, E. - 108, 183, 443
Schechter, H. - 31, 55
Scheeren, H. W. - 172
Scheffold, R. - 95, 228
Schlessinger, R. H. - 154, 168
Schleyer, P. v. R. - 206, 211
Schlosser, M. - 107
Schmidbaur, H. - 107
Schmidt, H. W. - 338
Schmidt, R. R. - 6, 121, 439
Schmidt, U. - 308
Schneider, D. F. - 110
Schneider, G. - 38
Schollkopf, U. - 16, 17, 45, 61, 105
Scholz, D. - 20
Schreiber, S. L. - 95, 271, 339
Schroth, W. - 359
Schuda, P. F. - 240
Schultz, A. G. - 2, 34, 350
Schulze, K. - 360
Schwartz, J. - 111

Scolastico, C. - 46, 55
Scott, F. - 149
Scriven, E. F. V. - 105, 452
Secrist, J. A., III - 114
Seebach, D. - 10, 17, 37, 60, 66, 67, 72, 106, 457
Seeman, J. I. - 452
Sekiya, M. - 11
Selve, C. - 309
Semmelhack, M. F. - 8, 24, 126, 202, 224, 231, 254
Sen, A. - 128
Sepiol, J. - 210
Sepulveda, J. - 60
Serratosa, F. - 426
Seto, H. - 94
Seyden-Penne, J. - 73
Seyferth, D. - 47, 60, 223
Shanzer, A. - 116, 364, 414
Sharma, R. P. - 95, 370
Sharp, J. T. - 353
Sharpless, K. B. - 29
Shastin, A. V. - 128
Shea, K. J. - 125, 166, 168, 186
Shechter, H. - 31
Sheppard, R. C. - 449
Shibuya, M. - 195
Shim, S. C. - 326
Shimizu, I. - 273
Shin, C. - 353
Shinozaki, H. - 426
Shiorri, T. - 97, 102, 361, 363
Shirahama, H. - 87
Shishoo, C. J. - 352
Shono, T. - 58, 66, 69, 128, 238, 253, 273, 424, 450
Sieler, J. - 227
Simchen, G. - 250
Simonet, J. - 49
Sinay, P. - 138
Singh, H. - 254, 440
Skattebol, L. - 179
Sliwa, W. - 364
Smadja, W. - 134, 452
Smart, B. E. - 146

AUTHOR INDEX

Smit, W. A. - 5
Smith, A. B., III - 122, 150, 172
Smith, E. H. - 183
Smith, F. X. - 45
Smith, J. G. - 30
Smith, J. R. L. - 245
Smith, K. - 59
Smith, P. A. S. - 143
Smith, R. F. - 362
Smithers, R. H. - 122
Snider, B. B. - 166, 179
Snieckus, V. - 207, 248, 250
Snowden, R. L. - 101
Soai, K. - 79, 296, 442
Soderquist, J. A. - 130
Solladie, G. - 55
Solladie-Cavallo, A. - 103
Sonawane, H. R. - 150
Sondengam, B. L. - 288
Soto, J. L. - 250, 362
Spencer, A. - 199
Stamm, H. - 14

Stammer, C. H. - 143
Stang, P. J. - 104, 143
Staunton, J. - 65
Stavber, S. - 245
Steglich, W. - 65, 114, 252
Stella, L. - 345
Sternbach, D. D. - 169
Stetter, H. - 91
Stevens, R. W. - 41
Still, I. W. J. - 265
Still, W. C. - 68, 112, 181,
Stillings, M. R. - 191
Stirling, C. J. M. - 21
Stoodley, R. J. - 157, 284
Stork, G. - 52, 75, 85, 127, 138, 330
Stothers, J. B. - 59
Strunz, G. M. - 61
Sturtz, G. - 260
Suarez, E. - 84

Subba-Rao, G. S. R. - 34
Suda, H. - 279
Suginome, H. - 131
Sukata, K. - 7
Sutherland, J. K. - 85, 88
Suzuki, A. - 121, 140, 219
Suzuki, H. - 111, 202, 291, 295, 402
Suzuki, K. - 410
Suzuki, N. - 289
Suzuki, S. - 24, 182
Suzuki, Y. - 192
Swenton, J. S. - 95, 208, 275
Tabbaa, I. - 92
Taber, D. F. - 34, 97
Taddei, M. - 90, 268
Takabe, K. - 251, 253
Takagi, K. - 131
Takahashi, H. - 61
Takahashi, K. - 8, 49
Takahashi, T. - 7
Takahashi, T. T. - 101
Takai, K. - 124, 237, 270, 393

Takaki, K. - 230
Takano, S. - 7, 438
Takayama, H. - 21, 120
Takeda, A. - 46, 112
Takeda, K. - 433
Takeda, T. - 20, 130
Takei, H. - 75
Takeshita, H. - 120
Takimoto, S. - 417
Tam, S. W. - 218
Tam, W. - 126
Tamao, K. - 23
Tamblyn, W. H. - 152
Tamm, C. - 157
Tamura, Y. - 215
Tanabe, Y. - 359
Tanaka, F. S. - 197
Tanaka, K. - 21, 335
Tanikaga, R. - 44, 334
Tanimoto, S. - 11, 22
Tanura, Y. - 46
Tashiro, M. - 199
Taticchi, A. - 165

AUTHOR INDEX

Tatsumi, T. - 15
Taylor, E. C. - 96, 101
Taylor, R. J. K. - 12
Taylor, S. K. - 189, 194
Teisseire, P. - 124
Teutsch, G. - 29
Tewari, R. S. - 312
Thakur, D. K. - 359
Thebtaranonth, Y. - 145
Thiem, J. - 54
Thomas, C. B. - 421
Thomas, E. J. - 168
Thomas, E. W. - 151, 152
Thomsen, I. - 358
Thompson, D. W. - 133
Thompson, W. J. - 70
Thyes, M. - 190
Tiecco, M. - 204, 390
Tietze, L. F. - 340
Tintel, C. - 205
Tipping, A. E. - 96
Tisler, M. - 364
Tius, M. A. - 123, 209

Tkatchenko, I. - 128
Tobinaga, S. - 210
Toda, F. - 174
Toda, T. - 178
Tokuda, M. - 95
Tokutake, N. - 322
Tolstikov, H. A. - 109
tom Dieck, H. - 160
Tomioka, H. - 118, 305
Tomoda, S. - 80
Toniolo, L. - 221
Torii, S. - 96, 164, 286
Torsell, K. B. G. - 51, 317, 343
Tou, J. S. - 102
Trimitsis, G. B. - 2
Trippett, S. - 207, 210
Trivedi, G. K. - 211
Trombini, C. - 47
Trost, B. M. - 14, 15, 16, 17, 21, 24, 106, 117, 149, 157, 175, 196, 404, 431
Tsien, Y. L. - 226

Tsuchihashi, G. I. - 217
Tsuchiya, T. - 353
Tsuge, O. - 161, 213, 314
Tsuji, J. - 6, 7, 17, 101
Tsujimoto, K. - 94
Turecek, F. - 176
Turner, J. S. - 209
Turner, J. V. - 159
Turner, R. W. - 214
Uda, H. - 56
Ueda, M. - 402
Uehara, A. - 288
Ueki, M. - 379
Uemura, M. - 118, 208
Ueno, Y. - 112, 232, 259, 292
Ugi, I. - 376, 377
Ullenius, C. - 78
Umani-Ronchi, A. - 47, 222
Uneyama, K. - 423
Ungaro, R. - 193
Uyehara, T. - 148
van Boeckel, C. C. A. - 372
Van De Mark, M. R. - 386

van der Gen, A. - 2, 112, 398
Vankar, Y. D. - 292
van Leeuwen, P. W. N. M. - 221
Van Saun, W. A. - 215
van Schaik, T. A. M. - 113
van Tamelen, E. E. - 85, 88, 195
Varvoglis, A. - 420
Vasi, I. G. - 60
Vasil'eva, L. L. - 57, 81
Vasil'eva, L. P. - 355
Vatele, J. M. - 183
Vaultier, M. - 310
Vedejs, E. - 160
Venkataramani, P. S. - 256
Venkateswaran, R. V. - 241
Venturello, P. - 108
Vermeer, P. - 124, 135
Viallefont, P. - 28
Viehe, H. G. - 93, 173
Vilieras, J. - 112
Villemin, D. - 45, 138, 227
Villenave, J. J. - 92

AUTHOR INDEX

Villieras, J. - 112
Vlad, P. F. - 88, 346
Vlietstra, E. J. - 375
Voelter, W. - 264, 373
Vogel, P. - 163
Vogtle, F. - 105
Volkmann, R. A. - 23, 348
Vollhardt, K. P. C. - 211, 224
Vo-Quang, Y. - 112
Vora, K. P. - 225
Vorbruggen, H. - 218
Wade, P. A. - 355
Waegell, B. - 142, 226
Wagner, P. J. - 94
Wakselman, C. - 13, 46, 367, 380
Walborsky, H. M. - 110
Walker, B. J. - 140
Wallace, T. W. - 212
Walling, C. - 188
Walter, D. - 458
Wamser, C. C. - 173
Wang, D. - 278

Wang, K. K. - 135
Ward, R. S. - 81
Warner, P. - 186
Warren, S. - 18, 19, 22, 54, 109
Warrener, R. N. - 128
Wartski, L. - 73
Watanabe, Y. - 285, 340
Watt, D. S. - 56, 62, 81, 168
Weber, G. - 272
Weber, L. - 457
Weedon, A. C. - 38
Weiler, L. - 85, 86
Weinreb, S. M. - 124, 406
Weller, D. D. - 157
Welzel, P. - 51
Wender, P. A. - 30, 94, 134
Wenkert, D. - 271
Wenkert, E. - 165, 204
Wennerstrom, O. - 186
Werstiuk, N. M. - 59, 448
Weyerstahl, P. - 141
Whalley, W. B. - 165

Wheeler, C. J. - 112
White, J. D. - 109, 199
Whitesell, J. K. - 4, 179, 426
Whitesides, G. M. - 37, 104
Wicha, J. - 17
Widdowson, D. A. - 208
Widmer, U. - 380
Wierenga, W. - 48, 306
Wightman, R. H. - 442
Williams, D. R. - 301, 371
Williams, J. R. - 186
Williams, R. M. - 190
Willner, I. - 118
Witczak, Z. J. - 449
Witkop, B. - 320
Wojcicki, A. - 175
Wolfe, J. F. - 204
Wolff, S. - 94
Wolfsberger, W. - 454
Wolinsky, J. - 179
Wolter, A. - 387
Wong, C. M. 193
Wong, H. N. C. - 212, 453

Wright, M. E. - 160
Wubbels, G. G. - 455
Wuest, J. D. - 93
Wulff, W. D. - 155
Wuts, P. G. M. - 68
Xu, Y. - 199
Yadav, J. S. - 17
Yakobson, G. G. - 35, 454
Yamada, H. - 441
Yamada, Y. - 226, 423
Yamaguchi, M. - 137, 138, 139
Yamakawa, K. - 443
Yamamoto, H. - 103, 109, 203
Yamamoto, K. - 41, 119
Yamamoto, S. - 393
Yamamoto, T. - 200
Yamamoto, Y. - 40, 57, 61, 62, 68
Yamamura, S. - 159
Yamanaka, H. - 43, 200
Yamashita, Y. - 146
Yamauchi, M. - 346
Yamauchi, T. - 217

AUTHOR INDEX

Yamazaki, Y. - 197

Yates, P. - 144, 146, 169

Yatsimirsky, A. K. - 198

Yoneda, N. - 399

Yoneda, S. - 44, 50

Yoshida, H. - 76

Yoshida, T. - 337

Yoshida, Z. - 11, 234

Yoshida, Z. I. - 99, 216

Yoshii, E. - 87, 216

Yuste, F. - 234

Zajac, W. W., Jr. - 247, 429

Zamboni, R. - 109

Zbiral, E. - 112, 113

Zeelen, F. J. - 106

Zefirov, N. S. - 13, 158

Zhelduborskaya, G. A. - 132

Ziegler, F. E. - 71, 74

Ziffer, H. - 396

Zimmer, H. - 114

Zwanenburg, B. - 55, 381

Zweifel, G. - 139, 140, 289

RAYMOND H. FOGLER LIBRARY
DATE DUE

BOOKS ARE SUBJECT TO RECALL